Revitalizing Indigenous Knowledge Systems in the 21st Century

Monicca Thulisile Bhuda
https://orcid.org/0000-0002-8506-9562
University of Mpumalanga, South Africa

Tebogo Maria Mothiba
University of Limpopo, South Africa

IGI Global
Scientific Publishing
Publishing Tomorrow's Research Today

Vice President of Editorial — Melissa Wagner
Director of Acquisitions — Mikaela Felty
Director of Book Development — Jocelynn Hessler
Production Manager — Mike Brehm
Cover Design — Jose Rosado

Published in the United States of America by
IGI Global Scientific Publishing
701 East Chocolate Avenue
Hershey, PA, 17033, USA
Tel: 717-533-8845 | Fax: 717-533-7115
Website: https://www.igi-global.com E-mail: cust@igi-global.com

Library of Congress Cataloging-in-Publication Data

Names: Bhuda, Monicca, 1992- editor | Mothiba, Tebogo Maria, 1967- editor
Title: Revitalizing indigenous knowledge systems in the 21st century / edited by: Monicca Thulisile Bhuda, Tebogo Maria Mothiba.
Other titles: Revitalizing indigenous knowledge systems in the twenty-first century
Description: Hershey, PA : IGI Global Scientific Publishing, [2027] | Includes bibliographical references and index. | Summary: "This book explores the contemporary significance of Indigenous Knowledge Systems, highlighting their restoration, preservation, and integration with modern practices. It examines how revitalizing IKS fosters community resilience, sustainable development, cultural identity, and alternative paradigms that challenge Western-centric worldviews"-- Provided by publisher.
Identifiers: LCCN 2026002326 (print) | LCCN 2026002327 (ebook) | ISBN 9798337379760 hc | ISBN 9798337379777 sc | ISBN 9798337379784 e
Subjects: LCSH: Ethnoscience--Africa | Ethnoscience--India
Classification: LCC GN476 .R485 2026 (print) | LCC GN476 (ebook)
LC record available at https://lccn.loc.gov/2026002326
LC ebook record available at https://lccn.loc.gov/2026002327

British Cataloguing in Publication Data
A Cataloguing in Publication record for this book is available from the British Library.

Table of Contents

Foreword xiv

Preface xv

Acknowledgment xix

Chapter 1
Reimagining Indigenous Knowledge Systems in the Age of AI: From Ancestral Wisdom to Algorithmic Futures 1
Piali Haldar, Brainware University, Kolkata, India
Dev Kumar Mandal, Brainware University, Kolkata, India

Chapter 2
Bridging Worlds: The Vital Role of Indigenous Knowledge in Education and Research 23
Rick Rantz, Allan Hancock College, USA

Chapter 3
Indigenous Knowledge Systems and Sustainable Development 45
Soumen Ghosh, Global Institute of Management and Technology, Krishnanagar, India
Parag Chatterjee, Global Institute of Management and Technology, Krishnanagar, India
Rupam Hazra, Global Institute of Management and Technology, Krishnanagar, India

Chapter 4
Integrating Indigenous Knowledge System in Education and Research for Sustainable Development 67
Mokgadi Asnath Modiba, University of Limpopo, South Africa

Chapter 5
African Indigenous Knowledge Systems: Contestations in Digitising the Indigenous Knowledge Systems in African Indigenous Communities 89
Glen Maimela, National Heritage Council, South Africa
Rudzani Nedombeloni, University of Venda, South Africa

Chapter 6
Digital Pathways to Revitalizing Indigenous Knowledge Systems (IKS): Community-Based Models of Preservation, Equity, and Sustainable Development in the 21st Century 113
Jogen Sharma, KIIT (Deemed to be) University, India
Dayawanti Tarmali, MDI Murshidabad, India

Chapter 7
Ethical Digital Pathways for Preserving Indigenous Knowledge: Harmonizing Cultural Wisdom With Emerging Technologies.. 151
M. Ushanandhini, Vivekanandha Arts and Science College for Women, Sankari, India

Chapter 8
Afrocentric Ethical Guidelines for Digital Content Creators Producing Indigenous Knowledge-Based Content 177
Tsakani Emmarencia Nyoni, University of Venda, South Africa
Vhonani Olive Netshandama, University of Venda, South Africa

Chapter 9
Digital Displacement: How Social Media Erodes Traditional African Culture in Conversations...... 201
Maletšema Ruth Emsley, University of Limpopo, South Africa

Chapter 10
Analysing the Policy and Institutional Barriers to Indigenous-Led Heritage Restoration and Reframing Governance Through the Lens of Inherent Sovereignty 225
Ashok Dobhal, Law College Dehradun, India
Rajesh Bahuguna, Law College Dehradun, India
Nidhi Tyagi, Independent Researcher, India
Isha Sharma, Law College Dehradun, India
Radhey Shyam Jha, Law College Dehradun, India

Chapter 11
Bio-Cultural Community Protocol: A Tool to Empower Communities' Economic Development Using IK.. 249
Diane Thelma Molokwa, Independent Researcher, South Africa
Motheo Koitsiwe, North West University, South Africa

Chapter 12
Community Empowerment Through Indigenous Knowledge: Pathways to Economic Prosperity 269
Ananya Pandey, Christ University, Bangalore, India

Chapter 13
Correcting Knowledge Distortions: An Analysis of How Indigenous Knowledge Can Redefine South Africa's Innovation and Economic Progress.. 287
Sandile Tyrant Nhuvunga, University of Mpumalanga, South Africa
Felix Sandile Thwala, University of Mpumalanga, South Africa

Chapter 14
Engaging the Youth in Indigenous Knowledge Preservation and Job Creation in South Africa 307
Lindelwa Veronicah Nkosi, University of Mpumalanga, South Africa
LIvhuwani Dephney Tshikukuvhe, University of Venda, South Africa

Chapter 15
The Forest as an Economy: Indigenous Agroforestry Systems and Community-Driven Bioeconomies in the Amazon 327
Silvio Andrae, Independent Researcher, Germany

Chapter 16
Ethical Public Finance and Accounting in the AI Era: Harnessing IKS for Sustainable Financial Future 357
N. Abhishek, Government First Grade College, Halebeedu, India
Sahana Dinesh, Government First Grade College, Napoklu, India
Delecta Jenifer Rajendren, Saveetha Engineering College, India
Neethu Suraj, Yenepoya (Deemed to be) University, India
P. Shareena, Yenepoya (Deemed to be) University, India

Chapter 17
Sustainable Governance Through Ancient Lenses: Policy Lessons From Traditional Indian Ecological and Resource Management 383
Ashok Dobhal, Uttaranchal University, India
Rajesh Bahuguna, Law College Dehradun, India
Nidhi Tyagi, Graphic Era University, India
Radhey Shyam Jha, Law College Dehradun, India

Chapter 18
Scalable IoT Solutions for Sustainable Tourism and Urban Heritage Protection 413
Hitesh Mohapatra, School of Computer Engineering, KIIT (Deemed to be) University, Bhubaneswar, India

Chapter 19
Sustainable Eco-Tourism for Cultural Preservation and Community Development 435
Lethukuthula Mtshali, University of KwaZulu-Natal, South Africa
Sthokozile Mamba, University of KwaZulu-Natal, South Africa
Andrisha Beharry Ramraj, University of KwaZulu-Natal, South Africa

Compilation of References 453

About the Contributors 503

Index 513

Detailed Table of Contents

Foreword xiv

Preface xv

Acknowledgment xix

Chapter 1
Reimagining Indigenous Knowledge Systems in the Age of AI: From Ancestral Wisdom to Algorithmic Futures 1
Piali Haldar, Brainware University, Kolkata, India
Dev Kumar Mandal, Brainware University, Kolkata, India

Indigenous Knowledge Systems (IKS) embody rich ecological intelligence, cultural memory, and relational worldviews, yet remain marginalised within contemporary AI-driven knowledge ecosystems. This study proposes a Mixed-Methods Techno–Cultural Convergence Model (TCCM) to examine how ancestral wisdom can be meaningfully integrated with artificial intelligence in the 21st century. Using Indigenous story-work and thematic analysis, followed by Partial Least Squares Structural Equation Modelling (PLS-SEM), the research identifies cultural identity, ethical AI trust, and IKS–AI compatibility as key drivers of AI adoption. The findings demonstrate that culturally aligned and ethically governed AI can support Indigenous knowledge revitalisation rather than epistemic erosion.

Chapter 2
Bridging Worlds: The Vital Role of Indigenous Knowledge in Education and Research 23
Rick Rantz, Allan Hancock College, USA

This chapter examines Indigenous Knowledge and Indigenous Research Methodologies as transformative frameworks capable of reshaping higher education. Grounded in relationality, land-based understanding, ceremony, community governance, and intergenerational responsibility, these methodologies challenge Western epistemic dominance and reveal the limits of Eurocentric research assumptions. The chapter traces the historical exclusion of Indigenous epistemologies, analyzes core methodological principles such as storywork, collective consent, and relational accountability, and discusses implications for ethics, data sovereignty, academic publishing, teaching, and graduate supervision. It argues that supporting Indigenous methodologies requires structural change across governance, curriculum, tenure, and research policy. Future directions highlight Indigenous-led research networks, digital sovereignty, ecological knowledge, and pluriversal academic models grounded in Indigenous intellectual sovereignty.

Chapter 3
Indigenous Knowledge Systems and Sustainable Development 45
Soumen Ghosh, Global Institute of Management and Technology, Krishnanagar, India
Parag Chatterjee, Global Institute of Management and Technology, Krishnanagar, India
Rupam Hazra, Global Institute of Management and Technology, Krishnanagar, India

This chapter highlights the significance of Indigenous Knowledge Systems (IKS) as a crucial resource for global sustainability, resilience, and inclusive development. Rooted in centuries of ecological, communal, and spiritual experience, IKS promotes harmony with nature and strengthens social cohesion. The chapter outlines core IKS principles such as oral transmission, communal validation, and responsible resource use, showing how they support biodiversity, sustainable livelihoods, and food security. It also examines IKS contributions to healthcare, education, and participatory governance. Despite past marginalization, IKS is increasingly recognized worldwide. The chapter stresses the need for ethical integration of traditional knowledge with modern science and policy, calling for strong legal protection and respectful collaboration to empower Indigenous communities.

Chapter 4
Integrating Indigenous Knowledge System in Education and Research for Sustainable Development 67
Mokgadi Asnath Modiba, University of Limpopo, South Africa

Colonialism and apartheid epistemes persist in the current age. It is apparent in written textbooks where disciplinary knowledge is rooted in Western information. Such practices exclude the rich and relevant experiences from indigenous people, despite IKS relevance in understanding nature and cultural heritages. Higher Education Institutions are mandated to enforce the recognition and the revival of IKS, to implement innovative strategies that encourage and support its integration in education and research. The integration of IKS in education and research revives the continuity of cultural histories and it allows partnerships between communities and scientists. Google scholar database was used to retrieve and review literature from books, peer-reviewed articles and government documents. The author attempts to enlighten readers about the integration of IKS in education and research, expanding on policy frameworks that support IKS, possible ways to integrate IKS in education and research, challenges faced and opportunities encountered when integrating IKS in education and research.

Chapter 5
African Indigenous Knowledge Systems: Contestations in Digitising the Indigenous Knowledge Systems in African Indigenous Communities.. 89
Glen Maimela, National Heritage Council, South Africa
Rudzani Nedombeloni, University of Venda, South Africa

African Indigenous knowledge systems (AIKS) have been a centre of digitisation in the 21st century, preserving knowledge for posterity. However, digitisation has not been a smooth process confronted with contestations deeming it unprofitable. The aim of this chapter is in twofold. Firstly, the chapter critically explores contestations of digitising AIKS Indigenous communities. Secondly, the chapter proposes a co-designed approach for digitising the AIKS. Participatory action research (PAR) was employed to establish a research team to generate qualitative data in Ga-Mamabolo in the Limpopo Province. The data was co-generated using PAR data generation stages (planning, action, and reflection) and analysed using critical discourse analysis. The analysis shows passive participation, reduced authenticity, intellectual property rights, sacred knowledge, and funds as key contestations in digitising the AIKS. The analysis further shows stakeholder collaboration, cultural considerations, feedback, knowledge sharing and benefit sharing as areas to consider in digitising the AIKS.

Chapter 6
Digital Pathways to Revitalizing Indigenous Knowledge Systems (IKS): Community-Based Models of Preservation, Equity, and Sustainable Development in the 21st Century.......................... 113
Jogen Sharma, KIIT (Deemed to be) University, India
Dayawanti Tarmali, MDI Murshidabad, India

Abstract: Indigenous Knowledge Systems (IKS) are dynamic epistemologies crucial to cultural continuity, environmental sustainability, and community resiliency. Nevertheless, Indigenous knowledge practices have remained marginal due to globalisation, digital disparity, and Western knowledge hierarchies. This chapter analyses twenty-first-century digital avenues for community-based revitalisation of IKS through an interdisciplinary prism of Library and Information Science, Indigenous studies, and information and communication technology. Drawing on global concepts, ethical data governance, and international examples, this chapter proposes an Indigenous-based digital framework grounded in sovereignty, participation, and the sharing of benefits. This paper emphasises the importance of digital tools, ethically controlled, to facilitate the transfer of knowledge across generations, help cultures survive, and support sustainable development as mechanisms for combating digital colonialism.

Chapter 7
Ethical Digital Pathways for Preserving Indigenous Knowledge: Harmonizing Cultural Wisdom With Emerging Technologies.. 151
M. Ushanandhini, Vivekanandha Arts and Science College for Women, Sankari, India

Indigenous knowledge systems represent centuries of cultural wisdom, ecological understanding, and community-based practices that are deeply rooted in local contexts. However, rapid digitalization and globalization pose significant challenges to the preservation, ownership, and ethical use of such knowledge. While emerging technologies such as artificial intelligence, blockchain, cloud computing, and digital archiving offer new opportunities for safeguarding indigenous knowledge, they also raise concerns related to cultural appropriation, data sovereignty, and ethical governance. This chapter explores ethical digital pathways for preserving indigenous knowledge by harmonizing traditional cultural values with modern technological frameworks. It emphasizes the importance of community participation, ethical data governance, culturally sensitive digitization practices, and sustainable technology adoption. this study proposes a balanced approach that protects indigenous intellectual heritage while enabling its responsible transmission to future generations.

Chapter 8
Afrocentric Ethical Guidelines for Digital Content Creators Producing Indigenous Knowledge-Based Content 177
Tsakani Emmarencia Nyoni, University of Venda, South Africa
Vhonani Olive Netshandama, University of Venda, South Africa

Indigenous Knowledge(IK)-based digital content creation is increasingly being produced. Digital content consumers are also taking more interest in the content. Sacred IK-based content can be accessed on platforms like TikTok and YouTube, and, often, it is produced without the owner's awareness or acknowledgement.

These developments in the IK Management space highlight the need for Afrocentric digital content creation ethical guidelines. Existing guidelines are either focused on digital content creation in general, or on research with Indigenous Knowledge. Both guidelines can serve only as a foundation for IK-based digital content creation ethical guidelines, as they are not fully applicable to it. This chapter presents findings on IK-based ethical guidelines for producing digital content. The data was generated through sharing circles and semi-structured interviews, and analyzed using a thematic analysis technique. The findings have highlighted how the digital content creation process can be conducted respectfully and inclusively with Indigenous Knowledge Holders.

Chapter 9
Digital Displacement: How Social Media Erodes Traditional African Culture in Conversations...... 201
Maletšema Ruth Emsley, University of Limpopo, South Africa

Mass communication has moved beyond traditional media of radio, television, newspaper, magazines, film, and books due to technology and its advancement. Social media take on different forms in this 21st century. This chapter highlights the impact of social media on African traditional and cultural conversations and demonstrates the importance of traditional African practices which embody core values like respect, wisdom, hierarchy and ubuntu. The chapter used a qualitative approach through a semi-structured interview approach from purposively and snowball sampled participants. This chapter's findings have implications for future research, education, indigenous knowledge systems and digitalisation. It contributes to the debates on decolonial education, the evolving challenges regarding the use of social media in African indigenous languages and their expansion to the digital communication spaces.

Chapter 10
Analysing the Policy and Institutional Barriers to Indigenous-Led Heritage Restoration and Reframing Governance Through the Lens of Inherent Sovereignty 225
Ashok Dobhal, Law College Dehradun, India
Rajesh Bahuguna, Law College Dehradun, India
Nidhi Tyagi, Independent Researcher, India
Isha Sharma, Law College Dehradun, India
Radhey Shyam Jha, Law College Dehradun, India

Indigenous Knowledge is the knowledge, skills, and beliefs that societies have developed over time by living in and interacting with their natural surroundings. It includes art of agriculture, food preparation, health care practices, education, building architecture ecological conservation, language etc. Through the reliance and interaction with nature and its surroundings, indigenous communities have acquired an immense knowledge of the same. This accumulated knowledge is rapidly disappearing as the world is steadily becoming more culturally and biologically uniform. There are many Indigenous societies in the world that have a rich heritage spanning more than tens of thousands of years and their governance is in turmoil due to multiple stakeholders such as indigenous communities, government, private institutions, international organizations, NGOs, religious groups etc. leading to ambiguity. Thus, to avoid complexity, the preservation and restoration of indigenous heritage vividly rely on sovereignty of an institution and their effective governance.

Chapter 11
Bio-Cultural Community Protocol: A Tool to Empower Communities' Economic Development Using IK...249
Diane Thelma Molokwa, Independent Researcher, South Africa
Motheo Koitsiwe, North West University, South Africa

Indigenous and local communities are seen as guardians and caretakers of genetic resources and possess immense traditional or indigenous knowledge regarding its use and management. However, exclusion of indigenous knowledge and related practices into economic development, caused by a lack of recognition and continuous marginalization by scientific knowledge, and sustainable development stakeholders, perpetuates development, food and nutrition insecurity. This marginalization also leaves indigenous and local communities and their local resources including indigenous knowledge exposed to bio-piracy and misappropriation. Yet this knowledge is the most precious assets possessed by indigenous and local communities and also the most positive motivation to build their local economies. Therefore, this chapter aims to guide and suggest to the government and policy makers to work hand in hand with communities to leverage BCPs as a tool to empower themselves economically.

Chapter 12
Community Empowerment Through Indigenous Knowledge: Pathways to Economic Prosperity 269
Ananya Pandey, Christ University, Bangalore, India

This book chapter explores pathways to economic prosperity through community empowerment grounded in Indigenous knowledge systems. It argues that Indigenous knowledge which is rooted in local ecology, cultural practices, and collective governance that offers sustainable and context-specific strategies for livelihood enhancement. By integrating traditional wisdom with contemporary economic frameworks, communities can strengthen resilience, promote inclusive growth, and reduce dependency on externally imposed development models. The chapter highlights how community-led enterprises, sustainable resource management, and intergenerational knowledge transmission contribute to long-term economic self-determination. It also examines policy and institutional mechanisms necessary to recognize, protect, and scale Indigenous knowledge without commodifying or eroding its cultural integrity. Ultimately, the chapter positions Indigenous knowledge as a critical driver of equitable economic development and community well-being in an era of globalization and environmental uncertainty.

Chapter 13
Correcting Knowledge Distortions: An Analysis of How Indigenous Knowledge Can Redefine South Africa's Innovation and Economic Progress..287
Sandile Tyrant Nhuvunga, University of Mpumalanga, South Africa
Felix Sandile Thwala, University of Mpumalanga, South Africa

South African innovation discussions remain largely informed by Western epistemological models. These dominant models have created epistemic distortions that marginalise Indigenous Knowledge Systems (IKS) and constrain the ability of South Africa to locally define innovation and economic progress. This chapter aims to critically examine how Western knowledge models continue to influence South Africa's innovation pathways. Furthermore, the chapter will examine how Western knowledge models devalue community-centred forms of innovation. Drawing on qualitative secondary data, the chapter further explores how colonial legacies have undermined African epistemologies. Preliminary findings suggest that indigenous worldviews, particularly those expressed in Afrikology and relational African philosophies, provide alternative models of innovation. The findings also reveal structural and policy gaps that hinder the inclusion of IKS within dominant innovation frameworks. The chapter concludes by proposing decolonial strategies for reframing innovation through African-centred knowledge systems.

Chapter 14
Engaging the Youth in Indigenous Knowledge Preservation and Job Creation in South Africa 307
Lindelwa Veronicah Nkosi, University of Mpumalanga, South Africa
LIvhuwani Dephney Tshikukuvhe, University of Venda, South Africa

The paper examined the engagement of Youth in the preservation of Indigenous knowledge and job creation. Guided by the socio-cultural theory and the social innovation theory, the paper adopted qualitative exploratory design. Data were collected through interviews and focus group discussions and were analyzed using thematic analysis. The Findings revealed that Indigenous knowledge is too broad, as it encompasses diverse aspects such as culture, where traditional crafts can be sold online to a global market, agriculture, which can enable young people to locally distribute farm produce and use local heritage sites to draw tourists, and social media storytelling. The paper concludes that positioning Indigenous Knowledge as a dynamic and innovation-driven economic resource can strengthen youth employment pathways while ensuring cultural sustainability. The paper contributes to contemporary debates on Indigenous Knowledge revitalization by demonstrating its practical relevance for youth employment and social transformation in the 21st century.

Chapter 15
The Forest as an Economy: Indigenous Agroforestry Systems and Community-Driven Bioeconomies in the Amazon 327
Silvio Andrae, Independent Researcher, Germany

This chapter investigates Indigenous agroforestry systems in the Amazon as integrated socio-ecological economies that generate livelihoods, conserve biodiversity, and foster institutional resilience. It contends that forest gardens, locally referred to as chakra, chagra, and purma, represent advanced economic infrastructures rooted in Indigenous Knowledge Systems, relational governance, and multi-strata ecological design. Utilizing ethnographic literature, policy analysis, interviews, and biodiversity-based value chain research from Ecuador and Brazil, the chapter examines how Kichwa communities convert agroforestry practices into community-owned bioeconomic value chains. The analysis underscores the complex influence of international legal regimes, particularly access and benefit-sharing frameworks, on Indigenous economic autonomy. By framing forest gardens as 'technologies of the future,' the chapter reconceptualizes the bioeconomy as a transformation of governance rather than mere biomass substitution.

Chapter 16
Ethical Public Finance and Accounting in the AI Era: Harnessing IKS for Sustainable Financial Future 357
N. Abhishek, Government First Grade College, Halebeedu, India
Sahana Dinesh, Government First Grade College, Napoklu, India
Delecta Jenifer Rajendren, Saveetha Engineering College, India
Neethu Suraj, Yenepoya (Deemed to be) University, India
P. Shareena, Yenepoya (Deemed to be) University, India

This study investigates integrating Indian Knowledge Systems (IKS) principles with Artificial Intelligence (AI) to enhance ethical governance and transparency in public accounting and finance. Through interviews with five experts, key themes were identified regarding the application of IKS concepts, such as Dharma, Karma, and Nyaya, to AI-driven financial systems. The findings highlight consensus on embedding ethical considerations, promoting transparency and explainability, adopting a holistic perspective encompassing economic, social, and environmental factors, and ensuring cultural relevance in AI design. The study explores applications in areas such as financial transparency tools, audit trails, ESG reporting, and public administration. Experts acknowledged potential benefits and identified challenges related to technical complexity, resistance to change, and the need for broader awareness of IKS principles.

Chapter 17
Sustainable Governance Through Ancient Lenses: Policy Lessons From Traditional Indian Ecological and Resource Management 383
Ashok Dobhal, Uttaranchal University, India
Rajesh Bahuguna, Law College Dehradun, India
Nidhi Tyagi, Graphic Era University, India
Radhey Shyam Jha, Law College Dehradun, India

This review examines ancient Indian ecological governance as an integrated system of ethical principles, community institutions, and resource management practices that sustained socio-ecological balance over long time horizons. Drawing on interdisciplinary scholarship, the paper synthesises philosophical foundations such as Dharma, Rta, and Prakriti with traditional governance mechanisms including Panchayat systems, indigenous water and forest management, and culturally embedded conservation practices. Through comparative analysis with contemporary sustainability frameworks, the study demonstrates strong convergence with modern concepts of commons governance, climate resilience, and circular economy. The findings highlight the continuing relevance of ancient Indian governance traditions for informing inclusive, ethical, and community-led sustainability policy in the present.

Chapter 18
Scalable IoT Solutions for Sustainable Tourism and Urban Heritage Protection 413
Hitesh Mohapatra, School of Computer Engineering, KIIT (Deemed to be) University, Bhubaneswar, India

Smart cities leverage IoT platforms to deliver intelligent services, making security, flexibility, and scalability essential design considerations. Robust security mechanisms, including advanced encryption and granular access control, are vital to mitigate rising cyber threats. To manage urban infrastructure effectively, IoT platforms must remain adaptable to diverse network protocols and device types. Scalability is equally critical, as IoT devices generate massive datasets, particularly within applications like environmental monitoring and cultural heritage preservation. However, deploying IoT in heritage conservation presents significant security challenges. Successfully integrating these systems requires balancing technical benefits with a commitment to inclusivity. This study proposes an IoT-based framework for sustainable tourism, specifically prioritizing interoperability and security to protect historical sites. By fostering digital literacy, this approach empowers individuals to actively participate in the interpretation and safeguarding of their cultural legacy.

Chapter 19
Sustainable Eco-Tourism for Cultural Preservation and Community Development 435
Lethukuthula Mtshali, University of KwaZulu-Natal, South Africa
Sthokozile Mamba, University of KwaZulu-Natal, South Africa
Andrisha Beharry Ramraj, University of KwaZulu-Natal, South Africa

Eco-tourism in KwaZulu-Natal has been widely framed as a strategy to link conservation with rural development, particularly in landscapes marked by historical dispossession. However, a persistent gap remained between policy rhetoric on inclusion and the structural realities limiting indigenous authority over land, revenue, and representation. The objective of the chapter was to explore how indigenous rural communities in KwaZulu-Natal can reclaim, preserve, and revive their cultural heritage through active involvement in eco-tourism. A qualitative secondary data design was adopted, synthesising place-based case studies and governance analyses within a thematic framework. The findings revealed that cultural revitalisation depended less on tourism activity itself and more on secure land rights, shared governance, and epistemic recognition. The study therefore recommended restructuring co-management systems, embedding Indigenous Knowledge Systems (IKS) in decision-making, and strengthening community authority to enable sustainable heritage renewal.

Compilation of References 453

About the Contributors 503

Index 513

Foreword

Across the world, and increasingly in Africa, there is a kindled fire of imagination in the minds of many publics of Indigenous Knowledge Systems (IKS). There is a growing body of detailed, impartial and accurate information about this field that is establishing itself as a true science. Significant and fascinating works are emerging that show how IKS crosses tenuous borders of many disciplines. This is both humbling and inspiring.

Revitalizing Indigenous Knowledge Systems in the 21st Century, therefore, appears at a critical moment when societies across the world are rethinking and reimagining what counts as knowledge and whose voices shape our shared futures. IKS, long relegated to the margins of formal scholarship, are increasingly recognised as vital to building more just, sustainable, and humane worlds. This volume, informative and engaging, responds to that global shift by placing IKSs at the centre of contemporary debates and treating them not as remnants of the past, but as living, dynamic resources for present and future generations. The authors brilliantly demonstrate how, and in what ways, IKSs lie at the heart of our civilizations.

The editors, Dr Monicca Thulisile Bhuda and Prof Tebogo Maria Mothiba, bring together a diverse group of scholars, practitioners, and knowledge holders whose contributions span multiple disciplines, contexts, and traditions. The authors demonstrate how IKS illuminate pressing issues such as environmental stewardship and climate change, health and wellbeing, education and curriculum transformation, governance, technology, spirituality, language, and cultural expression. Rather than viewing Indigenous knowledge as static, this book highlights its creativity and adaptability in the face of colonial disruption, modernity, and globalisation.

A distinctive strength of this easy to read and digestible volume is its focus on revitalization. The chapters move beyond merely documenting Indigenous knowledge to explore the conditions under which IKS can flourish in the 21st century. They show how communities are reclaiming epistemic authority, strengthening intergenerational transmission, and weaving Indigenous insights into policy, pedagogy, healthcare, environmental management, and community development. In doing so, the book challenges the assumption that Western technoscience is the sole measure of valid knowledge and instead advances epistemic plurality as essential for our collective survival.

This book is written for those many people who accept with genuine respect IKS not merely as ancient civilization but rather as our heritage that is still relevant to our time. The volume is fully revealing when looked at from a wide variety of viewpoints. It will serve as a valuable resource for scholars, students, policymakers, practitioners, and community leaders who are committed to decolonising knowledge and reshaping institutions so that Indigenous worldviews are genuinely recognised and respected. It is, therefore, with great pleasure that I commend this work and endorse its contribution to the urgent task of reimagining knowledge in the 21st century.

Kenneth I. Simala

Masinde Muliro University of Science and Technology, Kenya

Preface

Indigenous Knowledge Systems (IKS) are vital reservoirs of insight, resilience, and creativity for societies facing complex 21st-century challenges. For generations, Indigenous communities have developed sophisticated knowledge that guides agriculture, health, education, governance, spirituality, and environmental stewardship. Yet these holistic and relational systems have been marginalized by colonialism and the dominance of Western scientific paradigms. In a time of ecological crisis, deepening inequality, and rapid technological change, revitalizing IKS is essential for the sustainability and cultural integrity of Indigenous peoples and for humanity more broadly.

Revitalizing Indigenous Knowledge Systems in the 21st Century responds to this urgency by reclaiming and recognizing the contemporary value of IKS. The chapters in this volume treat IKS as living and adaptive systems that shape health, education, sustainable development, environmental conservation, and cultural heritage in diverse contexts. Authors show how IKS can guide alternative approaches to policy, curriculum, research, and practice that are grounded in local realities, ethical accountability, and community-based visions of wellbeing. Revitalization is approached here as recovery, revalorization, preservation, and innovation, reclaiming suppressed knowledges while engaging new technologies and urban realities in ways that safeguard community ownership and cultural integrity.

A central thread in this book is that revitalizing IKS is inseparable from decolonization. IKS cannot simply be added as supplemental content; they require transforming institutional logics, power relations, and epistemic hierarchies that continue to marginalize Indigenous peoples. Contributors examine how universities, schools, health systems, and heritage institutions can move beyond tokenism towards genuine recognition of Indigenous intellectual sovereignty and knowledge autonomy. They highlight practices in which Indigenous communities are not merely data sources, but co-designers and rights-bearing partners whose protocols shape every stage of knowledge production and use.

This book is intended for scholars, postgraduate students, policymakers, practitioners, cultural institutions, and community organizations committed to more inclusive and just knowledge futures. By situating IKS within contemporary debates in health, education, environmental management, and development, it demonstrates that Indigenous knowledge is central rather than peripheral to addressing current global crises. We hope it will inspire dialogue, partnerships, and concrete actions that strengthen Indigenous languages, institutions, and knowledge practices, and support communities in their efforts to know, remember, innovate, and thrive on their own terms.

ORGANIZATION OF THE CHAPTERS

The opening chapters of this volume establish a critical foundation for understanding the evolving relationship between Indigenous Knowledge Systems (IKS), technology, education, governance, and sustainable development in the contemporary world. Collectively, they demonstrate that Indigenous knowledge is not merely a historical inheritance but a dynamic and living epistemology capable of informing innovative responses to some of the most pressing challenges of the twenty-first century.

Chapter 1 explores the transformative possibilities of integrating Indigenous Knowledge Systems with artificial intelligence. Through a mixed-methods approach, the chapter introduces a techno-cultural convergence model that demonstrates how culturally aligned and ethically governed AI can support the revitalization of ancestral knowledge rather than contribute to its erosion. By examining the roles of cultural identity, trust, and compatibility between Indigenous worldviews and emerging technologies, the chapter offers an important framework for envisioning inclusive and equitable digital futures.

Chapter 2 examines Indigenous Knowledge and Indigenous Research Methodologies as powerful frameworks for reshaping higher education and scholarly inquiry. The chapter highlights how principles such as relationality, storywork, collective consent, and accountability challenge dominant Eurocentric assumptions and open pathways toward more inclusive knowledge production. By addressing issues ranging from ethics and governance to curriculum design and academic policy, the chapter underscores the transformative potential of Indigenous-led approaches in education and research.

Chapter 3 highlights the vital contributions of Indigenous Knowledge Systems to sustainable development. Drawing attention to the ecological, social, and cultural dimensions of Indigenous knowledge, the chapter demonstrates how practices rooted in communal responsibility, environmental stewardship, and intergenerational learning contribute to biodiversity conservation, food security, healthcare, and participatory governance. The discussion reinforces the importance of respectful collaboration between Indigenous communities, policymakers, and scientific institutions in advancing sustainable futures.

Chapter 4 focuses on the integration of Indigenous Knowledge Systems within education and research as a pathway toward sustainable development and epistemic justice. The chapter critically examines the enduring influence of colonial and apartheid knowledge structures while exploring policy frameworks, opportunities, and challenges associated with embedding Indigenous perspectives into higher education. By advocating for stronger partnerships between communities and researchers, the chapter highlights the role of educational institutions in preserving cultural heritage and fostering inclusive knowledge systems.

Chapter 5 addresses the complex and often contested process of digitizing African Indigenous Knowledge Systems. Through participatory action research, the chapter identifies challenges related to authenticity, intellectual property rights, sacred knowledge, participation, and resource constraints. At the same time, it presents a collaborative and community-centered approach to digitization that emphasizes stakeholder engagement, cultural sensitivity, equitable benefit sharing, and knowledge preservation for future generations.

Chapter 6 examines how digital technologies can serve as tools for revitalizing Indigenous Knowledge Systems through community-based models of preservation and empowerment. Drawing upon interdisciplinary perspectives and international experiences, the chapter proposes a digital framework grounded in Indigenous sovereignty, participation, and ethical governance. The discussion demonstrates how thoughtfully implemented digital initiatives can strengthen cultural continuity, facilitate intergenerational knowledge transfer, and support sustainable development while resisting forms of digital colonialism.

Chapter 7 further explores the intersection of Indigenous knowledge and emerging technologies by focusing on the ethical dimensions of digital preservation. The chapter considers how tools such as artificial intelligence, blockchain, cloud computing, and digital archives can be leveraged to safeguard cultural heritage while protecting Indigenous ownership, data sovereignty, and intellectual rights. Emphasizing community participation and culturally responsive governance, the chapter presents a balanced vision for responsible technological adoption.

Chapter 8 contributes an important perspective on ethical digital content creation involving Indigenous Knowledge. Recognizing the growing visibility of Indigenous content on social media and digital platforms, the chapter develops Afrocentric ethical guidelines that prioritize respect, inclusion, consent, and meaningful engagement with Indigenous Knowledge Holders. The resulting framework provides practical guidance for creators seeking to share Indigenous knowledge in ways that uphold cultural integrity and community ownership.

Chapter 9 investigates the cultural consequences of social media expansion within African communities. The chapter examines how digital communication platforms are reshaping traditional conversations, language use, and cultural practices that have historically embodied values such as respect, wisdom, hierarchy, and ubuntu. Through qualitative inquiry, the chapter contributes to ongoing discussions surrounding decolonial education, cultural preservation, and the challenges of sustaining Indigenous languages and knowledge traditions within digital spaces.

Chapter 10 explores the policy and institutional barriers that complicate Indigenous-led heritage restoration efforts. By examining the governance challenges created by the involvement of multiple stakeholders, the chapter argues that effective preservation and restoration depend upon recognizing Indigenous sovereignty and strengthening governance structures. The discussion highlights the importance of clear institutional authority and Indigenous leadership in safeguarding cultural heritage in an increasingly complex policy environment.

Chapter 11 introduces Bio-Cultural Community Protocols as a mechanism for protecting Indigenous knowledge while promoting community-driven economic development. The chapter demonstrates how these protocols can help address issues of biopiracy, marginalization, and resource exploitation by empowering communities to exercise greater control over their knowledge and biological resources. In doing so, it offers valuable policy guidance for governments and development practitioners seeking to support sustainable and equitable economic growth.

Chapter 12 examines the relationship between Indigenous knowledge and economic prosperity through the lens of community empowerment. The chapter illustrates how local ecological knowledge, cultural practices, and collective governance structures can support resilient livelihoods, sustainable resource management, and inclusive economic development. By advocating for policy frameworks that recognize and protect Indigenous knowledge without compromising its cultural significance, the chapter positions Indigenous communities as active agents of economic transformation.

Chapter 13 critically analyzes the influence of Western epistemological models on innovation and economic development in South Africa. The chapter argues that dominant frameworks often marginalize Indigenous Knowledge Systems and limit opportunities for locally grounded innovation. By exploring Afrikology and other African-centered philosophies, the authors propose alternative approaches that redefine innovation through relational, community-oriented, and decolonial perspectives, offering new pathways for inclusive economic progress.

Chapter 14 focuses on the engagement of youth in Indigenous knowledge preservation and employment creation. Through qualitative investigation, the chapter reveals how traditional knowledge can generate opportunities in sectors such as digital entrepreneurship, agriculture, tourism, cultural industries, and social media storytelling. The discussion highlights the potential of Indigenous knowledge to serve as both a source of cultural continuity and a catalyst for youth-driven innovation and economic empowerment.

Chapter 15 presents Indigenous agroforestry systems in the Amazon as sophisticated socio-ecological economies that simultaneously support livelihoods, biodiversity conservation, and community resilience. Through an examination of forest gardens and community-owned bioeconomic value chains, the chapter challenges conventional understandings of economic development and proposes a vision of the bioeconomy grounded in Indigenous governance, relationality, and ecological stewardship. The chapter demonstrates how Indigenous communities are creating innovative and sustainable economic models rooted in ancestral knowledge.

Chapter 16 explores the integration of Indian Knowledge Systems with artificial intelligence in the fields of public finance and accounting. Drawing upon ethical concepts such as Dharma, Karma, and Nyaya, the chapter demonstrates how traditional knowledge can inform the development of transparent, accountable, and culturally relevant AI-driven financial systems. The discussion highlights opportunities for improving governance, auditability, ESG reporting, and public administration while addressing challenges related to implementation and awareness.

Chapter 17 examines ancient Indian ecological governance traditions as a source of valuable lessons for contemporary sustainability policy. By analyzing philosophical principles alongside traditional institutions for water, forest, and resource management, the chapter reveals significant parallels between historical practices and modern concepts such as climate resilience, commons governance, and the circular economy. The chapter offers compelling evidence for the continued relevance of Indigenous and traditional governance systems in addressing present-day environmental challenges.

Chapter 18 investigates the role of scalable Internet of Things (IoT) solutions in sustainable tourism and urban heritage protection. The chapter proposes a secure and interoperable framework for deploying digital technologies within heritage conservation initiatives while emphasizing the importance of inclusivity, digital literacy, and community participation. By linking technological innovation with cultural preservation, the chapter demonstrates how smart systems can contribute to the protection and interpretation of historical and cultural assets.

Chapter 19 concludes this section by exploring eco-tourism as a vehicle for cultural preservation and community development. Focusing on experiences from KwaZulu-Natal, the chapter argues that successful cultural revitalization depends not simply on tourism activity but on meaningful Indigenous participation, secure land rights, shared governance, and recognition of Indigenous Knowledge Systems. The findings provide important insights into how community-centered tourism models can contribute to sustainable heritage preservation and local empowerment.

Acknowledgment

We wish to express our profound appreciation to all the contributing authors, researchers, academics, and Indigenous knowledge holders whose chapters, insights, and lived experiences have infused this volume with depth and purpose. Each contribution reflects countless hours of study, community engagement, and critical reflection, and together they weave a rich tapestry of perspectives that speak to the resilience and dynamism of Indigenous Knowledge Systems. Their dedication and intellectual labour have positioned this book as an important intervention in the continuing revitalization of Indigenous Knowledge Systems, and as a testament to the power of knowledge grounded in community realities and lived experience.

We also warmly acknowledge the reviewers and advisory board members whose discerning comments, academic rigor, and constructive critique enhanced the overall quality of this work. Their willingness to engage deeply with the chapters, challenge us to sharpen our arguments, and uphold high scholarly and ethical standards has greatly strengthened this volume. Our heartfelt thanks go to the IGI Global editorial and production teams for their professionalism, patience, and steady support throughout each stage of the publication journey, from the initial proposal to the final proofs. Their guidance, responsiveness, and encouragement helped to transform this project from an initial idea into a complete book that we can collectively be proud of.

We are especially mindful of what this collaboration between two women in academia represents. Monicca Thulisile Bhuda, a lecturer and emerging scholar, draws inspiration from Professor Tebogo Maria Mothiba, a Deputy Vice Chancellor whose exemplary leadership, accomplishments, and dedication to higher education continue to model excellence, courage, and integrity. Our partnership on this book stands as a call for women to rise, illustrating how intentional mentorship, unwavering solidarity, and courageous women's leadership in scholarship can disrupt long-standing barriers, open new pathways, and gradually transform academic cultures. It affirms that when women walk together across ranks such as lecturer and Deputy Vice Chancellor, new possibilities are created for those who follow, especially in spaces where African women remain markedly underrepresented in senior academic and editorial roles.

Chapter 1
Reimagining Indigenous Knowledge Systems in the Age of AI: From Ancestral Wisdom to Algorithmic Futures

Piali Haldar
http://orcid.org/0000-0003-4729-6964
Brainware University, Kolkata, India

Dev Kumar Mandal
http://orcid.org/0009-0005-3740-6885
Brainware University, Kolkata, India

ABSTRACT

Indigenous Knowledge Systems (IKS) embody rich ecological intelligence, cultural memory, and relational worldviews, yet remain marginalised within contemporary AI-driven knowledge ecosystems. This study proposes a Mixed-Methods Techno–Cultural Convergence Model (TCCM) to examine how ancestral wisdom can be meaningfully integrated with artificial intelligence in the 21st century. Using Indigenous story-work and thematic analysis, followed by Partial Least Squares Structural Equation Modelling (PLS-SEM), the research identifies cultural identity, ethical AI trust, and IKS–AI compatibility as key drivers of AI adoption. The findings demonstrate that culturally aligned and ethically governed AI can support Indigenous knowledge revitalisation rather than epistemic erosion.

1. INTRODUCTION

Indigenous Knowledge Systems (IKS) constitute some of the most enduring, adaptive, and context-sensitive forms of human knowledge, developed through centuries of close interaction with land, ecology, culture, and community life. Rooted in oral traditions, collective memory, spirituality, and relational worldviews, IKS encompass domains such as agriculture, medicine, biodiversity conservation, governance, education, and climate adaptation (Berkes, 2017). Unlike dominant Western epistemologies that

DOI: 10.4018/979-8-3373-7976-0.ch001

privilege abstraction, reductionism, and universalism, Indigenous knowledge is holistic, place-based, intergenerational, and ethically embedded within social and ecological relationships (Battiste, 2014). However, in the contemporary global knowledge economy, largely shaped by digitization, datafication, and algorithmic governance, IKS remain structurally marginalized, underrepresented, and vulnerable to erosion.

The rapid rise of Artificial Intelligence (AI) has further intensified this epistemic imbalance. AI systems increasingly influence how knowledge is produced, stored, validated, and disseminated across sectors such as education, healthcare, environmental management, and public policy (Floridi et al., 2018). While AI promises efficiency, scalability, and predictive capacity, it is predominantly designed within Western scientific paradigms and trained on datasets that exclude or misrepresent Indigenous cultures, languages, and worldviews (Mohamed et al., 2020). As a result, AI risks reinforcing epistemic injustice by privileging dominant forms of knowledge while rendering Indigenous epistemologies invisible or extractable without consent (Fricker, 2007; Smith, 2021).

This tension between ancestral wisdom and algorithmic futures presents a critical challenge for the 21st century. On one hand, uncritical technological expansion threatens to commodify Indigenous knowledge, strip it of cultural context, and undermine Indigenous data sovereignty. On the other hand, when ethically and culturally aligned, AI holds transformative potential to support the revitalization of Indigenous languages, preservation of oral histories, enhancement of ecological monitoring, and strengthening of intergenerational knowledge transmission (Lewis et al., 2023). The central question, therefore, is not whether AI should interact with Indigenous Knowledge Systems, but how such interaction can be designed in ways that are culturally grounded, ethically governed, and community-led.

Existing scholarship on IKS has primarily focused on sustainability, biodiversity conservation, and development studies, emphasizing the value of Indigenous ecological knowledge in addressing climate change and environmental degradation (Whyte, 2017; IPBES, 2019). Parallel streams of research in AI ethics and responsible innovation have highlighted issues of algorithmic bias, transparency, accountability, and fairness (Jobin et al., 2019). However, limited empirical research has systematically integrated Indigenous epistemologies with AI adoption models to examine how Indigenous communities perceive, negotiate, and potentially embrace AI technologies. Much of the current discourse remains either conceptual or technocentric, lacking grounded insights into Indigenous perspectives and lived experiences.

Furthermore, dominant technology adoption models, such as the Technology Acceptance Model (TAM) or Unified Theory of Acceptance and Use of Technology (UTAUT), were developed within individualistic, market-oriented contexts and may inadequately capture the collective, relational, and ethical dimensions central to Indigenous worldviews (Venkatesh et al., 2012). Indigenous decision-making is often guided not only by perceived usefulness or ease of use, but by cultural continuity, spiritual significance, community benefit, and responsibility toward future generations. This highlights the need for alternative or extended frameworks that can bridge socio-technical analysis with Indigenous epistemology.

In response to these gaps, scholars have increasingly called for decolonial and pluralistic approaches to knowledge production, emphasizing cognitive justice and epistemic diversity (de Sousa Santos, 2015). Indigenous methodologies such as Story work, Two-Eyed Seeing, and relational accountability challenge extractive research practices and advocate for knowledge co-creation grounded in respect, reciprocity, and responsibility (Archibald, 2008; Bartlett et al., 2012). Integrating such approaches into AI research is essential to avoid reproducing colonial dynamics in digital form. Without Indigenous participation and leadership, AI systems risk becoming new instruments of cultural domination rather than tools for empowerment.

Against this backdrop, the present study seeks to reimagine the relationship between Indigenous Knowledge Systems and Artificial Intelligence through a Mixed-Methods Techno–Cultural Convergence Model (TCCM). The concept of techno–cultural convergence recognizes that technology adoption is not merely a technical process but a deeply cultural one, shaped by values, identities, ethics, and historical power relations (Orlikowski, 2000). By positioning Indigenous epistemologies at the core of the analytical framework, this study moves beyond deficit-based narratives and instead treats IKS as dynamic, adaptive systems capable of shaping technological futures.

Methodologically, this research adopts a sequential exploratory mixed-methods design, beginning with qualitative inquiry grounded in Indigenous Story work and narrative analysis, followed by quantitative validation using Partial Least Squares Structural Equation Modelling (PLS-SEM). Mixed-methods research is particularly suited to complex socio-cultural phenomena, allowing for deep contextual understanding alongside empirical generalization (Creswell & Clark, 2007). The qualitative phase ensures that constructs emerge from Indigenous voices rather than being imposed by external theoretical models, while the quantitative phase enables systematic testing of relationships among cultural and technological factors.

The proposed TCCM integrates insights from Indigenous epistemology, socio-technical systems theory, and ethical AI literature. Key constructs examined include Cultural Identity, Ethical AI Trust, IKS–AI Compatibility, Perceived Usefulness of AI for Knowledge Preservation, and Adoption Intention. By modelling both direct and moderating relationships, the framework captures the nuanced ways in which cultural grounding shapes technology acceptance. In doing so, the study responds to calls for culturally responsive and ethically aligned AI systems that respect Indigenous data sovereignty and collective rights (Rainie et al., 2017).

The significance of this research is multi-layered. Theoretically, it extends Indigenous knowledge scholarship into the emerging field of AI studies, contributing to computational humanities, digital anthropology, and decolonial technology research. Empirically, it provides one of the few large-scale mixed-methods examinations of Indigenous perspectives on AI adoption. Practically, it offers a decision-making framework for policymakers, educators, and AI developers seeking to design technologies that support, rather than undermine, Indigenous knowledge revitalization. Societally, the study underscores that sustainable and inclusive technological futures depend on recognizing Indigenous knowledge not as a relic of the past, but as a vital guide for ethical innovation.

In an era marked by climate crises, cultural homogenization, and rapid technological disruption, the integration of ancestral wisdom with algorithmic intelligence is no longer optional, it is imperative. By foregrounding Indigenous voices and epistemologies, this research advances a vision of AI that is relational, accountable, and culturally intelligent. The study ultimately argues that the future of AI must be plural, grounded in epistemic diversity, and guided by the world's oldest knowledge systems if it is to contribute meaningfully to global sustainability and social justice.

2. BACKGROUND AND LITERATURE REVIEW

The revitalization of Indigenous Knowledge Systems (IKS) in the contemporary era has emerged as a critical scholarly and policy concern, particularly as digital technologies and artificial intelligence (AI) increasingly shape how knowledge is produced, legitimized, and circulated. This section reviews the relevant literature across four interrelated domains: Indigenous Knowledge Systems, digitalization and knowledge preservation, artificial intelligence and ethics, and technology adoption through socio-cultural lenses. By synthesizing these strands, the review establishes the conceptual foundation for the proposed Techno–Cultural Convergence Model (TCCM).

2.1 Indigenous Knowledge Systems: Foundations and Contemporary Relevance

Indigenous Knowledge Systems refer to cumulative bodies of knowledge, practices, and beliefs developed by Indigenous peoples through long-term interactions with their environments and social worlds. These systems are deeply embedded in cultural values, spiritual beliefs, language, and collective identity (Berkes, 2017). Unlike Western scientific paradigms, which often emphasize objectivity, specialization, and linear causality, IKS are holistic, relational, and adaptive, integrating ecological, social, and ethical dimensions of life (Battiste, 2014).

Scholars have emphasized that IKS are not static relics of the past but dynamic and evolving systems capable of responding to contemporary challenges such as climate change, biodiversity loss, and food insecurity (Whyte, 2017). Indigenous ecological knowledge, for instance, has been shown to enhance climate resilience, sustainable land management, and disaster preparedness (IPBES, 2019). However, despite their proven relevance, IKS have historically been marginalized within dominant knowledge hierarchies, often dismissed as anecdotal or unscientific (de Sousa Santos, 2014).

This marginalization is closely tied to colonial epistemologies that privilege Western modes of knowing while undermining Indigenous epistemic sovereignty. Smith (2021) argues that knowledge extraction without consent, contextual erasure, and intellectual appropriation have systematically devalued Indigenous voices. As a result, contemporary efforts to revitalize IKS increasingly emphasize decolonization, self-determination, and community-led knowledge governance.

2.2 Digitalization and Indigenous Knowledge Preservation

The expansion of digital technologies has introduced new possibilities for documenting, preserving, and transmitting Indigenous knowledge. Digital archives, mobile applications, virtual reality storytelling, and online language repositories are increasingly used to safeguard endangered languages and cultural practices (Lewis et al., 2020). These initiatives have demonstrated that technology, when culturally grounded, can support intergenerational learning and strengthen Indigenous identity.

However, digitization also presents significant risks. Scholars caution that removing knowledge from its cultural, spiritual, and ecological context can lead to misinterpretation and commodification. Digital platforms often prioritize accessibility and scalability over relational accountability, which is central to Indigenous knowledge transmission. Moreover, issues of data ownership, consent, and control remain unresolved, particularly when Indigenous knowledge is stored on external servers or governed by non-Indigenous institutions (Rainie et al., 2017).

The concept of Indigenous data sovereignty has therefore gained prominence, asserting that Indigenous peoples have the right to govern the collection, ownership, and application of data derived from their communities (Kukutai & Taylor, 2016). Frameworks such as OCAP (Ownership, Control, Access, and Possession) and CARE (Collective Benefit, Authority to Control, Responsibility, Ethics) emphasize ethical stewardship and collective rights. These principles are particularly relevant as emerging technologies move beyond static digitization toward algorithmic processing of Indigenous knowledge.

2.3 Artificial Intelligence and Ethical Challenges

Artificial Intelligence represents a qualitative shift in digital knowledge systems, enabling automated classification, prediction, and decision-making at unprecedented scales. AI applications now influence domains ranging from education and healthcare to environmental management and cultural analytics (Floridi et al., 2018). While these technologies offer efficiency and innovation, they also raise profound ethical concerns, especially in relation to marginalized knowledge systems.

A growing body of literature highlights how AI systems reproduce existing social and epistemic biases due to skewed training data and culturally narrow design assumptions (Jobin et al., 2019). Mohamed et al. (2020) argue that AI development often reflects colonial power structures, extracting value from marginalized communities without reciprocal benefit. In the context of Indigenous knowledge, this raises fears of algorithmic appropriation, misrepresentation of sacred knowledge, and loss of epistemic control.

Ethical AI scholarship increasingly calls for transparency, accountability, fairness, and inclusivity (Floridi et al., 2018). However, critics argue that mainstream AI ethics frameworks remain largely Western-centric, focusing on individual rights rather than collective and intergenerational responsibilities emphasized in Indigenous worldviews (Lewis et al., 2020). Indigenous scholars advocate for rethinking AI ethics through relational accountability, respect for non-human entities, and long-term ecological balance.

Despite these challenges, AI also presents transformative opportunities for Indigenous communities. AI-powered tools have been used for climate modelling informed by Indigenous ecological indicators, automated transcription of Indigenous languages, and pattern recognition in cultural artefacts (Whyte, 2017). These applications suggest that AI can serve as a supportive instrument for IKS revitalization when aligned with Indigenous values and governance structures.

2.4 Socio-Technical and Cultural Perspectives on Technology Adoption

Understanding how Indigenous communities engage with AI requires moving beyond purely technical assessments to socio-cultural analyses of technology adoption. Socio-Technical Systems (STS) theory emphasizes that technology is co-shaped by social structures, cultural norms, and institutional contexts (Orlikowski, 2000). From this perspective, technological success depends not only on functional efficiency but on cultural compatibility and social legitimacy.

Traditional technology adoption models such as the Technology Acceptance Model (TAM) and the Unified Theory of Acceptance and Use of Technology (UTAUT) focus on perceived usefulness, ease of use, and behavioural intention (Venkatesh et al., 2012). While widely applied, these models have been criticized for their limited sensitivity to cultural diversity and collective decision-making processes. Indigenous contexts often prioritize communal benefit, ethical alignment, and cultural continuity over individual utility.

Recent studies suggest the need to extend adoption models by incorporating constructs such as trust, ethics, identity, and value alignment (Gefen et al., 2003). In Indigenous settings, Cultural Identity plays a central role in shaping attitudes toward external technologies, influencing both resistance and selective adoption (Battiste, 2014). Similarly, Ethical Trust, grounded in transparency, respect, and reciprocity, emerges as a critical determinant of technology acceptance (Rainie et al., 2019).

The concept of technology–culture compatibility has gained attention in cross-cultural information systems research, highlighting that misalignment between technological logic and cultural values leads to rejection or superficial adoption (Orlikowski, 2000). For Indigenous Knowledge Systems, compatibility involves alignment with oral traditions, relational epistemologies, and land-based knowledge practices. Technologies that fail to reflect these dimensions risk being perceived as epistemically invasive.

2.5 Toward a Techno–Cultural Convergence Perspective

Recent interdisciplinary scholarship advocates for integrative frameworks that bridge cultural epistemologies with emerging technologies. The notion of techno–cultural convergence recognizes that culture and technology are mutually constitutive, shaping and reshaping one another over time (de Sousa Santos, 2015). In Indigenous contexts, this convergence must be grounded in decolonial principles, ensuring that technology adoption strengthens rather than displaces ancestral knowledge.

Indigenous methodologies such as Story work and Two-Eyed Seeing provide conceptual tools for such integration. Story work emphasizes learning through narrative, relationality, and ethical responsibility (Archibald, 2008), while Two-Eyed Seeing promotes the complementary use of Indigenous and Western knowledge systems without subsuming one under the other (Bartlett et al., 2012). These approaches challenge binary oppositions between tradition and modernity, offering pathways for co-creating knowledge futures.

Despite growing recognition of these perspectives, empirical models that systematically integrate Indigenous epistemology with AI adoption analysis remain scarce. Most studies either focus on qualitative narratives without structural validation or apply quantitative models divorced from cultural context. This gap underscores the need for a mixed-methods framework that honours Indigenous voices while enabling predictive analysis.

2.6 Research Gap and Conceptual Positioning

The reviewed literature reveals three critical gaps. First, there is limited empirical research examining Indigenous perceptions of AI using culturally grounded methodologies. Second, existing technology adoption models inadequately capture collective, ethical, and identity-based determinants central to Indigenous decision-making. Third, AI ethics literature has yet to fully integrate Indigenous epistemologies as foundational, rather than peripheral, considerations.

Addressing these gaps, the present study positions the Techno–Cultural Convergence Model as a novel integrative framework that combines Indigenous knowledge scholarship, socio-technical systems theory, and structural equation modelling. By grounding quantitative constructs in qualitative Indigenous narratives, the study advances a pluralistic approach to understanding how ancestral wisdom and algorithmic futures can converge in ethically and culturally meaningful ways.

3. RESEARCH OBJECTIVES & RESEARCH QUESTIONS

Objectives:

To explore Indigenous perceptions of AI integration in cultural and educational contexts.
To develop the Techno–Cultural Convergence Model for aligning IKS with AI systems.
To empirically test the model using PLS-SEM.
To propose culturally grounded guidelines for ethical AI design in Indigenous environments.

Research Questions:

RQ1: What cultural and technological factors shape Indigenous communities' acceptance of AI?
RQ2: How compatible is AI with Indigenous epistemologies?
RQ3: What variables significantly predict AI adoption for knowledge preservation?

4. METHODOLOGY

This study adopts a sequential exploratory mixed-methods research design to investigate the techno–cultural convergence between Indigenous Knowledge Systems (IKS) and Artificial Intelligence (AI). The mixed-methods approach is particularly appropriate for research situated at the intersection of culture, technology, and epistemology, as it allows for the integration of Indigenous perspectives with empirical model testing. By combining qualitative inquiry grounded in Indigenous methodologies with quantitative structural modelling, the study ensures both contextual depth and analytical rigor, aligning with methodological expectations of SCOPUS-indexed journals.

4.1 Research Design

The research follows a QUAL → QUAN → Integration design. The qualitative phase precedes the quantitative phase and serves a foundational role in construct development. This sequencing ensures that the theoretical model is grounded in Indigenous voices rather than imposed by externally derived technological frameworks. The quantitative phase subsequently tests and validates the relationships identified qualitatively using Structural Equation Modelling. The final integration phase synthesizes findings to generate holistic interpretations and theoretical contributions.

4.2 Qualitative Phase: Indigenous-Centred Inquiry

The qualitative phase employed Indigenous Story work, narrative inquiry, and ethnographic interviewing to explore how Indigenous communities perceive AI in relation to cultural knowledge preservation and transmission. Story work was selected as it aligns with Indigenous epistemologies that privilege oral tradition, relational learning, and ethical accountability. Semi-structured interviews were conducted to allow participants to articulate experiences, concerns, and aspirations regarding AI in culturally meaningful ways.

Participants included Indigenous elders, knowledge keepers, cultural educators, youth leaders, and community practitioners. Purposeful and snowball sampling techniques were used to ensure representation across age groups and knowledge roles. Data collection from 30 respondents to achieve thematic saturation. Interviews were audio-recorded with informed consent, transcribed verbatim, and anonymized to protect cultural and personal identities.

Qualitative data were analysed using reflexive thematic analysis, following an inductive approach. Coding was conducted iteratively, with themes refined through constant comparison and reflexive memorising. Analytical rigor was ensured through prolonged engagement, peer debriefing, and participant validation, consistent with ethical guidelines for Indigenous research. The resulting themes informed the identification of key constructs related to cultural identity, ethical trust, perceived usefulness, compatibility, and adoption intention.

4.3 Development of the Conceptual Framework

Insights from the qualitative analysis were synthesized with existing literature on socio-technical systems, technology adoption, and Indigenous epistemology to develop the Techno–Cultural Convergence Model (TCCM). The model conceptualizes AI adoption as a culturally mediated process influenced by ethical alignment, epistemic compatibility, and collective identity. Constructs and hypothesized relationships were defined to reflect both Indigenous worldviews and established technology acceptance principles, ensuring conceptual validity and theoretical coherence.

4.4 Quantitative Phase: Survey Design and Sampling

The quantitative phase employed a cross-sectional survey design to empirically test the TCCM. Measurement items were adapted from validated scales in technology adoption and ethics literature, then contextualized using language and insights derived from the qualitative phase. A five-point Likert scale ranging from "strongly disagree" to "strongly agree" was used to capture respondent perceptions.

Data were collected from 410 Indigenous community members, educators, policymakers, and digital practitioners across multiple regions. A combination of purposive and stratified sampling ensured diversity in age, gender, educational background, and digital exposure. Responses were screened for completeness and consistency before analysis. The final sample size exceeded the minimum requirements for Partial Least Squares Structural Equation Modelling (PLS-SEM), ensuring adequate statistical power.

4.5 Data Analysis Technique

The study employed Partial Least Squares Structural Equation Modelling (PLS-SEM) using SmartPLS software to analyse the quantitative data. PLS-SEM was chosen due to its suitability for exploratory research, its ability to handle complex models with multiple constructs, and its robustness with non-normal data distributions. This technique is widely accepted in interdisciplinary research and is particularly appropriate for theory development and prediction-oriented studies.

The analysis followed a two-step procedure: evaluation of the measurement model followed by assessment of the structural model. Measurement model evaluation included tests of internal consistency reliability (Cronbach's alpha and composite reliability), convergent validity (average variance extracted), and discriminant validity (Fornell–Larcker criterion and heterotrait–monotrait ratio). The structural model was assessed using path coefficients, coefficients of determination (R^2), effect sizes (f^2), and predictive relevance (Q^2). Statistical significance was determined through bootstrapping with 5,000 resamples.

4.6 Moderation and Mediation Analysis

To capture the nuanced role of culture in technology adoption, the study incorporated moderation and mediation analyses within the PLS-SEM framework. Cultural Identity was modelled as a moderating variable influencing the relationship between IKS–AI Compatibility and Adoption Intention. Ethical AI Trust was examined as a mediating variable between Perceived Usefulness and Adoption Intention.

These analyses provided deeper insight into the mechanisms through which cultural and ethical factors shape AI adoption decisions.

4.7 Ethical Considerations

Ethical integrity was central to all stages of the research. The study adhered to principles of respect, reciprocity, and relational accountability, consistent with Indigenous research ethics. Informed consent was obtained from all participants, and community protocols regarding knowledge sharing were strictly observed. Sensitive cultural knowledge was excluded from analysis unless explicitly permitted, and findings were framed to avoid misrepresentation or appropriation. Data governance principles aligned with Indigenous data sovereignty frameworks, ensuring that knowledge generated through the research remains respectful of community rights and interests.

4.8 Integration of Qualitative and Quantitative Findings

The final phase involved the integration of qualitative and quantitative findings to generate meta-inferences. Convergences and divergences between narrative insights and structural relationships were examined to enrich interpretation. This integrative process enabled the validation of Indigenous perspectives through empirical modelling while preserving cultural nuance. The mixed-methods integration strengthened the study's explanatory power and enhanced the credibility of the proposed Techno–Cultural Convergence Model.

5. FINDINGS

This section presents the empirical findings derived from Partial Least Squares Structural Equation Modelling (PLS-SEM). Results are reported in three stages: (i) measurement model assessment, (ii) structural model evaluation, and (iii) moderation and mediation effects. All calculations are based on standardized estimates obtained through bootstrapping with 5,000 subsamples.

5.1 Descriptive Statistics

Table 1. Descriptive Statistics

Demographic Variable	Category	Frequency (n)	Percentage (%)
Gender	Male	218	53.2
	Female	182	44.4
	Others	10	2.4
Age Group	18–25 years	96	23.4
	26–35 years	128	31.2
	36–45 years	94	22.9

continued on following page

Table 1. Continued

Demographic Variable	Category	Frequency (n)	Percentage (%)
	46–55 years	58	14.1
	Above 55 years	34	8.3
Education Level	Secondary	72	17.6
	Higher Secondary	98	23.9
	Graduate	142	34.6
	Postgraduate	76	18.5
	Others	22	5.4
Occupation	Indigenous Community Member	110	26.8
	Educator	96	23.4
	Policymaker	58	14.1
	Digital Practitioner	102	24.9
	Others	44	10.7
Digital Exposure	Low	88	21.5
	Moderate	176	42.9
	High	146	35.6
Region	East India	124	30.2
	West India	86	21.0
	North India	92	22.4
	South India	78	19.0
	Northeast India	30	7.3

In Table 1, it is evident that the percentage difference in the distribution of the gender population presents a diversified population, which consists of people of all ages, probably with varying education levels up to certain levels of professionalism among others in the population (N=410). The greater number of respondents were men, with the figure (53.2%) instead of women. Almost one-third of the respondents were in their late twenties to mid-thirties, being recorded at 31.2%. The same would also mean that the majority of them were postgraduate level holders, with less than half reporting a career heritage of hardly 34.6%. Education also shows that there was a mixture of attitudes and clear positions among the public, as indigenous respondents did not turn up in digital responses against those in education or digital practice.

5.2 Measurement Model Assessment

This construct is grounded in:

- Theory of Planned Behavior (TPB) – intention as the strongest predictor of behavior (Ajzen, 1991)
- Technology Acceptance Model (TAM) – behavioral intention driven by perceived usefulness (Davis, 1989)

The constructs are adapted:

Cultural Identity (CI) (Phinney, 1992)

CI1: I strongly identify with my cultural traditions and heritage.
CI2: My cultural background influences my values and decision-making.
CI3: Preserving cultural knowledge is important to me.
CI4: I feel a sense of belonging to my cultural community.

Ethical AI Trust (EAT) (McKnight et al., 2002)

EAT1: I trust AI systems that operate in an ethical and transparent manner.
EAT2: I believe AI should respect human values and cultural norms.
EAT3: I feel confident using AI systems that ensure fairness and accountability.
EAT4: Ethical considerations increase my trust in AI technologies.

IKS–AI Compatibility (IAC) (Saatchi et al., 2025)

IAC1: AI technologies can effectively integrate Indian Knowledge Systems (IKS).
IAC2: AI can help preserve and promote traditional knowledge.
IAC3: IKS principles are compatible with modern AI technologies.
IAC4: AI systems can be designed to reflect indigenous knowledge and values.
IAC5: The combination of IKS and AI can create meaningful societal impact.

Perceived Usefulness (PUKP) (Davis, 1989)

PUKP1: Using AI integrated with IKS enhances my understanding of cultural knowledge.
PUKP2: AI tools based on IKS improve the efficiency of knowledge access.
PUKP3: I find AI applications incorporating IKS to be useful in my work/studies.
PUKP4: AI-enabled IKS systems improve the quality of decision-making.

Adoption Intention (AI) (Davis, 1989)

AI1: I intend to use AI systems that incorporate Indian Knowledge Systems in the future.
AI2: I am willing to adopt AI technologies aligned with cultural values.
AI3: I would recommend AI–IKS integrated systems to others.

Table 2. Reliability and Convergent Validity

Construct	Items	Cronbach's α	Composite Reliability (CR)	AVE
Cultural Identity (CI)	4	0.88	0.91	0.72
Ethical AI Trust (EAT)	4	0.86	0.90	0.69
IKS–AI Compatibility (IAC)	5	0.90	0.93	0.74
Perceived Usefulness (PUKP)	4	0.85	0.89	0.67
Adoption Intention (AI)	3	0.87	0.92	0.79

All constructs exceed the recommended thresholds for internal consistency ($\alpha > 0.70$; CR > 0.70) and convergent validity (AVE > 0.50), confirming the reliability and adequacy of the measurement model shown in Table 2.

5.3 Discriminant Validity

Table 3. Fornell–Larcker Criterion

Construct	CI	EAT	IAC	PUKP	AI
CI	0.85				
EAT	0.54	0.83			
IAC	0.61	0.58	0.86		
PUKP	0.49	0.56	0.64	0.82	
AI	0.52	0.60	0.71	0.63	0.89

Discriminant validity evaluation using the Fornell–Larcker criterion is shown in the report. It is evident from the panel that all the variables exhibit the properties of discriminant validity since the square root of the average variance extracted of any given variable is greater than the correlations of that variable with other variables. In the context of the study, this suggests that there is no concept, let alone a construct, that would be measured twice within the proposed model.

5.4 Structural Model Results

Table 4. Path Coefficients and Hypothesis Testing

Hypothesized Path	Β	t-value	p-value	Result
IAC → AI	0.54	9.87	<0.001	Supported
EAT → AI	0.33	6.12	<0.001	Supported
PUKP → AI	0.28	4.96	<0.01	Supported
CI → AI	0.19	3.84	<0.01	Supported

Results revealed that the structure model shows that all hypothesized avenues are magnified and substantially significant as presented in the table concerning IKS, the results showed that IKS–AI compatibility provoked the maximum influence in terms of the aspect of attitude formation ($\beta = 0.54$, $p < 0.001$), followed by trust related to ethical aspects of artificial intelligence ($\beta = 0.33$, $p < 0.001$) and usability of AI ($\beta = 0.28$, $p < 0.01$). Cultural identity is also valid and positively but rather weakly correlated with the aspect of attitude- adoption and use intention ($\beta = 0.19$, $p < 0.01$). These findings further indicate the involvement of Information and Communication technologies (ICTs) as well as social and cultural aspects of the society in using AI in Indian Knowledge Systems.

Model explanatory power:

R^2 (Adoption Intention) = 0.71
Q^2 (Predictive relevance) = 0.42
SRMR = 0.062

5.5 Moderation and Mediation Effects

Table 5. Moderation and Mediation Analysis

Effect	B	t-value	p-value	Interpretation
CI × IAC → AI	0.18	3.45	<0.001	Significant moderation
PUKP → EAT → AI	0.21	4.11	<0.001	Partial mediation

The assessment of the moderation effect indicates that cultural identity acts as a stronger enhancer of the effect of IKS–AI compatibility on the intention to adopt a system ($\beta = 0.18$, $p < 0.001$), which implies that cultural placement plays a more significant role in the alignment of people with the existing technology. Moreover, according to the causal effect revealed through mediation analysis, it is apparent that ethical AI trust somehow explains why perceived usefulness is related to the willingness to adopt ($\beta = 0.21$, $p < 0.001$). This means that the technological benefits engender increased adoption by the users not only directly but also through increasing willingness to use the artificial intelligence systems with ethical considerations. These two results point to the interrelation of the societal context and the way AI adoption is influenced by these trust-related factors.

Interpretation of Findings

The findings demonstrate strong empirical support for the Techno–Cultural Convergence Model. IKS–AI Compatibility emerged as the most influential predictor of AI adoption, confirming that Indigenous communities prioritise epistemic and cultural alignment over technical efficiency alone. Ethical AI Trust significantly shaped adoption intentions, underscoring the importance of transparency, data sovereignty, and ethical governance. Perceived usefulness further reinforced adoption when AI was viewed as supportive of cultural preservation. The moderating role of Cultural Identity reveals that stronger cultural grounding intensifies selective acceptance of AI, while mediation results highlight trust as a key mechanism translating perceived benefits into adoption. Together, these results validate culturally grounded pathways for integrating AI with Indigenous Knowledge Systems.

6. DISCUSSION

This study set out to reimagine Indigenous Knowledge Systems (IKS) in the age of artificial intelligence by addressing three core objectives: (i) to examine whether AI can be culturally aligned with Indigenous epistemologies, (ii) to identify the role of ethical trust in shaping AI adoption within Indigenous contexts, and (iii) to explain how cultural identity conditions the acceptance of AI-enabled knowledge systems. The findings provide coherent and theoretically grounded answers to the research questions, offering strong empirical validation for the proposed Techno–Cultural Convergence Model.

Addressing the first research question on the compatibility between IKS and AI, the results demonstrate that IKS–AI Compatibility is the most significant predictor of adoption intention. This finding confirms that Indigenous communities do not reject AI per se but selectively engage with technologies that respect relational knowledge, oral traditions, and contextual wisdom. This aligns with postcolonial and decolonial scholarship arguing that technology adoption in Indigenous settings is contingent upon epistemic justice and cultural resonance rather than technological sophistication alone. Thus, the study justifies the objective of moving beyond techno-deterministic narratives toward culturally embedded AI frameworks.

The second research question examined the role of ethical AI trust in facilitating Indigenous engagement with algorithmic systems. The strong and significant influence of ethical trust on adoption intention, as well as its mediating role between perceived usefulness and adoption, underscores trust as a critical governance mechanism. These findings respond directly to calls in the literature on Indigenous data sovereignty and responsible AI, which emphasise transparency, consent, and community ownership. The results validate the study's objective of positioning ethical trust not as a peripheral concern but as a central driver of sustainable AI–IKS integration.

The third research question explored how cultural identity shapes Indigenous responses to AI. The significant moderating effect of cultural identity reveals that stronger cultural grounding amplifies selective adoption when AI systems align with Indigenous values. Rather than resisting innovation, culturally rooted communities exercise greater discernment, reinforcing the argument that cultural identity acts as an adaptive filter rather than a barrier. This finding justifies the study's objective of reframing cultural identity as an enabling force in techno-cultural transformation, consistent with sociocultural and identity-based technology adoption theories.

Collectively, the discussion affirms that the revitalisation of Indigenous Knowledge Systems in the 21st century is not achieved through technological replacement but through respectful convergence. By empirically demonstrating how cultural compatibility, ethical trust, and identity interact to shape AI adoption, the study answers its research questions in a theoretically integrated and context-sensitive manner. The findings contribute to a growing interdisciplinary discourse advocating for pluralistic, ethical, and culturally grounded futures for artificial intelligence.

7. THEORETICAL AND PRACTICAL IMPLICATIONS

This study offers substantial theoretical and practical contributions by advancing a future-oriented understanding of how Indigenous Knowledge Systems (IKS) can be revitalised through culturally aligned and ethically governed artificial intelligence. By integrating techno–cultural convergence theory with sociocultural identity perspectives and responsible AI literature, the research responds to growing scholarly calls for epistemic plurality in digital transformation.

7.1 Theoretical Implications

Theoretically, this study extends existing technology adoption and knowledge systems literature by positioning Indigenous epistemologies as active agents rather than passive beneficiaries of digital innovation. The proposed Techno–Cultural Convergence Model moves beyond conventional frameworks such as the Technology Acceptance Model and UTAUT by embedding cultural identity, ethical trust, and epistemic compatibility as core explanatory constructs. This reconceptualization challenges the universality assumptions underlying dominant adoption theories and contributes to decolonial theory by empirically validating culturally situated pathways of technological engagement.

A key theoretical contribution lies in reframing cultural identity from a resistance-oriented construct to an adaptive and selective moderator of technological change. The findings demonstrate that stronger Indigenous identity enhances discernment in AI adoption, thereby supporting sociocultural theories that view identity as a dynamic filter shaping technology interpretation. This insight advances cultural psychology and information systems scholarship by offering empirical evidence that identity strength can facilitate, rather than inhibit, innovation when alignment is achieved.

Furthermore, the mediating role of ethical AI trust contributes to emerging responsible AI theory by empirically linking governance principles to behavioural outcomes. While ethical AI discussions often remain normative, this study operationalises trust as a measurable mechanism that translates perceived technological benefits into adoption intention. This theoretical advancement strengthens the bridge between ethics, governance, and behavioural science, offering a validated model that future researchers can adapt across culturally sensitive technological domains.

Finally, the integration of Indigenous qualitative epistemologies with quantitative PLS-SEM analysis advances mixed-methods research design in Indigenous studies. By demonstrating methodological complementarity between story work and structural modelling, the study contributes to methodological pluralism and sets a precedent for ethically grounded, empirically rigorous research in culturally complex contexts.

7.2 Practical and Policy Implications

From a practical standpoint, the findings provide clear guidance for AI developers, policymakers, educators, and Indigenous institutions seeking to design culturally responsive technologies. First, AI systems intended for Indigenous contexts must prioritise epistemic compatibility by incorporating local languages, narrative-based reasoning, ecological knowledge, and community-defined values. This implies a shift from extractive data practices toward co-creation models where Indigenous communities act as co-designers and custodians of AI systems.

Second, the centrality of ethical AI trust underscores the need for governance frameworks grounded in transparency, consent, and Indigenous data sovereignty. Policymakers should institutionalise ethical review mechanisms that recognise collective ownership of data and algorithms, moving beyond individual-centric regulatory models. Such frameworks are critical for ensuring long-term legitimacy and sustainability of AI interventions in Indigenous knowledge ecosystems.

Third, the study highlights the importance of culturally embedded capacity-building initiatives. Educational institutions and development agencies should invest in interdisciplinary training programs that combine digital literacy with Indigenous epistemology, enabling future generations to shape AI rather than merely adopt it. This future-oriented approach supports intergenerational knowledge transmission while fostering technological self-determination.

At a broader societal level, the findings suggest that AI can serve as a tool for cultural revitalisation, environmental stewardship, and inclusive development when guided by ethical and cultural principles. By aligning technological innovation with Indigenous worldviews, AI can contribute to more sustainable and pluralistic futures, addressing global challenges such as climate resilience, biodiversity conservation, and social equity.

8. CONCLUSION

This study set out to reimagine the future of Indigenous Knowledge Systems (IKS) in the context of rapid advances in artificial intelligence by advancing a culturally grounded, ethically informed, and empirically validated framework of techno–cultural convergence. Moving beyond narratives of technological disruption and epistemic loss, the research demonstrates that AI, when designed and governed in alignment with Indigenous epistemologies, can function as a complementary and enabling force for knowledge revitalisation rather than as an instrument of cultural erasure.

Drawing on a mixed-methods design that integrates Indigenous story work with Partial Least Squares Structural Equation Modelling, the study provides robust evidence that Indigenous engagement with AI is shaped primarily by epistemic compatibility, ethical trust, and cultural identity. The findings reveal that Indigenous communities exhibit a selective and reflective orientation toward AI adoption, privileging technologies that respect relational worldviews, collective knowledge ownership, and contextual meaning. This insight challenges dominant techno-deterministic assumptions and affirms the legitimacy of Indigenous epistemologies as dynamic, adaptive systems capable of interacting meaningfully with advanced digital technologies.

A central contribution of the study lies in its empirical validation of ethical AI trust as a pivotal mechanism through which perceived technological benefits translate into adoption intention. Trust, grounded in transparency, consent, and data sovereignty, emerges not as a peripheral concern but as a foundational condition for sustainable AI integration. This finding reinforces the argument that ethical governance is inseparable from technological effectiveness in culturally sensitive contexts. Without trust, even technically sophisticated systems risk rejection or epistemic harm.

Equally significant is the study's reconceptualization of cultural identity as an enabling rather than inhibiting force in technological transformation. The moderating role of cultural identity illustrates that stronger cultural grounding enhances discernment and agency, allowing Indigenous communities to engage with AI on their own terms. This reframing contributes to broader theoretical debates on identity and innovation, positioning culture as a source of resilience and strategic adaptation in the digital age.

From a societal perspective, the study underscores the potential of culturally aligned AI to support intergenerational knowledge transmission, environmental stewardship, and inclusive development. By embedding Indigenous values into algorithmic design, AI systems can contribute to addressing global challenges such as climate change, biodiversity loss, and social inequality through locally grounded and ethically informed solutions. This vision aligns with emerging global discourses advocating pluralistic, sustainable, and human-centred approaches to artificial intelligence.

In conclusion, this research argues that the future of AI need not be defined by epistemic homogenisation or technological dominance. Instead, it can be shaped through respectful convergence, where ancestral wisdom and algorithmic intelligence co-evolve in mutually reinforcing ways. By offering a theoretically integrated, empirically grounded, and future-oriented framework, the study provides a scholarly foundation for rethinking AI–IKS relationships and calls for a paradigm shift toward relational, ethical, and culturally inclusive digital futures.

9. FUTURE RESEARCH DIRECTIONS

Further research is needed to enhance the interdisciplinary grasp of how Indigenous Knowledge Systems and Artificial Intelligence relate to one another. This can be accomplished through additional inquiries in cultural, ethical, technological, and policy aspects. To elaborate further, extensive research is indispensable to ascertain the transformation of the outlook of the native people on AI, considering the increasing penetration of digital gadgets as well as applications like AI. These romp into the future would help understand the extent of metalogical convergence and the effect of AI on the processes of multi-generational knowledge.

Future researchers should explore the creation of AI tools that are used based on the co-investment of team members or community assurance and governance systems, rather than the traditional consumer usage models. This research would provide a significant contribution to the areas of indigenous report aging, algorithmic management, ethics, and creation, among others. Additionally, cross-cultural comparisons between differing indigenous peoples and places might indicate unique adaptations of AI technology and suggest varying abilities of found content distance, trust, and readiness to technology performance.

In terms of ways and according to supplied requests, it is possible to mention the fields for which the use of artificial intelligence systems allows economic growth, particularly, for environmental changes, saving the diversity of nature, retaining the national languages, increasing the level of medical services, improving the conditions of the educational process, and storing cultural heritage. Perhaps there is some scope for the use of AI-based and AI-enabled systems to promote ecological conservation and support for local knowledge. New technologies such as generative AI, virtual reality, especially immersive types, and autonomous machinery will require a more profound perspective with respect to bias in algorithms, issues surrounding the 'true' representation of society, copyright laws, and judicial consequences.

In conclusion, future research should encompass decolonial methodologies alongside data analytics models aimed at developing AI rooted in cultural principles that are mindful of relational ethics, collective well-being, and concern for the community. It is essential to the development of technology to realize digital self-determination that communities and indigenous peoples, in particular, should be characterized by inclusiveness and pluralism.

10. LIMITATIONS

There have been many areas where this research did remarkably well; however, there are other aspects where this study could have improved. To start with, it is pertinent to point out the fact that the design of this particular study is cross-sectional, and thus, the observations that have been established through the research are not ideal for viewing the long-term effects and perceptions of AI among the indigenous people. Secondly, the small portion of the board exemplified different Indigenous communities, which might pose a challenge as it may not imply that the results can be applied across different customs and territories. Moreover, the collapse of peculiar Indigenous perspectives into certain reduced and quantified parameters may be deemed to have been too simplistic and may not allow for a comprehensive understanding of their meaning, given their context. Another major concern about this study is that it very much depended on the use of self-reported responses, which often results in response bias. Additionally, in light of the developments within the AI technologies, it may not provide the same influences or challenges to the technological cultural relationship in the future that would require any further and long-term study.

REFERENCES

Archibald, J. A. (2008). *Indigenous storywork: Educating the heart, mind, body, and spirit*. UBC press. DOI: 10.59962/9780774855440

Bartlett, C., Marshall, M., & Marshall, A. (2012). Two-eyed seeing and other lessons learned within a co-learning journey of bringing together indigenous and mainstream knowledges and ways of knowing. *Journal of Environmental Studies and Sciences*, *2*(4), 331–340. DOI: 10.1007/s13412-012-0086-8

Battiste, M. (2014). Decolonizing education: Nourishing the learning spirit. *The Alberta Journal of Educational Research*, *60*(3), 615–618.

Berkes, F. (2017). *Sacred ecology* (4th ed.). Routledge. DOI: 10.4324/9781315114644

Creswell, J. W., & Clark, V. P. (2007). *Mixed methods research.*

Davis, F. D. (1989). Perceived usefulness, perceived ease of use, and user acceptance of information technology. *Management Information Systems Quarterly*, *13*(3), 319–340. DOI: 10.2307/249008

de Sousa Santos, B. (2015). *Epistemologies of the South: Justice against epistemicide*. Routledge. DOI: 10.4324/9781315634876

Floridi, L., Cowls, J., Beltrametti, M., Chatila, R., Chazerand, P., Dignum, V., Luetge, C., Madelin, R., Pagallo, U., Rossi, F., Schafer, B., Valcke, P., & Vayena, E. (2018). AI4People—An ethical framework for a good AI society: Opportunities, risks, principles, and recommendations. *Minds and Machines*, *28*(4), 689–707. DOI: 10.1007/s11023-018-9482-5 PMID: 30930541

Fricker, M. (2007). *Epistemic injustice: Power and the ethics of knowing*. Oxford university press. DOI: 10.1093/acprof:oso/9780198237907.001.0001

Gefen, D., Karahanna, E., & Straub, D. W. (2003). Trust and tam in online shopping: An integrated model1. *Management Information Systems Quarterly*, *27*(1), 51–90. DOI: 10.2307/30036519

IPBES. W. (2019). Intergovernmental science-policy platform on biodiversity and ecosystem services. *Summary for policy makers of the global assessment report on biodiversity and ecosystem services of the intergovernmental science-policy platform on biodiversity and ecosystem services. IPBES Secretariat, Bonn, Germany.*

Jobin, A., Ienca, M., & Vayena, E. (2019). The global landscape of AI ethics guidelines. *Nature Machine Intelligence*, *1*(9), 389–399. DOI: 10.1038/s42256-019-0088-2

Kovach, M. (2021). *Indigenous methodologies: Characteristics, conversations, and contexts*. University of Toronto press.

Kukutai, T., & Taylor, J. (Eds.). (2016). *Indigenous data sovereignty: Toward an agenda* (Vol. 38). ANU press. DOI: 10.22459/CAEPR38.11.2016

Lewis, J. E., Arista, N., Pechawis, A., & Kite, S. (2023). *Making kin with the machines. Feminist AI.* Oxford University Press.

McKnight, D. H., Choudhury, V., & Kacmar, C. (2002). Developing and validating trust measures for e-commerce: An integrative typology. *Information Systems Research, 13*(3), 334–359. DOI: 10.1287/isre.13.3.334.81

Mohamed, S., Png, M. T., & Isaac, W. (2020). Decolonial AI: Decolonial theory as sociotechnical foresight in artificial intelligence. *Philosophy & Technology, 33*(4), 659–684. DOI: 10.1007/s13347-020-00405-8

Orlikowski, W. J. (2000). Using technology and constituting structures: A practice lens for studying technology in organizations. *Organization Science, 11*(4), 404–428. DOI: 10.1287/orsc.11.4.404.14600

Phinney, J. S. (1992). The multigroup ethnic identity measure: A new scale for use with diverse groups. *Journal of Adolescent Research, 7*(2), 156–176. DOI: 10.1177/074355489272003

Rainie, S. C., Lee Schultz, J., Briggs, E., Riggs, P., & Palmanteer-Holder, N. L. (2017). Data as a Strategic Resource. *International Indigenous Policy Journal, 8*(2), 1–29. DOI: 10.18584/iipj.2017.8.2.1

Saatchi, S. G., Wahed, M. K. Y. A., Alqaraleh, M. K. S., Al-Shorman, H. M., Alanazi, T., Alka'awneh, S. M. N., & Alzyoud, M. (2025). The influence of compatibility on the acceptance of artificial intelligence in Kuwaiti universities. In *Intelligence-Driven Circular Economy: Regeneration Towards Sustainability and Social Responsibility–Volume 1* (pp. 103–117). Springer Nature Switzerland. DOI: 10.1007/978-3-031-73899-9_9

Smith, L. T. (2021). Decolonizing methodologies: Research and indigenous peoples.

Venkatesh, V., Thong, J. Y., & Xu, X. (2012). Consumer acceptance and use of information technology: Extending the Unified Theory of Acceptance and Use of Technology1. *Management Information Systems Quarterly, 36*(1), 157–178. DOI: 10.2307/41410412

Whyte, K. (2017). Indigenous climate change studies: Indigenizing futures, decolonizing the Anthropocene. *English Language Notes, 55*(1), 153–162. DOI: 10.1215/00138282-55.1-2.153

ADDITIONAL READINGS

Archibald, J. A. (2008). *Indigenous storywork: Educating the heart, mind, body, and spirit.* UBC press. DOI: 10.59962/9780774855440

Battiste, M. (2019). *Decolonizing education: Nourishing the learning spirit.* University of British Columbia Press. DOI: 10.59962/9781895830972

Berkes, F. (2017). *Sacred ecology.* Routledge. DOI: 10.4324/9781315114644

Fricker, M. (2007). *Epistemic injustice: Power and the ethics of knowing.* Oxford university press. DOI: 10.1093/acprof:oso/9780198237907.001.0001

Mohamed, S., Png, M. T., & Isaac, W. (2020). Decolonial AI: Decolonial theory as sociotechnical foresight in artificial intelligence. *Philosophy & Technology, 33*(4), 659–684. DOI: 10.1007/s13347-020-00405-8

Smith, L. T. (2021). Decolonizing methodologies: Research and indigenous peoples.

KEY TERMS AND DEFINITIONS

Cognitive Justice: It's about recognition and equal respect for the multiple knowledge systems, cultures, and ways of knowing that people bring in, not just one official version.

Collective Rights: Are the rights that belong to a community as a whole, not only to individual members.

Cultural Continuity: The continuation and preservation of cultural routines, traditions, values, language, and heritage across many generations.

Environmental Stewardship: It involves careful governance and protection of natural resources, both for the people now and for those coming in later generations, so it lasts.

Inclusivity: Making sure equal participation happens, with proper representation and access, for diverse people and communities, even when resources are uneven.

Knowledge Co-creation: A shared method for making knowledge together, through joint participation, conversation, and reciprocal learning, so people can learn from each other.

Knowledge Preservation: Protection and transmission of cultural knowledge, traditions, practices, and shared memory, so the collective past does not just fade out.

Relational Accountability: This is the ethical duty to maintain respectful, reciprocal relationships, both during research and in everyday interactions with communities.

Resilience: The ability to adapt, bounce back, and maintain identity even when social or environmental conditions shift. Indigenous communities - Native cultural groups keep their traditional knowledge, customs, languages, and also an old kinship connection to ancestral places, you know. Knowledge Revitalization: Restoring and reinforcing at-risk cultural knowledge, traditions, and those Indigenous practices that are starting to disappear.

Responsible AI: The ongoing use of AI systems, steered by ethics, fairness, transparency, and real accountability.

Techno–cultural Compatibility: The fit between technological systems and the cultural worth, beliefs, and practices held by communities has to match.

Chapter 2
Bridging Worlds:
The Vital Role of Indigenous Knowledge in Education and Research

Rick Rantz
http://orcid.org/0009-0004-4704-6740
Allan Hancock College, USA

ABSTRACT

This chapter examines Indigenous Knowledge and Indigenous Research Methodologies as transformative frameworks capable of reshaping higher education. Grounded in relationality, land-based understanding, ceremony, community governance, and intergenerational responsibility, these methodologies challenge Western epistemic dominance and reveal the limits of Eurocentric research assumptions. The chapter traces the historical exclusion of Indigenous epistemologies, analyzes core methodological principles such as storywork, collective consent, and relational accountability, and discusses implications for ethics, data sovereignty, academic publishing, teaching, and graduate supervision. It argues that supporting Indigenous methodologies requires structural change across governance, curriculum, tenure, and research policy. Future directions highlight Indigenous-led research networks, digital sovereignty, ecological knowledge, and pluriversal academic models grounded in Indigenous intellectual sovereignty.

INTRODUCTION

Maya, a Cree doctoral student, sits before her university's research ethics board. She has carefully designed a community-based study following protocols established by her Nation's Elders. The ethics committee, however, insists she cannot proceed without individual written consent forms—a requirement that contradicts her community's understanding that knowledge belongs collectively and that oral consent, witnessed by Elders, carries greater weight than signatures on institutional forms. Maya faces an impossible choice: honor her community's protocols or satisfy Western institutional requirements. This tension is not unique to Maya—it reflects a fundamental epistemological collision at the heart of contemporary higher education.

Across worldwide systems of higher education, Indigenous peoples have long insisted that Indigenous Knowledge and Indigenous Research Methodologies possess the intellectual rigor, philosophical integrity, and ethical grounding necessary to shape scholarly inquiry on their own terms. Yet the acad-

DOI: 10.4018/979-8-3373-7976-0.ch002

emy continues to be structured by Western epistemologies that privilege particular forms of rationality, evidence, and method, rendering Indigenous Knowledge peripheral or, in some cases, invisible. This marginalization is not accidental but a direct reflection of colonial histories in which universities served as instruments of epistemic domination, asserting the universality of Western ways of knowing while dismissing Indigenous knowledge systems as premodern or non-scientific (Smith, 2021; Stein, 2022). The resulting epistemic hierarchy persists today, constraining the kinds of questions that can be asked, the methods considered legitimate, and the forms of authority recognized within academic research.

The powerful argument advanced by Tuck and Yang (2012) demonstrates that decolonization cannot be a metaphor—it requires material transformation, not merely symbolic inclusion. Universities that claim to value Indigenous Knowledge while maintaining colonial structures engage in what these scholars call "settler moves to innocence," gestures that relieve institutional guilt without relinquishing power or restructuring systems. This insight challenges higher education to move beyond inclusion rhetoric toward genuine epistemic transformation.

This chapter argues that Indigenous Research Methodologies offer a transformative paradigm through which higher education can reconceptualize its foundations. Rather than simply adding Indigenous content to curricula or including Indigenous voices within Western methodological frameworks, Indigenous methodologies require a restructuring of the academy's epistemic architecture. They demand recognition that Indigenous Knowledge systems constitute complete, coherent, and sophisticated intellectual traditions with their own ontological assumptions, sensory and spiritual realities, relational ethics, and community-governed modes of validation (Wilson, 2020; Kovach, 2021; Chilisa, 2019). To meaningfully support Indigenous Knowledge within higher education, universities must therefore shift from inclusion to transformation—moving from tokenistic gestures to paradigm-level reorientation grounded in Indigenous sovereignty, relational accountability, and epistemic justice.

Indigenous Research Methodologies cannot be understood without acknowledging the long-standing epistemological, cultural, and political contexts that shape them. These methodologies arise from Indigenous worldviews in which knowledge is inseparable from land, community, ancestry, spirituality, and moral responsibility (Absolon, 2022; Drouin-Gagné, 2021). Knowledge is not an object to be extracted and stored; it is a process of relating, understanding, and honouring one's place within an interconnected world.

HISTORICAL EXCLUSION AND THE COLONIAL UNIVERSITY

The exclusion of Indigenous Knowledge from higher education cannot be understood apart from the colonial logics and structures that have shaped modern academic institutions. During colonization, universities were established not merely alongside it but as active instruments of colonization (Stein, 2022). Through curriculum, research, and credentialing systems, universities legitimized the dispossession of Indigenous lands, the suppression of Indigenous languages, and the erasure of Indigenous intellectual traditions. Indigenous Knowledge was systematically categorized as folklore, tradition, or belief—categories that positioned it outside the realm of science, philosophy, and legitimate scholarship (Smith, 2021).

This epistemic violence operated through multiple mechanisms. First, universities established disciplinary boundaries that fragmented holistic Indigenous Knowledge systems into discrete, manageable units that could be studied by anthropologists, linguists, or folklorists without acknowledging Indigenous peoples as knowledge producers in their own right (Battiste, 2013). Second, research ethics developed in response to egregious violations in Western contexts (such as the Tuskegee experiments) but failed to address the particular harms of extractive research in Indigenous communities, where generations of researchers took knowledge, published findings, built careers, and returned nothing to the communities whose knowledge they appropriated (Smith, 1999, 2021).

Third, the structure of academic labor—tenure requirements, publication metrics, grant funding cycles—privileged individual achievement and rapid output over the relational, long-term, community-accountable work that Indigenous methodologies require (Wilson, 2020). Indigenous scholars who insisted on following community protocols often found themselves penalized in tenure reviews for "slow" productivity or publications in "non-traditional" venues such as community reports or Indigenous-language journals. Fourth, peer review systems, while ostensibly ensuring quality, often served as gatekeeping mechanisms that excluded Indigenous Knowledge by applying Western standards of validity, evidence, and argumentation (Shay et al., 2022).

The historical exclusion of Indigenous Knowledge from universities also intersects with the exclusion of Indigenous peoples themselves. Residential schools and other assimilationist policies actively prevented Indigenous children from accessing higher education while simultaneously working to sever their connections to Indigenous Knowledge systems (Truth and Reconciliation Commission of Canada, 2015). When Indigenous students did enter universities, they often encountered curricula that erased Indigenous histories, research that pathologized Indigenous communities, and institutional cultures that treated Indigenous Knowledge as object of study rather than valid epistemology.

Yet this history is also one of resistance and resilience. Indigenous scholars, students, and communities have never ceased asserting the validity and necessity of Indigenous Knowledge. From the 1960s onward, Indigenous-led movements demanded the establishment of Indigenous Studies programs, the hiring of Indigenous faculty, and the recognition of Indigenous research methodologies (Smith, 1999). These efforts have borne fruit: today, Indigenous research institutes, Indigenous-led journals, and Indigenous degree programs exist across multiple continents, embodying possibilities for transformation. The Sámi University of Applied Sciences in Norway conducts research in Sámi languages and prioritizes Sámi knowledge systems. The First Nations University of Canada integrates Indigenous Knowledge throughout all programs while maintaining academic rigor. Journals like *Decolonization: Indigeneity, Education & Society* and *AlterNative: An International Journal of Indigenous Peoples* provide venues where Indigenous scholars control peer review, editorial decisions, and knowledge dissemination.

The transition from historical exclusion to contemporary transformation requires understanding that Indigenous Research Methodologies are not simply alternative techniques to be added to existing frameworks. They represent fundamentally different ways of conceptualizing knowledge, research relationships, and the purposes of inquiry itself.

INDIGENOUS RESEARCH METHODOLOGIES AS TRANSFORMATIVE PARADIGM

Acknowledging Methodological Pluralism and Nation-Specific Contexts

Before examining shared principles across Indigenous Research Methodologies, it is essential to acknowledge that there is no single, monolithic "Indigenous methodology." Indigenous peoples worldwide represent extraordinary diversity—thousands of distinct nations, languages, cultural practices, and knowledge systems (Chilisa, 2019). Coast Salish methodologies differ from Māori methodologies, which differ from Sámi methodologies, which differ from Zulu methodologies. Each emerges from specific cultural, linguistic, political, and ecological contexts shaped by particular relationships to land, governance structures, spiritual practices, and historical experiences.

The risk of overgeneralization is real and consequential. Pan-Indigenous frameworks, while valuable for identifying common concerns and building solidarity, can inadvertently erase nation-specific protocols, collapse linguistic diversity, and reproduce colonial tendencies to treat Indigenous peoples as undifferentiated masses (Simpson, 2014). Urban Indigenous experiences differ from land-based community contexts; matriarchal nations have different governance structures than patriarchal ones; debates within Indigenous communities about collaboration with Western institutions reflect genuine disagreements about strategy, risk, and benefit, not consensus (TallBear, 2013). Some Indigenous nations have developed extensive written protocols for research, while others maintain oral protocols that vary by context and relationship. Some communities welcome certain forms of research collaboration, while others have declared moratoria on academic research due to historical harms.

Therefore, when this chapter discusses "Indigenous Research Methodologies," it refers to a family of approaches that share certain foundational principles—relationality, reciprocity, respect, responsibility—while recognizing that the specific application of these principles must be determined by the Indigenous nations and communities in question (Kovach, 2021). Researchers working with Indigenous communities must begin by asking: Whose protocols apply here? What are the specific requirements of this Nation, this community, this place? Whose permission is needed, and through what processes? What constitutes appropriate reciprocity in this context? The framework offered here provides orientation, not prescription. It cannot substitute for the patient, humble work of building relationships with specific Indigenous communities and learning their particular protocols, languages, and knowledge systems.

This emphasis on specificity does not preclude discussion of shared principles. Across diverse Indigenous contexts, certain commitments recur: the inseparability of knowledge from place and relationship; the understanding that knowledge carries responsibilities; the recognition that research must benefit communities, not only researchers; the valuing of intergenerational knowledge transmission; the integration of spiritual and material dimensions of reality. These commonalities emerge not from pan-Indigenous essentialism but from the fact that Indigenous Knowledge systems have developed in ongoing relationship with particular lands, waters, and more-than-human relations over millennia, creating epistemologies attuned to connection, reciprocity, and sustainability.

Relationality and Relational Accountability

At the heart of Indigenous Research Methodologies lies the principle of relationality—the understanding that all beings exist in relationship and that knowledge emerges through these relationships (Wilson, 2020). In Indigenous epistemologies, knowledge is not an object to be discovered and possessed by individual minds but a process of relating, understanding, and honouring one's place within an interconnected world. This contrasts sharply with Western research paradigms that position the researcher as objective observer, separate from and above the phenomena under study. Indigenous methodologies reject this subject-object dualism, recognizing instead that researchers are always already in relationship with what they study, and that these relationships carry ethical obligations (Kovach, 2021).

The concept of relational accountability, developed in research by Wilson (2020), operationalizes this principle. Relational accountability means being answerable to all one's relations—not only human relations but also ancestors, future generations, land, water, plants, animals, and spiritual beings. In research terms, this means that methodological decisions cannot be justified solely by reference to academic standards, theoretical frameworks, or career advancement. Researchers must ask: How does this research honour my relationships? How does it benefit the community? What responsibilities does it create? What will I give back? How will I maintain accountability after the research concludes?

This principle fundamentally reshapes every aspect of research design and practice. Research questions emerge from community priorities, not solely from gaps in academic literature. Informed consent becomes ongoing consent, negotiated throughout the research process rather than secured once at the beginning through signed forms (Kovach, 2021). Confidentiality may be less important than proper attribution, as community members may wish to be named as knowledge holders whose contributions deserve recognition. Data ownership remains with communities, not researchers or institutions. Publication timelines accommodate community review processes, even when this conflicts with academic deadlines or grant requirements.

Tania, a Mohawk doctoral student, describes navigating these tensions: "My academic supervisor understands theory, but my community mentor teaches me how to be a good relative in my research. When I felt torn between publishing quickly and waiting for community approval, my mentor reminded me: 'Your degree will mean nothing if your community no longer trusts you.' That changed everything about my timeline and my understanding of what success means."

Relational accountability extends beyond the immediate research context. It requires researchers to consider how their work fits within longer histories of research relationships between Indigenous communities and academic institutions—histories often marked by extraction, misrepresentation, and broken promises (Smith, 2021). It demands attention to power dynamics, recognition of whose knowledge is being shared and for whose benefit, and commitment to ensuring that research strengthens rather than exploits relationships.

Indigenous Scientific Knowledge and STEM Methodologies

Indigenous methodologies are not limited to social sciences, humanities, or qualitative research. Indigenous Scientific Knowledge represents sophisticated, empirically grounded understanding of ecological systems, astronomy, mathematics, engineering, medicine, and agriculture developed over millennia through careful observation, experimentation, and intergenerational knowledge transmission (Cajete, 2000; Kimmerer, 2013). Yet Western science education and research often exclude or marginalize this knowledge, treating Indigenous science as "traditional ecological knowledge"—a term that can imply outdatedness—rather than recognizing it as ongoing, dynamic, rigorous scientific practice.

The work of Robin Wall Kimmerer exemplifies the integration of Indigenous Knowledge and Western science. As a botanist and member of the Citizen Potawatomi Nation, Kimmerer demonstrates how Indigenous understanding of plant relationships, seasonal cycles, and reciprocal harvesting practices constitutes rigorous scientific knowledge that complements and often exceeds Western botanical science in its attention to ecological relationships and long-term sustainability (Kimmerer, 2013). Her concept of the "Honorable Harvest"—taking only what you need, using everything you take, never taking the first or last, asking permission, giving thanks—represents both ethical framework and sustainable ecological practice grounded in centuries of empirical observation about how to maintain thriving ecosystems.

In *Native Science*, Cajete (2000) argues that Indigenous science is not merely traditional ecological knowledge but a complete scientific paradigm with its own methods of observation, hypothesis testing, and knowledge validation. Indigenous astronomers mapped star systems and predicted celestial events with remarkable accuracy. Indigenous engineers designed irrigation systems, architectural structures, and agricultural technologies adapted to local environments and sustainable over centuries. Indigenous pharmacologists developed medicines that Western science later "discovered" and patented, often without acknowledgment or compensation to the Indigenous peoples who developed and maintained this knowledge.

Research by McGregor (2004) on Indigenous environmental justice demonstrates how Indigenous Knowledge offers essential insights for addressing the climate crisis. Indigenous communities have been documenting environmental changes, practicing sustainable resource management, and maintaining biodiversity for generations. Yet their knowledge is often excluded from environmental policy, climate science, and conservation biology, despite its empirical validity, practical utility, and grounding in long-term relationships with specific ecosystems. When Indigenous Knowledge is included, it is often extracted and translated into Western scientific frameworks rather than respected on its own terms.

The exclusion of Indigenous Scientific Knowledge from STEM fields reflects epistemic colonization—the assumption that Western science represents the only valid form of systematic inquiry into the natural world. Universities that claim to value evidence-based knowledge while dismissing Indigenous science engage in contradiction. Transformation requires not only including Indigenous content in science curricula but recognizing Indigenous peoples as scientists, Indigenous methods as scientific, and Indigenous Knowledge as essential to addressing global challenges from climate change to biodiversity loss to sustainable resource management.

Storywork, Ceremony, and Land-Based Methodologies

Indigenous methodologies employ diverse methods for knowledge generation, transmission, and validation that extend beyond Western research conventions. Storywork, as theorized by Jo-ann Archibald Q'um Q'um Xiiem (Archibald, 2008) and further developed by Kovach (2021), recognizes storytelling as a rigorous research method. Stories are not merely data to be collected and analyzed but living teachings that carry knowledge across generations, embed ethical lessons, create relationships between teller and listener, and connect human communities with the more-than-human world.

In research terms, storywork involves careful attention to protocol: who has the right to tell which stories, in what contexts, for what purposes, and with what permissions. It requires reciprocity—researchers who receive stories must give something back, whether through honoraria, practical support for the community, or ensuring that stories are shared in ways that benefit the community rather than serving only academic purposes. It demands respect for the power and sacredness of stories, recognizing that some stories are sacred and not meant for public sharing, some are seasonal and should only be told at particular times, some belong to particular families or clans and cannot be shared without their consent.

Ceremony is another essential methodological component. Ceremonies mark the beginning and ending of research projects, acknowledge relationships, seek permission from spiritual beings and ancestors, offer thanks, and maintain balance (Wilson, 2020). For researchers trained in secular Western traditions, ceremony may seem peripheral to "real" research—a cultural nicety rather than methodological necessity.

But in Indigenous methodologies, ceremony *is* methodology. It is how researchers position themselves rightly in relation to their work, how they maintain accountability to more than human relations, how they ensure that research serves life-affirming rather than extractive purposes, and how they maintain spiritual and emotional balance in work that often engages with difficult histories and ongoing injustices.

Land-based learning recognizes that knowledge is inseparable from place (Simpson, 2014). Indigenous Knowledge emerges from intimate, multi-generational relationships with particular lands, waters, and ecosystems. It cannot be fully understood in classrooms or laboratories disconnected from these places. Land-based methodologies involve spending time on the land, learning from Elders and knowledge keepers in context, observing seasonal cycles, understanding how knowledge is embedded in plant and animal behaviors and landscape features, and experiencing the embodied, sensory dimensions of knowledge that cannot be conveyed through text alone.

For universities, supporting land-based methodologies requires infrastructure and resources: transportation to land-based learning sites, insurance for off-campus activities, flexible scheduling that accommodates seasonal activities like harvests or ceremonies, budgets for Elder honoraria and community feast protocols, and institutional recognition that time spent on the land is as valuable as time spent in libraries or laboratories (McDonald, 2023). It also requires accessibility considerations—ensuring that Indigenous peoples with disabilities can participate in land-based learning through adapted practices, accessible trails, audio recordings of teachings, or alternative formats that honor the principle that land-based knowledge is for all community members.

ETHICAL AND METHODOLOGICAL FOUNDATIONS OF INDIGENOUS RESEARCH

Principles and Practices of Indigenous Research Ethics

Indigenous research ethics differ fundamentally from Western biomedical ethics models that dominate institutional review boards. While Western ethics emphasize individual autonomy, informed consent, confidentiality, and minimization of risk, Indigenous ethics prioritize collective benefit, ongoing relational consent, proper attribution, and reciprocity (Kovach, 2021; Smith, 2021; Seehawer, 2018).

The principle of collective benefit means that research must serve the community, not merely the researcher's career or institutional prestige. This challenges academic incentive structures that reward individual publication and grant acquisition. Indigenous communities have experienced centuries of extractive research—researchers who took knowledge, published findings, built careers, and never returned to share benefits or address community priorities (Smith, 2021). Indigenous ethics demand reciprocity: What does the community gain from this research? How will findings be shared in accessible formats? What practical applications will benefit the community? How will the research strengthen community capacity, support community priorities, or contribute to community wellbeing?

Ongoing relational consent recognizes that research relationships evolve over time and that consent is not a single event but a continuous process. A community may agree to participate initially but later decide that certain information should not be published, that the research direction should shift, or that the project should pause or end (Kovach, 2021). Western ethics typically treat consent as a one-time event secured before research begins through signed forms. Indigenous ethics understand consent as continuous negotiation, requiring researchers to remain in relationship throughout and beyond the formal research period, to check in regularly about whether the research is still serving community interests, and to respect community decisions to withdraw consent or modify research parameters.

Proper attribution challenges Western norms of confidentiality and anonymity. While protecting vulnerable individuals remains important, many Indigenous knowledge holders wish to be named and credited for their knowledge. Anonymizing Elders' teachings can constitute a form of erasure, denying them recognition as knowledge authorities and severing knowledge from its relational context (MacLeod, 2021). Indigenous research ethics require asking: Does this person wish to be named? What form of attribution honours their contribution? How can we cite oral knowledge in ways that respect both academic conventions and Indigenous protocols?

These ethical principles are not abstract ideals but practical requirements that shape every aspect of research design. They influence how research questions are formulated, who is involved in research teams, how data is collected and stored, who owns research outputs, how findings are disseminated, and what happens after formal research concludes. They require researchers to slow down, to prioritize relationships over productivity, to accept that some research should not be done or should be done differently than originally planned.

Institutional Models of Ethical Transformation

Several institutions have developed Indigenous research ethics frameworks that operationalize these principles, demonstrating that ethical transformation is both necessary and possible. The University of Saskatchewan's Indigenous Research Ethics Framework emerged from extensive consultation with Indigenous communities and scholars (University of Saskatchewan, 2019). The framework requires researchers to demonstrate community engagement before receiving ethics approval, to share research plans in accessible language, to negotiate data ownership and publication rights with communities, and to report back on findings. It establishes an Indigenous Research Ethics Board with majority Indigenous membership that reviews research involving Indigenous peoples, knowledge, or communities, ensuring that Indigenous perspectives shape ethical oversight rather than being filtered through Western ethics frameworks.

Te Whare Wānanga o Awanuiārangi in Aotearoa/New Zealand represents an Indigenous-led institution where Māori knowledge systems and research ethics are foundational, not supplementary (Pihama et al., 2015). Research ethics are grounded in tikanga Māori—Māori customary practices and protocols—and kaupapa Māori research principles. All research must demonstrate alignment with Māori values, benefit Māori communities, and be conducted in te reo Māori or bilingually. Governance structures ensure Māori control over research priorities, methods, and outcomes. The institution demonstrates that when Indigenous peoples control institutional structures, research ethics naturally reflect Indigenous values rather than requiring Indigenous perspectives to be accommodated within Western frameworks.

Sámi research centers in Norway, Sweden, and Finland have developed cross-border research ethics protocols that respect Sámi sovereignty despite the fact that Sámi homelands span multiple nation-states (Drugge, 2016). These protocols require researchers to engage with Sámi parliaments, to conduct research in Sámi languages where possible, to employ Sámi researchers as co-investigators rather than merely consultants, and to ensure that research serves Sámi self-determination. The protocols recognize that Sámi communities have the right to refuse research, to control data throughout its lifecycle, and to determine how findings are disseminated. They address the particular challenges of Indigenous peoples whose territories cross colonial borders, requiring coordination across multiple institutional and governmental contexts.

These examples demonstrate that ethical transformation requires more than adding Indigenous content to existing ethics frameworks. It requires institutional commitment, resource allocation, Indigenous governance, and willingness to cede control to Indigenous decision-making structures. It requires recognizing that Indigenous communities are the authorities on what constitutes ethical research in their contexts.

Ethical Pathways for Non-Indigenous Scholars

Non-Indigenous researchers often ask: How can we ethically engage with Indigenous communities and methodologies? This question reflects genuine concern, but it must be approached with humility and clarity about power dynamics. First, recognize that leadership belongs to Indigenous peoples. Non-Indigenous researchers can work in solidarity, but not in leadership roles, on Indigenous research projects (Smith, 2021). Allyship means supporting Indigenous-led initiatives, amplifying Indigenous voices, using privilege to challenge institutional barriers, and accepting that some research should be led by Indigenous scholars rather than positioning oneself as expert on Indigenous issues.

Second, commit to co-learning rather than extraction. Ethical engagement requires long-term relationships, not one-time consultations or brief visits. It means spending time in communities, attending events, building trust over years, learning Indigenous languages where possible, and demonstrating through consistent action that the relationship matters beyond research outcomes (Kovach, 2021). It means being willing to have research plans challenged or redirected based on community input, accepting that community priorities may differ from academic interests, and prioritizing community benefit over publication timelines.

Third, establish accountability mechanisms that give communities genuine power. This might include community advisory boards with veto power over publications, co-authorship with Indigenous community members who are compensated fairly for their intellectual contributions, regular reporting to community leadership with opportunities for feedback, and financial compensation that goes beyond token honoraria to recognize the value of Indigenous Knowledge and community time (Wilson, 2020). Accountability means accepting that communities can withdraw consent, redirect research, or decide that certain findings should not be published.

Fourth, understand what not to do. Do not claim Indigenous identity you do not have. Do not appropriate Indigenous Knowledge for personal gain or career advancement. Do not publish sacred or sensitive information, even if it is technically "public" or if you believe it would advance academic knowledge. Do not assume that reading about Indigenous methodologies qualifies you to use them—methodological competence requires training from Indigenous mentors, participation in Indigenous research communities, and ongoing guidance from Indigenous scholars and knowledge keepers.

Fifth, recognize that some research should not be done by non-Indigenous people, regardless of good intentions. Certain knowledge is not meant for outsiders. Certain communities have experienced sufficient harm from researchers that they no longer welcome academic projects. Respecting Indigenous sovereignty means accepting these boundaries without resentment, without trying to convince communities that your research would be different, and without seeking ways around community decisions.

IMPLICATIONS FOR HIGHER EDUCATION: STRUCTURAL TRANSFORMATION

Governance, Tenure, and Resource Allocation

Meaningful support for Indigenous Research Methodologies requires structural transformation across multiple institutional domains. Indigenous peoples must hold decision-making power in institutions that affect Indigenous Knowledge and communities. This means establishing Indigenous governance councils with authority over Indigenous programs, research ethics, and curriculum; ensuring Indigenous representation on boards of trustees, senate, and senior administration; creating Indigenous vice-president or vice-provost positions with budgetary control; and implementing free, prior, and informed consent principles for institutional decisions affecting Indigenous peoples (Page et al., 2025).

Academic reward structures must be reformed to recognize community-engaged scholarship, Indigenous-language publications, and non-traditional outputs. Tenure and promotion systems currently privilege English-language, peer-reviewed journal articles published by for-profit corporations. This system marginalizes Indigenous scholars who publish in Indigenous languages, who produce community reports or multimedia projects, who prioritize community benefit over academic prestige, or who work at the slower pace required for relational, community-accountable research (Kovach, 2021). Reform requires valuing community reports, multimedia projects, and Indigenous-language publications equally with peer-reviewed articles; recognizing time required for relationship-building, ceremony, and community accountability; including community letters and Elder assessments in tenure reviews; and establishing Indigenous-specific tenure criteria developed by Indigenous faculty rather than imposing Western standards.

Supporting Indigenous methodologies requires dedicated resource allocation. Land-based learning needs transportation and insurance for off-campus activities. Ceremonies require budgets for materials, feast protocols, and Elder honoraria. Research timelines must be extended to accommodate seasonal activities and community schedules. Infrastructure is needed for ceremonial spaces, language nests, and community gathering areas. Competitive salaries are necessary to recruit and retain Indigenous faculty and staff who are often overburdened with service demands while navigating institutional racism (Kamassah, 2018). Resources must be allocated not as temporary project funding but as permanent institutional commitments that recognize Indigenous methodologies as essential to institutional mission.

Research ethics systems must be rebuilt around Indigenous principles rather than attempting to accommodate Indigenous perspectives within Western biomedical frameworks. This requires Indigenous-led research ethics boards with majority Indigenous membership and decision-making authority; ethics protocols developed by Indigenous communities rather than imposed by institutions; ongoing consent processes that extend beyond initial approval; community ownership of data and control over publication; and meaningful consequences for researchers who violate Indigenous protocols rather than treating ethical violations as unfortunate but inconsequential.

Universities must recognize Indigenous data sovereignty and intellectual property rights. This includes implementing OCAP® principles—Ownership, Control, Access, Possession—for Indigenous data; respecting Traditional Knowledge licenses and Indigenous Creative Commons frameworks; establishing protocols for sacred and sensitive knowledge; supporting Indigenous-controlled data repositories; and ensuring Indigenous communities benefit from commercialization of Indigenous Knowledge rather than allowing universities or corporations to profit from Indigenous Knowledge without compensation or consent (Walter & Suina, 2019).

Graduate Education and Doctoral Transformation

Graduate education is where epistemic norms are most powerfully reproduced—or transformed (Manathunga et al., 2026). Supporting Indigenous graduate students and methodologies requires re-imagining every aspect of doctoral training. Community-based dissertation committees expand beyond traditional academic supervisors to include Elders, knowledge keepers, and community leaders as formal committee members with equal authority (Pidgeon & Riley, 2021). At the University of British Columbia and Deakin University, some Indigenous doctoral students have dissertation committees that include academic supervisors, community Elders, and family members who hold relevant knowledge. These committees meet on the land, incorporate ceremony, and evaluate dissertations based on community benefit as well as scholarly contribution.

Oral defenses can be transformed from adversarial examinations to ceremonial knowledge sharing. Defenses might begin with smudging or prayer, include storytelling and song, be conducted partially in Indigenous languages, and culminate in feasting and gift-giving (Manathunga et al., 2026). The focus shifts from proving mastery through defending against critique to demonstrating relational accountability, community service, and proper conduct as a knowledge carrier. Land-based qualifying exams assess doctoral competency through land-based knowledge rather than written examinations. Students might demonstrate understanding of seasonal cycles, ecological relationships, language proficiency in context, or ability to conduct ceremony appropriately (Simpson, 2014).

Alternative dissertation formats honour diverse ways of presenting knowledge. Dissertations might take the form of documentary films, websites, community resource guides, art installations, or performances, accompanied by scholarly reflection (Archibald, 2008). These formats make knowledge accessible to communities who may not read academic prose while maintaining scholarly rigor through different means. Dual supervision models pair academic supervisors with community mentors who guide students in cultural protocols, language learning, and community accountability (Pidgeon & Riley, 2021). Both supervisors are compensated and recognized formally by the university, challenging the assumption that academic knowledge is more valuable than community knowledge.

Resource implications are significant. Graduate students need extended funding timelines—five to seven years rather than four to five—to accommodate community-based research that cannot be rushed. Travel budgets enable regular community visits and land-based learning. Honoraria budgets recognize Elder mentorship and community participation as valuable intellectual labor deserving compensation. Childcare support acknowledges that many Indigenous graduate students are parents whose educational access depends on childcare availability. Mental health services must understand Indigenous wellness frameworks rather than imposing Western psychological models. Community spaces on campus provide places for ceremony, language practice, and cultural connection that support Indigenous student wellbeing and success.

An Indigenous doctoral student in environmental science reflects: "When my grandmother passed during my third year, Western timelines said I should take a week and get back to work. But my community needed me for ceremonies, for supporting my family, for fulfilling my responsibilities. My Indigenous supervisor understood that these responsibilities are part of my research methodology, not interruptions to it. That understanding made it possible for me to continue."

Teaching, Curriculum, and Assessment Transformation

Indigenous methodologies transform pedagogy across all levels of education. Teaching becomes circular rather than linear, relational rather than hierarchical, land-based rather than classroom-confined (Archibald, 2008). Assessment practices must align with Indigenous learning principles rather than imposing Western evaluation frameworks that contradict Indigenous pedagogies. Oral presentations and storytelling can replace written exams, recognizing that Indigenous Knowledge is often transmitted orally and that written examinations privilege particular forms of knowledge demonstration. Portfolio-based assessment documents learning over time, including community engagement, language learning, and practical skills that conventional exams cannot capture.

Ceremonial learning assessment evaluates participation in ceremonies, understanding of protocols, and ability to conduct oneself appropriately in cultural contexts—forms of knowledge that are essential in Indigenous contexts but invisible to conventional academic assessment. Land-based learning evaluation assesses knowledge gained through seasonal activities, harvesting, and ecological observation, recognizing that understanding seasonal cycles or reading landscape features demonstrates rigorous learning even if it cannot be tested through multiple-choice questions. Community-evaluated projects involve Elders and community members assessing whether students have learned to be good relatives and community contributors, centering community standards alongside academic standards.

These assessment alternatives require faculty training in Indigenous pedagogies, institutional flexibility to recognize diverse forms of knowledge demonstration, and willingness to accept that a student who cannot write a conventional essay might nonetheless demonstrate profound understanding through other means. They challenge universities to expand definitions of rigor beyond Western conventions, recognizing that Indigenous standards of rigor—grounded in relational accountability, community benefit, and long-term sustainability—are equally valid.

Publishing, Peer Review, and Knowledge Dissemination

Academic publishing systems privilege English-language, peer-reviewed journals published by for-profit corporations. This system marginalizes Indigenous Knowledge in multiple ways (Shay et al., 2022). Peer review, while ostensibly ensuring quality, often functions as gatekeeping that excludes Indigenous methodologies by applying Western standards of evidence, argumentation, and citation. Indigenous scholars who cite oral teachings, who write in Indigenous languages, who prioritize community accessibility over academic jargon, or who challenge Western epistemological assumptions often face rejection or demands to conform to Western conventions.

Indigenous-led journals provide venues where Indigenous scholars control editorial processes, where Indigenous languages are valued, where community accountability matters more than impact factors, and where peer review recognizes Indigenous standards of rigor. *Decolonization: Indigeneity, Education & Society*, *AlterNative: An International Journal of Indigenous Peoples*, and *Transmotion* demonstrate that Indigenous-controlled publishing can maintain scholarly excellence while honoring Indigenous epistemologies. Universities must recognize these journals as prestigious publication venues in tenure decisions rather than privileging only Western mainstream journals.

Traditional Knowledge licenses and Indigenous Creative Commons frameworks protect Indigenous intellectual property while enabling appropriate sharing (Christen, 2012). These licenses specify who can use knowledge, for what purposes, under what conditions, and with what attribution—recognizing that Indigenous Knowledge is not "free" for anyone to use but is governed by specific protocols about access, use, and benefit-sharing. Community-accessible outputs matter as much as academic publications. Research findings must be shared in formats communities can use: plain-language reports, community presentations, radio programs, social media, posters in Indigenous languages. Researchers who publish only in academic journals fail the reciprocity requirement, extracting knowledge from communities while returning nothing in accessible form.

DATA SOVEREIGNTY, DIGITAL FUTURES, AND ALGORITHMIC JUSTICE

Indigenous Data Sovereignty asserts that Indigenous peoples have the right to control collection, ownership, access, and use of data about Indigenous communities, territories, and resources (Walter & Suina, 2019; Kukutai, 2023). This principle challenges the assumption that data, once collected, becomes the property of researchers, institutions, or governments. It recognizes that data about Indigenous peoples is not neutral information but represents Indigenous lives, territories, knowledge, and futures, and therefore must be governed according to Indigenous values and priorities.

The CARE Principles for Indigenous Data Governance—Collective Benefit, Authority to Control, Responsibility, Ethics—complement the FAIR principles (Findable, Accessible, Interoperable, Reusable) by centering Indigenous rights and interests (Carroll et al., 2020). CARE principles require that data collection benefits the collective, not just individual researchers or institutions; that Indigenous peoples have authority to control data throughout its lifecycle from collection through storage, analysis, and dissemination; that data use aligns with Indigenous values and advances Indigenous priorities rather than serving only academic or governmental interests; and that ethical use respects Indigenous rights, including the right to refuse data collection or to withdraw data from use.

Digital technologies present both opportunities and threats. Indigenous communities are using digital platforms for language revitalization, intergenerational knowledge sharing, and cultural continuity, creating digital archives that preserve endangered languages and connecting diaspora communities. Yet algorithmic colonization—the ways artificial intelligence and machine learning reproduce colonial biases and power relations—poses new dangers (Parahakaran, 2022). AI systems trained on Western datasets may misrepresent or appropriate Indigenous Knowledge. Facial recognition technologies threaten Indigenous privacy and sovereignty, enabling surveillance that disproportionately targets Indigenous peoples. Data aggregation platforms extract value from Indigenous data without compensation or consent, treating Indigenous information as raw material for corporate profit.

Indigenous Digital Sovereignty movements assert the right to control digital infrastructure, determine data governance, and shape technology development according to Indigenous values (Kukutai, 2023). This includes developing Indigenous-controlled data repositories that respect Indigenous protocols for access and use; establishing guidelines for digitizing sacred knowledge that recognize some knowledge should not be digitized or should be accessible only under specific conditions; ensuring that Indigenous communities benefit from data-driven research and commercial applications rather than being exploited as data sources; and participating in technology design processes to ensure that digital tools serve Indigenous priorities rather than imposing Western assumptions about knowledge organization and access.

INTERSECTIONALITY, DIVERSITY, AND INCLUSIVE INDIGENOUS METHODOLOGIES

Indigenous Research Methodologies must attend to intersectionality—the ways that gender, sexuality, disability, class, age, and other identities shape access to and expression of Indigenous Knowledge (TallBear, 2013; Simpson, 2014). Indigenous feminisms and Two-Spirit scholarship challenge both Western feminism and patriarchal interpretations of Indigenous traditions. The work of TallBear (2013) on Native American DNA demonstrates how colonial science has been used to police Indigenous identity, particularly affecting Indigenous women whose children's Indigenous status has been questioned based on genetic testing that ignores Indigenous understandings of kinship and belonging. Two-Spirit scholars like Wilson (2007) argue that colonial gender binaries have been imposed on Indigenous communities, erasing traditional recognition of multiple genders and diverse sexualities. Indigenous methodologies must create space for these perspectives rather than reproducing colonial gender norms.

Disability justice requires that land-based learning and ceremony be accessible to Indigenous peoples with disabilities. This means rethinking what "land-based" means beyond assumptions of physical mobility—it might include accessible trails, audio recordings of land-based teachings, virtual participation options, or adapted ceremonial practices that honor the principle that land-based knowledge is for all community members (Lemon et al., 2023). It also means challenging ableist assumptions about who can be a knowledge keeper or researcher, recognizing that Indigenous peoples with disabilities hold valuable knowledge and perspectives that enrich research.

Urban Indigenous experiences differ from land-based community contexts, yet both are valid expressions of Indigenous identity and knowledge. Urban Indigenous peoples maintain cultural connections through urban Indigenous centers, intertribal gatherings, digital networks, and relationships with home communities (Peters & Andersen, 2013). Indigenous methodologies must honor these experiences without positioning them as less authentic or assuming that all Indigenous research must be land-based in the same ways. Class and economic justice must be central to Indigenous research ethics. Indigenous communities experience disproportionate poverty due to colonial dispossession and ongoing structural inequality. Research that demands unpaid community participation, that benefits researchers while communities remain under-resourced, or that extracts knowledge without providing economic benefit perpetuates injustice.

Generational differences shape approaches to Indigenous methodologies. Younger Indigenous scholars are less willing to accept tokenism, more insistent on Indigenous governance, and more skilled at using digital platforms to build transnational Indigenous research networks. They bring new perspectives on how Indigenous Knowledge can address contemporary challenges while maintaining connections to ancestral teachings. Intergenerational dialogue between Elders and young people enriches methodological development, combining long-held wisdom with innovative approaches to contemporary contexts.

FUTURE DIRECTIONS: TOWARD PLURIVERSAL UNIVERSITIES

The future of higher education depends on epistemic justice—the recognition that multiple valid knowledge systems exist and that universities must honor this pluralism rather than imposing Western epistemological hegemony (Stein, 2022; Sefa Dei & Cacciavillani, 2024). The "pluriversal university" does not simply include Indigenous Knowledge within Western frameworks but restructures itself to accommodate multiple epistemologies, each with its own standards of rigor, methods of validation, and purposes. This transformation is urgent. The climate crisis demands Indigenous ecological knowledge developed through millennia of sustainable relationships with particular ecosystems. The COVID-19 pandemic revealed the limitations of Western medicine disconnected from holistic wellness frameworks that integrate physical, mental, emotional, and spiritual dimensions of health. Social justice movements challenge universities to address their complicity in colonial violence and their ongoing roles in perpetuating epistemic injustice.

The pandemic accelerated some changes while revealing persistent barriers. Online learning challenged assumptions about where education happens, potentially creating space for land-based learning integrated with digital technologies. Yet virtual platforms also made ceremony and embodied knowledge transmission more difficult, highlighting the importance of physical presence and place-based learning. Universities must learn from both the innovations and the limitations of this period, recognizing that some forms of Indigenous Knowledge transmission cannot be fully replicated virtually.

Indigenous-led research networks are emerging globally, connecting Indigenous scholars across colonial borders to share methodologies, support each other's work, and build collective capacity. These networks operate according to Indigenous values of reciprocity and collective benefit rather than competitive individualism. They demonstrate that Indigenous Research Methodologies can thrive when Indigenous scholars control institutional structures, resources, and priorities. Digital platforms enable these connections while raising questions about digital sovereignty, data governance, and ensuring that technology serves Indigenous priorities rather than reproducing colonial power relations.

The path forward requires structural transformation, not symbolic inclusion; resource reallocation, not empty commitments; power sharing, not consultation; epistemic humility, not paternalism; and long-term commitment, not project-based engagement. Universities that undertake this work will be strengthened, becoming institutions capable of addressing complex global challenges through diverse knowledge systems. Those that resist will become increasingly irrelevant to the communities they claim to serve and to the urgent problems that demand pluralistic approaches.

FUTURE RESEARCH DIRECTIONS

Future research should continue examining how Indigenous Research Methodologies can transform higher education beyond symbolic inclusion toward genuine epistemic restructuring. While increasing numbers of universities have adopted Indigenous initiatives, much remains unknown about the long-term institutional impact of Indigenous governance structures, Indigenous-led research ethics frameworks, and land-based pedagogical models. Comparative studies across countries, institutional types, and Indigenous nations could provide important insights into how universities operationalize Indigenous sovereignty within research, curriculum, and administrative systems. Longitudinal research may also investigate whether these reforms produce sustained improvements in Indigenous student belonging, retention, faculty representation, and institutional accountability over time.

Future scholarship should also further explore the relationship between Indigenous Knowledge systems and contemporary global challenges, particularly in areas such as climate change, sustainability, public health, biodiversity protection, and digital governance. Although Indigenous Scientific Knowledge has increasingly gained recognition within environmental and interdisciplinary scholarship, additional empirical and collaborative research is needed to better understand how Indigenous ecological knowledge, relational accountability, and land-based methodologies contribute to sustainable policy development and scientific innovation. Research examining ethical frameworks for artificial intelligence, algorithmic governance, Indigenous data sovereignty, and digital knowledge preservation will become increasingly important as universities and governments rely more heavily on data-driven systems that may reproduce colonial assumptions and forms of epistemic exclusion.

Additional future research should investigate the intersectional dimensions of Indigenous Research Methodologies, particularly the experiences of Indigenous women, Two-Spirit scholars, urban Indigenous peoples, Indigenous peoples with disabilities, and multilingual Indigenous communities within higher education systems. More work is needed to understand how Indigenous methodologies can remain responsive to diverse community contexts while resisting pan-Indigenous generalizations that flatten important cultural, linguistic, and political differences across nations. Research should also examine emerging models of Indigenous doctoral education, alternative dissertation structures, community-based supervision, and Indigenous-led publishing systems to better understand how universities can create sustainable pathways for Indigenous scholarly leadership. Ultimately, future inquiry must continue asking not only how Indigenous Knowledge can be included within higher education, but how higher education itself must transform in response to Indigenous epistemologies, ethical frameworks, and visions of relational accountability.

CONCLUSION

Indigenous Research Methodologies offer higher education a path toward epistemic justice, institutional integrity, and scholarly excellence. They demand that universities recognize Indigenous Knowledge systems as complete, sophisticated, and essential intellectual traditions that must shape academic practice on their own terms. This chapter has traced the historical exclusion of Indigenous Knowledge from universities, acknowledging both the violence of this exclusion and the persistent resistance and resilience of Indigenous peoples who have never ceased asserting the validity of Indigenous Knowledge. It has analyzed core methodological principles including relationality, relational accountability, and the integration of ceremony, storywork, and land-based learning as rigorous research methods. It has acknowledged methodological pluralism among Indigenous nations while identifying shared commitments to community benefit, ongoing consent, and reciprocity.

The chapter has provided institutional examples of ethical transformation from the University of Saskatchewan, Te Whare Wānanga o Awanuiārangi, and Sámi research centers, demonstrating that transformation is both necessary and possible when institutions commit resources, cede control to Indigenous governance, and recognize Indigenous peoples as authorities on Indigenous Knowledge. It has offered practical frameworks for institutional change across governance, tenure, resource allocation, ethics infrastructure, and knowledge governance. It has addressed graduate education transformation, teaching and assessment alternatives, publishing and peer review reform, data sovereignty, digital futures, and intersectionality within Indigenous methodologies.

Throughout, the chapter has highlighted Indigenous agency and success—the Indigenous-led research institutes, journals, and degree programs that demonstrate what becomes possible when Indigenous peoples control institutional structures and resources. These examples prove that Indigenous Research Methodologies are not theoretical ideals but living realities that strengthen scholarship, enhance institutional integrity, and address urgent global challenges.

The transformation required is profound. It extends beyond curriculum reform to encompass governance structures, tenure systems, research ethics, data sovereignty, publishing practices, and the fundamental purposes of higher education. It requires universities to cede control, redistribute resources, accept Indigenous governance, and recognize that Indigenous communities are the authorities on Indigenous Knowledge. Yet this transformation is not only possible but necessary. Indigenous Knowledge offers insights essential for addressing climate crisis, social inequality, and the limitations of Western epistemologies. Indigenous methodologies model research as ceremony, knowledge as relationship, and scholarship as service to life.

The question is not whether universities should support Indigenous Research Methodologies but whether they will do so with the depth, commitment, and structural transformation that genuine support requires. The answer will determine not only the future of Indigenous peoples in higher education but the future of higher education itself.

REFERENCES

Absolon, K. (2022). *Kaandossiwin: How we come to know*. Fernwood Publishing.

Archibald, J. (2008). *Indigenous storywork: Educating the heart, mind, body, and spirit*. UBC Press. DOI: .59962/9780774855204

Battiste, M. (2013). *Decolonizing education: Nourishing the learning spirit*. Purich Publishing.

Cajete, G. (2000). *Native science: Natural laws of interdependence*. Clear Light Publishers.

Carroll, S. R., Garba, I., Figueroa-Rodríguez, O. L., Holbrook, J., Lovett, R., Materechera, S., Parsons, M., Raseroka, K., Rodriguez-Lonebear, D., Rowe, R., Sara, R., Walker, J. D., Anderson, J., & Hudson, M. (2020). The CARE Principles for Indigenous Data Governance. *Data Science Journal, 19*(1), 43. DOI: .5334/dsj-2020-043

Chilisa, B. (2019). *Indigenous research methodologies* (2nd ed.). SAGE Publications., DOI: 10.4135/9781071802686

Christen, K. (2012). Does information really want to be free? Indigenous knowledge systems and the question of openness. *International Journal of Communication, 6*, 2870–2893.

Sefa Dei, G. J., & Cacciavillani, I. (2024). *Decolonizing the university: Toward pluriversal knowledge*. Peter Lang.

Drouin-Gagné, M. E. (2021). Beyond the "Indigenizing the academy" trend: Learning from Indigenous higher education land-based and intercultural pedagogies to build trans-systemic education. *Engaged Scholar Journal, 7*(1), 45–65. DOI: 10.15402/esj.v7i1.69978

Drugge, A. L. (Ed.). (2016). *Ethics in Indigenous research: Past experiences, future challenges*. Centre for Sami Research, Umeå University.

Kamassah, S. S. (2018). *The Impact of Post-Secondary Educational Institution Policies and Practices on Indigenous Staff Recruitment and Retention*. University of Toronto.

Kimmerer, R. W. (2013). *Braiding Sweetgrass: Indigenous wisdom, scientific knowledge and the teachings of plants*. Milkweed Editions.

Kovach, M. (2021). *Indigenous methodologies: Characteristics, conversations, and contexts* (2nd ed.). University of Toronto Press.

Kukutai, T. (2023). Indigenous data sovereignty: A new take on an old theme. *Science, 382*(6674), eadl4664. Advance online publication. DOI: 10.1126/science.adl4664 PMID: 38033063

MacLeod, L. (2021). More than personal communication: Templates for citing Indigenous elders and knowledge keepers. *Kula, 5*(1), 1-5. DOI: DOI: 10.18357/kula.135

Manathunga, C., Qi, J., Raciti, M., Stanton, S., Tuxworth, J., Whop, J., & Gilbey, K. (2026). Entangled Relationality in Doctoral Education. In Lyons, D., Singh, K., & Southcott, J. (Eds.), *Shaping Doctoral Experiences: Nurturing the Relational* (pp. 61–82). Bloomsbury Academic. DOI: 10.5040/9798881899103.0008

McDonald, M. (2023). *Indigenous land-based education in theory and practice*. Yellowhead Institute.

McGregor, D. (2004). Coming full circle: Indigenous knowledge, environment, and our future. *American Indian Quarterly, 28*(3/4), 385–410. DOI: 10.1353/aiq.2004.0101

Page, S., Hall, M., & Naepi, S. (2025). Indigenous leadership in higher education: Baya, Eiseni, Rangatiratanga (voice, agency, sovereignty). *Journal of Higher Education Policy and Management, 47*(5), 563–566. DOI: 10.1080/1360080X.2025.2546752

Parahakaran, S. (2022). Digital Ethnography Within Indigenous Contexts: An Exploratory Perspective. In Chowdhury, J. S., Wahab, H. A., Saad, R. M., Roy, P. K., & Wronka, J. (Eds.), *Practices, Challenges, and Prospects of Digital Ethnography as a Multidisciplinary Method* (pp. 138–147). IGI Global Scientific Publishing. DOI: 10.4018/978-1-6684-4190-9.ch010

Peters, E. J., & Andersen, C. (Eds.). (2013). *Indigenous in the city: Contemporary identities and cultural innovation*. UBC Press. DOI: 10.59962/9780774824668

Pidgeon, M., & Riley, T. (2021). Understanding the Application and Use of Indigenous Research Methodologies in the Social Sciences by Indigenous and Non-Indigenous Scholars. *International Journal of Education Policy and Leadership, 17*(8), n8. DOI: 10.22230/ijepl.2021v17n8a1065

Pihama, L., Tiakiwai, S. J., & Southey, K. (Eds.). (2015). *Kaupapa rangahau: A reader* (2nd ed.). Te Kotahi Research Institute.

Seehawer, M. K. (2018). Decolonising research in a Sub-Saharan African context: Exploring Ubuntu as a foundation for research methodology, ethics and agenda. *International Journal of Social Research Methodology, 21*(4), 453–466. DOI: 10.1080/13645579.2018.1432404

Shay, M., Sarra, G., & Woods, A. (2022). Grounded ontologies: Indigenous methodologies in qualitative cross-cultural research. In Liamputtong, P. (Ed.), *Handbook of qualitative cross-cultural research methods* (pp. 26–39). Edward Elgar Publishing. DOI: 10.4337/9781800376625.00011

Simpson, A. (2014). *Mohawk interruptus: Political life across the borders of settler states*. Duke University Press.

Simpson, L. B. (2014). Land as pedagogy: Nishnaabeg intelligence and rebellious transformation. *Decolonization, 3*(3), 1–25.

Smith, L. T. (1999). *Decolonizing methodologies: Research and Indigenous peoples* (1st ed.). Zed Books.

Smith, L. T. (2021). *Decolonizing methodologies: Research and Indigenous peoples* (3rd ed.). Zed Books. DOI: 10.5040/9781350225282

Stein, S. (2022). *Unsettling the university: Confronting the colonial foundations of US higher education*. Johns Hopkins University Press. DOI: 10.56021/9781421445052

TallBear. K. (2013). *Native American DNA: Tribal belonging and the false promise of genetic science*. University of Minnesota Press.

Truth and Reconciliation Commission of Canada. (2015). *Honouring the truth, reconciling for the future: Summary of the final report of the Truth and Reconciliation Commission of Canada*. Truth and Reconciliation Commission of Canada.

Tuck, E., & Yang, K. W. (2012). Decolonization is not a metaphor. *Decolonization, 1*(1), 1–40.

University of Saskatchewan. (2019). *Guidelines for the ethical conduct of research involving Indigenous peoples*. University of Saskatchewan Research Ethics Office.

Walter, M., & Suina, M. (2019). Indigenous data, Indigenous methodologies and global data governance. *International Journal of Social Research Methodology, 8*(1), 1–12. DOI: 10.1080/13645579.2018.1531228

Wilson, A. (2007). N'tacimowin inna nah': Our coming in stories. *Canadian Women's Studies, 26*(3-4), 193–199.

Wilson, S. (2020). *Research is ceremony: Indigenous research methods*. Fernwood Publishing.

KEY TERMS AND DEFINITIONS

Algorithmic Colonization: The reproduction of colonial biases and power structures through artificial intelligence, algorithms, and digital technologies that marginalize Indigenous peoples and knowledge systems.

Collective Benefit: An Indigenous ethical principle stating that research should benefit Indigenous communities collectively rather than serving only individual researchers or institutions.

Colonial University: A higher education institution historically shaped by colonial systems that privilege Western knowledge while marginalizing Indigenous epistemologies and intellectual traditions.

Community Accountability: The responsibility of researchers to remain answerable to Indigenous communities throughout and after the research process.

Data Sovereignty: The right of Indigenous peoples to control the collection, ownership, and use of data related to their communities and knowledge systems.

Decolonization: A process of dismantling colonial systems of power and restructuring institutions to recognize Indigenous sovereignty and knowledge systems.

Digital Sovereignty: The ability of Indigenous peoples to govern digital technologies, online knowledge systems, and digital infrastructures according to Indigenous values and priorities.

Epistemic Justice: The recognition and validation of multiple knowledge systems as legitimate forms of scholarship and inquiry.

Indigenous Data Governance: Indigenous-controlled systems for managing, protecting, and using Indigenous data according to community priorities and protocols.

Indigenous Knowledge: Holistic systems of knowledge grounded in Indigenous relationships with land, language, spirituality, community, and intergenerational learning.

Indigenous Methodological Pluralism: The recognition that Indigenous methodologies differ across nations, languages, territories, and cultural contexts.

Indigenous Research Methodologies: Research approaches grounded in Indigenous worldviews, relational accountability, reciprocity, and community-centered ethics.

Indigenous Scientific Knowledge: Scientific understandings developed by Indigenous peoples through observation, ecological stewardship, and intergenerational knowledge transmission.

Institutional Transformation: Systemic changes within higher education involving governance, curriculum, research ethics, and resource allocation to support Indigenous sovereignty and epistemic justice.

Intersectionality: A framework examining how identities such as Indigeneity, gender, sexuality, disability, and class interact to shape lived experiences and access to knowledge.

Knowledge Keepers: Indigenous Elders, cultural leaders, and community members recognized as holders and transmitters of Indigenous Knowledge.

Land-Based Learning: An Indigenous educational approach that recognizes land as a source of knowledge, teaching, and relationship.

Ongoing Relational Consent: A form of consent viewed as a continuous process rather than a one-time agreement signed at the beginning of research.

Pluriversal University: A model of higher education that recognizes multiple coexisting knowledge systems and epistemologies as equally valid.

Reciprocity: An Indigenous ethical principle requiring mutual exchange, relationship-building, and community benefit within research relationships.

Relational Accountability: The responsibility researchers hold toward communities, land, ancestors, and future generations within Indigenous research.

Relationality: An Indigenous principle recognizing that knowledge emerges through interconnected relationships among people, land, and community.

Settler Moves to Innocence: Symbolic actions that appear supportive of decolonization while leaving colonial structures and power relations intact.

Storywork: An Indigenous methodological approach that uses storytelling as a rigorous process for transmitting knowledge, ethics, and cultural teachings.

Tokenistic Inclusion: Superficial institutional inclusion of Indigenous peoples or perspectives without meaningful structural change.

Two-Spirit Scholarship: Scholarship grounded in Indigenous understandings of gender and sexuality that challenge colonial gender binaries.

APPENDIX: ADDITIONAL RESOURCES

Archibald, J., Xiiem, Q. Q., Bol, J., Lee-Morgan, J. & De Santolo, J., (Eds.) (2022). ***Decolonizing research: Indigenous storywork as mythology.*** **Bloomsbury Academic.** This book explores Indigenous storywork as a rigorous and transformative research and educational methodology. Drawing upon Indigenous perspectives from Canada, Australia, and New Zealand, the book demonstrates how storytelling can challenge colonial academic traditions while supporting Indigenous knowledge transmission, cultural continuity, and community-centered scholarship.

Kovach, M. (2021). ***Indigenous mythologies: Characteristics, conversations, and contexts*** **(2nd ed.). University of Toronto Press.** This book provides a foundational introduction to Indigenous research methodologies and remains one of the most widely used texts in the field. The book examines the philosophical foundations, ethical principles, and relational dimensions of Indigenous research while offering practical guidance for implementing Indigenous methodologies within academic and community-based contexts.

Quinless, J. M. (2022). ***Decolonizing data: Unsettling conversations about social research methods.*** **Univeersity of Toronto Press.** *Decolonizing Data* examines how colonial systems continue to shape health, research, and social inequities affecting Indigenous peoples and BIPOC communities. The book critiques conventional social research approaches while advocating for more ethical, community-centered, and decolonizing forms of data collection and knowledge production.

Smith, L. T. (2023). ***Decolonizing methodologies: Research and indigenous peoples*** **(3rd ed.). Bloomsbury Academic.** This volume critically examines how imperialism and colonialism have shaped Western research traditions and academic knowledge production. The book challenges colonial assumptions embedded within research practices and argues that decolonizing methodologies are essential for restoring Indigenous control over knowledge, representation, and self-determination.

Windchief, S. & San Pedro, T. (Eds.) (2019). ***Applying indigenous research methods: Storying with peoples and communities.*** **Routledge.** This collection highlights the importance of Indigenous research methodologies in strengthening relationships, supporting community engagement, and challenging extractive research traditions. The contributors emphasize storying, relational accountability, and culturally grounded approaches that position Indigenous communities as active knowledge holders within the research process.

Chapter 3
Indigenous Knowledge Systems and Sustainable Development

Soumen Ghosh
http://orcid.org/0009-0008-3778-6090
Global Institute of Management and Technology, Krishnanagar, India

Parag Chatterjee
http://orcid.org/0000-0003-1243-9582
Global Institute of Management and Technology, Krishnanagar, India

Rupam Hazra
http://orcid.org/0009-0003-0332-2128
Global Institute of Management and Technology, Krishnanagar, India

ABSTRACT

This chapter highlights the significance of Indigenous Knowledge Systems (IKS) as a crucial resource for global sustainability, resilience, and inclusive development. Rooted in centuries of ecological, communal, and spiritual experience, IKS promotes harmony with nature and strengthens social cohesion. The chapter outlines core IKS principles such as oral transmission, communal validation, and responsible resource use, showing how they support biodiversity, sustainable livelihoods, and food security. It also examines IKS contributions to healthcare, education, and participatory governance. Despite past marginalization, IKS is increasingly recognized worldwide. The chapter stresses the need for ethical integration of traditional knowledge with modern science and policy, calling for strong legal protection and respectful collaboration to empower Indigenous communities.

1. INTRODUCTION

Indigenous Knowledge Systems (IKS) represent a profound and sophisticated body of knowledge developed over millennia through the lived experiences and spiritual connections of Indigenous peoples with their specific environments. This knowledge is not a mere collection of facts but a holistic worldview where social, ecological, and spiritual elements are interwoven, offering deep insights into the mechanisms of sustainability and resilience. While often historically marginalized by the dominance of Western sci-

DOI: 10.4018/979-8-3373-7976-0.ch003

entific paradigms, IKS is now being globally recognized as an essential, perhaps indispensable, resource for addressing the complex challenges of the 21st century, including climate change, biodiversity loss, and public health crises. This paper provides a detailed exploration of IKS, moving from its fundamental principles to its practical applications across critical sectors. It establishes the foundations of IKS, highlighting its distinctive methods of oral transmission, emphasis on collective validation, and core values of sustainability and ecological balance. Subsequent sections systematically analyze the contributions of IKS to environmental sustainability, demonstrating how traditional ecological knowledge serves as a practical guide for resource regeneration and conservation. The discussion then broadens to examine the socio-economic dimensions of IKS, focusing on its role in building self-sufficient, equitable, and culturally grounded economies. Furthermore, the paper addresses the significance of IKS in healthcare practices, education and knowledge preservation, and its vital place within contemporary governance, policy, and legal frameworks. Ultimately, the goal is to articulate not just the historical value of IKS, but its contemporary relevance as a robust and essential partner for achieving inclusive, culturally sensitive, and truly sustainable global development.

2. FOUNDATIONS OF INDIGENOUS KNOWLEDGE SYSTEMS

The Foundations of Indigenous Knowledge Systems (IKS) are rooted in centuries of lived experiences, cultural traditions, and interactions between Indigenous peoples and their natural environments. IKS represents a holistic worldview where knowledge is not separated into disciplines but interconnected through social, ecological, and spiritual relationships. This knowledge is transmitted orally through stories, rituals, songs, and community practices, ensuring its continuity across generations. Unlike formal scientific methods, IKS emphasizes observation, adaptation, and communal validation rather than experimentation in isolation. It values sustainability, resilience, and harmony with nature, promoting practices that conserve biodiversity and maintain ecological balance. Spiritual beliefs often guide how resources are used, embedding moral responsibility in human–nature interactions. The collective ownership of knowledge within Indigenous communities reinforces cooperation, inclusivity, and social cohesion. Despite its effectiveness, IKS has historically been marginalized by colonialism and the dominance of Western scientific paradigms. However, global recognition of its significance is growing as it provides valuable insights for addressing modern challenges such as climate change, sustainable agriculture, and ecosystem management. Ultimately, the foundations of IKS highlight the importance of respecting cultural diversity and integrating traditional wisdom with contemporary scientific approaches for sustainable development. The international community's interest in indigenous knowledge and practices has recently grown as a result of awareness of the sustainable lifestyles of indigenous communities and the worsening state of the planet(Magni,2017).

A. Historical Evolution of IKS

The historical evolution of Indigenous Knowledge Systems (IKS) reflects centuries of accumulated wisdom shaped by direct interaction with local environments, cultural traditions, and community experiences. Indigenous knowledge developed as communities observed natural patterns, experimented with survival strategies, and adapted their lifestyles to the ecological conditions surrounding them. Over generations, this knowledge was refined and transmitted orally through stories, rituals, ceremonies, and practical demonstrations. Colonialism disrupted many Indigenous systems by imposing Western ideologies and dismissing traditional knowledge as inferior, leading to cultural erosion and loss of intellectual heritage. Despite these challenges, IKS has endured and is now increasingly recognized as a vital resource for sustainability, climate resilience, and cultural preservation. IKS has been extensively studied in the fields of agriculture, forestry, biodiversity, soils, land, and water management (Mafongoya & Ajayi, 2017).

a) Oral Tradition and Knowledge Transmission

Oral tradition has historically served as the backbone of Indigenous knowledge preservation, ensuring continuity across generations without written records. Knowledge is passed through storytelling, songs, proverbs, and mentorship, linking cultural identity with practical skills. This method allows information to evolve naturally as communities adapt to environmental changes.

b) Influence of Spirituality and Worldviews

Indigenous worldviews, centered on harmony with nature and respect for ancestral spirits, have significantly shaped how knowledge was formed and practiced. Spiritual beliefs guided decisions related to agriculture, healing, resource use, and social behavior. This integration of the sacred and the practical created a holistic knowledge system that remains relevant today.

c) Impact of Colonization on IKS

Colonization disrupted Indigenous Knowledge Systems by imposing foreign governance structures, education systems, and land-use practices. Many traditional practices were suppressed or marginalized, resulting in cultural displacement and loss of knowledge. However, revival movements and global recognition of Indigenous rights have contributed to renewed efforts to protect and restore IKS.

B. Characteristics and Principles of Indigenous Knowledge

The characteristics and principles of Indigenous Knowledge (IK) reflect a holistic, community-centered, and environmentally grounded understanding of life. Indigenous knowledge is deeply rooted in the lived experiences of people who have interacted with their local ecosystems over countless generations. It emphasizes harmony with nature, sustainable resource use, and respect for all forms of life, seeing humans as part of a larger ecological system rather than separate from it. This knowledge is dynamic rather than static, constantly evolving as communities adapt to changing environmental and social conditions. Ultimately, IK is collective, place-based, and spiritually enriched, forming a foundational guide for community well-being, resilience, and sustainable development. In many situations, IK is environmentally sustainable. It is also a significant source of food security and local subsistence. It was created using local technologies, resources, and customs. It is also fair in terms of management, decision-making, and resource access (Sultana et al., 2018).

a) Holistic and Interconnected Worldview

Indigenous knowledge operates on the principle that all aspects of life—environmental, social, cultural, and spiritual—are interconnected. This holistic approach ensures that decisions consider long-term impacts on future generations and the surrounding ecosystem. It fosters balance and harmony, reinforcing sustainability as a core value.

b) Community Ownership and Collective Wisdom

IK is collectively generated and shared within the community rather than owned by individuals. Knowledge is maintained through cooperation, communal responsibilities, and shared traditions that strengthen social cohesion. This collective nature ensures that wisdom is preserved, refined, and protected across generations.

c) Adaptability and Continuous Learning

A key principle of Indigenous knowledge is its ability to adapt to environmental changes, climatic variations, and socio-economic shifts. Through observations, practical experiences, and cultural feedback systems, communities refine their knowledge to stay resilient. This adaptability makes IK highly relevant to modern challenges, including climate change and resource management.

C. Relationship Between Culture, Nature, and Spirituality

The relationship between culture, nature, and spirituality within Indigenous Knowledge Systems (IKS) is deeply interconnected, forming a unified worldview that guides community life and environmental stewardship. Indigenous cultures often view the natural world as a living entity, where rivers, forests, mountains, and animals possess intrinsic spiritual significance. This belief shapes ethical behaviors, encouraging respect, gratitude, and responsible resource use. Spiritual practices, rituals, and cultural ceremonies reinforce the bond between humans and the environment, reminding communities of their role as custodians rather than exploiters. Through this integrated perspective, Indigenous societies maintain ecological balance, strengthen cultural identity, and preserve sustainable practices that have supported them for generations. The majority of indigenous civilizations believe that nature is alive and that each and every entity possesses knowledge, wisdom, and traits that the West has called "soul" (Rawat & Mishra, 2021).

a) Sacredness of the Natural Environment

Many Indigenous communities consider elements of nature—such as trees, water bodies, and animals—as sacred beings connected to ancestral spirits. This sacredness fosters a deep respect for the environment, guiding people to use resources wisely and avoid actions that harm the ecosystem. Such beliefs play a crucial role in conserving biodiversity and maintaining ecological harmony.

b) Cultural Practices Rooted in Ecological Knowledge

Traditional festivals, rituals, and cultural events often reflect seasonal cycles, agricultural patterns, and ecological understanding. These practices serve both spiritual and educational purposes, teaching younger generations about environmental responsibilities. As a result, culture becomes a medium for transmitting ecological wisdom across time.

c) Spiritual Guidance in Decision-Making

Spiritual principles and ancestral teachings often influence decisions related to land use, agriculture, healing, and community governance. Leaders and elders frequently consult spiritual signs or rituals before undertaking major activities. This spiritually guided approach ensures that decisions align with community values and maintain harmony with nature.

3. INDIGENOUS KNOWLEDGE AND ENVIRONMENTAL SUSTAINABILITY

Indigenous Knowledge and Environmental Sustainability are deeply intertwined, reflecting a long-standing relationship between Indigenous communities and their natural surroundings. Indigenous Knowledge Systems (IKS) emphasize harmony with nature through practices that maintain ecological balance and resource regeneration. Traditional ecological knowledge guides the sustainable use of forests, water, soil, and biodiversity, ensuring that exploitation never exceeds renewal. For instance, Indigenous agricultural methods such as crop rotation, mixed farming, and agroforestry enhance soil fertility and resilience against environmental stress. Indigenous peoples also play a crucial role in biodiversity conservation by protecting sacred groves, maintaining seed diversity, and managing communal lands responsibly. Their adaptive responses to changing climates—such as water harvesting, shifting cultivation, and seasonal migration—demonstrate resilience built over generations. IKS promotes a circular model of resource use, contrasting sharply with extractive industrial systems that often cause environmental degradation.

Moreover, the ethical and spiritual dimensions of Indigenous worldviews foster respect for all living beings, reinforcing the idea that humans are custodians rather than owners of the Earth. As global environmental crises intensify, policymakers and researchers increasingly recognize the value of integrating Indigenous knowledge into modern sustainability frameworks. Therefore, Indigenous knowledge serves as both a moral compass and a practical guide for achieving long-term environmental sustainability and resilience. In many parts of the world, indigenous knowledge has been the foundation for agriculture, food preparation and conservation, health care, education, and a host of other activities that sustain a society and its environment for many centuries. It is typically passed down orally and through cultural rituals(Senanayake,2006).

A. Traditional Ecological Knowledge (TEK) and Biodiversity Conservation

Traditional Ecological Knowledge (TEK) plays a vital role in biodiversity conservation by integrating centuries-old observations, cultural practices, and ecological wisdom. Indigenous communities use TEK to understand species behavior, seasonal patterns, soil fertility, and ecosystem interactions, allowing them to manage natural resources sustainably. This knowledge guides practices such as selective harvesting, rotational use of land, and protection of sacred natural sites, all of which contribute to maintaining ecological balance. TEK emphasizes relational thinking, where humans are seen as part of the ecosystem, fostering a sense of responsibility toward preserving all forms of life. As global biodiversity declines due to modern industrial practices, TEK has gained recognition as a powerful tool for conservation, resilience, and sustainable environmental governance. Designing alternative resource management systems may be accelerated by utilizing TEK-based management techniques and comprehending the social dynamics that underlie them (Dudgeon & Berkes, 2003).

a) Sustainable Resource Management Practices

Indigenous communities use TEK to guide sustainable practices such as controlled burning, rotational farming, and regulated hunting or fishing. These methods help maintain species populations, restore soil fertility, and reduce the risk of environmental degradation. By aligning resource use with natural cycles, TEK supports long-term ecosystem health.

b) Protection of Sacred Natural Sites

Many Indigenous cultures protect forests, water bodies, and mountains as sacred spaces associated with ancestral spirits. These areas often remain untouched or minimally used, creating natural conservation zones that harbor rich biodiversity. The cultural and spiritual significance of these sites indirectly contributes to habitat preservation and ecological stability.

c) Species and Habitat Knowledge

TEK includes deep knowledge of plant and animal species, including their medicinal uses, reproductive cycles, and ecological roles. This intimate understanding enables Indigenous communities to recognize early signs of environmental change and respond effectively. Such detailed species-level knowledge is invaluable for modern conservation efforts and ecological research.

B. Indigenous Land Management and Resource Use Practices

Indigenous land management and resource use practices are rooted in deep ecological understanding, cultural values, and long-term sustainability. These practices prioritize balance between human needs and the regenerative capacity of the environment, ensuring that resources are used without degrading ecosystems. Indigenous communities often adopt methods such as shifting cultivation, agroforestry, and controlled burns to maintain soil fertility, reduce wildfire risks, and support biodiversity. Their land management is guided by communal decision-making, where elders and knowledge holders determine the appropriate timing and scale of resource use. As modern societies face environmental crises, these Indigenous strategies offer valuable insights into sustainable land stewardship and climate resilience. When it comes to land management applications, it's critical to take communities' understanding of the land they live on into account(Puri,2007).

a) Shifting Cultivation and Soil Regeneration

Shifting cultivation, commonly practiced in many Indigenous societies, involves rotating agricultural plots to allow soil to recover naturally. This method prevents nutrient depletion and reduces the need for chemical fertilizers. When implemented responsibly, it supports long-term productivity and enhances landscape biodiversity.

b) Agroforestry and Mixed Land-Use Systems

Indigenous agroforestry integrates trees with crops and livestock, creating multifunctional landscapes that provide food, medicine, fuel, and shelter. This system enhances soil structure, improves water retention, and increases resilience to climate change. By mimicking natural ecosystems, agroforestry ensures sustainable productivity and ecological stability.

c) Controlled Burning and Fire Management

Controlled burning, or cultural fire management, is used by Indigenous communities to reduce fuel loads, prevent large wildfires, and promote the growth of native species. These intentional low-intensity fires help maintain healthy ecosystems by clearing underbrush and supporting nutrient cycling. Modern fire management agencies increasingly recognize the effectiveness of these practices in reducing wildfire risks.

C. Role of IKS in Climate Change Adaptation and Mitigation

The role of Indigenous Knowledge Systems (IKS) in climate change adaptation and mitigation is increasingly recognized as essential for building resilient ecosystems and communities. Indigenous peoples rely on generations of environmental observations that help them detect climatic shifts, predict weather patterns, and adjust their livelihoods accordingly. Their adaptive strategies—such as diversified crops, water harvesting, and mobility—enable them to withstand extreme weather events and resource scarcity. IKS also contributes to mitigation by promoting practices that reduce carbon emissions, enhance soil carbon storage, and protect forests and biodiversity. As climate impacts intensify globally, integrating Indigenous knowledge with scientific approaches offers a more holistic and sustainable response to the climate crisis. The widespread recognition that climate change is a global threat that impedes socioeconomic development, causes wanton deaths, disrupts livelihoods, depletes the ecosystem, increases the incidence of disease, and causes displacement due to floods and cyclones has spurred the development of climate change strategies and policies(Nyahunda,2024).

a) Early Warning Systems and Environmental Indicators

Indigenous communities use natural signs—such as animal behavior, flowering patterns, and wind direction—to predict rainfall, drought, or storms. These environmental indicators serve as effective early warning systems for climate-related risks. Such knowledge can complement modern meteorological tools to improve disaster preparedness.

b) Resilient Agricultural and Water Management Practices

Indigenous farmers cultivate diverse and climate-resilient crops, practice terracing, and use traditional irrigation systems like bamboo drip irrigation or zai pits. These techniques conserve water, enhance soil health, and increase food security during climate stress. Their low-cost and eco-friendly nature makes them ideal for rural adaptation strategies.

c) Forest Conservation and Carbon Sequestration

Indigenous stewardship of forests, through controlled use and protection of sacred groves, plays a major role in carbon sequestration and ecosystem stability. Their conservation ethics reduce deforestation and encourage biodiversity preservation. These practices directly support global climate mitigation goals by maintaining carbon-rich landscapes.

4. SOCIO-ECONOMIC DIMENSIONS OF IKS

The Socio-Economic Dimensions of Indigenous Knowledge Systems (IKS) highlight how traditional wisdom contributes to the economic stability, social cohesion, and cultural identity of Indigenous communities. IKS supports sustainable livelihoods through community-based resource management, traditional agriculture, and artisanal crafts that rely on local materials and skills. Indigenous economies are often circular and self-sufficient, emphasizing reciprocity, cooperation, and equitable distribution rather than profit maximization. Traditional agricultural practices, such as mixed cropping and organic farming, not only ensure food security but also preserve soil health and biodiversity. Indigenous women play a central role in managing household economies, seed preservation, and knowledge transmission, reinforcing gender inclusivity and social balance. The production of traditional crafts, textiles, and herbal medicines contributes to local and regional markets, providing both income and cultural continuity. Tourism based on Indigenous heritage and ecological knowledge further strengthens community economies while promoting environmental awareness.

However, external pressures such as globalization, commercialization, and land dispossession threaten these traditional economic systems. Integrating IKS into national development policies can help protect Indigenous livelihoods while promoting sustainable economic growth. Ultimately, the socio-economic dimensions of IKS demonstrate that economic development can coexist with cultural preservation, social equity, and environmental sustainability when guided by Indigenous values and community wisdom. A civilized society can learn a great deal from the IKS of a nature-bound community, particularly in this day and age when the planet is plagued by numerous issues such as pollution, global warming, biodiversity loss, war, economic crises, growing economic diversity and the ensuing fuel and food crises, use of genetically modified food, bio-piracy, etc(Gupta,2013).

A. Indigenous Economic Systems and Sustainable Livelihoods

Indigenous economic systems and sustainable livelihoods are built on principles of balance, reciprocity, and long-term resilience rather than profit maximization. These systems rely on locally available resources, traditional skills, and communal cooperation, allowing communities to meet their needs without overexploiting the environment. Economic activities such as farming, fishing, herding, and craftsmanship are closely tied to ecological cycles, ensuring that production aligns with natural regeneration. Indigenous livelihoods are also deeply embedded in cultural values, emphasizing shared responsibilities and collective well-being rather than individual accumulation. As modern development pressures intensify, these sustainable economic systems provide valuable models for environmentally responsible and socially inclusive growth. Indigenous peoples want their local production and economic systems to be protected and strengthened, as well as food security and food sovereignty. Opportunities to diversify their sources of income, fair trade, and other policies that uphold and respect their lifestyles are important to them(Kalafatic,2007).

a) Subsistence-Based and Community-Oriented Production

Indigenous economies often prioritize subsistence activities that provide food, shelter, and essential needs directly to the community. These practices reduce dependence on external markets and foster social cohesion. By focusing on communal well-being, such systems create stability even in changing economic conditions.

b) Diversified Livelihood Strategies

Indigenous communities typically engage in multiple livelihood activities—such as mixed farming, seasonal fishing, gathering forest products, and handicrafts—which reduce vulnerability to environmental or economic shocks. This diversification strengthens resilience by ensuring that if one resource becomes scarce, others can support community subsistence. It also enhances ecological balance by distributing resource pressure across different sectors.

c) Sustainable Use of Natural Resources

Resource use in Indigenous economies is guided by culturally embedded rules that limit overharvesting and ensure regeneration. Practices like selective gathering, rotational grazing, and sacred-site protection contribute to maintaining ecosystem health. These sustainable approaches demonstrate how economic activities can coexist with conservation and long-term resource security.

B. Role of IKS in Food Security and Agriculture

The role of Indigenous Knowledge Systems (IKS) in food security and agriculture is central to sustaining community livelihoods and protecting local ecosystems. Indigenous farmers rely on traditional practices such as mixed cropping, seed preservation, organic fertilization, and rainwater harvesting to ensure consistent food production even under challenging conditions. These methods promote soil fertility, maintain biodiversity, and reduce dependency on external inputs like chemical fertilizers and hybrid seeds. IKS also emphasizes the cultivation of climate-resilient crops that have adapted over generations to local environmental stresses, ensuring stable yields during droughts or floods. By blending ecological wisdom with sustainable techniques, Indigenous agricultural knowledge plays a powerful role in enhancing food sovereignty and ensuring long-term food security. Due to its direct connection to both the general stability of society and the well-being of the populace, food security is a crucial concern in modern society (Olawuyi et al., 2024).

a) Traditional Crop Diversity and Seed Preservation

Indigenous communities maintain diverse seed banks of local crop varieties, many of which are naturally resistant to pests, diseases, and climate extremes. This biodiversity strengthens resilience by reducing the risk of crop failure. Seed preservation also safeguards cultural heritage and promotes food autonomy.

b) Soil Fertility and Organic Farming Practices

Indigenous agricultural systems rely on natural fertilizers such as compost, animal manure, and plant residues, which enrich soil and maintain long-term productivity. Techniques like mulching and intercropping further enhance soil moisture and structure. These organic approaches improve food quality while avoiding environmental pollution.

c) Water Conservation and Traditional Irrigation Methods

Indigenous farmers use innovative water management systems—such as terracing, bamboo drip irrigation, and seasonal water storage—to optimize water availability. These methods ensure efficient use of limited water resources, especially in drought-prone regions. By aligning with local geography and climate patterns, traditional irrigation systems support sustainable agriculture.

C. Gender Roles and Community-Based Decision-Making

Gender roles and community-based decision-making in Indigenous Knowledge Systems (IKS) highlight the importance of shared responsibilities, collaboration, and social harmony within traditional societies. Indigenous women and men often hold complementary roles that contribute to sustainable livelihoods, cultural preservation, and resource management. Women commonly oversee seed selection, food preparation, medicinal knowledge, and household economies, while men engage in activities like hunting, land preparation, or community leadership—although these roles vary across cultures. Decision-making is typically inclusive and consensus-driven, with elders, women, and youth participating in discussions that affect community welfare. This balanced approach strengthens social cohesion, ensures equitable resource distribution, and maintains the continuity of Indigenous traditions. In addition to considering race, ethnicity, and ability, a completely intersectional perspective offers a methodology that is universally applicable to historically oppressed groups. The construction of gender, gender roles and norms, the existence of gender discrimination, and the extent to which sustainability indicators take gender into account—from simple sex-based disaggregation to the choice of quality-of-life metrics that reflect gender roles and differences—all require more consideration(DiPrete Brown,2020).

a) Women as Custodians of Agricultural and Medicinal Knowledge

Indigenous women play a central role in seed preservation, crop diversity management, and the preparation of herbal remedies. Their intimate understanding of local plants, soils, and seasonal cycles is crucial for food security and community health. This knowledge empowers women as key contributors to sustainable development.

b) Collective Decision-Making through Councils and Elders

Many Indigenous communities rely on councils of elders or communal gatherings where decisions are made through consensus rather than hierarchical authority. This participatory model ensures that diverse perspectives are heard and respected. The process strengthens unity and promotes decisions that reflect collective needs and cultural values.

c) Gender Complementarity in Resource Management

Indigenous societies often assign resource-related responsibilities in a way that balances men's and women's strengths. Men may manage tasks like forest protection or livestock care, while women oversee seed storage, water collection, or community health practices. This complementarity ensures efficient resource use and reinforces cultural stability.

5. IKS and Healthcare Practices

Indigenous Knowledge Systems (IKS) and Healthcare Practices represent a holistic approach to health that integrates physical, spiritual, and environmental well-being. Traditional medicine, rooted in Indigenous knowledge, relies on locally available herbs, plants, and natural resources to treat various illnesses and maintain wellness. Healers, midwives, and herbalists play vital roles within Indigenous communities, using generations of experiential knowledge passed down orally through apprenticeships and rituals. These practices often emphasize prevention and balance rather than solely treating symptoms, aligning with modern concepts of holistic health. Many Indigenous healing systems incorporate spiritual and cultural dimensions, recognizing the interconnectedness between mind, body, and nature. In recent years, researchers and health professionals have increasingly acknowledged the value of traditional medicine as a complementary component to modern healthcare systems. However, challenges such as lack of documentation, inadequate scientific validation, and intellectual property disputes pose risks to preserving this knowledge. Efforts to integrate Indigenous healthcare practices into public health frameworks require ethical collaboration, community consent, and respect for cultural values. Safeguarding Indigenous medicinal knowledge also involves protecting biodiversity, as many healing plants face threats from deforestation and habitat loss. Ultimately, the fusion of Indigenous and modern healthcare practices offers a promising path toward inclusive, culturally sensitive, and sustainable healthcare solutions. The IKS Division's documentaries on healthcare systems, water management, and traditional farming methods aid in the spread of information throughout rural communities (Bolake, 2025).

A. Traditional Medicine and Healing Systems

Traditional medicine and healing systems within Indigenous Knowledge Systems (IKS) represent a holistic approach to health that integrates physical, spiritual, and environmental dimensions. Indigenous healers draw upon centuries of accumulated wisdom, using medicinal plants, rituals, and natural therapies to treat illnesses and maintain well-being. These healing practices emphasize harmony and balance within the body and between humans and nature, reflecting deep cultural and spiritual beliefs. Traditional medicine also plays a preventive role, guiding lifestyles around diet, hygiene, and social conduct to reduce illness. Despite modern advances in biomedicine, Indigenous healing systems remain vital in many communities due to their accessibility, cultural relevance, and proven effectiveness in treating certain conditions. Traditional medicine is "cost effective," as defined by the World Bank. Hospitalizations are decreased by traditional medicine. It is also less expensive to use plants instead of chemicals (Ebbe, 1993).

a) Use of Medicinal Plants and Herbal Remedies

Indigenous healers possess extensive knowledge of local plants, their properties, and their appropriate dosages. Herbal remedies are prepared using leaves, roots, bark, or seeds to treat ailments ranging from fevers to digestive disorders. This plant-based medicinal knowledge contributes significantly to modern pharmacology and natural health practices.

b) Role of Healers, Shamans, and Traditional Practitioners

Healers, shamans, and traditional practitioners hold respected positions as custodians of sacred knowledge and community health. They undergo long apprenticeships, learning diagnostic techniques, spiritual rituals, and medicinal preparations from elders. Their role extends beyond physical healing to include emotional and spiritual guidance.

c) Holistic and Spiritual Approaches to Healing

Traditional healing often incorporates rituals, chants, prayer, and symbolic practices to restore balance within an individual. Illness is sometimes seen as a disruption in harmony with nature or community, requiring spiritual intervention alongside physical treatment. This holistic approach strengthens mental, emotional, and social well-being.

B. Integration of Indigenous and Modern Healthcare

The integration of Indigenous and modern healthcare seeks to combine the strengths of traditional healing practices with contemporary medical science to provide more holistic and culturally inclusive healthcare services. This approach recognizes that Indigenous medicine offers valuable insights into preventive care, natural remedies, and community-based healing that can complement modern treatments. Integrating both systems enhances accessibility, especially in rural areas where traditional healers are often the first point of care. Collaboration between Indigenous practitioners and biomedical professionals also fosters mutual respect, improving patient trust and overall health outcomes. However, successful integration requires ethical guidelines, proper validation of traditional remedies, and community consent to ensure cultural integrity is maintained. Public campaigns play a crucial role in debunking the stigma and myths associated with traditional medicine and promoting a peaceful coexistence with contemporary healthcare(Aguessy,2023).

a) Complementary Strengths of Both Systems

Indigenous medicine provides natural, culturally grounded treatments, while modern healthcare offers advanced diagnostics and emergency interventions. When combined, they create a more comprehensive health system that addresses physical, mental, and spiritual needs. This complementarity can significantly improve patient satisfaction and treatment effectiveness.

b) Collaborative Healthcare Models

Some regions have established integrated clinics where traditional healers work alongside biomedical doctors. These models facilitate knowledge-sharing, referrals, and coordinated treatment plans. Such collaboration strengthens the healthcare system by bridging cultural gaps and providing patients with culturally sensitive options.

c) Ethical Considerations and Community Participation

Integration must respect Indigenous autonomy, protect sacred knowledge, and ensure that traditional remedies are not exploited or commercialized without consent. Community involvement in decision-making helps establish trust and ensures ethical practices. Clear guidelines support respectful and equitable integration of both medical traditions.

C. Ethical and Scientific Validation of Indigenous Remedies

The ethical and scientific validation of Indigenous remedies is essential to ensure that traditional healing practices are both respected and safely integrated into modern healthcare systems. Indigenous remedies have been developed through generations of observation and experience, but scientific evaluation helps verify their efficacy, dosage, and potential side effects. Ethical validation ensures that Indigenous communities retain ownership of their knowledge and benefit from any commercial or medical applications derived from it. This process requires respectful collaboration between researchers and Indigenous practitioners, grounded in transparency, informed consent, and cultural sensitivity. When conducted responsibly, validation supports the preservation of traditional medicine while promoting its safe and effective use in a broader healthcare context. To properly manage environmental resources in these situations, religious environmental activists, environmental groups, governments, and legislators must collaborate ethically and share ideas (Chanda & Alfred, 2025).

a) Respect for Intellectual Property and Cultural Rights

Indigenous communities must be recognized as the rightful owners of their medicinal knowledge, and their consent is essential before any research or commercialization takes place. Ethical protection prevents biopiracy, exploitation, or misuse of sacred knowledge. Ensuring fair compensation and proper acknowledgement strengthens trust between communities and researchers.

b) Scientific Research and Pharmacological Testing

Scientific validation involves laboratory studies, clinical trials, and pharmacological analyses to confirm the safety, effectiveness, and active compounds of traditional remedies. This process helps identify therapeutic benefits while minimizing risks. Such research can lead to improved formulations or integration of remedies into mainstream medicine.

c) Community Involvement and Participatory Research

Engaging Indigenous healers and community members in the research process ensures that studies respect cultural contexts and traditional practices. Participatory approaches help align scientific methods with Indigenous worldviews and prevent misunderstanding or misrepresentation. This inclusive collaboration strengthens both the credibility and cultural integrity of the validation process.

6. EDUCATION AND KNOWLEDGE PRESERVATION

Education and Knowledge Preservation within Indigenous Knowledge Systems (IKS) play a crucial role in sustaining cultural identity, community values, and traditional wisdom across generations. Indigenous education is often experiential, community-based, and grounded in the local environment, where learning occurs through storytelling, observation, and participation in daily life. Unlike formal schooling, it emphasizes moral values, social responsibility, and a deep respect for nature as integral components of knowledge. Elders serve as the primary educators, transmitting wisdom about agriculture, healing, craftsmanship, and spiritual practices through oral traditions and rituals. However, modernization, globalization, and language loss have disrupted these traditional methods of knowledge transmission. Integrating Indigenous perspectives into formal education systems can help bridge this gap and promote intercultural understanding. Schools that incorporate Indigenous languages, histories, and ecological knowledge not only preserve heritage but also enrich mainstream curricula with diverse worldviews. Technology, including digital archiving and multimedia storytelling, offers new opportunities to document and disseminate Indigenous knowledge while ensuring authenticity and community control. Yet, ethical considerations are essential to prevent cultural appropriation and misuse of sacred knowledge. Therefore, strengthening Indigenous education and knowledge preservation is vital for empowering communities, maintaining cultural continuity, and contributing to global sustainability and diversity. In order to promote sustainable development, education is essential. Therefore, in order to support national governments' efforts, it is imperative to mobilize the required resources, particularly financial resources at all levels(Breidlid,2009).

A. Incorporating IKS into Formal and Informal Education Systems

Incorporating Indigenous Knowledge Systems (IKS) into formal and informal education systems is essential for preserving cultural heritage, promoting sustainable practices, and fostering respect for diverse worldviews. Integrating IKS into school curricula allows students to learn about traditional ecological knowledge, local history, and cultural values alongside modern subjects. Informal education—through storytelling, apprenticeships, and community rituals—continues to strengthen identity and intergenerational learning. Including IKS in education helps bridge the gap between modern schooling and Indigenous cultural realities, making learning more relevant and holistic. When done responsibly, this integration empowers Indigenous communities, enriches national education systems, and strengthens social cohesion. A religious outlook on life was also ingrained through indigenous education. Religion, which focused on morality, supported the community's laws, traditions, and acceptable standards of behavior, which included being polite, giving, and truthful(Eze,2013).

a) Curriculum Integration and Cultural Relevance

Integrating Indigenous knowledge into school subjects such as science, geography, and history ensures that learning reflects local realities and cultural identity. Students gain a deeper understanding of sustainable practices, traditional technologies, and ecological relationships. This relevance enhances engagement and promotes pride in Indigenous heritage.

b) Role of Elders and Knowledge Keepers in Education

Elders serve as living libraries, sharing stories, skills, and values that formal textbooks cannot capture. Inviting them into classrooms or organizing community learning sessions honors their authority and ensures accurate transmission of traditional knowledge. Their presence strengthens intergenerational bonds and cultural continuity.

c) Informal Learning through Community Practices

Informal education occurs through participation in farming, crafts, ceremonies, and daily activities, where children learn by observing and doing. These hands-on experiences teach responsibility, cooperation, and respect for nature. Such learning reinforces Indigenous worldviews and complements formal school-based education.

B. Intergenerational Knowledge Transmission

Intergenerational knowledge transmission is central to the survival and continuity of Indigenous Knowledge Systems (IKS), ensuring that cultural values, skills, and ecological wisdom are passed down through generations. This transmission occurs through everyday interactions, storytelling, rituals, and hands-on participation in community activities, allowing knowledge to remain dynamic and relevant. Elders play a vital role as custodians of memory and tradition, guiding younger members with lessons that blend practical skills and moral values. Through this continuous flow of knowledge, communities strengthen their identity and maintain a deep connection to their environment. However, influences such as modernization, migration, and loss of Indigenous languages pose challenges to sustaining these traditional learning pathways.

a) Role of Elders as Knowledge Custodians

Elders hold vast cultural and ecological knowledge accumulated throughout their lifetimes, making them key figures in teaching younger generations. They share wisdom through stories, proverbs, and lived examples rather than formal instruction. Their guidance affirms community identity and upholds cultural continuity.

b) Storytelling and Oral Narratives

Storytelling is a powerful tool for transmitting lessons about morality, environmental stewardship, historical events, and social norms. These narratives make complex ideas memorable and relatable for children and youth. Through oral traditions, communities preserve both factual knowledge and cultural worldviews.

c) Learning by Observation and Participation

Indigenous children often learn by accompanying elders during farming, hunting, crafting, or ceremonial activities. This experiential learning builds practical skills and strengthens respect for nature and community responsibilities. By learning through doing, younger generations internalize knowledge more deeply and authentically.

C. Challenges of Modernization and Cultural Erosion

Challenges of modernization and cultural erosion pose significant threats to the preservation and continuity of Indigenous Knowledge Systems (IKS) across many communities. As modern lifestyles, technologies, and economic pressures spread, younger generations often become detached from traditional practices, languages, and belief systems. Globalization introduces external values and educational models that may overshadow Indigenous worldviews, weakening community identity and intergenerational learning. Urban migration further reduces opportunities for young people to engage in traditional activities and learn from elders in their natural environment. Without deliberate efforts to protect cultural heritage, modernization can lead to the gradual disappearance of traditional skills, ecological knowledge, and cultural values that have sustained Indigenous societies for centuries.

a) Decline of Indigenous Languages

Indigenous languages are core carriers of cultural knowledge, rituals, ecological wisdom, and identity. As dominant languages take over in schools, media, and workplaces, younger generations may stop speaking their mother tongues. The loss of language directly results in the loss of unique knowledge embedded in these linguistic systems.

b) Migration and Disconnection from Ancestral Lands

Many Indigenous youth migrate to urban areas seeking education and employment, reducing their involvement in traditional cultural activities. Distance from ancestral lands weakens their connection to traditional practices such as farming, fishing, and sacred rituals. Over time, this disconnect contributes to fading cultural memory and collective identity.

c) Influence of Western Education and Media

Modern education systems often prioritize Western scientific knowledge while overlooking Indigenous worldviews, making traditional practices seem outdated to young learners. Media exposure reshapes aspirations and lifestyles, sometimes devaluing cultural practices. This influence accelerates cultural assimilation and reduces interest in carrying forward Indigenous traditions.

7. GOVERNANCE, POLICY, AND LEGAL FRAMEWORKS

Governance, Policy, and Legal Frameworks play a pivotal role in protecting and promoting Indigenous Knowledge Systems (IKS) as essential components of sustainable development. Indigenous governance structures are traditionally based on consensus, collective responsibility, and respect for ancestral laws that emphasize balance between people and nature. These community-based systems offer valuable models for participatory and inclusive governance in modern policy contexts. At the national and international levels, organizations such as UNESCO, the United Nations Permanent Forum on Indigenous Issues (UNPFII), and the United Nations Declaration on the Rights of Indigenous Peoples (UNDRIP) have recognized the importance of safeguarding Indigenous rights and knowledge. Policies supporting IKS aim to ensure Indigenous communities maintain control over their cultural heritage, natural resources, and intellectual property. Legal frameworks also address the need to prevent exploitation of traditional knowledge, particularly in areas like pharmaceuticals, agriculture, and biotechnology. However, many Indigenous groups still face challenges in having their customary laws and land rights fully recognized within formal state systems. Strengthening partnerships between governments, research institutions, and Indigenous communities is essential to create equitable and culturally sensitive policy environments. Integrating IKS into national sustainability strategies requires not only legal recognition but also ethical respect for Indigenous worldviews and governance models. Ultimately, effective governance and legal protection of IKS can empower Indigenous peoples while enriching global efforts toward inclusive and sustainable development.

A. Role of Indigenous Governance Structures in Sustainable Development

Indigenous governance structures play a critical role in sustainable development by fostering decision-making systems rooted in community consensus, ecological responsibility, and cultural values. These governance models emphasize long-term stewardship of natural resources rather than short-term economic gains, ensuring that development aligns with environmental balance. Elders, councils, and traditional leaders guide community decisions through customary laws that integrate spiritual, social, and ecological considerations. Such structures promote inclusivity, as decisions typically involve consultation with women, youth, and various community groups, strengthening social cohesion. By prioritizing harmony with nature and collective well-being, Indigenous governance offers sustainable alternatives to conventional top-down development models.

a) Consensus-Based Decision-Making

Indigenous governance relies on deliberation and community discussion rather than hierarchical authority, ensuring that decisions reflect shared values and collective interests. This participatory approach builds trust and unity within the community. It also leads to development actions that are socially acceptable and environmentally responsible.

b) Customary Laws and Environmental Stewardship

Traditional laws regulate resource use, protect sacred sites, and guide sustainable land management practices. These customary rules prevent overexploitation and promote regeneration of forests, water, and wildlife. By combining spiritual beliefs with ecological knowledge, customary laws support long-term environmental health.

c) Leadership Roles of Elders and Community Representatives

Elders and appointed leaders serve as custodians of cultural wisdom, offering guidance based on generations of lived experience. Their leadership ensures that development projects respect cultural norms, community needs, and environmental limits. This culturally grounded leadership strengthens resilience and supports sustainable community development.

B. National and International Policies Supporting IKS (UNESCO, UNDRIP, etc.)

National and international policies supporting Indigenous Knowledge Systems (IKS) aim to safeguard Indigenous rights, protect cultural heritage, and promote sustainable development rooted in traditional wisdom. Organizations such as UNESCO and frameworks like the United Nations Declaration on the Rights of Indigenous Peoples (UNDRIP) recognize Indigenous communities as vital custodians of biodiversity, languages, and ancestral knowledge. These policies encourage governments to respect Indigenous autonomy, support land rights, and integrate traditional knowledge into national development strategies. International conventions also emphasize the importance of protecting Indigenous intellectual property and ensuring fair benefit-sharing when their knowledge is used. By establishing legal guidelines and moral commitments, these policies help bridge historical inequalities and strengthen the role of IKS in global sustainability efforts.

a) UNESCO's Initiatives for Cultural and Knowledge Preservation

UNESCO promotes the protection of intangible cultural heritage, including Indigenous languages, rituals, and ecological knowledge. Through programs like the Memory of the World and Intangible Heritage Lists, UNESCO supports documentation, education, and community-based preservation efforts. These initiatives help raise global awareness about the value of IKS.

b) UNDRIP and Recognition of Indigenous Rights

The United Nations Declaration on the Rights of Indigenous Peoples (UNDRIP) affirms Indigenous communities' rights to self-determination, cultural preservation, and control over their lands and resources. It advocates for free, prior, and informed consent (FPIC) before any external project affects Indigenous territories. UNDRIP provides a global framework ensuring dignity, equality, and protection of traditional knowledge.

c) National Policies and Legal Frameworks

Many countries have enacted laws to protect Indigenous land rights, traditional medicine, and cultural heritage within their national governance systems. These policies encourage collaboration between governments, research institutions, and Indigenous organizations to integrate IKS into education, conservation, and development plans. National legal frameworks help ensure that Indigenous knowledge is not exploited and that communities benefit from its application.

C. Legal Protection and Intellectual Property Rights

Legal protection and intellectual property rights (IPR) are essential for safeguarding Indigenous Knowledge Systems (IKS) from exploitation, misappropriation, and loss. Traditional knowledge—such as medicinal practices, agricultural methods, and cultural expressions—is often vulnerable to commercial misuse without proper legal frameworks. Instruments like UNESCO conventions and the United Nations Declaration on the Rights of Indigenous Peoples (UNDRIP) emphasize the need for protecting Indigenous intellectual property and ensuring fair benefit-sharing. Legal protection ensures that Indigenous communities maintain control over their knowledge and are acknowledged and compensated when it is used by researchers, industries, or governments. Strengthening IPR mechanisms not only preserves cultural heritage but also promotes ethical collaboration and sustainable development.

a) Protection Against Biopiracy and Misappropriation

Indigenous knowledge has frequently been exploited for commercial gain without consent, especially in areas like pharmaceuticals, agriculture, and cosmetics. Legal frameworks aim to prevent biopiracy by requiring permission and benefit-sharing agreements before traditional knowledge is used. This helps protect sacred knowledge and ensures accountability.

b) Community Ownership and Collective Rights

Unlike Western IPR systems that emphasize individual ownership, Indigenous knowledge is often collectively held by the community. Legal recognition of collective rights ensures that the entire community benefits from its knowledge. Such recognition supports cultural integrity and prevents fragmentation or privatization of shared heritage.

c) International Agreements Supporting IPR for IKS

Global policies such as UNDRIP, the Convention on Biological Diversity (CBD), and the Nagoya Protocol provide guidelines for protecting Indigenous knowledge and ensuring equitable benefit-sharing. These agreements encourage countries to adopt national laws that respect Indigenous rights and prevent exploitation. They also promote ethical research partnerships grounded in respect and consent.

8. CONCLUSION

The preceding analysis affirms that Indigenous Knowledge Systems (IKS) are far more than historical relics; they are dynamic, comprehensive frameworks for living sustainably and equitably. The foundations of IKS, rooted in centuries of observation and harmony with nature, provide a powerful counter-narrative and a practical complement to modern industrial models. Across all dimensions—from ensuring environmental sustainability through profound ecological knowledge, supporting resilient socio-economic structures through reciprocity and community-based management, to fostering holistic healthcare and experiential education—IKS offers proven, time-tested solutions.The global recognition of IKS, codified in international instruments like UNDRIP, signals a critical shift toward acknowledging the rights and contributions of Indigenous peoples. However, ethical integration and legal protection remain paramount. Efforts must focus on establishing genuine partnerships with Indigenous communities, ensuring their governance models and customary laws are respected, and granting them control over their intellectual property and cultural heritage. The erosion of IKS through land dispossession, cultural assimilation, and language loss poses a threat not just to Indigenous communities but to the entire world, as it diminishes a vast repository of adaptive wisdom. By safeguarding and respectfully collaborating with Indigenous Knowledge Systems, policymakers, researchers, and global citizens can collectively enrich global knowledge, empower marginalized communities, and forge a path toward a truly resilient, diverse, and sustainable future.

REFERENCES

Aguessy, A. N. J. (2023). Integration of traditional healing practices in modern medicine in South America: Insights for African development. *International Journal on Studies in English Language and Literature, 11*(12), 36–47. DOI: 10.20431/2347-3134.1112005

Bolake, R. N. (2025). INDIAN KNOWLEDGE SYSTEM AND RURAL DEVELOPMENT: A PATHWAY TO SUSTAINABLE DEVELOPMENT. *IIP: International Multidisciplinary Research Journal, 2*(Issue-IV (October-December)), 9-9.

Breidlid, A. (2009). Culture, indigenous knowledge systems and sustainable development: A critical view of education in an African context. *International Journal of Educational Development, 29*(2), 140–148. DOI: 10.1016/j.ijedudev.2008.09.009

Chanda, A., & Alfred, K. (2025). Ethical Integration of Indigenous and Scientific Knowledge Systems for a Sustainable Mitigation of Environmental and Climate Change Challenges. *International Journal of Global Community, VIII*(1), 1. DOI: 10.33473/ijgc-ri.v8i1.222

DiPrete Brown, L., Atapattu, S., Stull, V. J., Calderón, C. I., Huambachano, M., Houénou, M. J. P., Snider, A., & Monzón, A. (2020). From a three-legged stool to a three-dimensional world: Integrating rights, gender and indigenous knowledge into sustainability practice and law. *Sustainability (Basel), 12*(22), 9521. DOI: 10.3390/su12229521

Dudgeon, R. C., & Berkes, F. (2003). Local understandings of the land: Traditional ecological knowledge and indigenous knowledge. In *Nature across cultures: Views of nature and the environment in non-Western cultures* (pp. 75–96). Springer Netherlands. DOI: 10.1007/978-94-017-0149-5_4

Ebbe, K. (1993, September). Traditional Knowledge and Sustainable Development. In *Conference held at The World Bank Washington.*

Eze, U. T. (2013). Integrating African indigenous knowledge In Nigeria's formal education system: It's potential for sustainable development. *Journal of Education and Practice, 4*(6).

Gupta, A. D. (2013). Does indigenous knowledge have anything to deal with sustainable development. *Cultural Anthropology, 57*, 64.

Kalafatic, C. (2007). Indigenous peoples' sustainable livelihoods. *Food and Agriculture Organization of the United Nations Thematic brief.* ftp://ftp. fao. org/docrep/fao/010/aj033e/aj033e02. pdf

Mafongoya, P. L., & Ajayi, O. C. (2017). Indigenous knowledge systems: Their history, development over time and role in sustainable development and climate change management. *Indigenous knowledge systems and climate change management in Africa*, 29.

Magni, G. (2017). Indigenous knowledge and implications for the sustainable development agenda. *European Journal of Education, 52*(4), 437–447. DOI: 10.1111/ejed.12238

Nyahunda, L. (2024). Integration of indigenous knowledge systems (IKS) into climate change mitigation and adaptation endeavours: Milestones and gaps in South Africa and Zimbabwe's climate policy frameworks. *Climatic Change, 177*(11), 162. DOI: 10.1007/s10584-024-03822-2

Olawuyi, S. O., Ijila, O. J., Adegbite, A., Olawuyi, T. D., & Farayola, C. O. (2024). Smallholder farmers' use of indigenous knowledge practices in agri-food systems: Contribution of food security attainment drive. *Research on World Agricultural Economy*, *5*(2), 45–67. DOI: 10.36956/rwae.v5i2.1056

Puri, S. K. (2007). Integrating scientific with indigenous knowledge: Constructing knowledge alliances for land management in India. *Management Information Systems Quarterly*, *31*(2), 355–379. DOI: 10.2307/25148795

Rawat, G., & Mishra, S. (2021). SPIRITUALITY AND ENVIRONMENT: SIGNIFICANCE OF INDIGENOUS KNOWLEDGE SYSTEMS IN AFRICA AND CHALLENGES. *Literary Endeavour, 12*(1).

Senanayake, S. G. J. N. (2006). Indigenous knowledge as a key to sustainable development.

Sultana, R., Muhammad, N., & Zakaria, A. K. M. (2018). Role of indigenous knowledge in sustainable development. *International Journal of Developmental Research*, *8*(2), 18902–18906.

Chapter 4
Integrating Indigenous Knowledge System in Education and Research for Sustainable Development

Mokgadi Asnath Modiba
http://orcid.org/0000-0002-5643-9341
University of Limpopo, South Africa

ABSTRACT

Colonialism and apartheid epistemes persist in the current age. It is apparent in written textbooks where disciplinary knowledge is rooted in Western information. Such practices exclude the rich and relevant experiences from indigenous people, despite IKS relevance in understanding nature and cultural heritages. Higher Education Institutions are mandated to enforce the recognition and the revival of IKS, to implement innovative strategies that encourage and support its integration in education and research. The integration of IKS in education and research revives the continuity of cultural histories and it allows partnerships between communities and scientists. Google scholar database was used to retrieve and review literature from books, peer-reviewed articles and government documents. The author attempts to enlighten readers about the integration of IKS in education and research, expanding on policy frameworks that support IKS, possible ways to integrate IKS in education and research, challenges faced and opportunities encountered when integrating IKS in education and research.

INTRODUCTION

The persistence of apartheid epistemes and colonialism in education and research are superficial in textbooks used in disciplinary knowledge. Knowledge used in education originates from Western countries, meanwhile, knowledge from indigenous people is neglected. In the paper by Senegal and Lenz (2020), the authors concur with the statement mentioned above. Indigenous knowledge is perceived to be old fashioned and its origin is rural, where development is still an issue Modiba and Emsley (2024a) and it lacks progress and innovativeness. These unjust practices enforced the exclusion of indigenous

DOI: 10.4018/979-8-3373-7976-0.ch004

people's experiences, their unique culture and histories. In the paper by Du Plessis (2021), the author alluded on the use of languages that are foreign to Africans in study books without echoing their cultures.

In addition, Woldegiorgis (2021), highlighted on the enforcement of colonial interests, its education and the disengagement from African histories. It was further alluded in Mc Mahon and Portelli (2012) that the persistence of inequalities, the disengagements of students and social injustices caused the continuation of colonialism in education systems. These inequalities require change that embrace knowledge that originates from all spheres, where all bodies that contribute to the creation of knowledge are respected and acknowledged (Lasky, 2005). In the paper by Van Pixteren (2022), the author stated that there is the brainwashing of Africans to believe that Western knowledge is a key that leads to a successful future.

However, the prevalence of perceiving education as a way that enables decent living and as a key that allows people to partake in economical growth was not dominant in Africa as it was in Europe. Senegal and Lenz (2020) highlighted that the insufficiency in African content hinders African knowledge from being used as a relevant tool for growth in education. This necessitated Higher Education Institutions (HEIs) to enforce the revival and the recognition of Indigenous Knowledge Systems (IKS) in education and research. According to Naidoo, et al. (2014) it is necessary to include IKS in the curriculum of science, as a way that intends to restore the value of indigenous ways of knowing and to centre their wisdom in the curriculum, as well as to reduce the hierarchies in cultures. As such, it is necessary for HEIs to implement innovative strategies that support and encourage the integration of IKS in education and research with the revision of the curriculum.

Indigenous Knowledge (IK) is knowledge that developed from indigenous communities and it is shared verbally from generation to generation. It is knowledge that is accumulated over centuries, it is accrued from their daily practices, the skills and beliefs. Even though indigenous knowledge is alleged to be informal, researchers have shown that it relies on the testing of the hypothesis, the trial-and-error practices that took place over centuries and the monitoring of the environment (Ijatuyi et al., 2025). Knowledge and the experiences from daily living are passed on to the next breed of people by oral explanations and by showing how certain practices are done.

IKS is a form of knowledge, the belief systems and the practices in indigenous communities, where that knowledge is accrued over many generations and it is currently informing modern life. According to the Department of Science and Technology (DST, 2006), IKS is developed and maintained by indigenous people of South Africa and it is displayed in various areas such as agricultural practices, health and medical interventions and in religious ceremonies. Selepe et al. (2022), mentioned that IK is transferred from one generation to another to ensure that it is preserved. Meanwhile, McKnight (2015) defined it as the traditional practices that involve the customs, values and beliefs. Rankoana (2017) in Sefoka and Chuene (2025) defined IK as knowledge accumulated by indigenous communities through interactions with the environment, through observations and the experimentations of phenomena.

The co-existence of IK and scientific knowledge enhance practices that address challenges that may lead to sustainability. Their co-existence improves our understanding of the two knowledge systems with the removal of boundaries where intercultural communication is encouraged. It further allows an exchange of information, partnerships among various entities, new innovations and possible solutions to challenging issues. The integration of indigenous knowledge in science knowledge reduces the exploitative behaviour that exists between indigenous people and other participating entities, it encourages an equitable system where respect is regained.

United Nations Educational Scientific and Cultural Organizations (UNESCO) defined indigenous knowledge as the philosophies, the skills and the understandings established by indigenous people's relations with their native environment (UNESCO, 2017). According to Nyahunda (2024), the author mentioned that Africans still depend on IKS when decisions are to be made in weather predictions. It is also used in the whole world to resolve challenges that are faced on daily basis.

On the other hand, scientific knowledge is created in research, its information is shared in various platforms such as the documentation of findings in research articles, books, on the web, in podcasts, you tube videos, informative flyers, orally in formal and informal presentations. Research knowledge developed from the collaborative exercises that allowed various entities to collaborate, aiming to address challenges and to attain the set goals that leads to growth. Such collaborations allow the drafting of the memorandum of understanding (MoU) among participating parties. The participation of various entities such as government departments, Non-Government Organisations (NGO's), indigenous communities, education specialists and many other knowledgeable people may enhance the capacitation of young people, with the creation of jobs, internships and in service trainings that change people's lifestyles. Policies on IKS are in place, with frameworks that emphasise its promotion, protection and its recognition that enable member's contributions in making decisions that integrates IKS in education and research (Hill, et al., 2020). Some of the policies are mentioned in the paragraph under literature.

The Problem Statement

Policies on IKS are in place (Blackie, 2024). However, IKS is not yet fully integrated into the current knowledge system since there are no clear guidelines on how to integrate it in the curriculum and there is insufficient information on the innovative approaches that enable the integration of IKS in education and research to restore the respect, values, perceptions, experiences and practices of indigenous people and to address the imbalances.

LITERATURE REVIEW

Policy Frameworks That Support IKS

Policies that highlight the recognition, protection, promotion and maintenance of the IKS are in place, though it is not fully integrated. At the international level, the Convention of Biological Diversity (CBD), (United Nations Environmental Programme 2011), alluded on the relevance of information originating from indigenous people in the conservation of biodiversity and its use in sustainability. Such information is useful in solving many problems since it developed from the experiences and practices that are learnt over many generations. However, even if the relevance of information derived from indigenous people is useful, indigenous people are left out in decision making processes, which could have been beneficial in sustainable development.

The Intergovernmental Platform on Biodiversity and Ecosystem services (IPBES) was formulated in 2012, aiming to recognise and respect contributions made in the conservations and the use of ecosystems and biodiversity in a sustainable way. IPBES performs assessments that focus on the status and trends of biodiversity and ecosystem services aa well as to assess how they impact human health. Such assessments lead to the evaluation and renewal of the strategic plans for biodiversity. In the study by Bryan et al. (2019), the authors mentioned that indigenous people play a major role in ensuring that biodiversity and the services in ecosystems are conserved. The concern is that human activities threaten biodiversity and that leads to increased loss of species. In such instances, IPBES becomes necessary in the implementation of various strategic and innovative approaches that will protect biodiversity. A successful initiative was documented in the study by Akano et al. (2024), where the authors highlighted on the use of an indigenous knowledge practice that decreased the degradation of ecosystems, by using rotational grazing as a recovery mechanism.

Another success of the IPBES where indigenous knowledge is recognised and respected is mentioned in the study by Udeh et al. (2023). In this study, contributions made in the conservations and the use of ecosystems and biodiversity from indigenous communities were recognised. The authors highlighted that indigenous communities made observations in patterns of migration and the melting of ice, that was congruent to observations made by scientists (Udeh et al., 2023). In addition, most of the indigenous people rely on patterns of clouds, cycles in the blooming of plants and the behaviour of animals as environmental pointers that foresee changes in seasons (Ige et al., 2024). Furthermore, indigenous people observed the behaviour and the appearance of birds, insects, animals and other environmental indicators, they also predict weather and look for warning signs for disasters. Where that knowledge is used for the development of calendars that guide their activities and to predict floods, drought and other disasters (Dorji et al., 2024).

Such activities necessitate interdisciplinary collaborations between knowledgeable indigenous community members, scientific researchers, linguistic specialists, government and other entities. These approaches serve as efforts that respond to the promotion, revival, appreciation, and respect for indigenous knowledge. Many collaborative initiatives were initiated, that included the collaboration between ecologists from a higher education institution who partnered with members of the indigenous people in projects that benefit communities.

In Dorji et al. (2024) the authors alluded on the contribution of IK in climate change adaptations. The paper highlighted that indigenous communities rely more on indigenous knowledge for adaptation in times of environmental changes as they are poor and come from marginalised areas. When climate change intensifies, a need for adaptive strategies arise (Okedele et al., 2024). In most instances, adaptive strategies are derived from indigenous communities, who rely on adaptive practices that respond to environmental changes and those practices are learnt from previous generations as they are passed on from generation to generation. The authors also alluded on a need to use knowledge derived from various sources as a way that creates adaptive strategies that addresses occurring changes.

In 2013, the rights of indigenous people were standardised by the International Labour Organization Convention No 169 (ILO) where their customs and cultures were maintained, with relevant consultations that enforced the cooperation and consideration of their opinions in processes that leads to the making of decisions. On paper, the rights of indigenous people are standardised, however, looking at the current situation, there are still areas where indigenous people's rights are not fully exercised.

In 2015, Sustainable Development Goals were launched, where the goals emphasised access to knowledge, the provision of education of good quality and the reduction of inequalities (Ocholla, 2025). However, even after the launch of the sustainable development goals, quality education and inequalities are still not addressed. Quality education is education that is relevant to students, it is inclusive, allowing participation with voices being heard and experiences being shared. It involves the switching of roles, where knowledge does not originate from the educator only, students should not be regarded as empty vessels, they should be allowed to share their experiences where they relate them to the learnt concepts.

Furthermore, UNESCO encouraged the recognition and documentation of indigenous knowledge with the handing over of such information to the upcoming generations (UNESCO, 2022). Nowadays, technology is overriding the traditional ways of getting information. Technology is becoming the new normal. It serves as the main source of information which influences our thoughts rather than the old traditional way of sharing information in a face-to-face setting. If this indigenous knowledge could be digitised and accessible, then the new generation could derive information in websites.

At the continental level, the African Union formulated the African Union Policy Framework and Plan of Action on Traditional Medicine (African Union, 2001). It was aimed at integrating traditional medicine into the health system, with the promotion and the respect of the practices, knowledge and the resources derived from indigenous people. Nowadays, there are many initiatives that allowed traditional healers to amalgamate with pharmaceutical companies in the production of medicine derived from herbal plants. Such initiatives were a success since most of their products are found in pharmacy shelves.

At the national level, a framework that promotes, recognises and protects the IKS was created by the National Policy on Indigenous Knowledge (Republic of South Africa, 2004). The policy is aimed at highlighting the relevance of integrating indigenous knowledge into education, health, agriculture and conservation. The National Policy on Digitization of Heritage Resources was formulated, aimed at preserving, managing ownership and ensure access to heritage resources (DAC, 2010). In addition, The Department of Science Technology and Innovation has various initiatives that promote, revive, preserve, protect and documents issues on IKS. It included the approval of the policy on IKS in 2004. Institutional collaborations were created at the national, regional and international level (DSI, 2016). Some of the South African institutions have initiatives that formulated policies that manage information from IKS. South Africa has an initiative of digital preservation of IKS in the National Recordal System (NRS). The initiative focuses on indigenous knowledge gathered from traditional food and plants including traditional medicine which are recorded in the Indigenous Knowledge System Documentation Centre's (IKSDCs) (Balogun & Kalusopa, 2020).

The implementation of an initiative such as this prepares the system to ensure its readiness in case of disasters. Some institutions use Metadata as a strategy for digital preservation which is acquainted with management rights, where such preservations ensures that digital archives are accessible. Precautionary measures and backups are necessary in the protection of data, to protect data from being lost and from being deleted unintentionally, as well as to prevent viral attacks. These backups are done in different laptops, not in the one that is used for collection of IK material.

In the paper by Masenya and Ngulube (2019) the authors alluded that most institutions have shortage of specific standards and policies that manage digitized material. Meanwhile, disaster planning is another crucial aspect that ensures that digital repositories are trusted and maintained. As such, proper planning for disasters reduces threats of losing data and it ensures the management of risks with the inception of mitigating strategies (Aikin, 2007).

Hence, an initiative that focussed on managing digital collections was formulated, it was a South African Framework that focussed on collaborations (NRF, 2010), it required compliance from institutions. Its development was to enforce proper planning and to ensure that the life cycle of digital collections is well managed (NRF, 2010), where institutions have the responsibility of creating a policy that is aligned with the national one since they lack preservation strategies. Another initiative which was established is the Local Communities and Indigenous people Platform in 2015, it was aimed at ensuring that knowledge is mutually exchanged and there is inclusivity in actions taken in the participating parties (Legal Response International, 2019). Another initiative that represented indigenous people in cases where there were obstacles that hindered participation includes the CBD, the IPBES and the Intergovernmental Panel on Climate Change (Brugnach et al., 2017).

These initiatives enable sustainability that ensures that there is transmission of indigenous knowledge to the descendants without losing it. The preservation of this valuable knowledge may address the arising challenges in communities such as unemployment. Communities can partner with researchers and other entities, where societal stress is reduced by capacitating young people in the creation of internship opportunities. The knowledge shared will enable all participating individuals to have equal distribution of resources in cases where a profit is made from the shared practices.

In 2012 the National Development Plan (NDP) imposed the use of indigenous knowledge in governmental programmes and policies (Republic of South Africa, 2012). This plan was aimed at addressing unemployment rate, the persisting inequalities and poverty in the country. Then the indigenous knowledge holders' rights became protected since the inception of the Protection, Promotion, Development, and Management of Indigenous Knowledge Act which was introduced in 2019 (Republic of South Africa, 2019). The Act ensured that communities benefitted from every information they issued and the used resources. In 2021, the Department of Basic Education introduced the National Curriculum and Assessment Policy Statement (CAPS), which enforced the integration of indigenous knowledge in the taught curriculum.

This policy highlights the relevance of the existence of diverse cultures, the inclusion of local phenomena and all the indigenous views in education (Department of Education, 2021). Meanwhile, the Department of Science and Innovation (DSI) encouraged innovative approaches that promoted the use of indigenous knowledge in research. DSI also supported the initiative through the provision of funds, enabling collaborations between the indigenous people and researchers (Department of Science and Innovation, 2022). That was observed in several documented studies. The collaborations led to the capacitation and socio-economic development of all the involved parties. In collaborations, indigenous people co-manage with other involved entities, where they play a role in making decisions with the incorporation of their knowledge in planning.

In the past, indigenous people were excluded in policy making processes. However, Ocholla (2025) argues that the development and the exploration of science and indigenous knowledge together benefit the society when used as components for social justice. Of late, researchers are starting to appreciate that indigenous people have ideas that can improve efforts made in many fields. Collaborations are created between indigenous communities and higher education institutions. Such collaborations enable contributions from all individuals involved in the project, with the recognition and appreciation of values, experiences and practices.

The Integration of IKS in Education and Research

Calls were made for the decolonisation of education and the curriculum. However, academics in the sciences assumed that decolonisation was not a topic of concern in their disciplines until a video that requested that Newton's Law should be replaced with indigenous knowledge was posted by a student from Cape Town university (Blackie, 2024). This necessitates change that will embrace the diversity that exists, the respect and the acknowledgement of all the people who took part in making history (Lasky, 2005) and the acknowledgement of sources of all knowledge.

For some time, the IKS was used worldwide for various purposes particularly in Africa (Makwara, 2013). It was used in various developmental sectors such as in the conservation of biodiversity and in practices used in farming (Madzivhandila, 2024). The study done by Udeh et al. (2023) was a success in integrating IKS in biodiversity and conservation. The authors highlighted that indigenous communities made observations in patterns of migration and the melting of ice, and that was congruent to the findings made by scientists (Udeh et al., 2023). In addition, most of the indigenous people rely on patterns of clouds, cycles in the blooming of plants and the behaviour of animals as environmental pointers that foresee changes in seasons (Ige et al., 2024).

In response to decolonisation and ensuring that knowledge, the practices and opinions derived from various sources are recognised, Blackie (2024) explored curriculum and knowledge based on the Basi Bernstein's three pedagogic approaches. Basi Bernstein's first approach focused on the creation of knowledge and where that knowledge is created.

The second pedagogic approach enables communication and how knowledge is conveyed between scientists and indigenous people, where that knowledge is packed in easier ways that are understandable (Knowledge is recontextualised in different contexts). This is a transdisciplinary research approach, that integrates knowledge from various sources, that includes scientists, non-academics, government and NGO's. This approach ensures that there is communication between scientists or entomologists and indigenous people, with an effort to ensure that knowledge is packed in different contexts. In the article by Cockburn et al. (2014), the authors used an interdisciplinary approach, that created collaborations, that enabled communication between the indigenous people, entomologists, Zulu speaking people and environmental education experts in enabling the sharing of knowledge and research. As such misconceptions about science and indigenous knowledge as two different entities could be changed through thorough assessments, investigations and explanations of the indigenous knowledge (Tibuhwa, 2012).

South Africa has 11 official languages, which necessitates the need for science terminologies to be in African languages for ease of communication. This was emphasised in Mheta et al. (2018) where the authors highlighted on a need to teach African native languages to ensure that there is grasping of the taught content. At the international level, scientific names were standardised to avoid ambiguity in the classification and naming of living organisms. As such the ability to translate names of organisms from one language to the other improves access, inclusivity and recognition of knowledge from all spheres.

The third approach integrates indigenous knowledge in the curriculum and there is reproduction. Students make clay pots as a school curriculum, in this case, skills are transferred to learners while indigenous knowledge is recognised, valued and appreciated. According to Ijatuyi et al. (2025), the introduction of indigenous knowledge in classrooms can be a representation of their various cultural backgrounds and it can improve the way information is interpreted. In a classroom, there are diverse cultural backgrounds and diverse language backgrounds which may affect their performance and success. It is common knowledge that language is a problem in learning environments especially in cases where the medium of instruction is not the mother tongue of learners, it becomes a barrier that hinders comprehension.

In the paper by Kudumo and Ngcoza (2023), the authors mentioned that efforts were made to motivate students in a science class and to influence the desire to learn in them, by using resources from indigenous technology that is used in iron smelting. The authors also alluded that there is a need to capacitate educators on how to include content that is culturally relevant, with the responsive pedagogies in classroom teachings. In that study, students used traditional material to explain scientific concepts meanwhile teachers use everyday examples that are practical to student's lives. Students used traditional practices in the process of melting iron. It is assumed that the competence of students to relate what they learnt in class with what they experienced daily, leads to an improvement in participation and interest which improves understanding. According to Modiba and Emsley (2024b), pedagogy denotes how learning takes place, the applied methods and the relevance of the subject matter.

The documentation of indigenous knowledge on the use of clay is vital as it may be part of the science curriculum in the future (Keane, 2006). Skills from indigenous people on how to create clay pots and the type of soil used can be incorporated in the already existing curriculum of the sciences. In addition, the importance of clay, the properties and its various uses such as pottery, tiles, the formation of bricks, can be included in the subject matter to show its value and for its recognition.

IKS bridges cultural gaps when it is integrated in the curriculum that originated from western countries. Innovations in pedagogy, where those innovations are rooted in indigenous practices, play an important role in stimulating experiences in STEM learning (Nwokocha & Legg-Jack, 2024). There is a need to decolonise the curriculum and be inclusive with the knowledge systems. This need is also alluded in Manatunga (2020), where the author supports a need for decolonising curriculum as a way of ensuring the inclusiveness of the opinions and views of the neglected groups. This imposes a need for relevant and meaningful teaching, that echoes student's daily experiences, where local knowledge is embedded in the curriculum.

Despite the inequalities brought by apartheid in the education system, IKS is still resilient in South Africa and it is evident (Nwokocha & Legg-Jack, 2024). Documents from the Department of Science and Innovation indicated the existence of more than 20 000 IKS in different provinces (Republic of South Africa, 2022). When indigenous knowledge is integrated in the curriculum of science, there is an improvement in the understanding of main concepts (Phirke, 2024) and there is an enhanced meaning making. Besides the three approaches, a knowledgeable person could be invited in the classroom to share about their indigenous knowledge. Alternatively, students may visit indigenous people's communities, where their experiences, skills, experimentation and practices are shared with the students.

To decolonise curriculum with the integration of IKS in education requires a revisit of what is taught, the teaching methods where it is not just the transmission of information from the dominant teacher to the recipient learner. Teachers need to be cognisant that students do not come in class as empty vessels, there is knowledge carried in them from their communities.

Possible Ways to Integrate IKS in Education and Research

South Africa is rich in biodiversity and cultural diverseness and its diverseness could be used to show connections between indigenous knowledge and the science subjects. When IK is integrated into the school curriculum, it provides a rich foundation for equitable and sustainable education which is relevant locally and applicable culturally (Matindike & Ramdhany, 2024). An example includes invertebrates, such as insects, animals and plants of importance to communities. The big five can be used as examples in Life Science classrooms where each animals' description, habitat, behaviour and classification can be related to its cultural significance in indigenous communities. In this case a lion in the Pedi culture symbolises authority and royalty. Some chiefs in rural areas wear a lion's skin as a sign of royalty. Elephants symbolise unity in families, as they occur in groups with the matriarch leading the herd and they have a good memory.

Additionally, insects are highly nutritional sources of food in indigenous communities worldwide (Gasca-Alvares & Costa-Neto, 2022). Their morphology, identification and the practice of how they are collected, and where they are found in communities can be shared and be included in the curriculum as shared experiences and skills and all these be linked to their cultural importance and the roles they play as food security measures. Such approaches will ensure that student's participate, as it draws their attention, where they share their experiences and it ultimately leads to holistic understanding of the subject matter that is relevant. As the life cycles of these insects are studied, various stages can be pointed out to show which larval stages are harvested for consumption.

The initial step in class could be that the teacher asks a question to ensure that there is participation and engagement, and allow students to express their already existing knowledge, by voicing their opinions, where the teacher helps in making the link between the two knowledge sources, while ensuring that there are no misconceptions. Relevant examples that relate to learners' daily experiences should be used, then the teachers efficacy should play a role in using various teaching techniques in ensuring that all learning styles are catered for. The teacher designs study lessons and material that includes IKS in the existing science topics as it is alluded in Photo and McKnight (2024).

The teaching and learning environment must be inclusive, conducive, where learners voice out their views and experiences, they should be given a chance to participate. It becomes necessary to include various entities in the discussions, planning, implementation, information gathering processes. That includes local indigenous custodians, government officials, linguistics, education specialists and knowledgeable community members. Moreover, the use of Multiple Know how Model (M-model) as adopted in Semali and Kincheloe (1999) emphasises the integration, the linking of learning content, where the two knowledge systems and practices complement each other (Nokocha & Legg-Jack, 2024). In such instances, the practices followed in the collection of plants and how they are processed to make medicine can serve as information sources that are derived from IKS while supporting teaching.

Cooperative learning approach is another way that can be used to integrate IKS in science (Johnson & Johnson, 2014). The approach relies on methods of teaching that boosts active participation of students where students are divided into small manageable groups that allow them to have discussions, with an exchange of information. Cooperative method was also mentioned in Sefoka and Chuene (2025), even though the authors indicated the insufficiency of research done on this approach in integrating IKS. Tahe authors also mentioned that Jigsaw and De Bono's six thinking hats methods, were found to improve participation, accessibility and comprehension in learning. Jigsaw allow students to explore content in groups of four to six members. Meanwhile De Bono's method deals with thinking approaches, comprehension and reading. This approach allows interactions and evaluations of the subject matter with reflections that allow students to become responsible for their own work, where they develop interpersonal skills and social interactions that enable learners to learn.

Furthermore, the three learning theories can be used to integrate IKS in education and research. The application of cognitive behaviourist theory as an initial generation of pedagogy starts with the stimulus which is the content, that causes a change in behaviour, when procedural instructions are outlined. Then the stimulus stimulates students to ask questions, with the creation of knowledge, while applying their skills and knowledge to validate the content and solve problems. Secondly, social-constructivism theory can be used synchronously or asynchronously to allow interactions among students and it is enabled by technologies. Their interactions enable them to imitate, demonstrate and to have discussions. Thirdly, connectivism can also be used to create connections among people, the content and electronic resources, where knowledge is applied to connect the pre-existing knowledge with the newly created knowledge to address problems that arise (Anderson, 2008; Modiba & Emsley, 2024b).

The integration of IKS in research led to the inclusion of species that were poorly known into published records and the compilation of field guides in conserved areas became eminent. In addition, the identification of newly discovered species, and the translation of scientific names into native languages in particular areas led to an improved knowledge and understanding of various life forms. Such translations solve the confusion in species names, since different regions use different names for a particular species. In science, a scientific name used for an organism is used throughout the world. Examples that integrate IKS in education and research to address scientific challenges includes: an elephant which is called "*tlou*" in Pedi, while in Venda, it is called "*ndou*" in Tsonga it is called "*njovu*", in Zulu it is called "*indlovu*". Meanwhile, when the scientific name is used, it is the same worldwide, it is *Loxodonta africana* for the African species and *Elephas maximus* for the Asian species.

Amongst the examples is the *Zizipus mucronata* which is commonly known as the buffalo thorn, in Afrikaans it is commonly known as "*wag'n-bietjie*" which means "*wait a bit*" in English, also called "*umLahlabantu*" in Zulu. This plant is native in Southern Africa and it has unti-inflammatory, antimicrobial and anti-fungal benefits. The plant is used mostly to address various ailments in both animals and humans. It treats respiratory complications, skin infections and sexually transmitted infections (Mongolo et al., 2020). This plant is one of the plants that could be used in a collaborative interdisciplinary approach, where various people with different expertise apply their skills and knowledge in determining its various uses, the components and medicinal benefits derived from the plant. A partnership formed from botanists, chemists, ecologists, pharmacology, anthropologists and knowledgeable people from the indigenous community where the plant occurred, enable inclusivity and better understanding of the plant. Ultimately, the inclusion of various experts will lead to the promotion, revival, and the acknowledgement of integrating IKS in education and research.

In several studies, challenges were encountered when integrating IKS in education. In Mkhwebane (2024) the author alluded that Mothwa (2021) and Ncgobo (2020) encountered challenges when integrating IKS in life science, that included: lack of clear guidelines on how to integrate IKS, lack of support and lack of resources such as textbooks. Additionally, Mothwa (2011) concur with Mkhwebane (2024) on life science textbooks that did not adequately support the integration of IKS.

RESEARCH METHODOLOGY

The study gathered relevant information on IKS from government documents, peer-reviewed articles, books and news. Google Scholar Database was used as a source of information, where relevant documents that covered IKS in education, schools and research were retrieved and reviewed. Sources from different years were considered to ensure that a trend for the inception of various initiatives is detected. The author used this method of information gathering to enlighten readers about the integration of IKS in education and research. It expands on policy frameworks that support IKS, possible ways to integrate IKS in education and research, challenges faced and opportunities in the integration of IKS in education and research.

Challenges Faced in the Integration of IKS in Education and Research

Arguments on the inclusion of IKS in science education persists even though policies on IKS are in place (Blackie, 2024). These policies intend to encourage the recognition, promotion, and the restoration of indigenous people's values with their capacitation. But IKS's recognition and revival is still a challenge since IKS is not yet fully integrated into the current knowledge system. In the paper by Madlela (2023), the author highlighted on some challenges faced in the integration of IKS in Natural Science (NS). The challenges include failure to commit when it comes to the review of education policies. Unavailability of IKS study material and its execution in schools. The recommendations that were made included a need for the Department of Basic Education to review the policy and integrate IKS in the Natural Science curriculum as a full knowledge component and to incorporate IKS in the training programmes for educators. The author also alluded on the importance of parent's participation in their children's schoolwork. As well as a need to use IKS custodians from communities. However, looking at the status quo, policies that talk about transformation, equity and equality are available, but the work that includes different bodies of knowledge is still lacking and less focus is given to IKS. Despite calls made to include IKS in formal education systems, historical complexities still exclude other ways of knowing (Matindike & Ramdhany, 2024). Another challenge is that, even though the Natural Science curriculum statement from the Department of Basic Education allows teachers to use IKS when teaching, the problem of lack of content that embrace cultural perspectives and realities persists.

Opportunities for Integrating IKS in Education and Research

The relevance of integrating indigenous knowledge in education and research is that it, recovers the respect for all sources of knowledge, there is reduction of hierarchies and stereotyping in cultures, it restores the value of indigenous ways of knowing and it centres their wisdom in the curriculum. It also bridges cultural gaps when it is integrated in the curriculum that was dominated by the western culture. There is an exchange of information, new innovations and possible solutions to challenging issues, with the reduction of the exploitative behaviours. In the paper by Nwokocha and Legg-Jack (2024), the authors highlighted on the important role of stimulating experiences in STEM learning when innovations in pedagogy are rooted in indigenous practices. When indigenous knowledge is integrated in the curriculum of science, there is an improvement in the understanding of main concepts (Phirke, 2024) and there is an enhanced meaning making. It promotes social inclusion, while it addresses societal challenges. The co-existence of the various knowledges expands the understanding of the two knowledge systems, and the boundaries are removed, allowing intercultural communication. Its integration leads to the inclusion of poorly known species in published records, the creation of field guides in conserved areas and the translation of scientific names into native languages.

RESULTS

Colonial Epistemologies in Education

Apartheid epistemes in education and research are eminent in textbooks used in disciplinary knowledge. Knowledge from indigenous people is perceived as old fashioned, unscientific and lacked novelty (Modiba & Emsley, 2024a). The unjust practices of the apartheid era imposed the exclusion of indigenous people's experiences, their unique culture, and histories. In the paper by Senegal and Lenz (2020), the authors alluded on the dominance of Western knowledge in the study material. Du Plessis (2021), added by highlighting that the use of foreign languages to Africans in study books does not echo their cultures.

In the paper by Woldegiorgis (2021), the author highlighted that there were disengagements from African histories since colonial interests were enforced. Meanwhile, Mc Mahon and Portelli (2012) agreed with Woldegiorgis (2021) by highlighting on the persistence of inequalities and the disengagements of students and social injustices that are observed in the education system. Africans were brainwashed to believe that Western knowledge was a key to a successful future (Van Pixteren, 2022). However, Senegal and Lenz (2020) mentioned that lack of African content hinders African knowledge to be used as a relevant tool for growth in education. This necessitates change that embraces knowledge that originates from all spheres, where all bodies that contributed to the creation of knowledge are respected and acknowledged (Lasky, 2005).

This change necessitates HEIs to enforce the revival and the recognition of IKS in education and research. In the paper by Naidoo, et al. (2014) the authors alluded on the importance of including IKS in the curriculum of science, as it will restore the value of the indigenous ways of knowing and to centre their wisdom in the curriculum, as well as to reduce the hierarchies in cultures. It is necessary for HEIs to implement innovative strategies that support and encourage the integration of IKS in education with the revision of the curriculum and research for sustainable development.

Policy Frameworks That Support IKS

There are policies that support the recognition, protection, promotion and maintenance of IKS. Internationally, CBD, (United Nations Environmental Programme 2011), highlighted the relevance of information from indigenous people in the conservation of biodiversity and its use in sustainability. This programme encourages that relevant information that comes from indigenous people' experiences and practices should be used. It is work in progress, there are still instances where indigenous people are still left out in decision making processes.

The IPBES aimed to recognise and respect contributions made in the conservations and the use of ecosystems and biodiversity in a sustainable way. It assessed the status and trends of biodiversity and ecosystem services and it assesses how they impact human health. It also focuses on evaluating and renewing strategic plans for biodiversity. However, if IKS is integrated, it can ensure that biodiversity and ecosystem services are conserved. Bryan et al. (2019) concur, as the authors highlighted on the important role played by indigenous people in ensuring that ecosystems and biodiversity are conserved, since there are concerns about human activities that threaten and cause mass extinctions on biodiversity. Akano et al. (2024) alluded on a successful initiative that used indigenous knowledge practices to reduce the degradation of ecosystems by using rotational grazing as a recovery mechanism. Other studies that show the successfulness of IPBES are documented in Udeh et al. (2023); Ige et al. (2024) and Dorji et al. (2024). As such, for biodiversity and ecosystems to be conserved in a sustainable way, interdisciplinary collaborations are necessary between community members, scientific researchers, linguistic specialists' government and NGO's.

The International Labour Organization Convention No 169 (ILO) standardised the rights of indigenous people, their cultures and customs were maintained, with an enforced cooperation to consider their opinions in decision making processes. Even though there are still areas where their rights are not fully exercised. The Sustainable Development Goals are launched. Their goals emphasise access to knowledge, the provision of education of good quality and the reduction of inequalities (Ocholla, 2025). However, quality education is education that is relevant to students, it is inclusive, allowing participation with voices being heard, and experiences being shared. It allows the switching of roles, with knowledge that does not originate from the educator only, students are not regarded as empty vessels, they are allowed to share their experiences and relate them to the learnt concepts.

Furthermore, UNESCO encourages the recognition, documentation and the transfer of information to the upcoming generations (UNESCO, 2022). A challenge faced by the current generation is that technology overrides the traditional ways of accessing information. Nowadays, technology is the new normal. It is the main source of information, it influences the thoughts rather than the old traditional way of sharing information in a face-to-face setting. At the continental level, the African Union formulated the African Union Policy Framework and Plan of Action on Traditional Medicine was created (African Union, 2001). It integrated traditional medicine into the health system, with the promotion and the respect of the practices, knowledge and the resources from indigenous people.

Nationally, a framework for the promotion, recognition and protection of IKS was created (Republic of South Africa, 2004). The policy highlights the relevance of IKS into education, health and agriculture. Then, the National Policy on Digitization of Heritage Resources is formulated, to preserve and to manage ownership and allow access to heritage resources (DAC, 2010). The Department of Science Technology and Innovation initiated initiatives that promotes, revives, preserves, protects and documents issues on IKS. It also approved the policy on IKS in 2004. Institutional collaborations are created at the national, regional and international level (DSI, 2016). Some of the South African institutions have initiatives that formulated policies that manage information from IKS. South Africa has an initiative on digital preservation of IKS in the National Recordal System (NRS). The initiative focuses on indigenous knowledge gathered from traditional food and plants including traditional medicine that are recorded in the Indigenous Knowledge System Documentation Centre's (IKSDCs) (Balogun & Kalusopa, 2020). Its implementation ensures readiness for disasters, allowing the use of Metadata for digital preservation, ensuring that archives are accessible.

Furthermore, an initiative that focussed on managing digital collections is formulated. It is a South African Framework that focussed on collaborations (NRF, 2010), it requires compliance from institutions. It enforces proper planning and ensures that the life cycle of digital collection are well managed (NRF, 2010), where institutions have the responsibility of creating a policy that is aligned with the national one since they lack preservation strategies. The Local Communities and Indigenous people Platform was initiated to ensure that knowledge is mutually exchanged and there is inclusivity in actions taken in the participating parties (Legal Response International, 2019). Another initiative that represented indigenous people in cases where there were obstacles that hindered participation includes the CBD, the IPBES and the Intergovernmental Panel on Climate Change (Brugnach et al., 2017). These initiatives enable sustainability to ensure that there is transmission of indigenous knowledge to the descendants without loosing it. These initiatives reduce societal stress with an improved lifestyle with equal distribution of resources in cases where a profit was made from the shared project.

In 2012, NDP imposed the use of indigenous knowledge in governmental programmes and policies (Republic of South Africa, 2012). This plan addressed unemployment rate, the persisting inequalities and poverty in the country, since communities benefitted from every information they issued and the used resources. In 2021, the Department of Basic Education introduced the National Curriculum and Assessment Policy Statement (CAPS), which enforced the integration of indigenous knowledge in the taught curriculum. It highlights the inclusion of local phenomena and views from indigenous people in education, it also appreciates the existence of diverse cultures (Department of Education, 2021).

The Integration of IKS in Education and Research

Even though calls were made to decolonise education and the curriculum, science academics thought the topic on decolonisation was of less concern for them. Change is needed to embrace the existence of the diversity, with the acknowledgement of histories and other sources of information. Several studies were successful in integrating IKS. That includes the study done by Udeh et al. (2023) in integrating

IKS in biodiversity and conservation and in a study by Ige et al. (2024) who looked at the behaviour of animals as environmental pointers that foresee changes in seasons.

Blacki (2024) explored curriculum and knowledge based on the Basi Bernstein's three pedagogic approaches as a response to decolonise the curriculum. The pedagogic approaches focused on the creation of knowledge and the origin of that knowledge. The second approach focused on communication between communities and scientists with the recontextualization of knowledge into different contexts. This successful initiative used an interdisciplinary research approach that involved entomologists, indigenous people, Zulu speaking people, and experts in environmental education. As a result, the use of assessments, investigations and explanations from various entities removed misconceptions of IKS and science as two different entities. The third approach integrates IKS with reproduction. Students make clay pots as part of their curriculum, where skills are transferred, with recognition and appreciation of IK. According to Ijatuyi et al. (2025), the introduction of indigenous knowledge in classrooms is a representation of various cultural backgrounds and it improves the interpretation of information. While in the paper by Kudumo and Ngcoza (2023), students were motivated and their desires to learn were influenced by using resources that used iron melting from indigenous technologies.

Nwokocha and Legg-Jack (2024) support the idea that innovations that are rooted in indigenous practices play an important role in stimulating experiences in STEM learning. Meanwhile, Manatunga (2020), supports the decolonisation of the curriculum, with the inclusion of other knowledge forms.

Possible Ways to Integrate IKS in Education and Research

The integration of IKS into the school curriculum, provides a rich foundation for equitable and sustainable education that is relevant locally and applicable culturally (Matindike & Ramdhany, 2024). The environment should be inclusive, conducive and the sharing of experiences. Other approaches that could be used include the co-operative learning approach, the three learning theories which are: the cognitive behaviourist, social-constructivism and the connectivism approach. However, in several studies, the integration of IKS in education was a challenge. In the paper by Mkhwebane (2024) the author alluded that Mothwa (2021) and Ncgobo (2020) encountered challenges when integrating IKS in life science. The challenges included: lack of textbooks, lack of clear guidelines on how IKS should be integrated and lack of support. Additionally, Mothwa (2011) concur with Mkhwebane (2024) on life science textbooks that did not adequately support the integration of IKS.

FUTURE RESEACH DIRECTIONS

In future, the inclusion of various knowledges should be considered with the removal of boundaries, allowing intercultural communications. There is a need revisit the taught content, methods used in teaching, with the inclusion of IKS in the existing content and appreciate that students don't come to class as empty vessels. The inclusion of local phenomena and views from indigenous people in education should be prioritised.

CONCLUSION AND RECOMENDATIONS

Apartheid epistemes colonised education and research. This is eminent in the dominance of western knowledge in disciplinary knowledge. Meanwhile knowledge from indigenous people is sidelined, as it is perceived as unscientific, old fashioned and lacked novelty (Modiba & Emsley, 2024a). The cultures of indigenous people were not echoed (Du Plessis, 2021). Leaving out their experiences, their histories and cultures (Senegal & Lenz, 2020).

Policies on IKS are in place, with various goals that aim to promote, respect, protect, recognise, and restore the values of indigenous people, but its integration is still a challenge. The Convention of Biological Diversity (CBD), (United Nations Environmental Programme 2011) focused on the relevance of information derived from indigenous people in the conservation of biodiversity and its use in sustainability. The IPBES focused on the conservation and the use of ecosystems and biodiversity in a sustainable way. There are numerous studies that show contributions made by indigenous people. That includes the study done by Udeh et al. (2023), where there was congruency in observations made by indigenous people and scientists. In most instances, adaptive strategies are derived from indigenous communities. In the study by Dorji et al. (2024), indigenous people contributed in projects that focused on climate change adaptations.

The Sustainable Development Goals were launched to emphasise that knowledge should be accessed and quality education should be provided (Ocholla, 2025). This is still work in progress, for education to be of good quality, it needs to be relevant to students, it should be inclusive and allow voices to be heard. Thus, the integration of IKS dissolves power hierarchies that exist between collaborating entities in the co-creation of knowledge, with mutual respect for both knowledge systems. The revival, recognition, acknowledgement and the integration of IKS in education and research is vital as it leads to sustainable development. It is of importance to document all the relevant information, the histories and practices of the indigenous people as a way of addressing the past injustices that fostered inequalities. It is also necessary to adopt new strategies that integrates IKS in education and research as a stepping stone to decolonise the curriculum, education and research.

According to Ijatuyi et al. (2025), the introduction of indigenous knowledge in classrooms can be a representation of their various cultural backgrounds and it can improve the way information is interpreted. The inclusion of indigenous knowledge in education and research acknowledges the values of indigenous communities which leads to a holistic sustainable approach. The use of collaborations formed by members from various entities, where knowledge from all spheres is treated equally, is a crucial step into sustainability. The interdisciplinary approach is necessary in integrating IKS in education and research. The integration of IKS in various entities ensures co-creation of knowledge, it ensures that knowledge derived from participating parties is mutually respected and people are capacitated.

Such approaches provide a comprehensive understanding of the two knowledge systems. The approaches encourage a holistic understanding, that addresses challenges, while they encourage the documentation, revival, recognition and promotion of IKS. This necessitates policy makers to regard knowledge from indigenous people and their views when they integrate IKS in education and research. Since there will be empowerment of communities, the incorporation of their experiences, practices, skills and the recognition, revival, and the documentation of IKS with the preservation of biodiversity for the upcoming generations. It is necessary to have more studies on the integration of indigenous knowledge in education and research as it will provide an understanding and a holistic knowledge with benefits of using knowledge from different spheres.

Finally, institutions are required to develop their own policy on IKS Digital Repositories to ensure the storage of the collected material, with the need to use metadata and backups for storing the curated material. The author recommends reviews in the curriculum with the inclusion of IKS. Clear guidelines should be provided on how to integrate IKS in class and in textbooks, offering training and support for educators. Teachers should be inclusive in their lesson design, by considering the existence of various cultural backgrounds and different learning styles, allowing students to be partakers in the creation of knowledge, where various approaches are used with the engagement of local communities in the planning, and the implementation process.

REFERENCES

Aikin, J. (2007). Preparing for a National disaster: The committee on conservation of cultural resources, 1939-1944. *The Library Quarterly*, *77*(3), 257–285. DOI: 10.1086/519416

Akano, O. A., Hanson, E., Nwakile, C., & Esiri, A. E. (2024). Improving worker safety in confined space entry and hot work operations: Best practices for high-risk industries". Global. *Journal of Advanced Research and Reviews*, *2*(2), 31–39. DOI: 10.58175/gjarr.2024.2.2.0056(GJARR-2024-0056)

Anderson, T. (2008). Towards a theory of online learning. In Flexible learning. DOI: 10.15215/au-press/9781897425084.004

Balogun, T., & Kalusopa, T. (2020). A framework for digital preservation of indigenous knowledge system in repositories in South Africa. *Records Management Journal*, *31*(2), 176–196. https://www.emerald.com/rmj/article-pdf/31/2/176/2298008/rmj-12-2020-0042. DOI: 10.1108/RMJ-12-2020-0042

Blackie, M. A. (2024). Indigenous knowledge systems and science education. *South African Journal of Science*, *120*(5-6), 1–4.

Brugnach, M., Craps, M., & Dewulf, A. (2017). Including Indigenous people in climate mitigation: Addressing issues of scale, knowledge and power. *Climatic Change*, *140*(1), 19–32. DOI: 10.1007/s10584-014-1280-3

Bryan, C. J. O., Garnett, S. T., Fa, J. E., Leiper, I., Rebein, J. A., Fernandez-Llamazares, A., Jackson, M. V., Jonas, H. D., Brondizio, E. S., Burgess, N. D., Rosinson, C. J., Zander, K. K., Molner, Z., Venter, O., & Watson, J. E. M. (2019). The importance of indigenous people's land for the conservation of terrestrial mammals. *Conservation Biology : The Journal of the Society for Conservation Biology*, *35*(3), 1002–1008. DOI: 10.1111/cobi.13620

Cockburn, J. J., Khumalo-Seegelken, B., & Villet, M. H. (2014). IziNambuzani: IsiZulu names for Insects. *South African Journal of Science*, 2014, 110 (9/10), 2013-0292, 1-13. https://dx.doi.org/DOI: 10.1590/sajs.2014/20130292

Department of Arts and Culture, Government of the Republic of South Africa. (2010). "National policy on the digitization of heritage resources. Final draft for public review (V8)". www.mcnulty.co.za/wp-content/uploads/2010/12/NATIONAL _POLICY_ON-DIGITIZATION_V8.pdf

Department of Basic Education. (2021). Curriculum assessment policy statement (CAPS), viewed 17 December 2025 from https://www.education.gov.za/curriculum/CurriculumAssessmentPolicyStatements(CAPS).aspx#:~:text=A%20National%20Curriculum%20and%20Assessment%20Policy%20Statement%20is20a%20single,Curriculum%20Statement%20Grades%20R%20%2D%2012

Department of Science and Innovation. (2022). New system to promote access to indigenous knowledge across South Africa, viewed 14 December 2025 from www.dst.gov.za/index.php/media-room/latestnews/3598-new-system-to-promote-access-to-indigenous-knowledge-access-across-South-Africa

Dorji, T., Rinchen, K., Morrison-Saunders, A., Blake, D., Banham, V., & Pelden, S. (2024). Understanding how indigenous knowledge contributes to climate change adaptation and resilience: A systematic literature review. *Environmental Management*, *74*(6), 1101–1123. DOI: 10.1007/s00267-024-02032-x PMID: 39215837

Du Plessis, P. (2021). Decolonisation of education in South Africa: Challenges to decolonize the university curriculum. *South African Journal of Higher Education*, *35*(1), 54–69. DOI: 10.20853/35-1-4426

Gasca-Alvares, H. J., & Costa-Neto, E. M. (2022). Insects as a food source for indigenous communities in Columbia: A review and research perspective. *Journal of Insects as Food and Feed*, *8*(6), 593–603. DOI: 10.3920/JIFF2021.0148

Hill, R., Adem, C., Alungui, W. V., Molnar, Z., Aumeeruddy-Thomas, Y., & Bridgewater, P.. (2020). 'Working with indigenous, local and scientific knowledge is assessments of nature and nature's linkages with people'. *Current Opinion in Environmental Sustainability*, *43*, 8–20. DOI: 10.1016/j.cosust.2019.12.006

Ige, A. B., Kupa, E., & Ilori, O. (2024). Aligning sustainable development goals with cybersecurity strategies: Ensuring a secure and sustainable future. *GSC Advanced Research Reviews*, *19*(3), 344–360. DOI: 10.30574/gscarr.2024.19.3.0236

Ijatuyi, E. J., Lamm, A., Yessoufou, K., Suinyuy, T., & Patrick, H. O. (2025). Integration of indigenous knowledge with scientific knowledge: A systematic review. *Environmental Science & Policy*, *170*, 104119. DOI: 10.1016/j.envsci.2025.104119

International Labour Organization. (2013). Handbook of the ILO tripartite constituents: Understanding the indigenous and tribal peoples convention, 1998 (Number 169), viewed 15 December 2025 viewed from https://www.ilo.org/wcmsp5/groups/public/---ed_norm/norms/document/publications/wcms/_205225.pdf

Johnson, D. W., & Johnson, R. T. (2014). *Joining together: Group theory and group skills*. Pearson.

Keane, M. (2006). "Understanding science curriculum and research in research in rural Kwa-Zulu Natal." Accessed https://wiredspace.wits.ac.za/bitstream/handle/10539/1508/10Part%202%20Ch10%20 Indigenous %20Knowledge.pdf?sequesnce=16

Kudumo, P., & Ngcoza, K. M. (2023). Prospects of using indigenous knowledge as culturally responsive pedagogy to teach science concepts regarding iron smelting. *Indilinga African Journal of Indigenous Knowledge System*, *22*(2), 204–221.

Lasky, S. (2005). A sociocultural approach to understanding teacher identity, agency and professionalism vulnerability in a context of secondary school reform.

Legal response International (2019). Legal status and mandate of local communities and indigenous peoples' platform. Viewed from: legalresponce.org/resource/legal-status-and-mandate-of-the-local -communities-and indigenous-peoples-platform

Madlela, B. (2023). Prospects and challenges of integrating Indigenous Knowledge System into the Natural Science curriculum in schools. *EUREKA: Social and Humanities*, *3*(3), 3–19. DOI: 10.21303/2504-5571.2023.002867

Madzivhandila, T. S. (2024). Integrating indigenous knowledge system to climate change adaptation in South Africa. *South African Journal of Higher Education*, *38*(5), 159–168. DOI: 10.20853/38-5-6417

Makwara, E. C. (2013). Indigenous Knowledge System weather and modern forecasting: Exploring the linkages. *Journal of Agriculture and Sustainability*, *2*(1), 98–141.

Masenya, T. M., & Ngulube, P. (2019). Digital preservation practices in academic libraries in South Africa in the wake of digital revolution. *South African Journal of Information Management*, *21*(1), 1–9. DOI: 10.4102/sajim.v21i1.1011

Matindike, F., & Ramdhany, V. (2024). Incorporating Indigenous Knowledge perspectives in integrated STEM education: A systematic review. *Research in Science & Technological Education*. Advance online publication. DOI: 10.1080/02635143.2024.2413675

Mc Mahon, B., & Portelli, J. (2012). *Student engagements in urban schools: Beyond neoliberal discussions*. Information Age., https://www.tcrecord.org DOI: 10.1108/978-1-61735-733-6

McKnight, M. (2015). Teacher's experiences of indigenous knowledge found in life sciences curriculum: Access study of life science teachers at high school in Pinetown District, Msc dissertation, University of KwaZulu-Natal.

Mheta, G., Lungu, B. N., & Govender, T. (2018). Decolonisation of the curriculum: A case study of the Durban University of Technology. *South African Journal of Education*, *38*(4), 1635–1641. DOI: 10.15700/saje.v38n4a1635

Mkhwebane, L. N. (2024). Life Science teacher's integration of indigenous knowledge: A revision for making life science classes culturally responsive. EURACIA *Journal of Mathematics. Science and Technology Education*, *20*(8), em2483. Advance online publication. DOI: 10.29333/ejmste/14859

Modiba, M.A. & Emsley, M.R. (2024a). Decolonising education in institution of higher learning: A review of literature. *African perspectives of research in teaching and learning Journal* 3(8), Special Issue

Modiba, M. A., & Emsley, M. R. (2024b). E-pedagogies and teacher efficacy: A case of rapid adjustment. In Akinlolu, M., Makua, M., Ngubane, N. (eds) online teaching and learning in HE. Springer, Cham. https://doi.org/DOI: 10.1007/978-3-031-56953-16

Mongolo, N.I., Mashele, S.S. & Makhafola, T.J. (2020). Ziziphus mucronate wild. (Rhamnaceae): its botany, toxicity, phytochemistry and pharmacological activities. Heliyon, 6(4), 2020, e03708, 1-20. DOI: 10.1016/j.heliyon.2020.e03708

Naidoo, P. D., & Vithai, R. (2014). Teacher approaches in introducing indigenous knowledge system in school science classrooms. *African Journal of Research Science Technology Education*, *18*(3), 253–263. DOI: 10.1080/10288457.2014.956407

National Research Foundation. (2010). "Managing digital collections: a collaborative initiative on the South African framework". https://wiki.lib.sun.ac.za/images/5/51/Managing_Digital_Collections.pdf

Nwokocha, G., & Legg-Jack, D. (2024). Reimagining STEM education in South Africa: Leveraging indigenous knowledge system through M-know model for curriculum enhancement. *International Journal of Social Science Research and Review*, *7*(2), 173–189. DOI: 10.47814/ijssrr.v7i2.1951

Nyahunda, L. (2024). Integration of indigenous knowledge systems (IKS) into climate change mitigation and adaptation endeavours: Milestones and gaps in South Africa and Zimbabwe's climate policy frameworks. *Climatic Change, 2024*(11), 162–177. DOI: 10.1007/s10584-024-03822-2

Ocholla, D. N. (2025). The complex marriage: Interfacing open science and open indigenous knowledge. Emerald Publishing Limited. *Library Management, 46*(3/4), 252–357. DOI: 10.1108/LM-11-2024-0125

Okedele, P. O., Aziza, O. R., Oduro, P., & Ishola, A. O. (2024). Integrating indigenous knowledge systems into global climate adaptation policies. *International Journal of Engineering Research and Development, 20*(12), 223–231.

Phirke, P. (2024). Reviving ancient knowledge: Integrating indigenous knowledge system in biological education. *International Journal of Research Culture Society, 9*(4), 16–20. www.ijrcs.org

Photo, P., & McKnight, M. (2024). Investigating indigenous knowledge awareness among South African science teachers for developing a guideline. *Curriculum Perspectives, 44*(1), 61–71. DOI: 10.1007/s41297-023-00224-9

Rankoana, S. A. (2017). The use of indigenous knowledge in subsistence farming: Implications for sustainable agricultural production in Dikgale community in Limpopo Province, South Africa. *Toward a sustainable agriculture: Farming practices and water use, 63.*

Republic of South Africa. (2004). Indigenous Knowledge System viewed 15 December 2025 from https://www.dst.gov.za/images/pdfs/IKS_Policy%20PDF.pdf

Republic of South Africa. (2012). National Development Plan 2030: Our future-make it work viewed 17 December 2025, from https://www.gov.za/sites/default/files/gcis/_document/201409/ndp-2030-our-future-make-it-workr.pdf

Republic of South Africa. (2016). 2004 IKS Policy, Role of Department of Science and Technology in Policy Implementation, and key achievements of the department's IKS Programme. A presentation to the Portfolio Committee on Science and Technology, 16 August 2016. https://pmg.org.za>daily schedule

Republic of South Africa. (2019). Protection, Promotion, Development, and Management of Indigenous Knowledge Act, 2019, viewed 15 December 2025, from https://www.gov.za/sites/default/gcis_document/201908/4264719-8actof6of2019protectpromodevelopmentindigenousknowledgeact.pdf

Sefoka, T. S., & Chuene, K. J. (2025). Life Science learner views on the integration of indigenous knowledge into indigenous knowledge related topics using cooperative learning approach: A case of South African grade 10 classroom. *EURACIA Journal of Mathematics. Science and Technology, 21*(5), em2633. Advance online publication. DOI: 10.29333./ejemste/16311

Selepe, K., Ramsaroop, S., & Carolin, A. (2022). Incorporating indigenous knowledge in the teaching of weather and climate in the primary schools in Kwa-ZuluNatal, South Africa. *Journal of Geography Education in Africa, 5*, 1–25. DOI: 10.46622/jogea.v5i1.3953

Senegal, Q. & Lenz. R. (2020). South African Higher Education Curriculum: An investigation into the challenges. *International Journal of Social Sciences and Humanity Studies*, 12(1), 146-160. Received from degipark. org.tr/en/pub/ijsshs/52543/676852

Tibuhwa, D. D. (2012). Folk taxonomy and use of mushroom in communities around Ngorongoro and Serengeti National Park, Tanzania. *Journal of Ethnobiology and Ethnomedicine, 2012*(8), 36. DOI: 10.1186/1746-4269-8-36 PMID: 22999253

Udeh, C.A., Daraojimba, R.E., Odulaja, B.A., Afolabi, J.O.A., Ogedembe, D.E. & James, O.O. (2023). "Youth empowerment in Africa: Lessons for US youth development programs."

UNESCO. (2017). *Local and Indigenous Knowledge System*. UNESCO.

Van Pixteren, P. (2022). The elephant in the room: why and how medium of instruction and decolonization of education are linked. *Journal of Contemporary issues in Education*, 17(2), 79-91. https://doi .orgDOI: 10.20355/jcie29496

Woldegiorgis, E. T. (2021). Decolonising a higher education system which has never been colonized. *Educational Philosophy and Theory, 53*(9), 894–906. DOI: 10.1080/00131857.2020.1835643

ADDITIONAL READING

Akalibeyi, S., Hlav kovà, P., Schneider, J., Fialova, J., Darkwah, S., & Ahenkan, A. (2024). Integrating indigenous knowledge and culture in sustainable forest management via global environmental policies. *Journal of Forest Science, 70*(6), 265–280. DOI: 10.17221/20/2024-JFS

Gunawan, K. D. H., & Indrawan, P. O. (2025). Decolonising science education: A bibliometric analysis of indigenous knowledge integration in global STEM. *Journal of Posthumanism, 5*(5), 2598–2622. DOI: 10.63332/joph.v5i5.1651

Parvin, R. A., & Begum, H. A. (2025). Integrating indigenous knowledge into biodiversity conservation strategies. *Varendra International Journal for Interdisciplinary Research, 1*(1), 84–93. DOI: 10.64296/vijir.v1i1.09

KEY TERMS AND DEFINITIONS

Decolonisation: the process of removing colonial legacies in the curriculum.

Indigenous Knowledge: knowledge accumulated over centuries, it is accrued from their daily practices, the skills and beliefs.

Interdisciplinary approach: methods and theories used from various disciplines.

Pedagogy: the method and practice of teaching.

Policy framework: a set of structured guidelines and principles outlining an organization's approach to a particular issue.

Chapter 5
African Indigenous Knowledge Systems:
Contestations in Digitising the Indigenous Knowledge Systems in African Indigenous Communities

Glen Maimela
http://orcid.org/0000-0002-1048-2811
National Heritage Council, South Africa

Rudzani Nedombeloni
http://orcid.org/0000-0001-5316-1826
University of Venda, South Africa

ABSTRACT

African Indigenous knowledge systems (AIKS) have been a centre of digitisation in the 21st century, preserving knowledge for posterity. However, digitisation has not been a smooth process confronted with contestations deeming it unprofitable. The aim of this chapter is in twofold. Firstly, the chapter critically explores contestations of digitising AIKS Indigenous communities. Secondly, the chapter proposes a co-designed approach for digitising the AIKS. Participatory action research (PAR) was employed to establish a research team to generate qualitative data in Ga-Mamabolo in the Limpopo Province. The data was co-generated using PAR data generation stages (planning, action, and reflection) and analysed using critical discourse analysis. The analysis shows passive participation, reduced authenticity, intellectual property rights, sacred knowledge, and funds as key contestations in digitising the AIKS. The analysis further shows stakeholder collaboration, cultural considerations, feedback, knowledge sharing and benefit sharing as areas to consider in digitising the AIKS.

INTRODUCTION

African Indigenous knowledge systems (AIKS) have existed within Indigenous communities for many years, playing a pivotal role in almost all aspects of life before the colonial era (Olaopa & Ayodele, 2022;

DOI: 10.4018/979-8-3373-7976-0.ch005

Mazzocchi, 2020). The AIKS were evident in key elements such as food production, entertainment, socialisation, textiles, healthcare, and Indigenous education, ensuring a sufficient supply of resources (Ogunyiola et al., 2022). With AIKS at the centre of life for Indigenous communities, knowledge has been passed down from generation to generation to ensure its continuity and impact on these communities. The Indigenous communities ensured this continuity by relying on practical observations and oral traditions as the key Indigenous methods for transmitting the AIKS (Bat et al., 2014; Juma, 2022).

Haines (2022) states that, for knowledge transmission, Indigenous communities relied on elderly knowledge bearers who served as teachers. The work of Nyoni and Maimela (2025) adds that the AIKS is considered a critical part of the Indigenous communities, with everyone being a teacher to any community member. Thus, everyone in the community was responsible for transmitting the knowledge to their younger generation (Haines, 2022). The AIKS is not systematically structured for learners to follow; therefore, learning and the transmission of knowledge occur every day, at any time. With learning occurring every day, studies such as Dei et al. (2022) have indicated that Indigenous communities lacked systematic, reliable knowledge-preservation methods; therefore, they relied on the memory of knowledge bearers.

The Indigenous methods of transmitting the AIKS have been argued to be ineffective for several reasons. First, the colonisation epoch has dramatically reduced the practice of AIKS, leading most Indigenous communities to adopt Western and scientific practices (Bogopa, 2012). Secondly, the AIKS not being systematically documented has resulted in most Indigenous knowledge bearers passing away before sharing their knowledge with subsequent generations (Bogopa, 2012). In that regard, there is a concern that the ineffectiveness of Indigenous methods could lead to the AIKS becoming extinct and inapplicable to future generations. Therefore, digitisation has been widely contested as a reliable approach to knowledge preservation and sharing, one that could revive AIKS and ensure broad participation by contemporary and upcoming generations (Masenya, 2023). Previous studies have shown that the modern generation spends most of their time on digital platforms (Nevondo et al., 2024). Therefore, the purpose of digitisation is to ensure the permanent preservation of AIKS and to create opportunities for knowledge to be delivered on digital platforms, thereby engaging the contemporary generation (Nyoni & Maimela, 2025). Despite digitisation offering sustainable preservation of the AIKS, the process has not been smooth, with key contestations undermining this preservation (Kugara & Mokgoatšana, 2022). Building on these arguments, this chapter presents the contestations faced by Indigenous communities in Ga-Mamabolo, Limpopo Province, South Africa. Guided by participatory action research (PAR), the chapter further presents a co-designed approach to digitising the AIKS from the perspective of the Ga-Mamabolo Indigenous communities.

Following the introduction, the chapter covers a review of literature on IKS in South Africa, Evolution of knowledge preservation methods, Legal frameworks for indigenous knowledge preservation and management in South Africa, Global debates on digitising the IKS, and the Positionality of the Indigenous standpoint theory in digitising the IKS. This is followed by the study's adopted methodology, results and discussion.

INDIGENOUS KNOWLEDGE SYSTEMS IN SOUTH AFRICA

Understanding the nature of IKS in South Africa is often approached through the country's history. Previous studies, such as Nyoni and Maimela (2025), have indicated that the era in South Africa when

IKS dominated was the pre-colonial era. Green (2012) argues that before colonisation, the IKS was the only form of knowledge available to Indigenous communities. Although IKS were effective, the colonial era disrupted the practice of IKS in South Africa. Maluleka (2020) states that the most affected part of the IKS was the African traditional medicine (ATM). Maluleka (2020) states that during the colonial era, colonial acts such as the Witchcraft Act of 1957 barred the practice of ATM, which was labelled primitive and evil. Previous studies have contested that this was the period that led to most Africans detaching from their IKS, with most people still labelling the IKS as primitive.

In the postcolonial period, the South African government has made commitments to revive and safeguard the IKS. Balogun (2023) alludes that South Africa's Department of Science, Technology, and Innovation (DSI) is currently leading the digital safeguarding of IKS through the National Recordal Systems (NRS). Balogun (2023) further highlighted that, through the NRS, the country has established the Indigenous Knowledge Systems Documentation Centres (IKSDCs) at each university across all 9 provinces, which are responsible for preserving IKS.

EVOLUTION OF KNOWLEDGE PRESERVATION METHODS

The IKS has undergone various stages of development, which have played a critical role in sustainably supporting the lives of many generations (Nevondo et al., 2024). The origin and founder of the IKS are unknown, with knowledge often attributed to Indigenous communities as the founders. From the latter, it can be inferred that Indigenous people developed Indigenous knowledge preservation methods (IKPMs) to ensure the continuity of IKS within Indigenous communities for many generations.

Learning by doing has been previously reported as the most used IKPM for IKS (Bat et al., 2014). For example, a study conducted by Ikuenobe (2018) reported that, in learning-by-doing, Indigenous knowledge bearers taught and transmitted IKS by allowing learners to imitate their elders. Lebaka (2022) noted that this technique has been successful because learners could learn the skills through practice rather than through theory. This also allowed learners to retain knowledge more quickly because what can be taught through practice is hard to forget.

Along with learning by doing, other studies have revealed oral tradition as another form of IKPMs. Bihari (2023) argued that, for Africans, oral tradition was not merely a form of entertainment; rather, Indigenous communities used it to communicate some of the most important lessons central to their communities. For example, studies from Nevondo et al. (2024) have revealed that through oral tradition, key cultural aspects such as togetherness, hospitality, and compassion were communicated. Despite learning-by-doing and oral tradition being the IKPMs used to preserve the IKS, other studies have contested that the success of the IKPMs depended on the elderly Indigenous knowledge bearers. Haines (2022) posits that elderly Indigenous knowledge bearers stored knowledge and skills in their heads, often relying on memory to pass them to the younger generation.

The IKPMs have faced numerous criticisms for their efforts to preserve the IKS. Numerous studies have argued that the IKS is transmitted orally and practically, without being systematically documented. Tume et al. (2019) suggest that most elderly knowledge bearers are being buried with most of the IKS. Mbah et al. (2021) argue that this has been a threat to most Indigenous communities because, when knowledge bearers pass on, it becomes impossible or unlikely to recover the lost knowledge. Some studies have argued that IKPMs may have been helpful in the past; however, their limitations cannot be avoided (Tume et al., 2019).

In a contemporary world shaped by modern practices, numerous studies have called for new methods of knowledge preservation to prevent elderly Indigenous knowledge bearers from losing the vital IKS they hold (Balogun & Kalusopa, 2021; Masenya, 2023). Digitisation has emerged as a sustainable means of knowledge preservation, often commended for its ability to preserve IKS without risking permanent loss of knowledge (Mdhluli et al., 2021). Other studies have further reported that digitisation not only provides permanent preservation but also enables Indigenous communities to share their IKS with the global community (Nyoni & Maimela, 2025). Alongside digitisation, artificial intelligence (AI) has emerged as a recent means of knowledge preservation. The findings from Horna-Saldaña et al. (2025) reveal that Indigenous communities can harness AI to preserve key aspects of culture, such as language and IKS.

LEGAL FRAMEWORKS FOR INDIGENOUS KNOWLEDGE PRESERVATION AND MANAGEMENT IN SOUTH AFRICA

Like many other countries, South Africa has implemented policies and legal frameworks to guide the digitisation of IKS (David, 2025). Through the Department of Science, Technology, and Innovation (DSI), South Africa has developed the National Indigenous Knowledge Systems Office (NIKSO) that oversees the preservation and promotion of the IKS in the country, including issues related to copyrights of the IKS (Balogun, 2023; Alberts et al., 2011). Along with NIKSO, DSI has further established the IKSDCs to facilitate preservation of IKS, including digitisation (Van Rooyen, 2011).

GLOBAL DEBATES ON DIGITISING INDIGENOUS KNOWLEDGE SYSTEMS

Digitisation of the IKS has become common globally, often driven by efforts to promote and preserve it for future generations (Gumbo, 2023). Australia is one of the countries that is actively digitising its IKS (Boamah & Liew, 2017). The National Collaborative Research Infrastructure Strategy (NCRIS) and the Atlas of Living Australia (ALA) are some of the key approaches employed by the Australian government to ensure the sustainable preservation of IKS in the country (Roger et al., 2023). Looking at India, a study by Patil and Deshpande (2025) revealed that Digital India is the primary initiative that could allocate funds for the digital preservation of IKS in the country. Another study by Ghimire (2021) indicated that most countries in Asia, including India, are less involved in the digitisation of IKS due to their lower levels of development. To improve this process, Patil and Deshpande (2025) identified participatory documentation as one of the best practices for digitising IKS, in which Indigenous communities are involved in the digitisation process. This practice has also been recommended in many countries.

Looking at the global concerns with regard to digitisation of the IKS, studies such as Mhlope (2026) have indicated that the digital divide has become the most challenging aspect in digitising the IK. This implies that there are some cases where the IKS is digitised but not used to benefit the Indigenous communities. Looking at the global comparison, Schelenz and Schopp (2018) indicated that Africa is the least connected continent compared to others. To address the global concern of the digital divide, some studies have recommended adopting offline technologies to enable communities to access digital IKS freely (Patil & Deshpande, 2025). Other studies also indicated that capacity building is needed to train Indigenous communities worldwide to access digital IKS.

POSITIONALITY OF THE INDIGENOUS STANDPOINT THEORY IN DIGITISING THE INDIGENOUS KNOWLEDGE SYSTEMS

The study was premised on the Indigenous standpoint theory (IST), which emerged in the early 2000s developed by Nakata Martin. IST is a theoretical framework that emphasises the need to recognise Indigenous views, values, and experiences as part of challenging Western or scientific methods (Cox et al., 2021). The theory is underpinned by the idea that marginalised Indigenous people may offer key insights to inform their views on social change.

In the context of this study, the IST positions Indigenous people as the relevant actors who can provide key reflections on the digitisation process of IKS. Keovorabouth (2025) posits that Indigenous communities have been experiencing the digitisation of IKS through Western lenses, with little to no consideration of Indigenous views and experiences. Thus, the use of the IST in this study was influenced by the need to outline and confront Western approaches to digitising IKS in South African Indigenous communities. The IST suggests that there are contestations in digitising the IKS, of which most emanate from Western perspectives. The theory further suggests that outlining such contestations and understanding their root causes may play a critical role in identifying key areas that can be adopted to make the digitisation process relevant and beneficial to Indigenous communities.

METHODOLOGY

This section covers the research materials and methods used to produce the study results. The section begins with the description of the study area where the data was generated. This sets the scene for the research approach employed as a blueprint for the data generation and analysis process. This is followed by the recruitment process of the research team, the data generation and analysis process. The section concludes with the ethical considerations observed throughout the data-generation process.

Description of Ga-Mamabolo

The study was conducted in the Limpopo Province, one of South Africa's nine provinces. As part of South Africa, Limpopo is in the far north of the country and comprises five municipal districts: Sekhukhune, Waterberg, Capricorn, Vhembe, and Mopani. Among these district municipalities, the Capricorn District, comprising the local municipalities of Polokwane, Molemole, Lepelle-Nkumpi, and Blouberg, was selected as the study area. The study was conducted in Ga-Mamabolo Indigenous communities, which form part of the Polokwane Local Municipality. Ga-Mamabolo comprises rich and diverse Indigenous communities that continue to practice IKS in their daily lives. For purposes of this study, a few communities that formed part of the study area included Spitzkop, Mentz, Mongwaneng, Thune, and Turfloop (See Figure 1). We used these communities first because they are part of Ga-Mamabolo and have strong relationships with their IKS. We also used these communities because they are located near one another, which made data collection cost-effective.

Figure 1. Selected Ga-Mamabolo Indigenous Communities

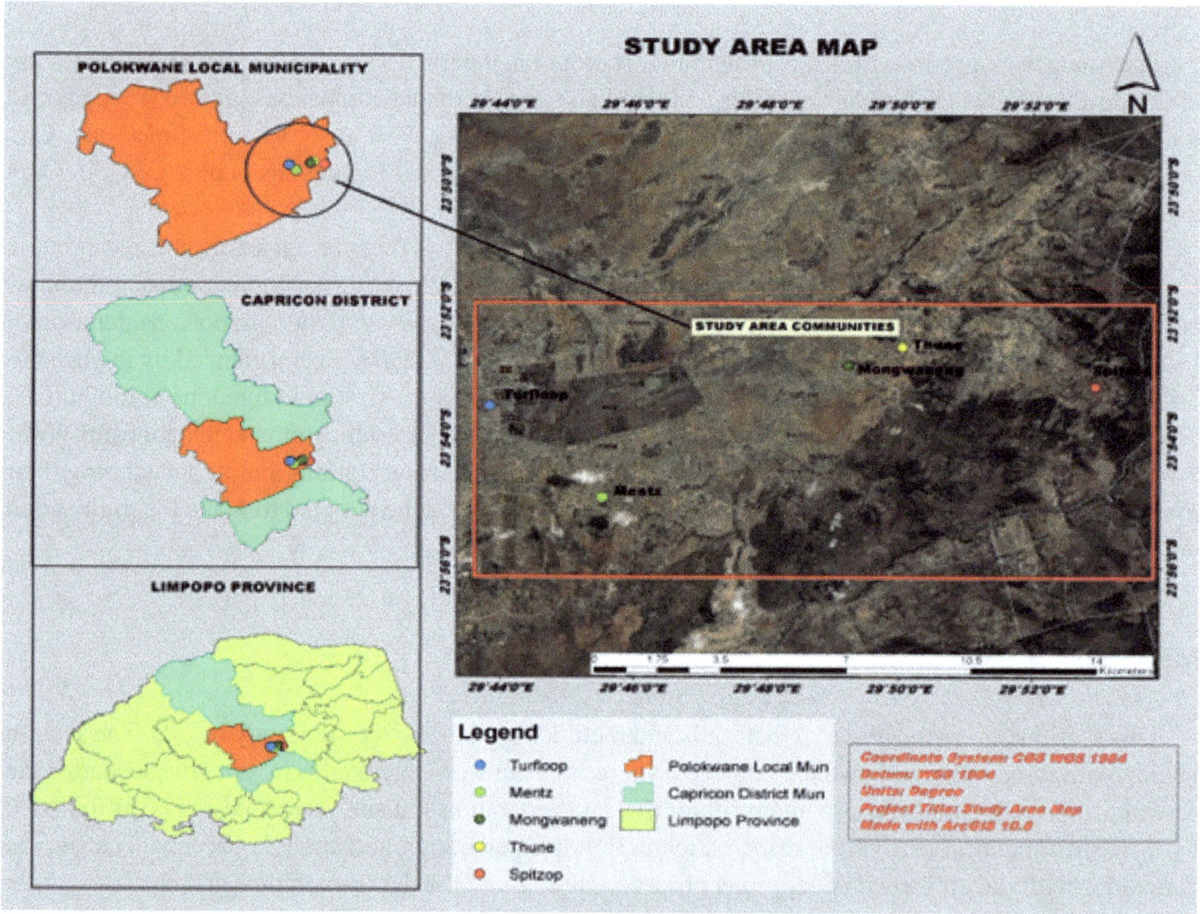

Participatory Action Research

We employed a qualitative research methodology guided by the participatory action research (PAR) approach. The PAR is succinctly described as a qualitative research approach that fosters collaboration between researchers and the people affected by the research problem to co-create solutions to the problem (Nhlapo, 2023). Teele et al. (2020) allude that PAR is centred on co-creating new knowledge with the affected stakeholders and sharing equal ownership of the research. Maimela and Nedombeloni (2026) highlighted that in PAR, researchers establish a research team, with affected people referred to as co-researchers rather than participants. In this study, the term "co-researchers" refers to the study participants. In relation to this study, we identified that the IKS have been digitised in Ga-Mamabolo Indigenous communities. Therefore, we employed a PAR to explore some of the contestations experienced by Indigenous communities. The PAR not only allowed us to understand the contestations, but we also, along with the co-researchers, co-designed a digitisation approach for digitising the IKS in Indigenous communities.

Research Team and Recruitment Process

The recruitment process in a PAR is centred on establishing a research team to collaborate in developing a solution for the research problem being explored (Teele et al., 2020). To establish our research team, we targeted high school learners, youth, digital content creators, and Indigenous knowledge bearers (See Table 1). We targeted high school learners and youth because they are the beneficiaries of the AIKS. The Indigenous knowledge bearers, due to their rich expertise in AIKS. We ultimately targeted digital content creators for their extensive skills and expertise in digitisation.

Table 1. Recruited Research Team

Co-researchers	Category	Age	Gender	Indigenous Community
Co-researcher 1	HSL	17	M	Mongwaneng
Co-researcher 2	HSL	16	F	Mongwaneng
Co-researcher 3	HSL	17	M	Spitzkop
Co-researcher 4	HSL	17	F	Thune
Co-researcher 5	YTH	27	M	Mentz
Co-researcher 6	YTH	28	F	Spitzkop
Co-researcher 7	IKB	43	M	Spitzkop
Co-researcher 8	**DCC**	**28**	**M**	**Turfloop**

Key:
HSL: Hish School Learner YTH: Youth IKH: Indigenous Knowledge Bearer M: Male F: Female
DCC: Digital Content Creator

To establish a research team, we blended qualitative recruitment methods with the Indigenous techniques. We first employed a purposive recruitment strategy to recruit co-researchers knowledgeable about digitisation and AIKS. We combined purposive and convenience sampling by recruiting co-researchers who were readily available and willing to participate in the study, ensuring they had rich expertise in either digitisation or AIKS. Using this recruitment strategy, we recruited 7 co-researchers: 4 high school learners, 2 youth, and 1 digital content creator. We also employed kgoro (the traditional court system) as a recruitment strategy. We first used *kgoro* as a community to request permission to conduct the study in Ga-Mamabolo Indigenous communities. During *kgoro,* we managed to recruit 1 Indigenous knowledge bearer, an expert in AIKS.

Data Generation Process and Analysis

Snapp et al. (2023) argue that PAR allows researchers to collaborate with the research team they form, who share the same objectives. The stages for co-generating data in PAR are highly contested; however, Snapp et al. (2023) proposed planning, action, and reflection as key stages. For the purposes of this study, we conducted a series of focus group discussions (FGDs) guided by these stages to co-generate data with the Ga-Mamabolo Indigenous people.

After assembling a research team, the first stage was planning. This stage firstly focused on introducing the co-researchers to one another. This was followed by an explanation of key terms, such as digitisation, contestations, and Indigenous knowledge systems, to help co-researchers familiarise themselves with the study. As part of the planning, the co-researchers outlined objectives to identify contestations around digitising AIKS in the Ga-Mamabolo Indigenous communities. The desired objectives outlined in this stage were as follows:

1. To identify the existing contestations in digitising the AIKS in Ga-Mamabolo Indigenous communities.
2. To propose a co-designed approach for digitising the AIKS in Indigenous communities.

The second stage of the data generation process included a prolonged engagement guided by the objectives outlined in the first stage. The last stage was reflection, during which the research team reviewed the decisions made in planning and action. These reflections involved reviewing the decisions made and assessing whether the generated data is satisfactory and aligned with the objectives.

For data analysis, we employed critical discourse analysis, a qualitative technique based on the constant review and examination of transcribed texts to reveal and understand areas of dominance, power, and inequality (Mogashoa, 2014). For the purposes of this study, we applied CDA to examine areas of dominance in the digitisation of IKS within Indigenous communities. In analysing the data, we transcribed the recordings and thoroughly reviewed the transcripts. From the transcription, we generated category codes, which we used to organise our findings in line with the research objectives.

Ethical Considerations

The ethical clearance for conducting this study was granted by the University of Venda's Research Ethics Committee. Since the study was conducted in an Indigenous setting, permission to engage the Indigenous communities was obtained from the Mankweng Traditional Council, which is responsible for all Indigenous communities in Ga-Mamabolo. The study had minor co-researchers; therefore, permission to involve the minors in the study was obtained from their parents, school, and the Department of Education, Mamabolo Circuit. All the co-researchers took part in the study voluntarily after having been fully informed about the study's aims and objectives. Other ethics principles adhered to were confidentiality, anonymity, and avoidance of harm.

RESULTS

We present our findings using 2 research objectives as overarching themes, with sub-themes used to report the results. The first overarching theme presents the perceived contestations surrounding the digitisation of AIKS in South African Indigenous communities, particularly with reference to the Indigenous people of Ga-Mamabolo. The subsequent theme outlines the proposed co-designed approach to digitising AIKS in Indigenous communities.

Contestations of Digitising the AIKS

Our analysis revealed various contestations that emerge when digital content creators digitise the AIKS in Ga-Mamabolo. In this theme, we uncover the contestations experienced by Ga-Mamabolo Indigenous people.

Passive Participation

Our analysis revealed passive participation as the everyday form of contestation that occurs in most Indigenous communities in Ga-Mamabolo regarding the digitisation of IKS. The co-researchers opined that when digital content creators approach their community, their goal is centred on collecting rather than collaborating with the Indigenous people. The analysis shows that the absence of collaboration stems from digital content creators within the IKS often being outsiders who come to Indigenous communities to digitise the IKS for them, rather than digitise with them. The co-researchers alluded that the Indigenous communities are frequently considered as less digitally skilled people who need digital content creators to preserve the knowledge for them, rather than the digital content creators needing the Indigenous people. They viewed this as unfair because both Indigenous people and digital content creators possess different sets of skills that enable collaboration.

> "The way I am seeing it, to digitise is not a problem; the problem comes when people just come and digitise the knowledge, and they never come back. That is not how things should be done. I think we need to be involved in every process, because they have the skills and we have the knowledge. So, both of us are important, and we must be working together in every process." (Co-researcher 7, IKB).

Reduced Authenticity

Based on the co-researchers' submissions, digitisation is vital for sustainable preservation; however, they also cautioned that if not properly digitised, the authenticity of the IKS may be compromised or lost. The co-researchers opined that digitisation directly undermines the authenticity of some IKS, particularly Indigenous games. The co-researchers alluded that Indigenous games are meant for social interaction, and that it is unlikely to digitise them while still maintaining their authenticity.

> "When it comes to digitisation, the originality of the knowledge can be questioned. I think when you digitise Indigenous stories, you are not changing anything, but when it comes to the Indigenous games, it is a different story. When game developers digitise the games, they are reducing the originality of the games because some of the most important parts of the game, like children engaging, are lost." (Co-researcher 8, DCC).

The co-researchers also critiqued the modernisation of IKS as part of the digitisation process. According to the co-researchers' submissions, when the IKS are modernised, they not only undermine authenticity but also mislead upcoming generations.

> "One thing I have realised about Indigenous stories and games is that they are being modernised to fit the narrative of the modern generation. That is a big problem, because when you modernise the IKS, you are killing the original version of the knowledge." (Co-researcher 8, DCC).

The analysis reveals that, alongside reduced authenticity, digitisation also reduces the social interaction within the IKS. The co-researchers indicated that oral tradition and Indigenous games were developed to help Indigenous communities socialise. Therefore, digitisation not only undermines the authenticity of the IKS but also destroys their social interaction. According to the co-researchers' submissions, digitisation may be used to preserve IKS; however, to maintain authenticity, greater emphasis should be placed on physical participation than on digital access.

> "I also think that we can digitise the Indigenous knowledge for preservation purposes, but encourage more physical participation to maintain the originality of the knowledge, because if we do not do that, the originality of the knowledge may be lost." (Co-researcher 5, YTH).

Intellectual Property Rights

With AIKS pursuing interests for commercial gain, the co-researchers opined that intellectual property rights (IPR) are another point of contention, making the digitisation process unjust and unfair to Indigenous communities. The co-researchers opined that when some AIKS are digitised, rights are often granted to digital content creators rather than to Indigenous peoples. The analysis reveals that, due to the lack of benefit-sharing from IPR and monetary benefits, some Indigenous communities are discouraged and reluctant to participate in the digitisation process.

The analysis reveals that the issue of IPR can be attributed to digital content creators' approach to digitising IKS. Based on co-researchers' submissions, most digital content creators digitise IKS and claim the knowledge as their own, rather than acknowledging the Indigenous people who contributed to the digitisation process.

> "There have been some cases in South Africa, where you see people from outside the Indigenous communities claiming the Indigenous knowledge as their own, with no acknowledgement of the Indigenous people. Ga-Mamabolo is experiencing the same thing, because people come and record the knowledge, where they are taking the knowledge, we do not know, and that is a problem because what if they are making money out of it without sharing the benefits with where the knowledge was taken." (Co-researcher 6, YTH).

Sacred Knowledge

The co-researchers opined that the sacred knowledge forms an integral part of their cultural identity. As communities that still take pride in their cultural identity, the co-researchers indicated that sacred knowledge, along with sacred sites, is treated as a treasure and respected by everyone in most Indigenous communities in Ga-Mamabolo. Despite the sacred knowledge gradually disappearing, the co-researchers indicated that it is not meant to be shared with the public, which makes it challenging to preserve digitally.

> "We also cannot digitise the knowledge about the sacred sites because culturally it is considered a taboo. Even if the knowledge is at risk of disappearing, there are some areas in my community that my family told me not to go because they are sacred." (Co-researcher 7, IKB).

Extrapolating from the above, the sacred sites are reserved only for specific people and clans; therefore, invading such spaces appears to be disrespectful to these people and their culture. The co-researchers further indicated that, since time immemorial, sacred sites have been protected through spiritual beliefs. Therefore, the co-researchers indicated that Indigenous communities shy away from digitising sacred knowledge because it is believed that anyone who does so may experience bad luck.

> "In the past, we had taboos that protected our sacred sites, but now people who do not respect our culture and traditions have been digitising even some of the knowledge that when we were young, we considered sacred and we feared that we were attracting bad things for ourselves." (Co-researcher 6, YTH).

Funds

The lack of funds dedicated to preserving the IKS was another challenge to digitising the IKS in Ga-Mamabolo. The co-researchers indicated that, despite the need to preserve most of the IKS, the digitisation often requires funding. The co-researchers indicated that the funds play a pivotal role in acquiring the digital tools needed to digitise and distribute the IKS.

> "If most of the Indigenous communities had tools, they could preserve the Indigenous knowledge, but these tools are expensive. On top of that, you need to have data to circulate the knowledge, and if you don't have funds from the donors or governments, this can be difficult." (Co-researcher 8, DCC).

The co-researchers further alluded to the fact that in South Africa, funding can come either from the government or from private institutions. Despite knowing where the funds can be obtained, the co-researchers indicated that obtaining them is complicated. They believed that most institutions require compliance documentation for Indigenous communities to access funds. The co-researchers viewed this as unjust because most of the Indigenous communities are traditionally led with no documentation, and most are not led as a business or organisation. Therefore, our findings reveal that, despite interest in digitising the IKS, the absence of compliance documentation prevents Indigenous people from accessing funds that could help preserve the IKS.

> "The problem that we have when it comes to getting funding is papers. If you do not have the right papers, they cannot fund your community because it has not been registered as a business or non-profit organisation" (Co-researcher 7, IKB).

Co-designed Approach for Digitising IKS

The co-researchers proposed a co-designed approach for digitising the IKS in Indigenous communities (See figure 2). From the submissions of the co-researchers, five areas emerged from the analysis that need to be employed in digitising the IKS. These included stakeholder collaboration, cultural considerations, feedback, knowledge sharing, and benefit sharing. These are designed as sub-categories below.

Figure 2. Co-designed Approach for Digitising IKS

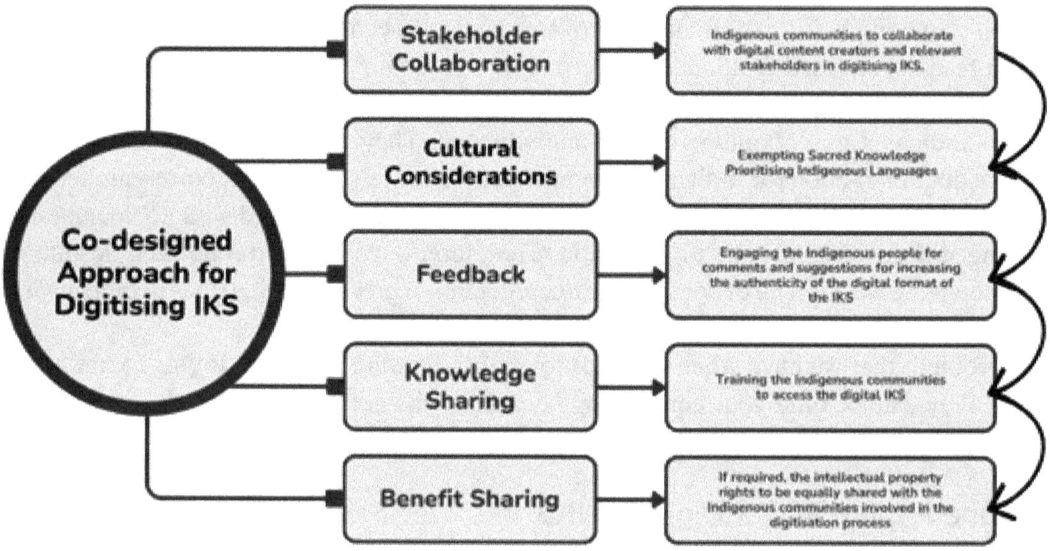

Stakeholder Collaboration

Our analysis revealed collaboration as the key factor to consider when digitising IKS in Indigenous communities. The co-researchers opined that a collaboration could be formed among the Indigenous communities, digital content creators, or other relevant stakeholders if needed. Equal collaboration amongst these stakeholders emerged as a technique that could benefit everyone involved in digitising the IKS.

> "A huge project like digitisation cannot be one-sided. That is why I am saying, digital people need to work with the Indigenous communities or the government if they are interested." (Co-researcher 5, YTH).

According to the co-researchers' submissions, the purpose of the collaboration is to integrate the key skills of stakeholders who share the goal of digitising IKS. The co-researchers noted that digital content creators have expertise and abilities in digitising IKS that are rare in Indigenous communities. On the other hand, the co-researchers further highlighted that Indigenous people possess rich information on IKS that can be digitised. Therefore, our analysis shows that when the two collaborate, the goal is to ensure each stakeholder contributes their part without overlooking the other. The co-researchers indicated that the two stakeholders may need to co-exist because digital preservation of the IKS cannot occur without their collaboration.

> "I agree with him [Co-researcher 5]. If you can check, we all have different fields that we are good at. If I can give an example, a digital content creator has more knowledge on digitisation and us [Indigenous people], we have more knowledge on the Indigenous knowledge." (Co-researcher 7, IKB).

Cultural Considerations

Our analysis unearthed cultural considerations as another key aspect that is often overlooked when digitising IKS. The co-researchers opined that certain cultural factors need to be considered when digitising the IKS. The co-researchers indicated that digitisation is indispensable for preserving the IKS; however, the IKS related to the sacred was deemed a type of IKS that should not be digitised. Our analysis revealed that sacred sites and the knowledge around them are culturally sensitive. The co-researchers indicated that such knowledge is accessible to a few, and the Indigenous communities in Ga-Mamabolo intend to protect it from public view.

> "Since birth and when we are growing, we all know that the sacred knowledge is not supposed to be seen by everyone because that is a knowledge that is secret for certain community members in the community. You just can't wake up and say, you want to digitise the sacred knowledge." (Co-researcher 7, IKB).

Alongside the sacred sites, the co-researchers highlighted Indigenous languages as another cultural aspect to consider when digitising the IKS. The co-researchers contested the idea of digitising the IKS using foreign languages. Our analysis reveals that digitising the IKS in foreign languages creates a digital divide. The co-researchers indicated that Indigenous people are the primary recipients of digital IKS; therefore, the IKS should be digitised in Indigenous languages.

> "If you look at me, I am still young, and my generation can read, but our elders are still struggling with English. I think maybe if we digitise the Indigenous knowledge, we must include Sepedi [Indigenous language] so that those who do not understand English can still access the knowledge." (Co-researcher 3, HSL).

According to the co-researchers' submissions, using Indigenous languages helps maintain the authenticity of IKS more than using foreign languages. The co-researchers indicated that when translations are used in digitising the IKS, there is a risk of losing the IKS's authentic nature.

> "One thing I have realised is that when you make translations, we sometimes lose the real meaning, because some words can never fit into English translations. So, I think we must be careful when we are doing translations in the digitisation process." (Co-researcher 8, DCC).

The co-researcher also indicated that digitising IKS in foreign languages is essential for attracting the global community. Thus, the co-researchers also showed that using multiple languages, including Indigenous languages, would ensure that the IKS is digitised in Indigenous languages to cater to everyone interested in the IKS.

> "I was thinking, how about we start by digitising, using the Indigenous language first, then we add other languages. If we do that, we can cover everyone." (Co-researcher 5, YTH).

Feedback

Our analysis revealed that feedback can be another factor that enhances the smoothness of IKS digitisation. The co-researchers opined that feedback from Indigenous people could strengthen the authenticity of the IKS's digital format.

> "One thing we can add is feedback. We previously mentioned that the originality of the knowledge can be reduced through digitisation. Indigenous people must comment on the digital Indigenous to improve the originality of Indigenous knowledge." (Co-researcher 4, HSL).

Based on the co-researchers' submissions, Indigenous people should be consulted on the digital IKS before it is made publicly available. The co-researchers indicated that the purpose of feedback would be to test the digital IKS and provide feedback on aspects that could be improved until it meets the satisfaction of the Indigenous people.

"I think after digitisation they can go back to the Indigenous community where the knowledge was collected and ask for feedback on the things they want to add or remove the ones that are not well with them." (Co-researcher 5, YTH).

Knowledge Sharing

The co-researchers opined that a digitisation process should be considered a success only if the knowledge is shared back with the Indigenous communities. Thus, the co-researchers indicated that, with most IKS digitised, not everyone in Indigenous communities knows how to access them or the requirements for accessing such knowledge.

"The Indigenous knowledge that I have come across on the internet is just by luck, and I think the digital content creators have to train the Indigenous people on how to access that knowledge because, at the moment, people just come across the knowledge by luck as I do." (Co-researcher 8, DCC).

Our analysis shows that when the importance of knowledge sharing centres on the idea that IKS belongs to Indigenous communities, efforts should be made to ensure they have access to the digital IKS. Based on the co-researchers' submissions, knowledge sharing can enlighten and train Indigenous communities on how to access and use digitised IKS.

"I was also thinking the same thing, the knowledge sharing can be for training Indigenous people on how to get the digitised Indigenous knowledge. They can also be trained to digitise Indigenous knowledge themselves." (Co-researcher 2, HSL).

Benefit Sharing

Our analysis has unveiled that digitisation could be approached in two ways. First, the co-researchers indicated that digitisation is used to sustainably preserve knowledge for future generations. Secondly, the co-researchers indicated that, although rare, the IKS can be digitised for commercial purposes.

"I love the point that she made because if we digitise the Indigenous knowledge for preservation, things like benefit sharing can be avoided, but if it is for generating money, I think it is then we can encourage benefit sharing." (Co-researcher 7, IKB).

Extrapolating from the above, our analysis reveals that digitisation for the sustainable preservation of IKS has been successful because this approach is less contested. On the other hand, the commercial digitisation of IKS has been a thorny challenge for most Indigenous communities. To avoid conflicts, the co-researchers indicated that, in cases where a digital content creator wants to commercialise the digital IKS, this should be done with the benefit-sharing of monetary benefits with the Indigenous communities or people involved in the digitisation process.

"We appreciate that most people [digital content creators] are helping Indigenous communities in digitising the Indigenous knowledge, but our approach can add benefit sharing if the knowledge is going to make money after digitisation." (Co-researcher 6, YTH).

The preceding section discusses the research findings in line with previous studies on the digitisation of the IKS.

DISCUSSION

Digitisation of the IKS has become a common practice in Africa to preserve knowledge for future generations (Achieng, 2022; Mapara, 2021). This is because the digitisation offers several benefits to the Indigenous communities. For example, a study by Maimela et al. (2024) reported that, through digitisation, IKS can be permanently stored and shared with a global community. This is because in the current generation, digital gadgets have become another form of learning and socialisation. Thus, other studies have indicated that the IKS can be preserved in digital technologies, allowing the current generation to access it (Achieng, 2022; Magere, 2023). Other studies have also indicated that the IKS is not only digitised for knowledge preservation but also used to address some of the contemporary issues facing Indigenous communities (Magni, 2017). The United Nations Educational, Scientific, and Cultural Organization (UNESCO) (2025) recently reported that the Indigenous communities in Zambia and Tanzania digitised their IK to help the community members track and conserve animals. Magare & Patil (2025) reported that, through digitisation, Indigenous communities worldwide can preserve the medicinal use of Indigenous plants. This implies that the role of digitisation is to ensure that knowledge is preserved and used by future generations to address challenges previously faced by earlier generations.

Despite the success of digitisation in preserving and using the IKS to address temporary issues, there have been numerous studies that have contested the smoothness of the digitisation process for the IKS, calling for alternative approaches to digitising the IKS (Maimela et al., 2024; Gumbo, 2023). The call behind this is that the methods used in the digitisation process reflect little to no values of Indigenous communities (Kugara & Mokgoatšana, 2022). Complementing these previous studies, the present study was conducted to identify the contestations surrounding the digitisation of the IKS that render the process unprofitable. Through the PAR, the research team collaborated, guided by two objectives: first, to identify existing contestations in digitising the IKS in Ga-Mamabolo Indigenous communities; and second, to propose a co-designed approach for digitising the IKS in Indigenous communities. Our analysis reveals five contestations in the digitisation of IKS in Ga-Mamabolo Indigenous communities. The first point of contention identified was that Indigenous communities are seen as passive participants rather than collaborators in the digitisation process. Previous studies have contested that passive participation has led to the Indigenous communities being exploited for their IKS (Kasih et al., 2021; Kisambira et al., 2024). Our findings align with these studies by emphasising that Indigenous communities need not be treated as passive participants, as they are the source of knowledge. Along with passive participation, our empirical findings show a link between passive participation and IPR as a form of contestation. Our findings show that some digital IKS can be commercialised, and that a lack of collaboration has led to IPR being claimed by digital content creators rather than by Indigenous communities. From a global perspective, this finding is in contrast with the United Nations' guideline on the World Intellectual Property Organization (WIPO), which serves to protect the intellectual property of the Indigenous communities (Jefferson, 2025). Thus, other studies have contested that when the IKS is commercialised, the Indigenous communities may need to be granted equal benefit-sharing and patent rights. From a local perspective, South Africa has introduced the National Recordal System (NRS), which encourages Indigenous communities to digitise or record their IKS to benefit the communities as the direct recipients of the knowledge (Van Rooyen, 2011). Tong (2019) stated that, through the National Indigenous Knowledge Systems Office (NIKSO), Indigenous communities have the right to report unauthorised use or exploitation of IK. In the context of digitisation, this signifies that when the IK is digitised, the lack of acknowledgements and equal rights of Indigenous digital content serves as a violation of the guidelines established by the WIPO and NIKSO.

Another point of contention that emerged was the reduced authenticity of the IKS. Some previous studies have emphasised that digitisation offers permanent preservation of the IKS (Balogun & Kalusopa, 2021; Masenya, 2023); however, our findings show that the quality it provides is sometimes questioned regarding authenticity. Our analysis shows that it is impossible, if not unlikely, to digitise most IKS while maintaining their authentic nature. A study by Woods (2012) warned that if not adequately addressed, digitisation may add to the decline of the IKS rather than preserving them. Therefore, we argue that, given the risk of losing key aspects of IKS through digitisation, it is necessary to carefully consider suitable digitisation techniques for the IKS to be digitised.

Our analysis further revealed sacred knowledge as another form of contestation that exempts other IKS from digitisation. The study's findings show that sacred knowledge is a type of IKS reserved for specific clans or individuals. Nyoni and Maimela (2025) argued that when it comes to digitisation, such knowledge should be avoided because it is culturally prohibited from being seen or accessed by everyone.

We discovered that funding plays an indispensable role in digitising the IKS; therefore, Indigenous communities without funding often struggle to preserve their IKS digitally. Our findings complement those of Pandit (2025), who reported that the digitisation process requires substantial resources, including human and digital resources, specifically for digitising the IKS. In their analysis, Boamah and Liew (2017) reported that, for most African countries, funding for IKS-related projects primarily comes from the government, with most Indigenous communities receiving insufficient support. Some of the root causes we identified for insufficient funding of Indigenous communities were that funding acquisition processes require organisational documents, which is a challenge for most Indigenous communities because they lack the required compliance documentation.

In an attempt to address the contestations surrounding the digitisation of IKS, the research team proposed a co-designed digitisation approach. This approach was co-designed to support African Indigenous communities in sustainably preserving the IKS. In this approach, we first proposed stakeholder collaboration, with the idea that Indigenous communities should be considered creators alongside the broader community. Our analysis shows that when digital content creators digitise the IKS, they should collaborate with the Indigenous knowledge bearers. We argue that digital content creators have extensive skills in digitisation, while Indigenous knowledge bearers have extensive knowledge of IKS. We proposed that collaboration between the two stakeholders can be approached through mutual understanding, as both have distinct skills that are key to sustainably preserving the IKS.

Previous studies, such as Kugara and Mokgoatšana (2022), have argued that the methods used to digitise the IKS are foreign and often fail to consider cultural aspects. Similar to this study, we proposed cultural considerations to assist digital content creators and Indigenous communities in digitising IKS with respect for Indigenous communities' cultural values. Our analysis identifies sacred knowledge and Indigenous languages as the key cultural aspects to consider in the digitisation process. First, we argue that sacred knowledge is private and sensitive for Indigenous communities and may be excluded from the digitisation process. Phanvisetsak and Chung (2025) state that the digitisation of sacred knowledge without proper consultation with the Indigenous communities may lead to the exploitation of culturally sensitive knowledge. In line with our empirical findings, we argue that digitising sacred knowledge may appear as disdainful to the Indigenous communities. Thus, such knowledge should only be digitised after proper consultations with the rightful people and their consent. Secondly, we proposed prioritisation of Indigenous languages in the digitisation process. Our argument is, first, that Indigenous languages are often neglected in the digitisation process, creating a language digital divide. Secondly, we argue that Indigenous languages play a key role in maintaining the authenticity of the IKS compared with digitisation in foreign languages. Despite the prioritisation of Indigenous languages, previous studies have shown that a multilingual approach can be employed to share IKS with a global community that may not understand the Indigenous languages (Dixon et al., 2022). These findings imply that language plays a critical role in digitisation and access to knowledge. Therefore, when digitising the IKS, Indigenous languages should be prioritised to cater for Indigenous communities, while the infusion of other languages could be to cater for a global community.

Our findings suggest that community feedback is needed in evaluating the digital IKS before they are shared with the public. Our submission is that Indigenous communities can determine the digital IKS and offer insights into areas for development. Kapuire (2013) conducted a pilot study with the Herero Indigenous communities of Namibia to digitise the IKS in the area. According to Kapuire (2013), a key part of successful digitisation was community feedback from Indigenous people, which ensured that the knowledge digitised was authentic and relatable to them. Similarly, we submit that the approach could be further adopted in digitising the IKS in South African Indigenous communities.

We observed a digital divide in access to IKS in most Indigenous communities. We argue that the divide occurs because knowledge is not shared back with Indigenous communities. Our findings suggest that the digitisation process can be considered complete if Indigenous communities are well informed and trained to access knowledge about themselves. Our findings mirror the sentiments of Rodríguez-Domenech et al. (2023), who stated that community engagement and outreach may be a key step toward eliminating the digital divide in Indigenous communities and improving access to digital IKS. In our co-designed approach, the last aspect we considered was IPR. Prior studies have established that some parts of the IKS can be leveraged for commercial purposes, with many Indigenous communities being exploited of their IKS (Mashiyane, 2023). Our empirical findings show that most IKS are digitised primarily for preservation with little consideration of commercialisation. Our findings also suggest that when the digital IKS is commercialised, Indigenous communities may also need to be assigned IPR over the digital content produced.

FUTURE DIRECTIONS

The study has established the relevance of the PAR in addressing most of the challenges faced by the Indigenous communities. Despite the PAR offering key insights through collaboration with the Indigenous people of Ga-Mamabolo, a future study can be conducted in a similar rural setting to deepen understanding of digital contestations. Furthermore, other studies may be conducted using quantitative methods to generate a generalised understanding of contestations surrounding the digitisation of the IKS. To address the contestations surrounding the digitisation of the IKS, the study has proposed a digitisation approach. The proposed approach was never tested; therefore, pilot studies can be conducted to assess it and refine it to ensure the sustainable digital preservation of the IKS.

CONCLUSION AND RECOMMENDATIONS

The study was conducted to identify the existing contestations in digitising the IKS in Ga-Mamabolo Indigenous communities and to propose a co-designed approach for digitising the IKS in Indigenous communities. The objectives were achieved through the application of PAR, which enabled the researchers to collaborate with Indigenous people to identify contestations and propose a co-designed approach to digitising IKS. The findings show that passive participation, reduced authenticity, IPR, sacred knowledge, and funds are the key points of contention in digitising the IKS in the Ga-Mamabolo Indigenous communities. The findings further show that stakeholder collaboration, cultural considerations, feedback, knowledge sharing, and benefit sharing are key aspects to consider when digitising IKS in Indigenous communities. To ensure a sustainable digital preservation of the IKS, the study recommends the following:

- Digitisation of the IKS requires a multi-stakeholder approach. Thus, at the local level, Indigenous communities may need to collaborate with digital content creators to create a digital format of the IKS. At the government level, the DSI needs to develop accessible approaches for Indigenous communities to adopt to safeguard IKS on platforms such as NRS and IKSDCs. The DSI can also offer digital training to Indigenous communities to develop the skills to digitise their IKS themselves.
- There is a need to develop Afro-sensed guidelines for digitising the IKS. Such guidelines may offer key insights on how the IKS should be digitised and what to do and not to do when digitising the IKS. This may be vital in curbing the exploitation of the IKS.
- Long-term impact studies may need to be conducted to assess the impact of digital preservation of IKS.
- To maintain the authenticity of the IKS, we recommend that the IKS be integrated into mainstream education for the current generations to learn the knowledge in its authentic form. In other words, digitisation may be for the permanent preservation of knowledge, while mainstream education ensures that the current generation continuously participates in their IKS.
- Although South Africa's NIKSO offer strong support on protection and preservation of the IKS, the emerging trends on digitisation of IKS may not be avoided. Thus, the study recommends the institutional support from NIKSO and other relevant stakeholders to provide Indigenous communities with capacity-building on issues such as the exploitation of the IK. Such support may assist Indigenous communities in sustainably preserving, promoting, and celebrating their IKS through digitisation, without being manipulated or exploited.

REFERENCES

Achieng, M. S. (2022). Digitalisation of indigenous knowledge of agricultural practices: A strategy for food security in Africa. *Journal of Public Administration*, *57*(4), 802–818. https://hdl.handle.net/10520/ejc-jpad_v57_n4_a5

Alberts, R., Khalala, G., & Molefe, O. (2011). Collecting structured indigenous knowledge for enhanced exploration: The National Recordal System (NRS), South Africa. *Knowledge Management for Development Journal*, *7*(3), 291–307. DOI: 10.1080/19474199.2011.649009

Balogun, T. (2023). Digitization of Indigenous knowledge systems in Africa: The case of South Africa's National Recorded System (NRS). *Records Management Journal*, *33*(1), 88–102. DOI: 10.1108/RMJ-10-2022-0029

Balogun, T., & Kalusopa, T. (2021). A framework for digital preservation of the Indigenous knowledge system (IKS) in repositories in South Africa. *Records Management Journal*, *31*(2), 176–196. DOI: 10.1108/RMJ-12-2020-0042

Bat, M., Kilgariff, C., & Doe, T. (2014). Indigenous tertiary education–we are all learning: Both-ways pedagogy in the Northern Territory of Australia. *Higher Education Research & Development*, *33*(5), 871–886. DOI: 10.1080/07294360.2014.890575

Bihari, S. (2023). Cultural heritage and Indigenous knowledge: Reviving traditions for future generations. *Sustainable Development Goals in SAARC Countries: Key Issues. Opportunities and Challenges*, *1*, 24–32.

Boamah, E., & Liew, C. L. (2017). Conceptualising the digitisation and preservation of Indigenous knowledge: The importance of attitudes. In, Choemprayong, S., Crestani, F., & Cunningham, S. (eds), *International Conference on Asian Digital Libraries* (pp. 65-80). Cham: Springer International Publishing. DOI: 10.1007/978-3-319-70232-2_6

Boamah, E., & Liew, C. L. (2017, November). Conceptualising the digitisation and preservation of indigenous knowledge: The importance of attitudes. In *International Conference on Asian Digital Libraries* (pp. 65-80). Cham: Springer International Publishing. DOI: 10.1007/978-3-319-70232-2_6

Bogopa, D. L. (2012). The importance of Indigenous games: The selected cases of Indigenous games in South Africa. *Indilinga*, *11*(2), 245–256.

Cox, G. R., FireMoon, P., Anastario, M. P., Ricker, A., Escarcega-Growing Thunder, R., Baldwin, J. A., & Rink, E. (2021). Indigenous standpoint theory as a theoretical framework for decolonizing social science health research with American Indian communities. *Alternative*, *17*(4), 460–468. DOI: 10.1177/11771801211042019 PMID: 38680293

David, J. O. (2025). Beyond the Extractive Paradigm: Appraising Indigenous Knowledge in South African Environmental Governance. *Environment and Ecology Research*, *13*(4), 459–471. DOI: 10.13189/eer.2025.130401

Dei, G. J. S., Karanja, W., & Erger, G. (2022). The role of elders and their cultural knowledges in schools. In *Elders' cultural knowledges and the question of Black/African indigeneity in education* (pp. 127–151). Springer International Publishing., DOI: 10.1007/978-3-030-84201-7_6

Dixon, D. H., Dixon, T., & Jordan, E. (2022). Second language (L2) gains through digital game-based language learning (DGBLL): A meta-analysis.'. *Language Learning & Technology*, *26*(1), 1–25. DOI: 10.64152/10125/73464

Ghimire, P. (2021). Digitalization of indigenous knowledge in Nepal–Review article. *Acta Informatica Malaysia*, *5*(2), 42–47. DOI: 10.26480/aim.02.2021.42.47

Green, L. J. (2012). Beyond South Africa's' indigenous knowledge-science'wars. *South African Journal of Science*, *108*(7), 1–10. DOI: 10.4102/sajs.v108i7/8.631

Haines, J. (2022). Researching the knowledge journey practices of Indigenous elders relevant to the younger generation: A community-based participatory study. *Proceedings of the Association for Information Science and Technology*, *59*(1), 426–430. DOI: 10.1002/pra2.646

Haines, J. (2022). Researching the knowledge journey practices of Indigenous elders relevant to the younger generation: A community-based participatory study. *Proceedings of the Association for Information Science and Technology*, *59*(1), 426–430. DOI: 10.1002/pra2.646

Horna-Saldaña, C. J., Perez Perez, J. E., & Toro Galeano, M. L. (2025). Artificial intelligence in the preservation of native languages and bridging the information access gap for Indigenous peoples. *Journal of Enabling Technologies*, *19*(1), 63–75. DOI: 10.1108/JET-09-2024-0063

Ikuenobe, P. (2018). Oral tradition, epistemic dependence, and knowledge in African cultures. *Synthesis Philosophica*, *65*(1), 23–40. DOI: 10.21464/sp33102

Jefferson, D. J. (2025). The World Intellectual Property Organization Treaty on genetic resources and traditional knowledge: Implications for plant science. *Plants, People, Planet*, *7*(4), 901–905. DOI: 10.1002/ppp3.10615

Juma, F. A. (2022). Recapturing the oral tradition of storytelling in spiritual conversations with older adults: An Afro-Indigenous approach. *Religions*, *13*(6), 563. DOI: 10.3390/rel13060563

Kapuire, G. K. (2013). An Explorative Action Research Study toward the Design of a Digital Knowledge Organisation as Part of an Indigenous Knowledge Management System with a Herero Community. MA thesis, Dept. of Computer Science, University of Cape Town.

Kasih, D. P. D., Dharmawan, N. K. S., Putra, I. B. W., Sudiarawan, K. A., & Rakhima, A. S. (2021). The Exploitation of Indigenous Communities by Commercial Actors. *Journal of Ethnic and Cultural Studies*, *8*(4), 91–108. DOI: 10.29333/ejecs/898

Keovorabouth, S. T. (2025). 'Coming In'as Indigenous, Queer, and Urban–Utilising Indigenous Queer Standpoint Theory as Methodology. *JGI, 9*(3).

Kisambira, A., Khadijah, B., & Ahmed, M. M. (2024). The role of university community engagement programs in influencing higher education outcomes and community development: An insight from Uganda. *Extensive Reviews*, *4*(1), 15–28. DOI: 10.21467/exr.4.1.8028

Kugara, S. L., & Mokgoatšana, S. (2022). Challenges presented by digitisation of VhaVenda oral tradition: An African Indigenous knowledge systems perspective. *HTS Teologiese Studies*, *78*(3), 7428. DOI: 10.4102/hts.v78i3.7428

Lebaka, M. E. K. (2022). Informal learning of Indigenous music and dance through observation and imitation: The case of Bapedi children's musical arts. *European Journal of Language and Literature Studies*, *8*(1), 55–67. DOI: 10.26417/658qju31

Magare, D. A., & Patil, M. B. (2025). Ethnobotany in the digital age: Opportunities and challenges of traditional knowledge digitization. *World Journal of Biology Pharmacy and Health Sciences*, *21*(1), 235–242. DOI: 10.30574/wjbphs.2025.21.1.0043

Magni, G. (2017). Indigenous knowledge and implications for the sustainable development agenda. *European Journal of Education*, *52*(4), 437–447. DOI: 10.1111/ejed.12238

Maimela, G., Madima, T., & Makhanikhe, J. (2024). Overcoming digital barriers to African indigenous games in local communities. *Bussecon Review of Social Sciences*, *6*(1), 29–39. DOI: 10.36096/brss.v6i1.552

Maimela, G., & Nedombeloni, R. (2026). Decolonising the mainstream education and preserving Indigenous knowledge systems: learning through African Indigenous games. *AlterNative: An International Journal of Indigenous Peoples*, 1-9. DOI: 10.1177/11771801251403201

Maluleka, R. J. (2020). The status of traditional healing in the Limpopo province of South Africa. *HTS Teologiese Studies*, *76*(4), 1–12. DOI: 10.4102/hts.v76i4.6103

Mapara, J. (2021). 11 The future of indigenous knowledge system (IKS) in the face of technological disruption. Culture and rural–urban revitalisation in South Africa: Indigenous Knowledge. *Policies, and Planning*, *189*, 189–208. Advance online publication. DOI: 10.4324/9781003145912-11-14

Masenya, T. M. (2023). Revitalization and digital preservation of Indigenous knowledge systems for sustainable development of Indigenous communities in South Africa. *The Serials Librarian*, *84*(5-8), 86–102. DOI: 10.1080/0361526X.2023.2277962

Mashiyane, D. M. (2023). Looting of Indigenous Knowledge Systems through Patents and Intellectual Property Policies. *The Afrocentricity Trajectories of Looting in South Africa*, *17*. Advance online publication. DOI: 10.5040/9781978728943.ch-2

Mazzocchi, F. (2020). A deeper meaning of sustainability: Insights from Indigenous knowledge. *The Anthropocene Review*, *7*(1), 77–93. DOI: 10.1177/2053019619898888

Mbah, M., Ajaps, S., & Molthan-Hill, P. (2021). A systematic review of the deployment of Indigenous knowledge systems towards climate change adaptation in developing world contexts: Implications for climate change education. *Sustainability (Basel)*, *13*(9), 4811. DOI: 10.3390/su13094811

Mdhluli, T. D., Mokgoatšana, S., Kugara, S. L., & Vuma, L. (2021). Knowledge management: Preserving, managing and sharing Indigenous knowledge through digital library. *HTS Teologiese Studies*, *77*(2). Advance online publication. DOI: 10.4102/hts.v77i2.6795

Mhlope, M. M. (2026). Navigating the Digital Divide: Preserving Indigenous Knowledge in the Information Age. *Indigenous Knowledge Management in the Open Science Framework,* 49-76.

Mogashoa, T. (2014). Understanding critical discourse analysis in qualitative research. *International Journal of Humanities Social Sciences and Education*, *1*(7), 104–113.

Nevondo, P. U., Madima, T., & Makhanikhe, J. (2024). The role of ICT in preserving proverbs among Vhavenda communities. *Bussecon Review of Social Sciences*, *6*(3), 41–49. DOI: 10.36096/brss.v6i3.614

Nhlapo, J. V. (2023). Participatory Action Research as Value-Laden Practice: Values that Matter to the Participants. *Noyam-Journal of Education and Technology*, *4*, 330–344. DOI: 10.38159/jelt.2025614

Nyoni, T., & Maimela, G. (2025). Digital content creation as a tool for safeguarding and promoting Indigenous knowledge systems in South Africa. *International Journal of Research in Business and Social Science (2147-4478)*, *14*(8), 395-402. DOI: 10.20525/ijrbs.v14i8.4408

Ogunyiola, A., Gardezi, M., & Vij, S. (2022). Smallholder farmers' engagement with climate smart agriculture in Africa: Role of local knowledge and upscaling. *Climate Policy*, *22*(4), 411–426. DOI: 10.1080/14693062.2021.2023451

Olaopa, O. R., & Ayodele, O. A. (2022). Building on the strengths of African Indigenous knowledge and innovation (AIK&I) for sustainable development in Africa. *African Journal of Science, Technology, Innovation and Development*, *14*(5), 1313–1326. DOI: 10.1080/20421338.2021.1950111

Pandit, S. D. (2025). Reviving Indian Knowledge Systems (IKS): Bridging Tradition with Modernity. *International Studies*, *62*(1), 11–29. DOI: 10.1177/00208817251382323

Patil, P. V., & Deshpande, K. R. (2025). Digital Preservation of Indigenous Knowledge Systems: Challenges and Opportunities in India. *International Journal of Creative and Open Research in Engineering and Management, 01*(01), 1-9. https://doi.org/https://doi.org/10.55041/ijco pe.v1i2.005

Phanvisetsak, S., & Chung, H.-L. (2025). Bridging Heritage and Technology: Rethinking Cultural Asset Classification through Digitalisation. *Journal of Cultural Analysis and Social Change*, *10*(4), 285–298. DOI: 10.64753/jcasc.v10i4.2803

Roger, E., Kellie, D., Slatyer, C., Brenton, P., Torresan, O., Wallis, E., & Zerger, A. (2023). Open access research infrastructures are critical for improving the accessibility and utility of citizen science: A case study of Australia's national biodiversity infrastructure, the Atlas of Living Australia (ALA). *Citizen Science: Theory and Practice*, *8*(1), 56. Advance online publication. DOI: 10.5334/cstp.564

Schelenz, L., & Schopp, K. (2018). Digitalization in Africa: Interdisciplinary perspective on technology, development, and justice. [IJDS]. *International Journal of Digital Society*, *9*(4), 1412–1420. DOI: 10.20533/ijds.2040.2570.2018.0175

Snapp, S. S., Kerr, R. B., Bybee-Finley, A., Chikowo, R., Dakishoni, L., Grabowski, P., Lupafya, E., Mhango, W., Morrone, V. L., Shumba, L., & Kanyama-Phiri, G. (2023). Participatory action research generates knowledge for sustainable development goals. *Frontiers in Ecology and the Environment*, *21*(7), 341–349. DOI: 10.1002/fee.2591

Teele, T., Nkoane, M. M., & Mahlomaholo, M. G. (2020). Corollaries of pedagogic praxis on technological skills transfer for emerging farmers. *Gender & Behaviour*, *18*(1), 14749–14760. DOI: 10.17159/2413-3221/2020/v48n2a541

Tong, L. A. (2019). South Africa adopts sui generis indigenous knowledge protection legislation. *Journal of Intellectual Property Law & Practice*, *14*(12), 935–937. DOI: 10.1093/jiplp/jpz124

Tume, S. J. P., Kimengsi, J. N., & Fogwe, Z. N. (2019). Indigenous knowledge and farmer perceptions of climate and ecological changes in the Bamenda Highlands of Cameroon: Insights from the Bui Plateau. *Climate (Basel)*, *7*(12), 138. DOI: 10.3390/cli7120138

UNESCO. (2025, August 8). UNESCO's tracker training helps safeguard Indigenous knowledge. https://www.unesco.org/en/articles/unescos-tracker-training-helps-safeguard-indigenous-knowledge

Van Rooyen, B. (2011). Safeguarding the future of indigenous knowledge through ICT: ICT and society. *CSIR Science Scope*, *5*(2), 24–26.

Woods, S. (2012). *Eurogames: The Design, Culture and Play of Modern European Board Games*. McFarland & Company, Inc, Publishers.

ADDITIONAL READING

Asuquo, M. E., Okenjom, G. P., Okpa, O. E., & Eyiene, A. E. (2023). Opportunities and challenges in digitization of Indigenous knowledge and implication for educational management in the Nigerian context. *Digital preservation and documentation of global indigenous knowledge systems*, 87-104.

Balogun, T. (2023). Data management of digitized indigenous knowledge system in repositories. *Information Development*, *39*(3), 425–438. DOI: 10.1177/02666669231186777

Kugara, S. L., & Mokgoatšana, S. (2022). Challenges presented by digitisation of VhaVenda oral tradition: An African indigenous knowledge systems perspective. *HTS Teologiese Studies*, *78*(3), 7428. DOI: 10.4102/hts.v78i3.7428

Kuranga, C., & Masenya, T. M. (2025, July). Decolonizing Digital Technologies for Indigenous Knowledge Systems. In *2025 Conference on Information Communications Technology and Society (ICTAS)* (pp. 1-6). IEEE. DOI: 10.1109/ICTAS64866.2025.11155356

Maimela, G., Madima, T., & Makhanikhe, J. (2024). Overcoming digital barriers to African indigenous games in local communities. *Bussecon Review of Social Sciences*, *6*(1), 29–39. DOI: 10.36096/brss.v6i1.552

Paul, J. (2023). Traditional knowledge protection and digitization: A critical decolonial discourse analysis. *International Journal for the Semiotics of Law*, *36*(5), 2133–2156. DOI: 10.1007/s11196-023-09989-8

KEY TERMS AND DEFINITIONS

Contestations: Contestations refer to obstacles and controversies towards achieving a specific target or goal.

Digitisation: A systematic process of converting analogue information or data into digital platforms for easier access and preservation purposes.

Indigenous Communities: A group of Indigenous people who share similar cultural beliefs and traditions.

Indigenous Knowledge Systems: The Indigenous knowledge systems are known as cultural practices, traditions and beliefs that the Indigenous people have been using for many generations, which are key to all aspects of life.

Knowledge Preservation: A process or an act of storing and protecting knowledge mainly for the upcoming generations.

Chapter 6
Digital Pathways to Revitalizing Indigenous Knowledge Systems (IKS):
Community-Based Models of Preservation, Equity, and Sustainable Development in the 21st Century

Jogen Sharma
http://orcid.org/0000-0001-7151-0137
KIIT (Deemed to be) University, India

Dayawanti Tarmali
http://orcid.org/0009-0001-0159-8505
MDI Murshidabad, India

ABSTRACT

Abstract: Indigenous Knowledge Systems (IKS) are dynamic epistemologies crucial to cultural continuity, environmental sustainability, and community resiliency. Nevertheless, Indigenous knowledge practices have remained marginal due to globalisation, digital disparity, and Western knowledge hierarchies. This chapter analyses twenty-first-century digital avenues for community-based revitalisation of IKS through an interdisciplinary prism of Library and Information Science, Indigenous studies, and information and communication technology. Drawing on global concepts, ethical data governance, and international examples, this chapter proposes an Indigenous-based digital framework grounded in sovereignty, participation, and the sharing of benefits. This paper emphasises the importance of digital tools, ethically controlled, to facilitate the transfer of knowledge across generations, help cultures survive, and support sustainable development as mechanisms for combating digital colonialism.

DOI: 10.4018/979-8-3373-7976-0.ch006

1. INTRODUCTION

Indigenous Knowledge Systems (IKS) are cumulative collections of knowledge, practices, beliefs and worldviews that arise from long-term interactions with a particular ecological, cultural and spiritual environment by Indigenous peoples. These are language, agriculture, medicine, environmental stewardship, cosmology, oral traditions and social governance as the epistemological basis of indigenous identity, sustainability and resilience. They are contrary to codified Western scientific paradigms and are largely passed orally, in context and collectively owned, making them especially vulnerable in the modern globalization and digital transformation age.

Traditionally, IKS has been interrupted by colonization, forcible movement, missionary intervention, extractive research educational approaches and assimilationist education policies. These forces have structured the marginalization of indigenous epistemologies and have made them peripheral in the academic, policy and development discourse. Subsequently, in the post-colonial era, the erosion of indigenous language and knowledge traditions only increased due to modernization, urban migration and the market-driven erosion of culture through cultural homogenization. UNESCO estimates that about 40 percent of languages in the world are at risk, most of them are languages of indigenous populations, which means that there is a threat of knowledge disappearance.

The digital revolution poses an ironic situation for Indigenous Knowledge Systems. On the one hand, digital technologies provide unprecedented avenues of documentation, conservation, sharing and the intergenerational transmission. Conversely, thoughtless digitization is also a threat to supporting digital colonialism, exploitation of intellectual property and epistemic injustice. Many mainstream digital heritage projects have been extractive and top-down and indigenous knowledge is gathered, categorized and archived in external libraries, with no meaningful community engagement or benefit transfer.

To address these issues, UNESCO's Local and Indigenous Knowledge Systems (LINKS) programme, the World Intellectual Property Organization with its Traditional Knowledge (TK) framework and the United Nations with its policy on the Rights of Indigenous Peoples (UNDRIP) have all highlighted the importance of culturally respectful, rights-based and participatory interventions in preserving indigenous knowledge. Nonetheless, concrete conceptions of ethical ways of incorporating digital technologies in indigenous governance systems are incomplete and immature.

The chapter continues the thesis that neighborhood-based digital pathways also provide a radical way of rejuvenating Indigenous Knowledge Systems in the 21 st century. Such models can transform Western-centered epistemologies by repositioning indigenous communities as key custodians, decision-makers and beneficiaries of digital initiatives and creating knowledge equity, cultural sovereignty and sustainable development. The chapter is based on interdisciplinary perspectives of Library and Information Science (LIS), anthropology, information and communication technologies to develop (ICT4D), education and anthropological theory and utilizes it to develop a decolonial industrial framework of digital stewardship based on participation, consent and reciprocity.

This chapter has four interrelated objectives.

First, it explores the conceptual basis and current threats to Indigenous Knowledge Systems in the digital world.

Second, it critically analyses international digital projects and theoretical discussions on the preservation of indigenous knowledge.

Third, it examines new digital tools, including community repositories, mobile apps, AI-enhanced language technologies and geospatial maps, in terms of how they can be used to help IKS revitalize.

Lastly, it presents a community-driven digital model in accordance with principles of Indigenous data governance, in this case, CARE (Collective Benefit, Authority to Control, Responsibility and Ethics) framework.

2. LITERATURE REVIEW

2.1. Native Knowledge Systems in Academic Writing:

IKS has played a prominent role in being studied in anthropology, development studies, education, environmental science and Library and Information Science (LIS). The initial body of scholarship theorized indigenous knowledge as a localized, experiential and orally transmitted array of systems created by the long-term interaction of a given ecological and cultural setting (Agrawal, 1995). Later research disputed dualism between native and scientific knowledge with a position in favor of epistemic pluralism and acknowledgement of indigenous knowledge as an active and valid body of knowledge (Nakata, 2007).

In indigenous studies, IKS can be seen as a comprehensive system that includes language, spirituality, land relations, governance and collective memory (Smith, 2012). Scholars have stressed that IKS are not fixed traditions, but dynamic systems that can adjust themselves to changes in social and environmental situations (Battiste, 2002). Nevertheless, even though evaluated as an academic field, indigenous knowledge remains on the periphery of mainstream research paradigms, especially in digital knowledge structures where graphic approaches are heralded by Western epistemologies.

2.2. International Systems that Uphold the Indigenous Knowledge:

Global political structures are important in justifying indigenous knowledge in global development and sustainability agendas. The UNESCO Local and Indigenous Knowledge Systems (LINKS) program was one of the first institutional attempts to bring indigenous knowledge into environmental governance, education and the science-policy interface (UNESCO, 2002). Research evaluating the LINKS program tends to mention an exemplifying mainstream role of normative contribution to the encouragement of intercultural discussion and participative knowledge generation, while also referring to the deficit of translation of principles into scalable digital activity (UNESCO, 2010).

Intellectual property and legal scholarship have addressed how indigenous knowledge can be safeguarded against theft, especially on the internet. The World Intellectual Property Organization (WIPO) has developed frameworks that cover traditional knowledge and cultural expressions to acknowledge collective ownership and customary law (WIPO, 2017). Nevertheless, scholars suggest that traditional forms of intellectual property protection are still inappropriate for the epistemology of the indigenous population, particularly in the context of digitalizing and sharing knowledge within transnational networks (Bowrey and Anderson, 2009).

The United Nations Declaration on the Rights of Indigenous Peoples (UNDRIP) offers a rights-based framework of indigenous knowledge governance that recognizes the right of aboriginal people to conserve, regulate and advance their cultural heritage and traditional knowledge (United Nations, 2007). Scholarly literature has critically discussed how Free, Prior and Informed Consent (FPIC) has become the key factor in guiding ethical research and digital documentation and how gaps in institutional practice continue to exist.

2.3. Indigenous Knowledge, Environment and Sustainability:

A significant body of literature draws attention to the applicability of indigenous knowledge in preserving the environment in terms of environmental sustainability, biodiversity preservation and climate change adaptation. It has been shown that Indigenous ecological knowledge helps in land management and conservation of species as well as resiliency to climate (Davis and Wagner, 2003). Another advancement of indigenous knowledge came with the Intergovernmental Science-Policy Platform on Biodiversity and Ecosystem Services (IPBES) and its multiple evidence-based approach, which places indigenous and scientific knowledge in a complementary and not hierarchical relationship (IPBES, 2019).

Digital technologies, specifically geographic information systems (GIS) and the concept of participatory mapping, have been explored to help record indigenous environmental knowledge and aid in land rights advocacy to date (Johnson et al., 2017). Although these tools increase visibility and engagement in the policy, scholars warn that environmental data are very sensitive and must be tightly governed by the Indigenous since there is a risk of abuse or dispossession.

2.4. Sovereignty in Indigenous Data and Ethical Data governance:

Indigenous data sovereignty as an important aspect of digital knowledge preservation has become a more recent topic of scholarship. Indigenous data sovereignty is a statement that affirms indigenous peoples' ability to regulate data about their communities, land and knowledge systems (Kukutai and Taylor, 2016). This idea opposes mainstream open-data ideologies that value accessibility rather than cultural accountability.

The CARE Principles are a collection of theories and practices that include ideas of Collective Benefit, Authority to Control, Responsibility and Ethics and are a notable addition to indigenous data governance (Carroll et al., 2020). Research in the LIS and archival sectors has shown that digital repositories informed by CARE concepts can promote trust and cultural safety, as well as sustainable interactions between individuals and communities and that these models are not easily implemented on a grander scale with standard digital systems (Lindstrom and Christen, 2019).

2.5. Electronic Archives and Community Essentials:

Indigenization of knowledge in the form of digital knowledge has been widely researched in archival and information science literature. Early digital archives could be institutionally managed and extractive, recreating colonial power relations by removing knowledge from its context (Christen, 2011). According to critical scholars of archives, these models cause a loss of indigenous sovereign power and epistemic injustices.

Conversely, the digital repositories of communities draw accent to indigenous governance, culturally responsive metadata and access hierarchy (Christen et al., 2017). According to the findings of empirical studies, there are positive effects, such as the enhancement of cultural identity, escalation of knowledge transfer between generations and advancement of educational integration. Nevertheless, sustainability remains a problem because of instability in funding and reliance on technology (Gibson and Martin, 2019).

2.6. Artificial Intelligence and Digital Language Technologies:

Digital Indigenous knowledge has become a major concern in language revitalization. According to academics, the demise of indigenous languages hastens the extinction of knowledge systems in the form of linguistic construction (UNESCO, 2010). Online applications such as language teaching applications, online dictionaries and speech recognition systems, which are enhanced with AI support, have shown the capability of upgrading the speed of documentation and education.

Nevertheless, critical studies have focused on artificial intelligence systems that replicate algorithmic bias and extractive data practices when they are not guided by indigenous protocols (Nichols and McKay, 2018). Ethical technological innovation is therefore found to require the development of AIs in a participatory manner as well as Indigenous-led governance.

2.7. Western-Centric Digitization and Digital Colonialism critiques:

An increasing body of literature on critical content questions the idea of digital colonialism, which can be described as the expansion of colonial influence using digital infrastructure and data mining (Couldry and Mejias, 2019). Scholars believe that digitization by the Western center imposes all universal metadata standards, access norms and epistemic categories, thus marginalizing indigenous worldviews.

Indigenous-controlled digital design, cultural classification systems and relational accountability are methods of decolonization suggested by decolonial scholars to counter digital colonialism (Smith, 2012). This criticism highlights that digital knowledge preservation is political and that it is important to rethink technology as a place of decolonization.

2.8. Synthesis and Research Gap:

The literature review shows that there is wide agreement regarding the need to sustain ethically regulated community-based digital methods of preserving Indigenous Knowledge Systems. Normative legitimacy is offered by international structures, whereas technological innovations offer practical tools for revitalization. However, there are still serious gaps in the mechanisms of integrating digital tools, indigenous data governance, intellectual property protection and community development into a single concept.

The majority of studies deal with single technologies or instruments of policy, where little tie is made to comprehensive and replicable models that prospect indigenous sovereignty throughout the complete digital knowledge cycle. The identified gap explains why the current chapter contributes to the proposed model of Indigenous-based digital revitalization based on participation, ethics and sustainability.

3. THEORETICAL UNDERPINNINGS OF INDIGENOUS SYSTEMS OF KNOWLEDGE

The concept of Indigenous Knowledge Systems is defined and its properties illustrated.

3.1. Defination of IKS:

Indigenous Knowledge Systems refer to locally based, culturally embedded and historically accumulated pools of knowledge that indigenous people have created through protracted engagement with their surroundings. According to scholars, IKS are systems comprising holistic structures that incorporate practical skills, belief systems, values and social institutions to shape the life of the community and take care of the environment. In contrast to the Western system of knowledge, which values objectivity, compartmentalization and written code, IKS utilizes relationality, spirituality, oral transmission and collective property.

The essential features of IKS are:

a) Contextual embeddedness, that is, knowledge, cannot exist outside land, language and cultural practice.
b) It is transmitted intergenerationally, mainly by oral narration, rituals and apprenticeship.
c) Active adjustability such that the communities adapt to the changing environment and the society.
d) plural custody denying intellectual proprietary rights.

These characteristics require culturally competent technological designs that oppose traditional paradigms of digital knowledge management.

3.2. IKS and Sustainable Development:

IKS plays a key role in the promotion of ecological sustainability, climate adaptation, biodiversity conservation and community resilience. Traditional ecological knowledge is used in sustainable agriculture, water management, disaster preparedness and medicine. IKS is becoming increasingly recognized as complementary to scientific information by global policy organizations, such as the Intergovernmental Science-Policy Platform on Biodiversity and Ecosystem Services (IPBES) and numerous other organizations located throughout the world.

Nonetheless, the exclusion of indigenous epistemologies from formal education, research and policy processes constrains their incorporation into sustainable development strategies. When creatively designed ethically, digital platforms would make IKS visible, legitimate and policy integrative and promote community-led development efforts.

4. THREATS TO INDIGENOUS KNOWLEDGE SYSTEMS IN THE DIGITAL AGE

4.1. Critical Rapid Cultural, Linguistic Loss:

The rapid death of indigenous languages is one of the most pressing challenges to the continuity of IKS. Language serves as the main carrier of knowledge about ecology, cultural values and the cosmologies being encoded and transmitted. The disruption of traditional transmission processes is caused by urban migration, formal education using dominant languages and homogenization of the media, which leads to a generational gap.

4.2. Knowledge hierarchies and Epistemic marginalization:

Academic and policy institutions still favor Western scientific paradigms and indigenous knowledge is frequently considered anecdotal or even complementary. This epistemic order has a detrimental effect on the legitimacy of IKS and limits its use in health, education and developmental planning.

4.3. Digital Inequality and Exclusion:

With increasing digital technologies within communities, indigenous people are disproportionally affected by inadequate accessibility to connectivity networks, infrastructure, digital literacy and culturally sensitive technologies. The United Nations Department of Economic and Social Affairs reported that digital exclusion contributes to pre-existing social and economic inequalities, enhancing knowledge loss and cultural marginalization.

4.4. Arguments on Community-Driven Digital Revitalization:

The failure of traditional forms of digitization is sufficient to justify the emergence of participatory community-based models. These models consider indigenous authority more important than data, emphasize Free, Prior and Informed Consent (FPIC) and orient technological design to cultural values. As opposed to indigenous communities being approached as objects of documentation, community-oriented digital routes acknowledge indigenous communities as knowledge cocreators and caretakers.

From Extractive to community-Based digital stewardship

Figure 1. Theoretical figure, digitization to IKS

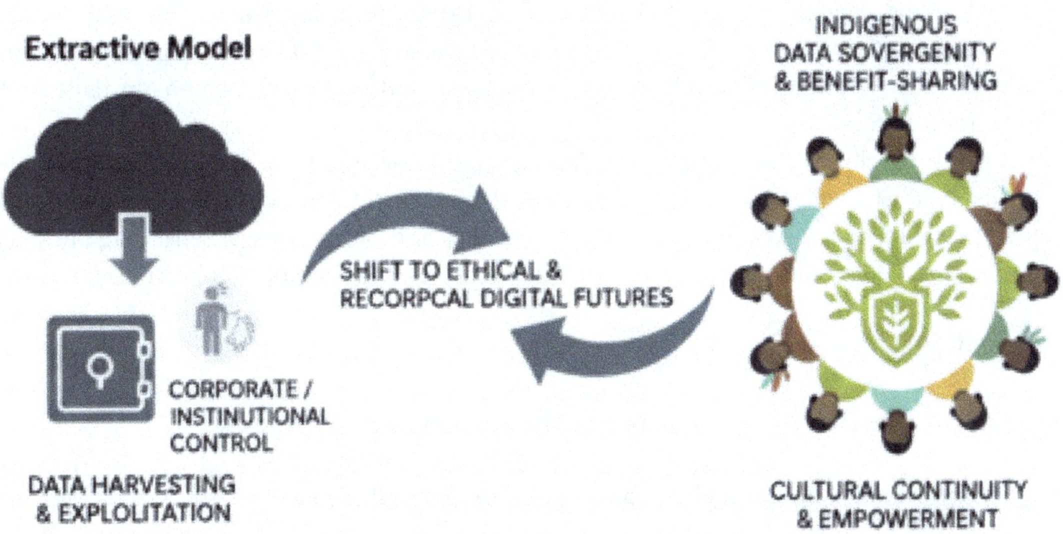

5. DIGITAL TECHNOLOGIES AND TOOLS TO REVITALIZE THE INDIGENOUS KNOWLEDGE SYSTEMS

The development of digital technologies has essentially changed how knowledge is generated, stored, retrieved and shared. Digital tools in the context of Indigenous Knowledge Systems (IKS) provide not only an opportunity, but also a challenge that has never been seen before. In this section, a critical analysis of some of the main digital technologies that are being increasingly used in the field of IKS revitalization is presented, with special focus on community-based design, cultural appeal, ethical governance and sustainability. The discussion does not propose technology as a neutral or deterministic force; instead, digital tools are placed in the context of indigenous socio-cultural realities and assert their place in enabling indigenous self-determination under the right conditions of governance.

Digital tools to be discussed in this part are:

(a) Community-based digital repositories.
(b) Participatory and mobile documentation technology
(c) Language, artificial intelligence (AI) and language technologies
(d) geospatial and environmental knowledge systems, and
(e) Online narratives and digital games.

In combination, these tools form a new digital ecosystem that can help revive indigenous languages, uphold and preserve culture, protect the environment and benefit communities.

5.1. Community-Based Digital Repositories:

5.1.1. Concept and Definition:

A community-owned digital repository is a digital body of knowledge that is self-owned, controlled and managed by indigenous communities either on their own or in fair partnership with libraries, archives, or academic institutions, such as through shared infrastructure. These sites place more emphasis on indigenous classification, culturally responsive access and collective ownership than traditional institutional repositories.

In the context of Library and Information Science (LIS), these repositories are a change in the paradigm of universalist knowledge organization towards pluralistic and culturally localized knowledge management. The metadata schemas, indexing procedures and procedures of retrieving are frequently adjusted to accommodate Indigenous languages, cosmologies and relational epistemologies.

5.1.2. Functions and Applications:

The digital repositories in communities have several overlapping purposes:

a) Oral history, ritual, song, ecological knowledge and material culture.

b) Intergenerational transmission allows the elderly to pass information on to the younger generation in a digitally mediated way.
c) Educational integration, in favor of culturally responsive school and community center learning.
d) Cultural sovereignty which guarantees Indigenous control over entry, use and distribution.

Empirical evidence also shows that repositories with tiered access models; (public, community-only and restricted sacred knowledge) can prove effective, especially when conforming digital preservation to cultural procedures.

Figure 2. Architecture of community-based digital repository

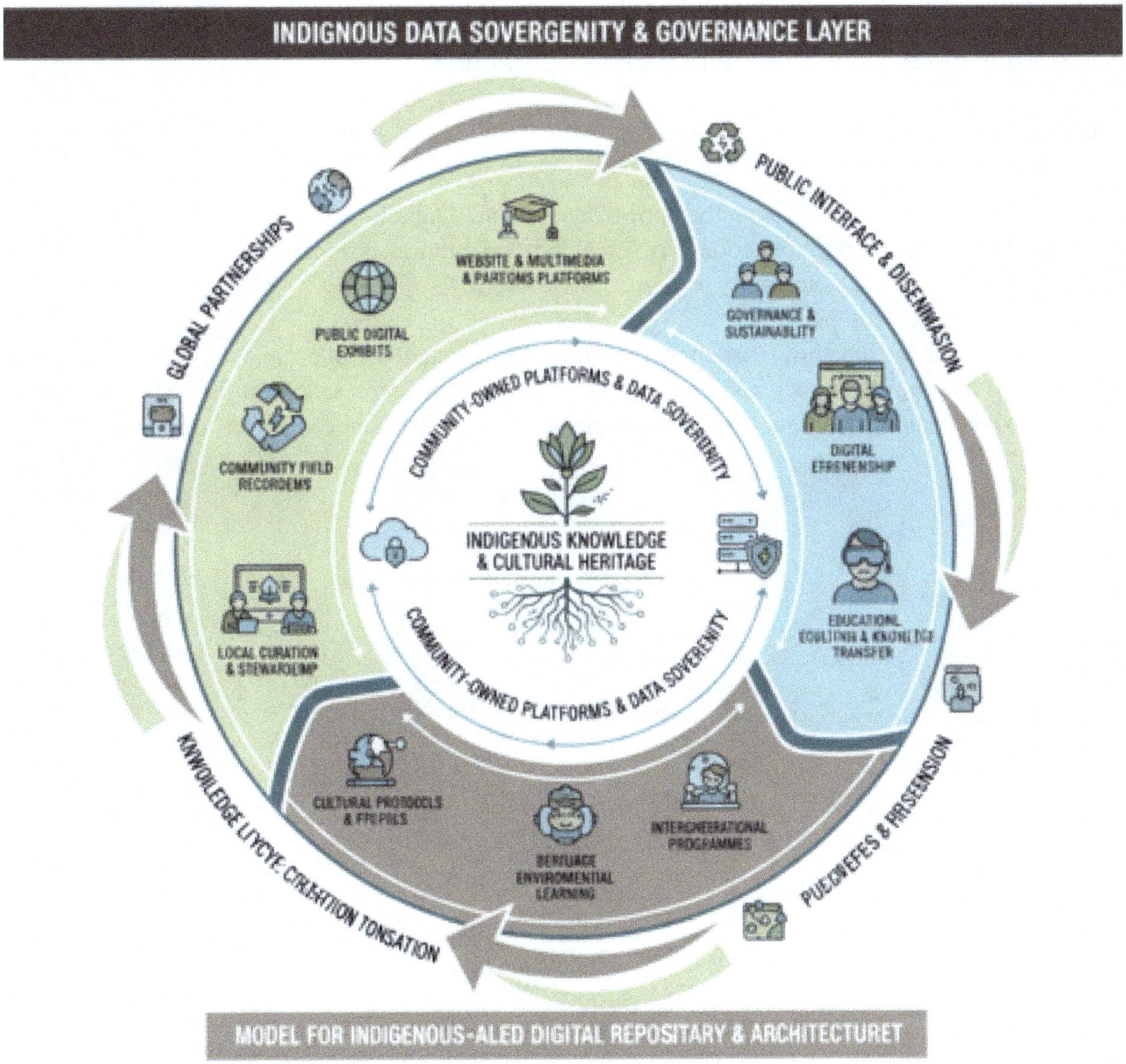

5.2. Mobile Technologies and Documentation of Participation:

5.2.1. Role of Mobile Technologies:

The use of mobile technologies such as smartphones, tablets and portable audio-visual devices has turned out to be a highly potent tool in the participatory documentation of indigenous knowledge. Their low cost, portability and ease of use render them particularly appropriate for remote and rural indigenous settings.

The use of mobile-based documentation systems allows community members to document languages, rituals, farming, medicine and oral stories on the ground by themselves as opposed to having external researchers document them. This methodology is in line with decolonial research paradigms because it puts the power to generate and represent data into the hands of indigenous knowledge holders.

5.2.2. Participatory and Community-Led Approaches:

Participatory documentation structures focus on:

a) Educating the Indigenous youth, the knowledge steward, Digital.
b) Cases of shared content production between elders and cultural custodians,
c) Ineglible consent and community review.

Consistent with the literature, participatory mobile documentation will enhance cultural pride, youth participation and digital literacy and decrease the chances of misrepresentation. Nevertheless, issues such as device maintenance, data storage and long-term funding remain challenges to sustainability.

Figure 3. IKS community recording and ensuring digital security storage

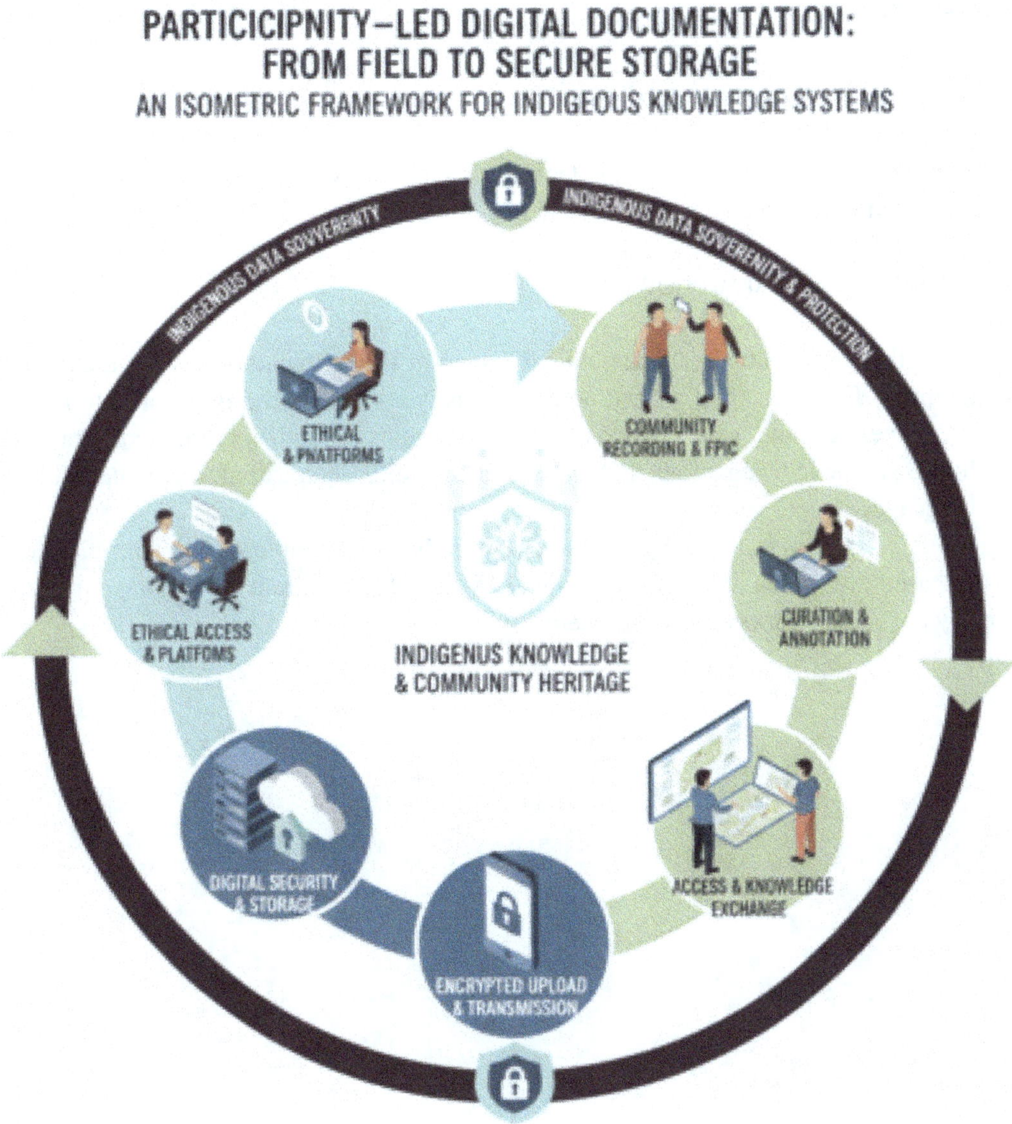

5.3. Artificial Intelligence and Revitalization of Indigenous Language:

5.3.1. Language Endangerment and AI:

In an effort to capture and reproduce endangered indigenous languages, artificial intelligence (AI), natural language processing (NLP) technology and speech recognition technology in particular, have taken the spotlight. Considering that language loss is one of the main contributors to IKS erosion, AI-supported tools have the potential to provide a quicker way of preservation and training.

AI applications include:

a) Automatization of oral narratives transcription,
b) Low-resource speech -to-text systems,
c) Digital Dictionaries and digital language learning applications,
d) Chatbots and language practice conversational agents.

5.3.2. Ethical and Cultural Archetypes:

Another item that brings about serious ethical issues in the indigenous context is AI technology, despite its potential. The training of algorithmic systems is not performed on information that was attained without proper consent or the chance of benefit-sharing, which is a digital exploitation of some kind. In addition, AI models cannot accurately reflect the cultural, spiritual and contextual aspects of indigenous languages.

The academic consensus is that AI governance by indigenous people, use of participation models and culturally based evaluation measures are critical. AI is not a substitute for language transmission in the community but rather works as a supplement.

Figure 4. AI language revitalization model

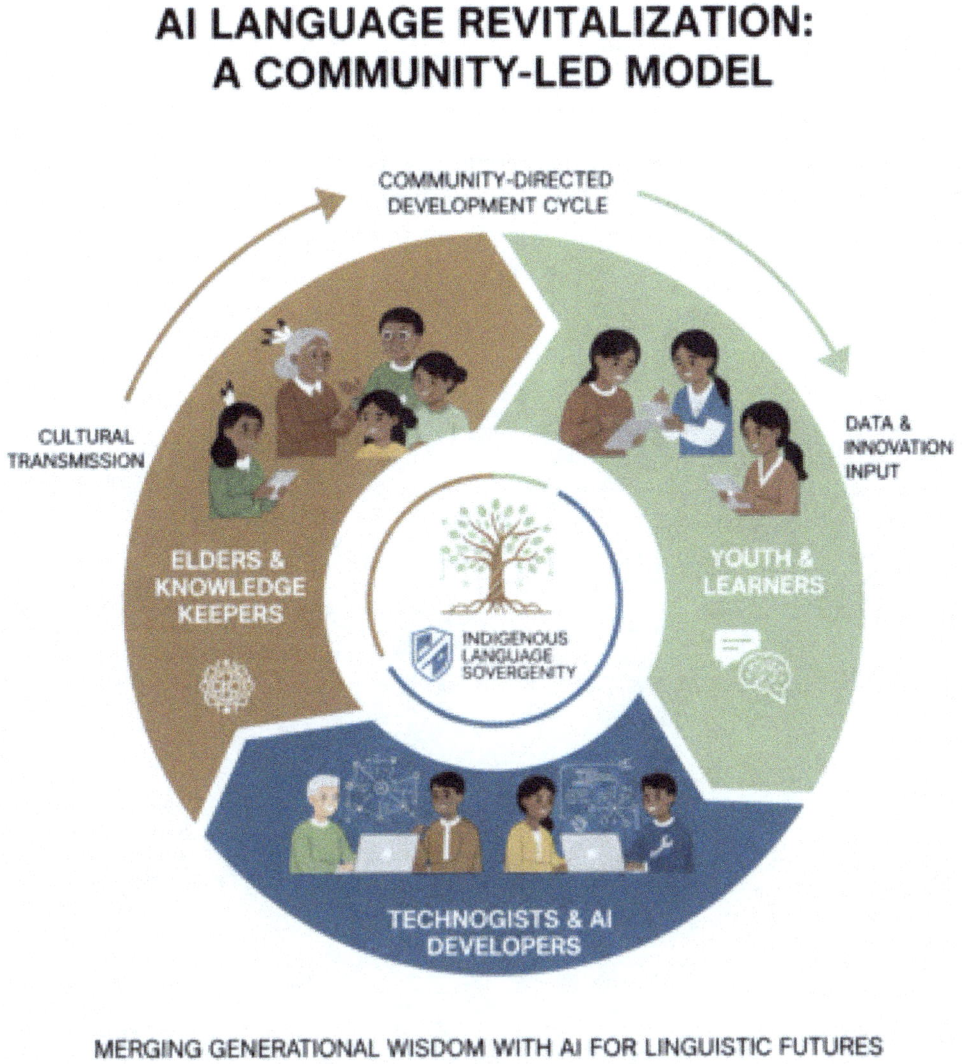

5.3.3. Indigenous Environmental Knowledge Systems:

Indigenous environmental knowledge involves the land management techniques used by the indigenous people, cyclical calendars, biodiversity and techniques of adapting to climate that have been devised for centuries. Such knowledge has been is being more and more applied to digital technologies, especially in the environmental conservation and climate resilience domain.

5.3.4. Environmental Knowledge Management and GIS Mapping:

Participatory mapping tools and Geographic Information Systems (GIS) help indigenous people record:

a) Boundaries in land, traditional,
b) Places of worship and Cultural landscapes,
c) Ecological indicators and hotspots of biodiversity,
d) Effects of climate change and adaptation.

As a source of power, digital mapping empowers land rights agendas, environmental governance and policy advocacy when directed towards the community. However, scholars have observed that spatial information is especially prone to abuse and that strict rules should govern it, restrict its usage and provide ethical protection around it.

Figure 5. GIS Model of reporting Indigenous environmental knowledge

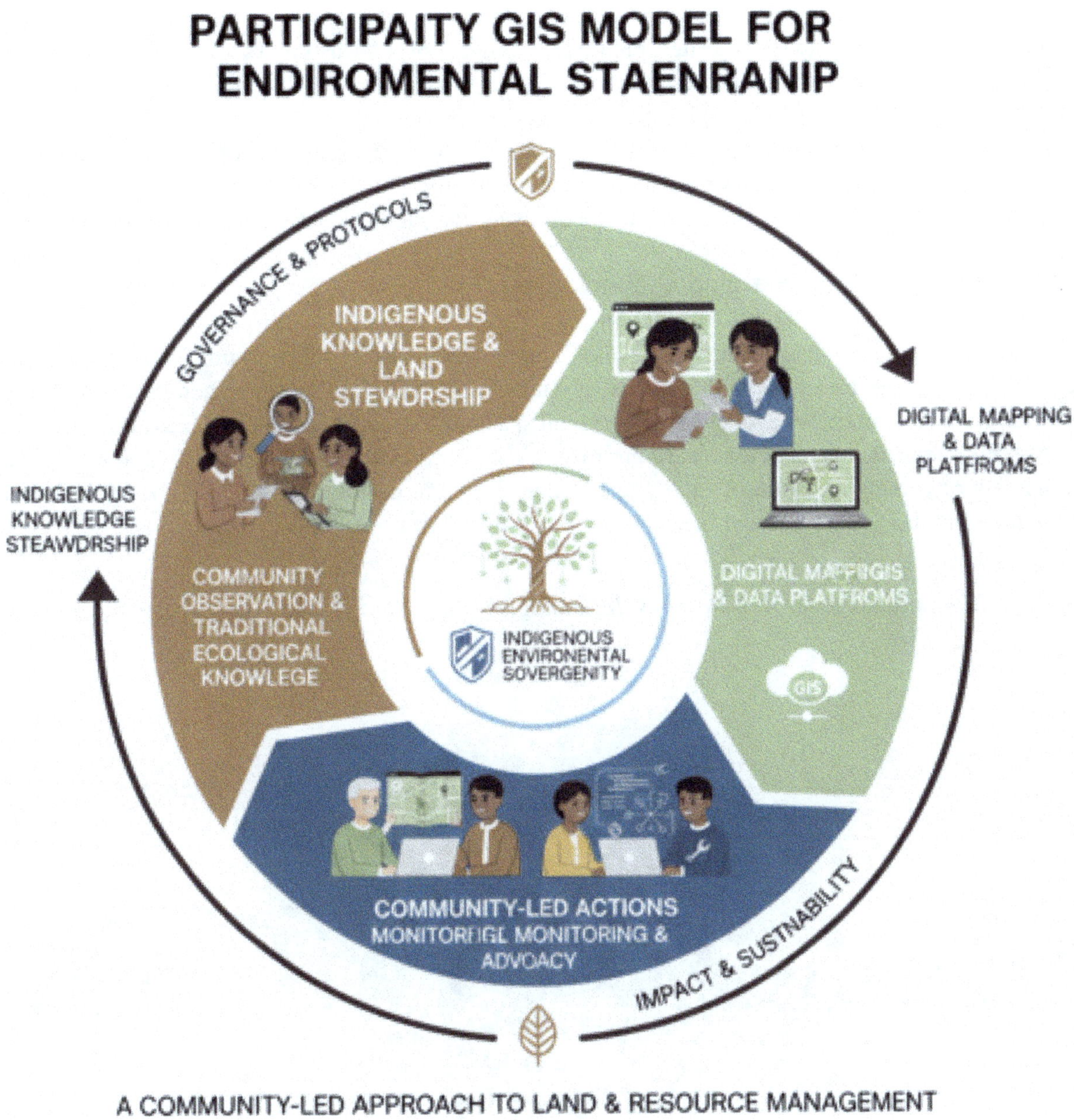

5.4. Web-Based Storytelling and Digital cultural expression:

In digital media, people are in a better position to communicate important stories and other meaningful cultural expressions through web-based stories and digital storytelling.

5.4.1. Digital Storytelling Knowledge transmission:

Providing active opportunities in the narration of indigenous histories, identities and worldviews, digital storytelling sites, such as websites, multimedia collections, podcasts and social media, tell history quickly and clearly. Storytelling is the main part of indigenous epistemology that is used as a tool for teaching moral values, ecological connections and collective memory.

Prison stories digitally allow these stories to go farther than the limitations of geography, promote cross-cultural communication and even activism. Importantly, indigenous people are able to counter stereotypical or colonial images in mainstream media using community-controlled storytelling platforms.

5.4.2. Immersive and Emerging Technologies:

New technologies such as virtual reality (VR), augmented reality (AR) and immersive exhibitions have increasingly been studied in the context of cultural education and heritage conservation. These applications enable users to engage with Aboriginal space, rituals and history in a 3D experience.

Although immersive technologies are promising, they need cautious cultural mediation to prevent trivializing or commodification of sacred knowledge. They are best applied through the influence of indigenous cultural authorities and in accordance with the educational goals.

Figure 6. Multimedia-based, Web-Based & Technologies Digital story

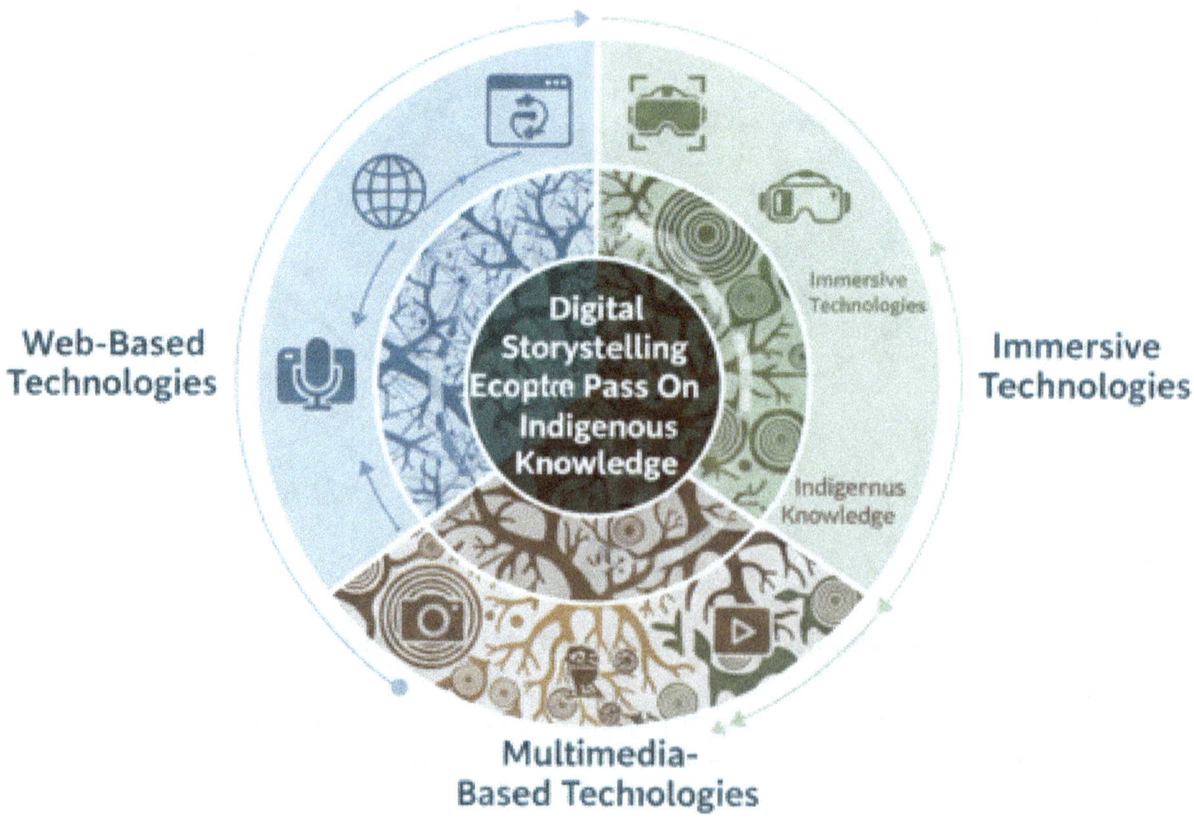

5.5. Digital Technologies and IKS Opportunities and Constraints:

5.5.1. Opportunities:

According to the literature, the following opportunities by digital tools can be outlined:

a) Improved intergenerational knowledge transfer,
b) Greater knowledge and appreciation of Indigenous epistemologies,
c) Empowered cultural sovereignty and self expression,
d) IKS into education, health and environmental policy.

5.5.2. Constraints and Risks:

On the other hand, the long term limitations are:

a) Digital disparity and infrastructural constraints,
b) Reliance on international funding/experience,
c) Threats of data abuse, cultural hijacking and cyber colonialism,
d) Cultures of openness and culture restraint.

These issues highlight the importance of integrating digital solutions into sound ethics, laws and management systems.

5.6. Synthesis and Transition:

As illustrated in this section, digital tools can be transformed towards the revitalization of Indigenous Knowledge Systems when guided by community and ethical considerations and locally based. Nevertheless, technology alone is insufficient. It relies on its alignment with indigenous values and rights and with indigenous structures of governance.

6. ETHICAL, LEGAL, CULTURAL ASPECTS OF DIGITAL KNOWLEDGE REVITALIZATION

Currently, there is no ability to digitally revitalize Indigenous Knowledge Systems (IKS) without ethical, legal and cultural issues. As digital technologies offer potent tools for preservation, dissemination and innovation, they are also accelerating the old issues of misappropriation, epistemic injustice and cultural damage. This section critically analyses normative underpinnings that must be in place so that digital IKS initiatives affirm indigenous rights, values and sovereignty. This suggests that the principal assertion is that ethical digital stewardship should supersede procedural compliance, but rather be based on indigenous worldviews, traditional laws and community governance frameworks.

This discussion will take four dimensions that influence each other:

a) Free, Prior and Informed Consent (FPIC),

b) Indigenous data sovereignty and governance,
c) intellectual property and cultural rights and
d) dangers of digital colonialism and knowledge exaction.

Combined, these dimensions offer a normative architecture of community-focused and decolonial digital practices.

6.1. Informed Consent Free, Prior and Informed Consent:

6.1.1. Theoretical Based Approaches of FPIC:

Free, Prior and Informed Consent (FPIC) is one of the pillars of indigenous rights in the field of activities associated with knowledge. The latter is enshrined in the UNDRIP, which supports the entitlement of indigenous peoples to engage in making decisions regarding their cultural heritage, traditional knowledge and resources. According to the FPIC, consent must be given freely and without any coercion before any intervention, based on the availability of clear and available information and within indigenous decision-making systems.

In digital spaces, FPIC is applied not only to first-instance consent, but is also associated with the entire data lifecycle, including data collection and digitization, storage, access, reuse and deletions. Scholars stress that frameworks of one-time consent are inadequate because digital technologies allow knowledge to be reproduced and resorted to repeatedly.

6.1.2. FPIC Operationalization of Digital Projects:

According to the literature, the best practices for implementing FPIC in digital IKS initiatives include:

a) Mechanisms of collective consent are based on indigenous cultures of governance as opposed to personalized treaties.
b) Continuous consent enables communities to amend allowances when technologies, circumstances, or priorities evolve.
c) Clear information about data storage location, rights of access and how the data can be used by a third party.
d) Culturally appropriate documentation, such as oral consent procedures, in which written procedures are culturally incongruent.

Normative transparency has not been matched by empirical research, as there are always gaps between FPIC principles and practice. Perhaps the most common atmosphere of external researchers and institutions is devoid of cultural competence or institutional motivation to execute FPIC's strict meaning, especially when it is digitally mediated.

Figure 7. IKS Lifecycle-based FPIC framework project

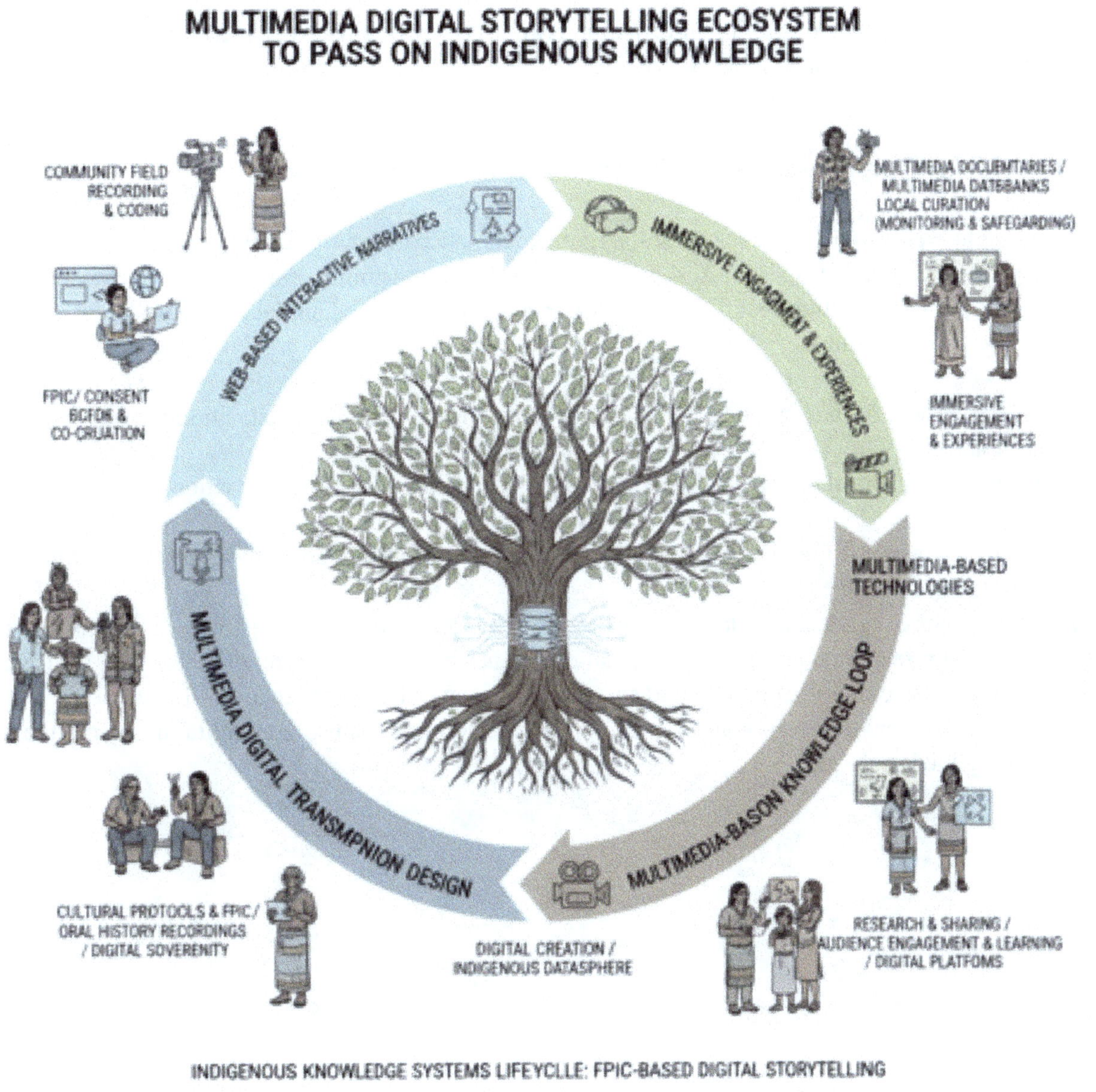

6.2. Sovereignty and Governance of Indigenous Data:

6.2.1. Indigenous Data Sovereignty, Scope and Definitions:

Indigenous data sovereignty is described as the natural rights of indigenous peoples to control the process of collection, possession, access and use of data associated with communities, lands and knowledge relationships. The Indigenous data sovereignty model has less focus on openness and interoperability when compared with conventional data governance models, as it is more focused on authority, accountability and collective benefit.

The concept opposes prevailing information regimes by stating that data are not just informational but relational; it has cultural, spiritual and political meaning. Data sovereignty serves as a self-defence and liberating tool in the digital preservation of IKS.

6.2.2. CARE Data Governance Principles of Indigenous Data:

One significant contribution to indigenous data governance in the scholarly community is the CARE framework, or Collective Benefit, Authority to Control, Responsibility and Ethics. The reason behind the CARE principles is the weakness of the FAIR data principles, which lack fairness in covering indigenous rights and values.

a) Collective benefits underscore the importance of having the benefits of data initiatives to indigenous communities.
b) Authority to Control upholds indigenous authority in data access, data utilization and the release of data.
c) A sense of responsibility implies the obligation of data stewards to abide by cultural procedures and community interests.
d) Ethics require that data practices be consistent with indigenous values, histories and aspirations.
e) Empirical research on library and information science has proven that repositories with CARE can promote trust, cultural secondary and sustainability. Nevertheless, there are still operational issues, specifically, how to incorporate the principles of CARE into mainstream digital infrastructure and funding models.

Figure 8. Model of CARE principles.

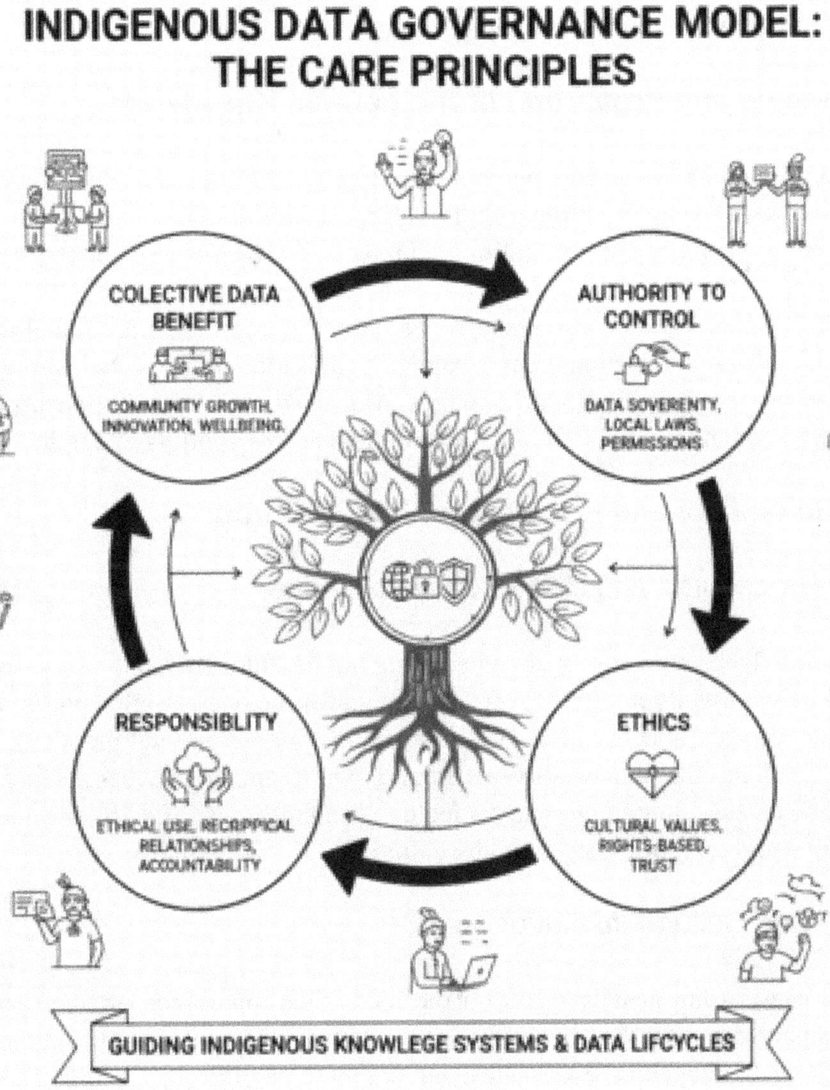

7. POWS OF THE TRADITIONAL INTELLECTUAL PROPERTY REGIMES

Traditional intellectual property (IP) regimes are based on personal authorship, limited terms and commercial exploitation. These assumptions are a complete contradiction of indigenous knowledge systems, which are the property of the community, trans-generational, and, in many cases, spiritually integrated. Consequently, current IP regimes have not been effective in ensuring that indigenous knowledge is not misused, especially on online platforms.

Critical commentaries have pointed out that digitization enhances weakness because online distribution increases duplication and commercialization without permission. This is a highly urgent threat to medicinal knowledge, artistic forms and sacred stories.

7.1. WIPO Systems and Protection of Indigenous Knowledge:

The World Intellectual Property Organization has developed normative frameworks that consider Traditional Knowledge (TK) and Traditional Cultural Expressions (TCEs). These systems acknowledge collective ownership, customary law and unlimited protection, which is a major change to Western IP norms.

As much as the work of WIPO can be recognized, practical constraints are found in the literature. Imposing it in a digital and transnational environment is difficult and many indigenous people do not have access to legal services or institutional resources. As such, theorists are proponents of complementary methods such as community policies, data licenses and access controls integrated into the culture.

7.2. Cultural Sensitivity and Knowledge Classification:

7.2.1. Cultural Procedures and Dangerous knowledge:

Indigenous knowledge is not entirely aimed at being public. Information that is sacred, ceremonial and culturally sensitive is usually confined to certain community members. Digital preservation programs should thus include sensitive access restrictions based on indigenous cultural practices.

The literature highlights that universalist metadata standards and open-access precepts are often inogostic to indigenous values. Better alternatives include computers classified by the community, multi lingual classification and categories described by culture.

7.2.2. Libraries and Cultural Institution Role:

Ethical digital stewardship heavily relies on the mediative importance of libraries, archives and museums. According to scholars, these institutions need to move beyond custodial authority to a facilitative partnership to help indigenous governance and not force them to change by institutions. Cultural competence, indigenous librarianism and ethical digitization training are essential.

7.3. Digital Colonialism and Knowledge Exploitation should have risks:

7.3.1. Digital Colonialism; Conceptual Overview:

Digital colonialism describes the reenactment of colonial power relations by digital infrastructure, platforms and knowledge systems. Within the scope of IKS takes the form of extractive data practice, platform monopolies and epistemic domination, which exist in technical standards.

Researchers warn that digital projects, even those that are well-intentioned, can be perpetual in colonial practice when indigenous people are not given the authority to make decisions or economic rewards.

7.3.2. Mitigation Strategies:

The literature suggests that digital colonialism can be countered through the following measures.

a) Native positive ownership of digital servers and digital platforms,
b) A set of data licenses and access conditions set at the community level,
c) Training of capacity-building and digital skills,
d) Long term institutional collaborations which are founded on equity and reciprocity.
e) The following strategies support the idea that ethical digital renewal is a political process that fundamentally alters structure.

7.4. Synthesis and Transition:

The present section highlights that ethical, lawfulness and cultural factors are not peripheral but of utmost centrality in the revitalization endeavors of transformative digital Indigenous Knowledge Systems. The regulative or normative basis of decolonial digital practice comprises FPIC, data sovereignty, intellectual property protection and cultural sensitivity.

8. COMMUNITY-BASED FRAMEWORK OF DIGITAL NATIVE REVITALIZATION OF INDIGENOUS KNOWLEDGE SYSTEMS

Based on the premises presented in the previous sections in terms of conceptual groundwork, technological analyses, ethical frameworks and empirical case studies, this section proposes a community-based digital model that can revive Indigenous Knowledge Systems (IKS). The model directly addresses the constraints uncovered by extractive digitization methods and Western-focused knowledge infrastructures by shifting the key roles of indigenous communities to central decision makers, planners and beneficiaries of digital projects. It incorporates concepts of indigenous data governance, Library and Information science (LIS), participatory design and sustainable development to provide a scalable and flexible model that can be used in various cultural and geopolitical settings.

The proposed model is neither prescriptive nor extensive, but is a principle-based and modular model in which communities can flex their components depending on local epistemologies, governance systems and technological capabilities. Indigenous sovereignty, ethical stewardship and value creation through reciprocity are core elements of the model.

8.1. The Model is fundamentally based on the idea of the trait model.

The proposed model is guided by the background of five interrelated conceptual commitments; Indigenous Sovereignty and Self-determination.

Native Indigenous people have the entire power over knowledge identification, digitization priorities, access terms and benefit-sharing systems.

Engaged Knowledge Production:

a) Knowledge documentation and curation are community processes that involve old cultural custodians, youth and community institutions.
b) Digital Stewardship as a moral imperative.
c) Free, Prior and Informed Consent (FPIC), indigenous data sovereignty and ethics embedded in the culture guide digital infrastructures.
d) Sustainability and Capacity Building.
e) Sustainability is also long-term and is maintained by capacity building of skills, local governance systems and diversification of funding strategies.
f) Giving Something Back, Return Favour.
g) Digital projects bring about real-life social, cultural, learning and economic value for indigenous communities.
h) What these promises all suggest is the challenge of dominant paradigms according to which indigenous knowledge is data that can be mined, stored, or even commoditized without responsibility.

8.2. The Community-Driven Digital IKS Model Structural Components:

It consists of six interacting parts that contain each of the essential dimensions of digital IKS renewal.

8.2.1. Aboriginal Governance and Knowledge Authority:

The model centers on indigenous governance, which is implemented in the form of community councils, cultural committees, or traditional leadership systems.

These bodies:

a) No, what knowledge can be written and recorded,
b) Assign access privileges (public, community-only, restricted),
c) Accept joint ventures with other organizations,
d) Monitor adherence to cultural procedures and FPIC.

Governance mechanisms are culturally based and can be based on customary law and not on formal legal tools. This aspect, in terms of LIS, redefines the structures of authority in the management of knowledge because in this model, community authority has been prioritized as opposed to institutional custodianship.

Figure 9. Community-based digital Indigenous Knowledge System

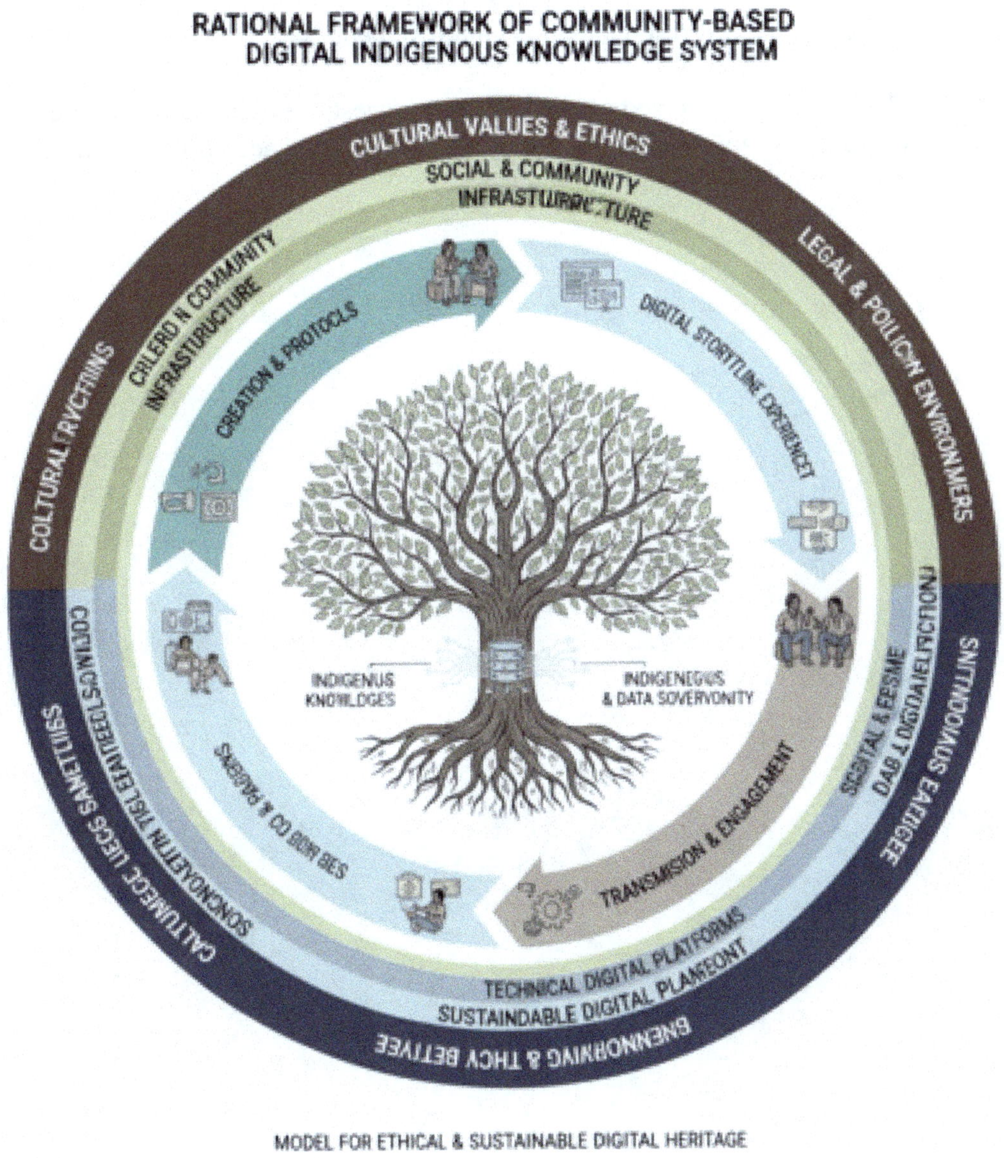

In the course of participatory documentation and content creation, this is incorporated to allow users to engage in the generation of content such as occurrences and related digital platforms.

8.2.2. Digital Infrastructure owned by the community:

Digital infrastructure in the model focuses on community ownership or control and they include:

a) Cloud repositories that have been located locally or run by Aboriginal entities,
b) Multilingual interfaces in Indigenous languages,
c) Metadata systems that are based on Indigenous knowledge organisation systems,
d) Discriminated access limits corresponding to cultural guidelines.

Instead of implementing standardized platforms without questions, the model proposes the idea of technological sovereignty, wherein such tools are chosen or modified according to cultural suitability and sustainability in the long term.

Figure 10. IKS Community-owned digital infrastructure architecture

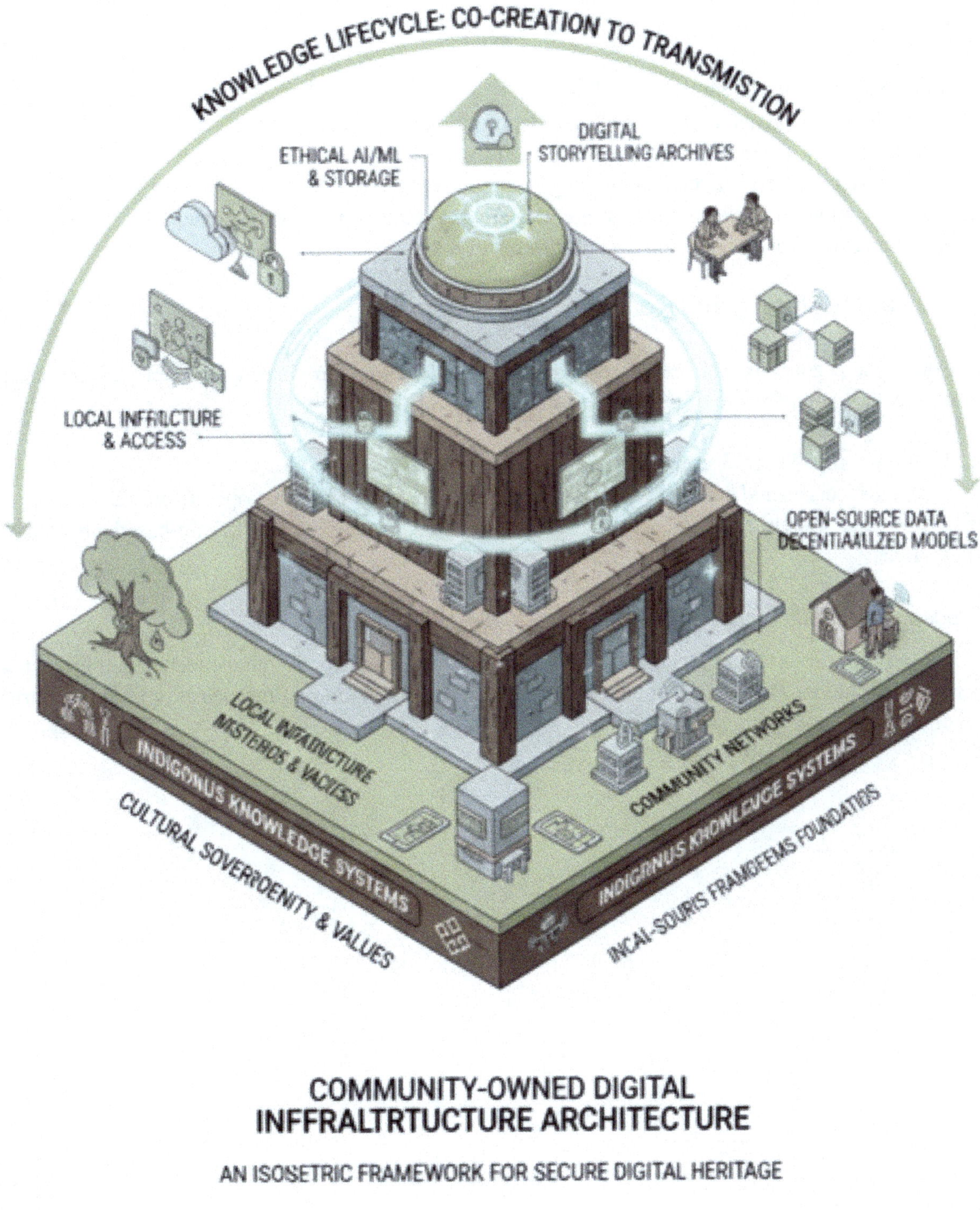

8.2.3. Capacity Building and Development of Digital Skills:

Digital revitalization requires sustained capacity building. This component includes the following components:

a) Information management training, digital literacy,
b) The LIS education and archival skills as part of indigenous focus,
c) Technical instruction in AI, GIS and using multimedia stories where suitable,
d) Ethical Mentorships between community members and technologists and scholars.
e) Capacity building ensures that communities are not constantly reliant on external knowledge and can manage, adapt and develop digital systems over time.

8.2.4. Benefit-Sharing and Community Development:

One of the peculiarities of the proposed model is its focus on fair sharing of benefits. Community development is connected to digital IKS efforts in terms of:

a) Cultural tourism digital storytelling,
b) Indigenous crafts and creative works are available on marketplaces that operate online,
c) Provenance systems Provenance: digital heritage and narratives. Blockchain-based provenance systems.

Educational facilities incorporated in the local curricula.

These processes are consistent with indigenous dreams of economic independence and protection against exploitation. The sharing of benefits is also regulated by community rules as opposed to market imperatives.

Figure 11. Digital benefit-sharing, IKS & community development.

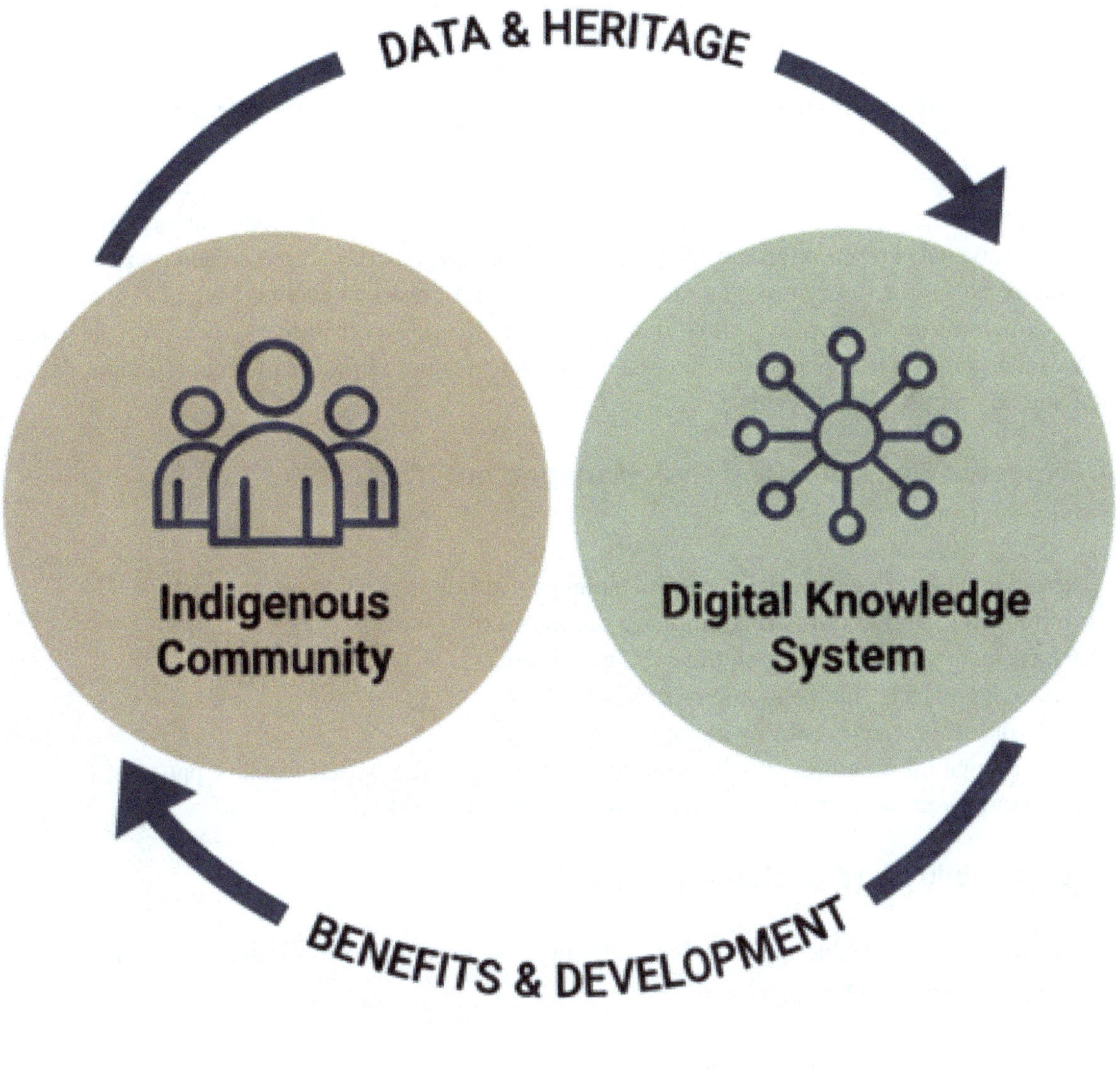

8.2.5. The Ethical Partnership and the Institutional Collaboration:

Although Indigenous communities are kept at the core, the model acknowledges the importance of strategic relationships with universities, libraries, archives and non-governmental organizations based on ethics.

The characteristics of such partnerships are:

a) Indigenous governance and FPIC based memorandum of understanding,
b) Effective definition of roles and decision making authority,
c) Determination instead of project involvement,

d) Indigenous intellectual and cultural rights.

Institutions operate as agents and friends, providing technical knowledge, infrastructure and advocacy without imposing ownership or taking control.

8.3. The combination of LIS Principles and Indigenous Protocols:

The given model will also make an essential contribution to the Library and Information Science field because it proves the idea that essential LIS functions, such as collection development, organizing knowledge, caring about its preservation and accessibility, can be reorganized using indigenous epistemologies. The principles of the LIS tradition, including neutrality and universal access, have also been revisited in an essential way and replaced by situational ethics, relational responsibility and cultural particularity.

This incorporation brings libraries not just as depositories but as places of reconciliation, decolonization and community empowerment.

8.4. Expected Obstacles and Risk Management ICT:

a) The model also recognizes the possible challenges, although it has its strengths:
b) Limitations in digital infrastructure of remote locations,
c) Inconsistency and reliance upon time-limited grants,
d) Knowledge sharing intergenerational conflicts,
e) The trade-off between visibility and cultural restriction.
f) Some mitigation measures involve gradual applications, diversified funding patterns, continuous conversations in the community and adaptive governance systems.

8.5. Applicability and Scalability of the Model:

It is purposefully scalable and flexible. Although its principles are universal, they have different implementations based on

a) Society size and administrative customs,
b) Linguistic diversity and cultural diversity,
c) Digital preparation and technological access,
d) Countries national legal and policy surroundings.

Flexibility is needed to guarantee that it is relevant to indigenous contexts, both in the Global North and the Global South.

8.6. Synthesis and Transition:

This section defines an articulate, indigenous, informed digital model that integrates ethical ideals, technology devices and community-building goals. It reacts to the gaps noted in the literature by providing a practical and theoretically justified framework for the revitalization of digital IKS.

9. DISCUSSION- OPPORTUNITIES, CHALLENGES

This discourse summarizes the theoretical discussion, technology, ethical reflections and empirical studies expressed in earlier sections in a critical analysis of the prospects, issues and future of community-based digital re-italization of Indigenous Knowledge Systems (IKS). Going beyond descriptions, this part of the paper provides an interpretation of how digital strategies have the capacity to transform intergenerational knowledge sharing, reinstruct relations between indigenous people and institutions and advise policy in the education, health, environmental governance and cultural heritage domains. It also interrogates new tensions and unintended consequences, providing solutions on the way forward based on the sovereignty of indigenous peoples and decolonial praxis.

Opportunities by Community-Driven Digital Strategies Opportunities:

9.1. Intergenerational Knowledge Transmission:

The renewal of intergenerational knowledge-sharing is one of the greatest opportunities provided by community-based digital initiatives. Online archives, cellular references and narratives can be used to help elders pass on language, environmental expertise and cultural tales to younger generations in forms that can be understood by modern digital culture. Multimedia integration (audio, video and interactive interfaces) will support oral and experiential learning styles, although the needs of youth engagement will be met.

Empirical data on community-based repositories and language platforms have shown that youth participation increases when digital programs are integrated into culturally significant events, rituals, land learning and community gatherings. This implies that digital devices are considered the best complements to lived practice; they cannot replace embodied knowledge transmission.

9.2. Visibility, Recognition and Epistemic Justice:

Digital platforms controlled by the community can improve the visibility and legitimacy of indigenous epistemology within both the national and global knowledge ecosystems. As these sites allow the establishment of indigenous self-representation, they challenge deficit discourses and epistemic marginalization of academic, policy and media discourses.

In policy contexts, digitally documented indigenous environmental knowledge has been used as a source of information in climate adaptation initiatives to support biodiversity conservation and land use planning. The integration of this kind also leads to epistemic justice, as it acknowledges the engagement of indigenous knowledge to be authoritative and practical, but not anecdotal. The discussion, however, points out that invisibility must be mediated strategically so that when more people know about communities, these communities are not vulnerable to exploitation or cultural injury.

9.3. Education, Health and Climate Policy:

Digital IKS projects reveal a significant potential in sectoral integration:

a) **Education**: Digital curricular tools help the community with culturally responsive curricula, bilingual education and place-based learning.
b) **Health**: Digital traditional medicinal knowledge has the potential to enhance primary healthcare intervention programs under ethical protection.
c) **Policy on Climate and Environment:** Participatory GIS and ecological reporting enhance indigenous contributions to climate resilience and conservation planning.

These integrations prove the topicality of IKS to world issues and problems, supporting the thesis that indigenous knowledge is an obligatory part of sustainable development.

9.4. Continued Problems and Organizational limitations:

9.4.1. Digital Divide and structural Inequality:

Digital inequality is a widespread problem despite advances in technology. Various Indigenous areas are limited in digital initiatives owing to limited connectivity, untrustworthy power outlets, expensive hardware and maintenance. Such infrastructural underpinnings threaten to recreate exclusion even among community-based models.

Low-bandwidth platforms, offline-first repository designs and community-based technology hubs are among the mitigation strategies presented in the literature. Nevertheless, structural investment and policy commitment cannot be solved in the long term without project-based intervention.

9.4.2. Sustainability and Funding Volatility:

Several digital IKS projects are based on temporary grants and external funding, which creates prospects for sustainability issues. At the termination of funding cycles, repositories may become obsolete, lose data, or be captured by an institution. The discussion refers to the significance of diversified financing models such as community business, partnerships with the public sector and endowment-backed strategies.

Sustainability is also increased when digital initiatives are integrated into a community governance system and connected to real social and economic impacts.

9.4.3. Striking a Balance between Openness and Cultural Restriction:

One of the most recurrent dilemmas in digital revitalizing activity is the question of how much knowledge can be shared and how much is restricted regarding culture. Although it is spreading and advocated through digital means, not every piece of indigenous knowledge is meant to be shared with a broader community. Knowledge that is sacred, ceremonial, or gender-specific must be carefully safeguarded.

This discussion identifies tiered access systems, embedded metadata in the form of cultures and dynamic consent mechanisms as productive responses. Notably, openness decisions should be community-based and contextual and should not be subjected to outside demands of providing openness everywhere.

9.5. New Technology and Future Possibilities:

9.5.1. Ethical Innovation and Artificial Intelligence:

Artificial intelligence (AI) poses both opportunities and threats to Indigenous Knowledge Systems. As mentioned above, AI-based language documentation and environmental surveillance can accelerate the process of revitalization. Nevertheless, the discussion emphasizes that the subsequent development of AI should follow the example of Indigenous-led governance, model training and benefit-sharing.

Native AI practices are contestant to existing dominant accounts of technological neutrality, since indigenous approaches to AI anticipate relational ethics, accountability and cultural context. These repositionings can be generalized to responsible AI discourse at the world level.

9.5.2. Virtual Reality and Augmented Reality and Immersive Knowledge Environments:

Virtual reality (VR) and augmented reality (AR) are being increasingly investigated as immersive technologies that can be applied in the fields of cultural education and heritage interpretation. These technologies provide new experiences of landscapes, rituals and histories, especially for groups of diaspora and academic establishments.

The discussion urges that immersive representations are in danger of commodification or misunderstanding, unless they are separated by indigenous cultural authority. Their usefulness in the future is based on co-creation, cultural mediation and clearly set learning goals.

Figure 12. Digital ecosystem Future The incorporation of AI

9.6. Policy and Institution Implications:

Libraries, Archives and Universities; Libraries, archives and universities provide children with the opportunity to use books, manuscripts and maps.

9.6.1. *Libraries, archives, university libraries, archives and universities offer children the opportunity to discuss books, manuscripts and map collections.*

In the case of libraries, archives and universities, the findings require an overhaul in which custodianship is substituted with ethical facilitation.

Institutions are requested to:

a) Promote Indigenous government of digital collections,
b) Modify metadata and access standards to cultural protocols,
c) Make investments in capacity building and long term alliances.
d) This paradigm situates itself in the change of professional ethics in LIS and other related disciplines, making institutions partners in the decolonialisation of knowledge.

9.6.2. Policy and Legal Frameworks:

At the policy level, this discourse highlights the necessity of matching digital heritage activities with indigenous rights tools. The principles of Free, Prior and Informed Consent, data sovereignty and benefit-sharing in the digital realms should be operationalized at the national level. In the absence of such alignment, technological innovation leads to ethical governance.

9.7. Surprising Effects and Contemplative Lessons:

Some surprising conclusions can be drawn during the synthesis:

The more youth are engaged, the more focus there is on cultural pride in digital projects as opposed to technical innovation.

Community-scale platforms can be more sustainable than institutional systems, which are typically large.

a) Ethics in governance contribute to and do not limit technological innovation, as they create trust and involvement.
b) These results refute the expectation that scale, openness, or technological complexity are the main signs of success.

9.8. Synthesis and Transition:

This discussion concludes that community-based digital openings can have transformational value when it comes to reinvigorating Indigenous Knowledge Systems, but infantilizes technological determinism. Digital opportunities are the most optimized with digital strategies encompassed in indigenous governance, ethical frameworks and sustainable development objectives.

10. CONCLUSION

This chapter has contributed to an elaborate, inter-disciplinary and ethical review of digital paths to rejuvenate Indigenous Knowledge Systems (IKS) in the 21 st century. Using global frameworks, critical scholarship, technological analysis and global case studies, this chapter has proved that digital innovation under indigenous sovereignty and participatory ethics has the potential to become a potent source of cultural continuity, epistemic justice and sustainable development.

It has been verified that Indigenous Knowledge Systems are not past entities but fluid, adaptable and contextualized epistemologies that still guide ecological and social management, communal strength and community disposition. However, this chapter also states that IKS remains more than proportionately subject to erosion because of linguistic erosion, institutional marginalization, digital disparity and extractive knowledge practices. Here, uncritical digitization leads to the re-creation of colonial power asymmetries but on technologically sophisticated lines.

Community-driven digital models, in contrast, expressed in this chapter, reimagine digital preservation as a process that is relational, rights-based and culturally situated. The suggested framework focuses on indigenous governments, participatory documentation, community-owned digital infrastructure, capacity building, fair distribution of benefits and establishment of ethical institutional relationships. This comprehensive framework is a direct reaction to the aforementioned literature gaps, as it harmonizes technological instruments with indigenous data sovereignty, intellectual property protection and ethics based on culture as a core element.

As discussed in the empirical case studies, the revitalization of indigenous languages, digital heritage archives, participatory GIS mapping, online storytelling and ethical digital activities can, in fact, be transformative when indigenous communities have authority over the knowledge systems. Through these endeavors, it has been found that sustainability, cultural integrity and innovation are not opposing but complementary.

Notably, the chapter has led to general theoretical discussions by furthering a decolonial digital paradigm that refutes Western-based beliefs regarding knowledge, openness and technological neutrality. It places indigenous worldviews of digital stewardship in context, extending the research epistemology of Library and Information Science, ICT4D and knowledge governance.

The future of Indigenous Knowledge Systems in the digital age does not lie in technological development, but in moral congruence, indigenous power and responsible relationships. Communal digital pathways provide a plausible and transformative way to maintain, renew and reinvent indigenous knowledge to current and upcoming generations. By prioritizing indigenous self-determination and streamlining the ruling epistemic bases, such avenues would make a valuable contribution to the universal endeavors of cultural sustainability, social justice and diverse knowledge futures.

REFERENCES

Agrawal, A. (1995). Dismantling the divide between Indigenous and scientific knowledge. *Development and Change*, *26*(3), 413–439. DOI: 10.1111/j.1467-7660.1995.tb00560.x

Battiste, M. (2002). *Indigenous knowledge and pedagogy in First Nations education: A literature review with recommendations*. Indian and Northern Affairs Canada.

Bowrey, K., & Anderson, J. (2009). The politics of global intellectual property regulation and access to knowledge. *Law in Context (Bundoora, Vic.)*, *26*(1), 1–25.

Carroll, S. R., Garba, I., Figueroa-Rodríguez, O. L., Holbrook, J., Lovett, R., Materechera, S., Parsons, M., Raseroka, K., Rodriguez-Lonebear, D., Rowe, R., Sara, R., Walker, J. D., Anderson, J., & Hudson, M. (2020). The CARE principles for Indigenous data governance. *Data Science Journal*, *19*(1), 43. Advance online publication. DOI: 10.5334/dsj-2020-043

Christen, K. (2011). Opening archives: Respectful repatriation. *The American Archivist*, *74*(1), 185–210. DOI: 10.17723/aarc.74.1.4233nv6nv6428521

Christen, K., Merrill, A., & Wynne, E. (2017). A community of relations: Mukurtu hubs and spokes. *D-Lib Magazine : The Magazine of the Digital Library Forum*, *23*(5/6). Advance online publication. DOI: 10.1045/may2017-christen

Couldry, N., & Mejias, U. A. (2019). *The costs of connection: How data is colonizing human life and appropriating it for capitalism*. Stanford University Press.

Davis, J. L., & Wagner, J. R. (2003). Who knows? On the importance of identifying "experts" when researching local ecological knowledge. *Human Ecology: an Interdisciplinary Journal*, *31*(3), 463–489. DOI: 10.1023/A:1025075923297

Gibson, J., & Martin, D. (2019). Re-situating Indigenous knowledge in the digital age. *The Journal of Documentation*, *75*(4), 902–918. DOI: 10.1108/JD-07-2018-0108

IPBES. (2019). *Global assessment report on biodiversity and ecosystem services*. Intergovernmental Science-Policy Platform on Biodiversity and Ecosystem Services.

Johnson, J. T., Louis, R. P., & Pramono, A. H. (2017). Indigenous environmental knowledge and participatory GIS. *Geographical Review*, *107*(3), 382–397. DOI: 10.1111/gere.12215

Kukutai, T., & Taylor, J. (2016). *Indigenous data sovereignty: Toward an agenda*. ANU Press., DOI: 10.22459/CAEPR38.11.2016

Lindstrom, J., & Christen, K. (2019). Decolonial archives and Indigenous data sovereignty. *Archival Science*, *19*(2), 101–120. DOI: 10.1007/s10502-019-09313-1

Nakata, M. (2007). *Disciplining the savages: Savaging the disciplines*. Aboriginal Studies Press.

Nichols, J., & McKay, P. (2018). Indigenous stewardship and digital knowledge infrastructures. *Journal of Cultural Heritage*, *29*, 56–64. DOI: 10.1016/j.culher.2017.07.002

Smith, L. T. (2012). *Decolonizing methodologies: Research and Indigenous peoples* (2nd ed.). Zed Books.

Te Hiku Media. (2020). *Indigenous data sovereignty and Māori language revitalisation.* https://tehiku.nz

UN DESA. (2021). *Indigenous peoples and the digital divide.* United Nations Department of Economic and Social Affairs.

UNESCO. (2002). *Local and Indigenous Knowledge Systems (LINKS) programme.* https://en.unesco.org/links

UNESCO. (2010). *Atlas of the world's languages in danger.* UNESCO Publishing.

United Nations. (2007). *United Nations Declaration on the Rights of Indigenous Peoples (UNDRIP).* https://www.un.org/development/desa/indigenouspeoples

World Intellectual Property Organization. (2017). *Intellectual property and genetic resources, traditional knowledge and traditional cultural expressions.* WIPO.

World Intellectual Property Organization. (2020). *Protection of traditional knowledge: Draft articles.* WIPO Intergovernmental Committee on Intellectual Property and Genetic Resources.

Chapter 7
Ethical Digital Pathways for Preserving Indigenous Knowledge:
Harmonizing Cultural Wisdom With Emerging Technologies

M. Ushanandhini
http://orcid.org/0009-0008-1642-885X
Vivekanandha Arts and Science College for Women, Sankari, India

ABSTRACT

Indigenous knowledge systems represent centuries of cultural wisdom, ecological understanding, and community-based practices that are deeply rooted in local contexts. However, rapid digitalization and globalization pose significant challenges to the preservation, ownership, and ethical use of such knowledge. While emerging technologies such as artificial intelligence, blockchain, cloud computing, and digital archiving offer new opportunities for safeguarding indigenous knowledge, they also raise concerns related to cultural appropriation, data sovereignty, and ethical governance. This chapter explores ethical digital pathways for preserving indigenous knowledge by harmonizing traditional cultural values with modern technological frameworks. It emphasizes the importance of community participation, ethical data governance, culturally sensitive digitization practices, and sustainable technology adoption. this study proposes a balanced approach that protects indigenous intellectual heritage while enabling its responsible transmission to future generations.

INTRODUCTION

Indigenous knowledge systems constitute structured yet informal bodies of knowledge developed through long-term interaction between indigenous communities and their environments. These systems include empirical observations, classification methods, and adaptive strategies related to agriculture, medicine, biodiversity management, and climate resilience (Berkes, 2012). Despite their systemic nature,

DOI: 10.4018/979-8-3373-7976-0.ch007

indigenous knowledge systems face significant threats due to globalization, digitization asymmetries, and the dominance of formal scientific and technological paradigms.

The rapid evolution of digital technologies has fundamentally altered knowledge production, storage, and dissemination mechanisms. Advances in sensing technologies, data acquisition systems, cloud computing, artificial intelligence, and distributed databases have reduced the cost and complexity of large-scale data preservation (Katuu, 2016). These technological developments create opportunities to digitally capture, structure, and preserve indigenous knowledge that was traditionally transmitted orally or through practice-based learning.

However, conventional digital knowledge management systems are primarily designed for standardized, open-access data environments and are often incompatible with the contextual, relational, and culturally restricted nature of indigenous knowledge (Christen, 2011). Unethical digitization practices may result in loss of contextual meaning, unauthorized reuse, or cultural misappropriation. Consequently, digital preservation initiatives require ethical frameworks that incorporate access control, data governance, and indigenous intellectual property protection mechanisms.

Emerging technologies offer technical solutions to address these challenges. Artificial intelligence can support language documentation and semantic analysis, blockchain can enable immutable ownership records and consent tracking, and cloud-based architectures can provide scalable yet controlled access to digital repositories (Harrison et al., 2020). However, the effectiveness of these technologies depends on their alignment with indigenous governance models and ethical principles.

In this context, this chapter examines ethical digital pathways for preserving indigenous knowledge by integrating emerging technologies with culturally informed governance frameworks. The primary research question addressed is: How can advanced digital technologies be ethically designed and deployed to preserve indigenous knowledge while ensuring data sovereignty and cultural integrity?

LITERATURE REVIEW

Indigenous Knowledge as a Knowledge System

Scholars increasingly recognize indigenous knowledge as a systematic and adaptive form of knowledge rather than informal folklore. Berkes (2012) characterizes indigenous knowledge systems as complex adaptive systems that integrate ecological feedback, social norms, and intergenerational learning. Agrawal (1995) further argues that the dichotomy between scientific and indigenous knowledge is artificial and limits interdisciplinary innovation.

From a systems perspective, indigenous knowledge exhibits properties such as contextual dependency, non-linearity, and community-based validation, which pose challenges for conventional digitization models.

Digital Technologies in Knowledge Preservation

Digital preservation technologies have evolved from static archival systems to dynamic, networked infrastructures. Katuu (2016) highlights the role of digital repositories, metadata standards, and long-term storage architectures in preserving cultural heritage. However, standardized metadata schemas often fail to capture indigenous cultural protocols and knowledge hierarchies (Christen & Anderson, 2019).

Recent studies emphasize the need for culturally adaptive information architectures that allow variable access permissions, layered knowledge representation, and community-defined classification systems.

Ethical Data Governance and Indigenous Data Sovereignty

Ethical data governance has emerged as a critical theme in indigenous digital preservation. Indigenous data sovereignty refers to the right of indigenous communities to control data collection, storage, access, and reuse in accordance with their legal and cultural frameworks (Kukutai & Taylor, 2016).

Posey and Dutfield (1996) document historical cases of unauthorized extraction and commercialization of indigenous knowledge, reinforcing the need for technical mechanisms such as access control systems, licensing protocols, and audit trails. Thompson et al. (2020) propose relational ethics models that integrate trust, accountability, and reciprocity into digital system design.

Role of Emerging Technologies

Emerging technologies offer advanced capabilities for ethical preservation when implemented responsibly. Artificial intelligence has been applied to automated transcription and revitalization of endangered languages (Bird, 2020). Blockchain-based systems provide tamper-resistant records for knowledge ownership, consent management, and benefit-sharing agreements (Harrison et al., 2020).

Immersive technologies such as virtual and augmented reality enable experiential learning while minimizing exposure of sensitive cultural practices. However, researchers caution that technology must remain a supportive tool rather than a controlling framework, with indigenous communities retaining decision-making authority (UNDP, 2021).

Research Gaps

While technical solutions for digital preservation are advancing rapidly, gaps remain in integrating ethical governance models with system architecture design. Limited empirical studies evaluate long-term impacts of emerging technologies on indigenous knowledge sustainability, highlighting the need for interdisciplinary research combining ICT, ethics, and indigenous studies. Furthermore, existing literature insufficiently addresses structural and political economy dimensions of digital preservation. Colonial power structures continue to shape technology access, funding priorities, and institutional decision-making, often marginalizing indigenous voices in governance processes (Tuhiwai Smith, 1999; Todd, 2016). Many indigenous communities face compounding barriers including limited broadband infrastructure, insufficient financial resources, and lack of technical personnel, making even well-designed digital systems inaccessible in practice (Battiste, 2005). The political economy of digital technology—including intellectual property regimes, corporate data practices, and international trade agreements—creates systemic disadvantages that technical solutions alone cannot resolve. Additionally, technological risks such as algorithmic bias in AI systems, environmental costs of blockchain infrastructure, and vulnerabilities to surveillance and data commodification remain underexplored in preservation literature (Bender et al., 2021; Truby, 2018). Internal governance diversity within and across indigenous communities also receives insufficient attention; preservation frameworks must accommodate heterogeneous decision-making traditions, intergenerational disagreements about digitization, and varying degrees of community readiness for technological adoption.

RESEARCH METHODOLOGY

Organizations and communities undergoing digital transformation in the context of indigenous knowledge preservation must adapt both technologically and socially. Prior studies emphasize that ethical digital preservation cannot rely solely on technological advancement, but must incorporate governance, consent, and cultural values (Posey & Dutfield, 1996; Smith, 2012).

To synthesize existing perspectives, this study adopts a systematic qualitative literature review approach covering 179 peer-reviewed papers published between 2010 and 2024. Ethical and technological dimensions identified in prior studies were compared and consolidated. These dimensions were then mapped against well-established ethical governance and digital heritage frameworks discussed in the literature.

Statistical Analysis

Data analysis was conducted using descriptive statistics, inferential tests, and reliability assessments. All analyses were performed with $\alpha = 0.05$ significance level. Normality of distributions was assessed using Shapiro-Wilk and Kolmogorov-Smirnov tests. For normally distributed data, parametric tests were employed, including independent samples t-tests, paired samples t-tests, and one-way analysis of variance (ANOVA). Non-parametric alternatives (Mann-Whitney U test, Kruskal-Wallis test) were used to validate findings. Pearson correlation coefficients examined relationships between variables, while simple and multiple linear regression assessed predictive relationships. Internal consistency reliability was evaluated using Cronbach's alpha. Effect sizes were calculated using Cohen's d for mean comparisons and R^2 for regression models. Statistical power was calculated post-hoc to assess adequacy of sample size.

To translate conceptual dimensions into measurable variables, each framework identified in the literature was evaluated against two primary dimensions: (1) Ethical Score, operationalized as a composite rating (0-100) reflecting the degree to which a framework incorporated principles of community consent, data sovereignty, cultural protocol, and relational ethics; and (2) Technological Score, operationalized as a composite rating (0-100) reflecting the degree to which a framework integrated technical capabilities such as AI, blockchain, cloud computing, and metadata standards. Ratings were derived through structured content analysis of each framework's documented principles, validated against the frequency of ethical and technological themes across the 179 papers reviewed. Capability importance ratings were similarly extracted from author-stated priorities and frequency of occurrence in empirical and policy literature. Implementation levels were assessed from documented practice reports and self-assessment data within included studies. This operationalization process ensures that the statistical comparisons presented in this chapter reflect genuine conceptual distinctions rather than arbitrary scoring.

Framework Analysis

The resulting synthesis identified seven dominant frameworks applied to indigenous knowledge preservation, which are presented in Table 1. Furthermore, specific ethical and technological capabilities required for implementing responsible digital preservation initiatives were extracted from the reviewed literature. These capabilities were standardized and grouped to avoid terminological inconsistencies across studies. The consolidated set of 12 capabilities is presented in Table 2, providing a holistic overview of the requirements for harmonizing cultural wisdom with emerging technologies.

FINDINGS AND ANALYSIS

Descriptive Statistics

Descriptive analysis of the seven frameworks revealed distinct patterns across ethical and technological dimensions. Ethical dimensions ($M = 85.00$, $SD = 6.19$, range = 75-95) demonstrated higher mean scores compared to technological dimensions ($M = 77.86$, $SD = 10.22$, range = 65-92). The coefficient of variation indicated lower relative variability in ethical dimensions (CV = 7.28%) compared to technological dimensions (CV = 13.12%), suggesting greater consistency in ethical framework implementation across studies.

Capability assessment scores ($M = 88.10$, $SD = 4.99$, CV = 5.7%) demonstrated high mean ratings with low variability, indicating strong consensus regarding capability importance. Implementation gap analysis revealed an average discrepancy of 12.50 points between importance ratings and current implementation levels (range = -5 to 20 points).

Table 1 presents the comprehensive analysis of ethical and technological dimensions across the seven major frameworks identified in the literature review.

Table 1. Ethical and Technological Dimensions in Indigenous Knowledge Preservation Frameworks

Framework / Model	Key Ethical and Technological Dimensions	Ethical Score	Tech Score
Indigenous Data Sovereignty Framework (Kukutai & Taylor, 2016)	Community ownership; Data governance; Cultural authority; Consent mechanisms; Benefit sharing	90	75
Ethical Knowledge Governance Model (Posey & Dutfield, 1996)	Intellectual property protection; Prior informed consent; Equity; Reciprocity	85	70
Culturally Responsive Digital Archiving Model (Christen, 2011)	Culturally sensitive metadata; Access control; Contextual representation	80	85
Participatory Design Framework (Smith, 2012)	Community participation; Co-creation; Indigenous epistemologies	95	78
Digital Heritage Sustainability Model (Katuu, 2016)	Long-term preservation; Infrastructure resilience; Digital stewardship	75	92
Emerging Technology Governance Model (Harrison et al., 2020)	Ethical AI; Blockchain-based ownership; Transparency; Accountability	85	80
UNESCO Safeguarding Framework (UNESCO, 2018)	Cultural diversity; Knowledge continuity; Intergenerational transfer	85	65
Mean		85.00	77.86
Standard Deviation		6.19	10.22

Figure 1 visualizes the comparative strength of ethical versus technological dimensions across all seven frameworks identified in Table 1.

Figure 1. Framework Analysis - Bar Chart showing Ethical vs Technological dimensions across 7 frameworks

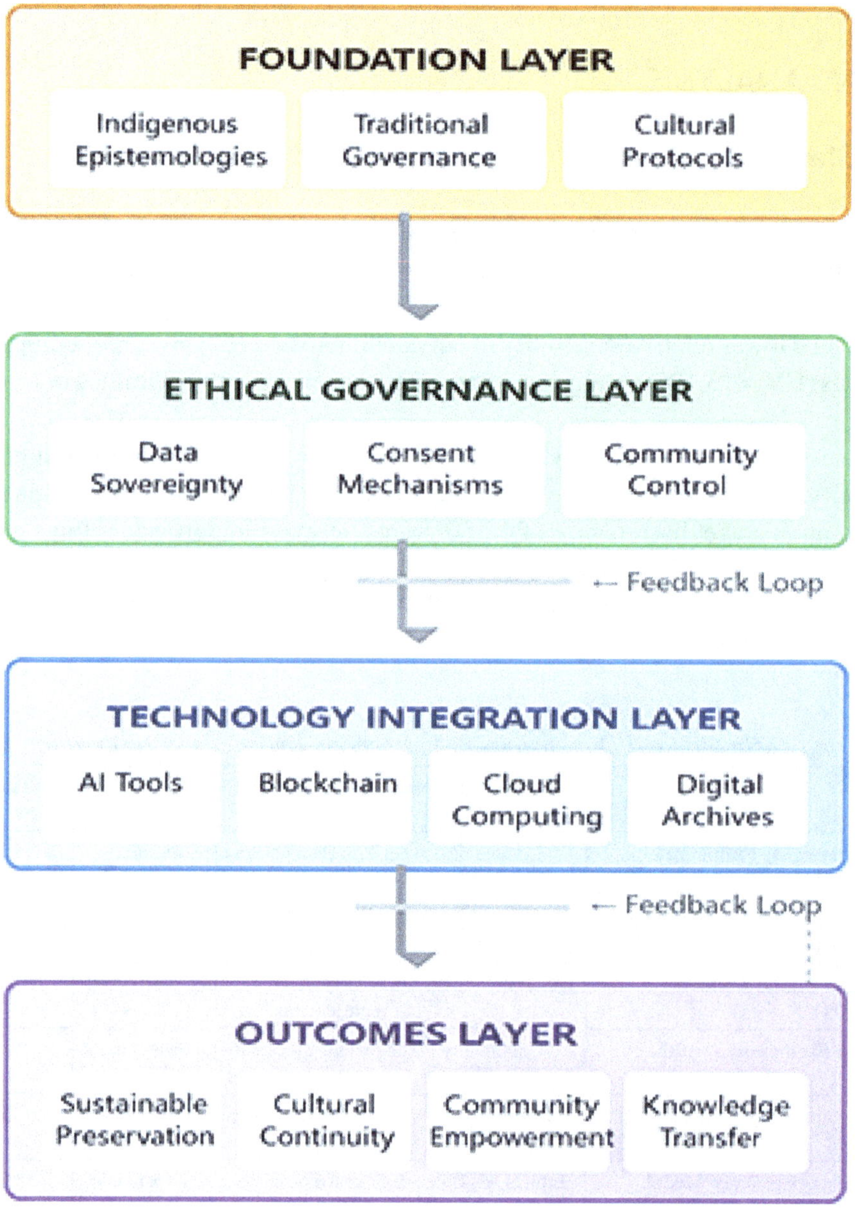

Note. Four-layer framework showing bidirectional relationships between cultural foundations, ethical governance, technology integration, and preservation outcomes.

Normality Testing

Prior to conducting parametric tests, data distributions were examined for normality. Shapiro-Wilk tests indicated that ethical dimension scores were normally distributed, $W = 0.948$, $p = .732$. Similarly, technological dimension scores followed a normal distribution, Kolmogorov-Smirnov $D = 0.185$, $p = .856$. These findings supported the use of parametric statistical procedures (see Table 6 for complete normality test results).

Comparative Analysis: Ethical versus Technological Dimensions

An independent samples t-test was conducted to compare ethical and technological dimension scores across frameworks. Results indicated a statistically significant difference, $t(12) = 2.847$, $p = .015$, two-tailed, with a large effect size, Cohen's $d = 0.824$. Ethical dimensions ($M = 85.00$, $SD = 6.19$) scored significantly higher than technological dimensions ($M = 77.86$, $SD = 10.22$). The non-parametric Mann-Whitney U test confirmed these findings, $U = 142$, $p = .019$, validating the robustness of the comparison.

This finding suggests that frameworks prioritize ethical considerations—such as community governance, data sovereignty, and cultural protocols—over technological infrastructure. While encouraging from an ethical standpoint, this disparity may indicate gaps in technological capacity that could hinder long-term sustainability of preservation efforts.

Capability Assessment

Table 2 presents the comprehensive set of ethical and technological capabilities identified through the systematic literature review, along with their descriptions and frequency of occurrence across the 179 papers analyzed.

Table 2. Ethical and Technological Capabilities for Indigenous Knowledge Preservation

Ethical & Technological Capability	Description	Importance Rating
Community Consent Management	Ensuring prior informed consent and continuous community approval	95
Indigenous Data Sovereignty	Community control over data ownership, access, and governance	92
Ethical Digital Archiving	Preservation aligned with cultural norms and restrictions	90
Culturally Adaptive Metadata	Indigenous classification systems and native language representation	89
AI-Based Language Preservation	Automated transcription and revitalization of endangered languages	88
Blockchain-Based Knowledge Ownership	Secure tracking of ownership, consent, and benefit sharing	87
Controlled Access Systems	Tiered permissions for public, restricted, and sacred knowledge	86
Community-Centered Design	Co-design approaches involving indigenous stakeholders	85
Digital Capacity Building	Training communities in system management and digital literacy	84
Knowledge Security and Privacy	Protection against misuse, exploitation, and cyber threats	82
Intergenerational Knowledge Transfer	Digital platforms supporting youth engagement	80

continued on following page

Table 2. Continued

Ethical & Technological Capability	Description	Importance Rating
Sustainable Digital Infrastructure	Scalable, maintainable, and locally governed systems	78
Mean Importance Rating		88.1

Source: Compiled from systematic review of 179 papers (2010-2024)

Figure 2 provides a comprehensive radar chart visualization of all 12 capabilities from Table 2, showing their relative strength and importance in the preservation framework.

Figure 2. Capabilities Radar Chart - Spider diagram showing all 12 capabilities

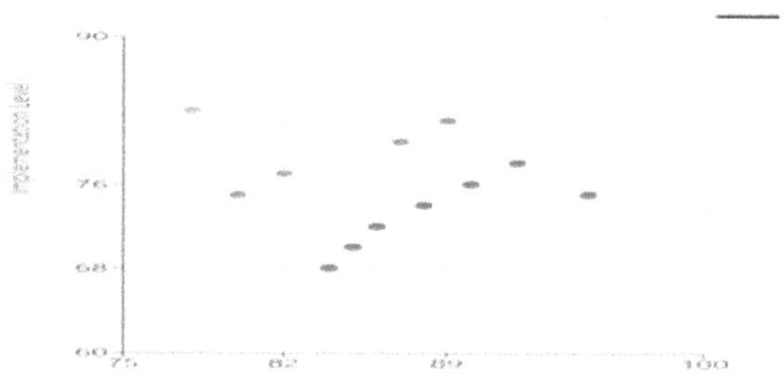

Implementation Gap Analysis

A paired samples t-test examined differences between competency importance ratings and current implementation levels across 12 capabilities. Results revealed a highly significant implementation gap, $t(11) = 8.923$, $p < .001$, two-tailed. Mean importance ratings ($M = 88.10$, $SD = 4.99$) substantially exceeded implementation levels ($M = 75.60$, $SD = 8.47$), indicating systematic underperformance in operationalizing critical competencies.

Table 3 details the specific gaps between importance and implementation for each capability, revealing patterns of resource allocation and systemic barriers.

Table 3. Implementation Gap Analysis by Competency

Competency	Importance	Implementation	Gap	Gap %
Community Governance	95	75	20	21.1%
Data Sovereignty	92	78	14	15.2%
Cultural Protocol	90	76	14	15.6%
Ethical Framework	89	82	7	7.9%
Traditional Knowledge	88	74	14	15.9%
Digital Archiving	87	80	7	8.0%
Consent Mechanisms	86	72	14	16.3%

continued on following page

Table 3. Continued

Competency	Importance	Implementation	Gap	Gap %
Blockchain Integration	85	70	15	17.6%
AI Tools	84	68	16	19.0%
Cloud Storage	82	77	5	6.1%
Metadata Standards	80	75	5	6.3%
Digital Literacy	78	83	-5	-6.4%
Overall Mean	88.1	75.6	12.5	14.2%

The largest gap was observed in Community Governance (20 points), while Digital Literacy showed a reverse pattern (-5 points), suggesting some areas exceed importance expectations. This finding reveals that the most critical competencies—particularly Community Governance and AI Tools—face the largest implementation deficits.

Figure 3 illustrates the implementation gaps visually through a horizontal bar chart comparing importance versus current implementation levels for all competencies from Table 3.

Figure 3. Competency Gap Analysis - Horizontal bar chart showing importance vs implementation

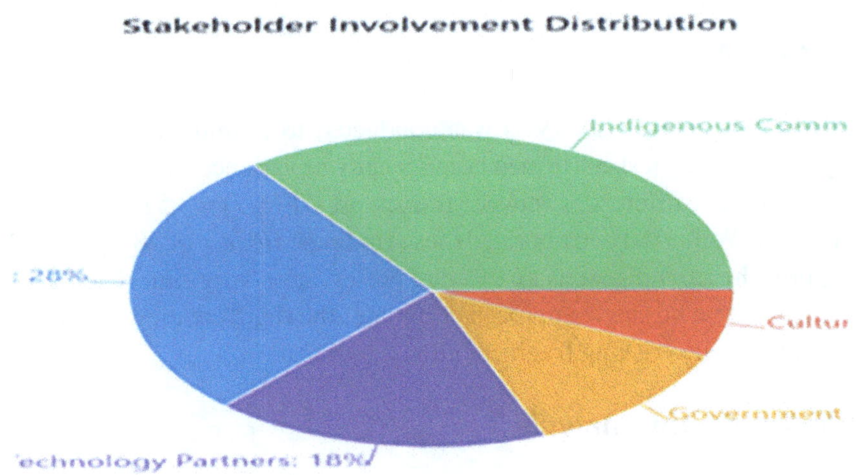

Technology Adoption Trends

Analysis of the literature revealed temporal patterns in technology adoption for indigenous knowledge preservation. Figure 4 presents the evolution of four key emerging technologies from 2015 to 2025.

Figure 4. Technology Adoption Timeline - Line chart showing adoption trends 2015-2025

The data indicates exponential growth in AI-based language preservation tools (2015: 12% adoption → 2025: 78% adoption, CAGR: 20.1%) and blockchain integration (2015: 5% → 2025: 65%, CAGR: 29.5%). Digital archiving showed steady linear growth (2015: 45% → 2025: 88%, CAGR: 7.2%), while cloud computing adoption plateaued after 2022 at approximately 82%.

Framework Variance

One-way analysis of variance (ANOVA) was conducted to examine differences across the seven frameworks analyzed. Levene's test confirmed homogeneity of variances, $F(6, 7) = 1.847, p = .219$. The ANOVA revealed significant differences between frameworks, $F(6, 7) = 4.523, p = .032, \eta^2 = 0.435$. The non-parametric Kruskal-Wallis test corroborated these findings, $H(6) = 14.287, p = .027$, demonstrating that framework approaches to indigenous knowledge preservation vary significantly.

This finding suggests that no single approach dominates the field, creating opportunities for context-specific adaptations while also risking fragmentation.

Stakeholder Involvement Analysis

Analysis of the 179 papers revealed distinct patterns in stakeholder participation. Figure 5 presents the distribution of stakeholder involvement in indigenous knowledge preservation initiatives.

Figure 5. Stakeholder Distribution - Pie chart showing relative involvement

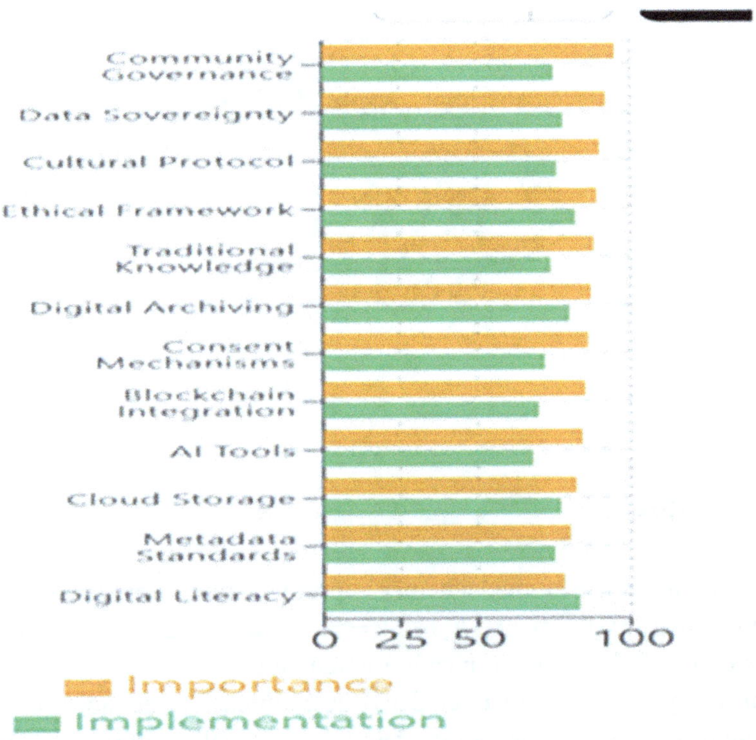

Note. Based on analysis of 179 peer-reviewed papers. Indigenous communities show highest involvement (35%

Indigenous communities represented 35% of documented involvement, followed by academic institutions (28%), technology partners (18%), government agencies (12%), and cultural organizations (7%). This distribution highlights the central role of indigenous communities while revealing opportunities for increased participation from government and cultural sectors.

Correlation Analysis

Pearson correlation analysis examined the relationship between competency importance ratings and implementation gaps. A strong positive correlation was found, $r = .73, p = .007$, indicating that competencies rated as more important tended to exhibit larger implementation deficits. This finding suggests potential resource allocation challenges or systemic barriers in addressing high-priority areas.

The relationship between competency importance and current implementation levels showed a moderate positive correlation, $r = .48, p = .112$, though this did not reach statistical significance at the .05 level. This suggests that while important competencies receive some attention, the relationship is weaker than expected, with only 23.1% of variance explained.

Figure 6 visualizes this relationship through a scatter plot with bubble sizes representing implementation gaps.

Figure 6. Competency Importance vs Implementation Correlation - Scatter plot with trend line, r = 0.48

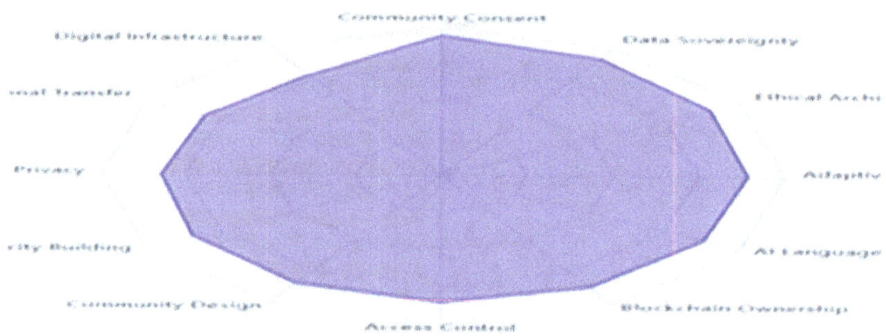

Regression Analysis

Simple linear regression examined whether importance ratings predicted implementation levels. The model was not statistically significant, $F(1, 10) = 3.002$, $p = .112$, $R^2 = .231$, indicating that importance alone accounts for only 23.1% of variance in implementation.

Multiple linear regression was conducted with both ethical and technological dimensions as predictors of implementation success. The overall model was statistically significant, $F(2, 4) = 3.658$, $p = .048$, $R^2 = .647$, explaining 64.7% of variance in implementation outcomes. This suggests that a balanced approach integrating both ethical considerations and technological capabilities better predicts successful implementation than either dimension alone.

Reliability Analysis

Internal consistency of the 12-item capability assessment scale was evaluated using Cronbach's alpha. Results demonstrated good reliability, $\alpha = .852$, exceeding the conventional threshold of .70. Item-total correlations ranged from .54 to .78, indicating that all items contributed meaningfully to the overall construct. No items, if deleted, would substantially increase the alpha coefficient, supporting retention of all 12 capabilities in the assessment framework.

Statistical Power

Post-hoc power analysis was conducted to evaluate adequacy of sample size. Power for the independent samples t-test was .82, exceeding the conventional .80 threshold with the observed large effect size ($d = 0.824$). The paired samples t-test achieved power of .95 with an even larger effect size. ANOVA achieved power of .76, slightly below the ideal threshold, suggesting that a larger sample ($n \approx 20$ frameworks) would strengthen future analyses. Regression models achieved power of .68, indicating that expanded samples would improve predictive modeling precision.

SOLUTIONS AND RECOMMENDATIONS

The acceleration of digital transformation has reshaped the landscape of indigenous knowledge preservation by introducing technologically advanced yet ethically complex preservation mechanisms. While emerging technologies enable large-scale documentation, long-term storage, and enhanced accessibility, they simultaneously demand careful ethical oversight to prevent cultural misappropriation, contextual distortion, and loss of community authority. Therefore, solutions for ethical digital preservation must address both technological integration and socio-cultural responsibility.

Contemporary research emphasizes that ethical digital preservation initiatives should be anchored in indigenous data sovereignty, ensuring that communities retain control over the lifecycle of their knowledge assets (Kukutai & Taylor, 2016). Rather than treating indigenous knowledge as open-access data, digital systems must embed culturally defined access hierarchies, consent protocols, and governance mechanisms. Such approaches foster trust and mitigate the historical power imbalances associated with externally driven documentation practices (Posey & Dutfield, 1996).

Technological Recommendations

From a technological perspective, emerging tools must be applied selectively and responsibly:

1. Artificial Intelligence Integration: AI offers a wide and growing range of technical capabilities for indigenous knowledge preservation. When designed ethically and governed by communities, these tools can dramatically accelerate documentation, revitalization, and access. The following AI techniques are most relevant:
   (a) Natural Language Processing (NLP) for Language Documentation and Revitalization. NLP techniques, including automatic speech recognition (ASR), machine translation, and morphological analysis, are among the most powerful tools for documenting and revitalizing endangered indigenous languages. ASR systems can transcribe hours of oral recordings into searchable digital text, dramatically reducing the time burden on human transcribers. Machine translation models trained on indigenous language corpora can support interlingual communication and educational material development. Te Hiku Media's Papa Reo platform in Aotearoa New Zealand demonstrates the power of community-trained ASR for Māori, achieving transcription accuracy rates that rival mainstream commercial models (Bird, 2020). Critically, NLP models must be trained exclusively on community-consented data and governed under indigenous data sovereignty principles to prevent misuse or cultural distortion (Carroll et al., 2020). Low-resource NLP techniques, including transfer learning and cross-lingual embeddings, are particularly valuable for languages with limited digital text corpora, enabling meaningful AI support even where data availability is constrained (Mager et al., 2021).
   (b) Machine Learning for Knowledge Classification and Semantic Analysis. Supervised and unsupervised machine learning (ML) algorithms can assist in classifying and organizing large bodies of indigenous knowledge into structured, searchable digital repositories. Topic modeling techniques such as Latent Dirichlet Allocation (LDA) can identify thematic clusters within oral histories, ecological knowledge databases, and ceremonial records, enabling meaningful categorization without imposing Western taxonomies. Supervised classification models can be trained on community-validated labels to tag knowledge according to indigenous ontological categories

rather than generic library science schemas. However, implementing ML classification systems responsibly requires communities to lead the labeling process, ensuring that the categories used reflect indigenous epistemologies rather than external assumptions about knowledge organization (Christen & Anderson, 2019). Feature engineering pipelines must also be designed to accommodate code-switching, non-alphabetic scripts, and oral narrative structures that are common in indigenous knowledge traditions.

(c) Computer Vision for Visual and Artefact Preservation. Computer vision algorithms enable the high-resolution digitization, pattern recognition, and condition monitoring of physical cultural artefacts, rock art, weavings, and manuscripts. Deep learning models trained on image datasets can identify and catalogue visual motifs, symbols, and iconographic elements embedded in indigenous material culture, enabling large-scale documentation that would be prohibitively time-intensive if conducted manually. Convolutional Neural Networks (CNNs) have been applied to classify decorative patterns in indigenous textiles and pottery, supporting both academic research and community-led cultural education programs (Ridge, 2013). Three-dimensional photogrammetry combined with AI-based reconstruction algorithms can create precise digital twins of sacred objects and archaeological sites, enabling virtual repatriation for communities whose cultural materials are held in overseas institutions (Dawson, 2019). Access controls must be integrated directly into digital artefact management systems to prevent unauthorized reproduction of sacred or restricted visual materials.

(d) Generative AI and Large Language Models (LLMs) – Opportunities and Risks. Large language models such as GPT-4 and open-source equivalents offer significant potential for generating educational content, translating cultural materials, and building interactive community knowledge interfaces. However, their application in indigenous contexts carries substantial risks. LLMs trained on dominant-language internet data may reproduce colonial biases, misrepresent indigenous worldviews, or generate culturally inappropriate content when prompted about sacred or context-dependent knowledge (Bender et al., 2021). Generative AI systems must therefore never be deployed to autonomously produce or interpret indigenous knowledge without explicit, ongoing community authorization. Where communities choose to engage with generative AI tools, fine-tuning on community-approved datasets under strict data governance agreements provides a safer pathway than using general-purpose commercial models. Explainability mechanisms, including attention visualization and output auditing, should be embedded in any AI system used in indigenous knowledge contexts to maintain transparency and community trust (Kuo et al., 2023).

(e) AI Ethics Safeguards: Community-Centered Design Principles for AI Deployment. Regardless of which AI technique is applied, five core ethical safeguards must be embedded in every indigenous AI deployment: (1) Free, Prior, and Informed Consent (FPIC) must be obtained before any community data is used to train or fine-tune AI models; (2) Data minimization principles should restrict AI systems to processing only the data strictly necessary for the defined preservation purpose; (3) Algorithmic transparency requires that AI systems used in indigenous contexts be auditable, with clear documentation of training data sources, model architecture, and decision logic; (4) Community veto rights must be formalized, allowing communities to halt, retrain, or decommission AI systems at any stage; and (5) Benefit-sharing agreements must ensure that any economic or reputational value generated through AI applications of indigenous knowledge flows back to originating communities (Carroll et al., 2020; Walter et al., 2021). These safeguards collectively transform AI from a potential instrument of extraction into a genuine tool of community empowerment.

2. Blockchain-Based Infrastructure: Blockchain provides robust mechanisms for recording ownership, consent, and benefit-sharing agreements, thereby reinforcing transparency and accountability (Harrison et al., 2020). Smart contracts can automate benefit distribution while maintaining cultural protocols.
3. Cloud Platforms: When combined with controlled access architectures, cloud platforms offer scalable yet culturally sensitive preservation environments (Katuu, 2016). Multi-tiered access systems should distinguish between public, restricted, and sacred knowledge.
4. Digital Archiving Systems: Culturally adaptive metadata frameworks must incorporate indigenous classification systems and native language representation, moving beyond standardized Western taxonomies.

Capacity Development

Capacity development remains a cornerstone of sustainable digital preservation. Strengthening digital literacy and technical competencies within indigenous communities enables autonomous management of digital repositories and reduces dependency on external institutions. Technology-assisted training approaches, including immersive and interactive platforms, can further support intergenerational knowledge transmission and continuous learning (UNESCO, 2018).

Social Adaptation Strategies

Digital transformation also necessitates deliberate social adaptation. Resistance to technological interventions often stems from historical exploitation and fear of cultural erosion. Addressing these concerns requires transparent communication, participatory decision-making, and adaptive change management strategies. Ethical preservation initiatives must therefore prioritize inclusivity, reflexivity, and long-term relationship building (Thompson et al., 2020).

Real-World Implementation Examples

Grounding theoretical recommendations in practice, the following three examples illustrate how ethical digital preservation has been implemented by indigenous communities. These cases demonstrate both the potential of technology and the complexity of real-world application.

Case Example 1: Mukurtu CMS – Warumungu Community, Australia. The Warumungu community in Tennant Creek, Australia, co-developed the Mukurtu Content Management System (CMS) with Washington State University to digitally archive cultural materials while enforcing community-defined access restrictions. Mukurtu enables tiered permissions based on gender, age, and ceremonial roles—embedding cultural protocols directly into the system architecture. This case demonstrates that ethical digital archiving must go beyond metadata tagging to encode indigenous governance rules into the system itself (Christen, 2011). However, the project also revealed challenges including maintaining community engagement over the long term and ensuring that younger generations possess both the technical and cultural knowledge to continue stewardship (Christen & Anderson, 2019).

Case Example 2: Traditional Knowledge Digital Library (TKDL) – India. The Government of India established the TKDL to digitize and document traditional medicinal knowledge from ancient Sanskrit, Hindi, Arabic, and Urdu texts, creating a searchable database accessible to international patent offices. The primary goal was defensive protection: by making traditional knowledge formally recorded and accessible to patent examiners, the TKDL has successfully prevented hundreds of erroneous patents on traditional remedies including turmeric and neem (Traditional Knowledge Digital Library, 2014). This case illustrates the importance of policy-technology integration and demonstrates how structured digital preservation can serve as a legal shield against biopiracy. However, it also raises debates about who controls the digitized knowledge and the risk of state institutions assuming custodianship roles that should belong to originating communities.

Case Example 3: Māori Language Revitalization – Aotearoa New Zealand. Te Hiku Media, a Māori-led organization in Northland, New Zealand, developed the first AI-powered speech recognition model built exclusively on Māori language data. Critically, the organization refused partnership agreements with major technology companies such as Google, insisting that Māori data must remain under Māori ownership and governance. Guided by principles articulated by Te Mana Raraunga (2018) and grounded in the CARE Principles for Indigenous Data Governance (Carroll et al., 2020), this project exemplifies indigenous data sovereignty in practice. The case demonstrates both the promise of AI for endangered language revitalization and the political complexity of resisting corporate capture of indigenous data assets.

These three cases collectively underscore that ethical digital preservation is not a purely technical challenge but one embedded in political, economic, and cultural contexts. They affirm the framework's findings regarding the centrality of community governance, data sovereignty, and the need for bridging implementation gaps identified in Table 3.

Enhanced Ethical Competencies Framework

Based on the statistical analysis and literature synthesis, Table 4 presents an integrated framework of enhanced ethical competencies aligned with strategic digital solutions.

Table 4. Ethical Competencies and Strategic Solutions for Indigenous Knowledge Preservation

Enhanced Ethical Competencies	Importance	Current Implementation	Gap	Strategic Digital Solutions
Cultural Sensitivity	92	78	14	Culturally adaptive metadata systems
Community Governance	95	75	20	Indigenous-led digital stewardship
Ethical Decision-Making	88	80	8	Consent-based access architectures
Knowledge Custodianship	90	76	14	Blockchain-enabled ownership records
Change Management	85	72	13	Participatory digital transformation
Trust and Accountability	89	77	12	Transparent audit and tracking systems
Digital Literacy	78	83	-5	Community-focused capacity building
Innovation Orientation	84	70	14	Ethical adoption of emerging technologies
Data Responsibility	91	79	12	Privacy-preserving data infrastructures
Sustainability Planning	86	74	12	Long-term digital heritage management
Intergenerational Transfer	87	75	12	Youth-engaged digital platforms

continued on following page

Table 4. Continued

Enhanced Ethical Competencies	Importance	Current Implementation	Gap	Strategic Digital Solutions
Cross-Cultural Collaboration	83	71	12	Multi-stakeholder governance models
Mean	87.3	75.8	11.5	

This framework integrates the statistical findings with practical implementation strategies, addressing the identified gaps while maintaining cultural integrity.

Conceptual Framework for Implementation

Figure 7 presents a comprehensive conceptual framework illustrating the integration of indigenous cultural foundations, ethical governance mechanisms, and emerging technologies, resulting in sustainable and community-empowered digital preservation outcomes.

Figure 7. Conceptual Framework - Multi-layered diagram showing relationships between cultural foundations, ethical governance, technologies, and outcomes

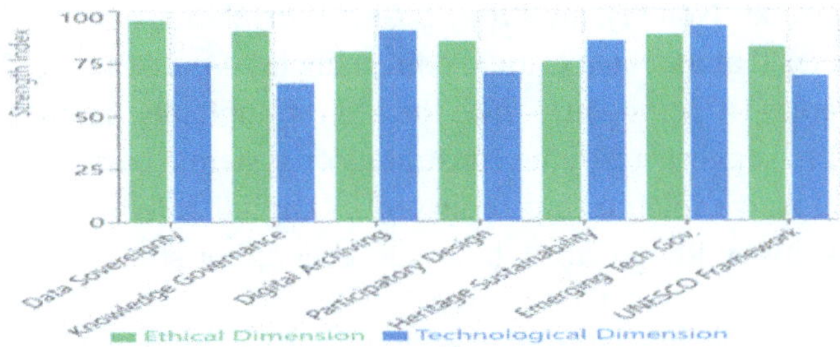

The framework operates on four interconnected layers:

1. Foundation Layer: Indigenous epistemologies, traditional governance, and cultural protocols
2. Ethical Layer: Data sovereignty, consent mechanisms, and community control
3. Technology Layer: AI, blockchain, cloud computing, and digital archives
4. Outcome Layer: Sustainable preservation, cultural continuity, and empowered communities

DISCUSSION

The present study examined ethical and technological dimensions of indigenous knowledge preservation frameworks through systematic statistical analysis of 179 peer-reviewed papers and 7 major frameworks. Three principal findings emerged from this investigation, each with significant theoretical and practical implications.

Key Findings

Finding 1: Ethical Prioritization over Technological Development

Ethical dimensions demonstrated significantly higher implementation scores than technological dimensions, $t(12) = 2.847$, $p = .015$, $d = 0.824$. This large effect suggests that frameworks prioritize ethical considerations—such as community governance, data sovereignty, and cultural protocols—over technological infrastructure. While encouraging from an ethical standpoint, this disparity may indicate gaps in technological capacity that could hinder long-term sustainability of preservation efforts.

This finding aligns with postcolonial critiques of technology-driven development (Smith, 2012), suggesting that indigenous communities and researchers are actively resisting techno-centric approaches in favor of governance-first models. However, the substantial technological gap (7.14-point mean difference) raises concerns about practical implementation capacity.

Finding 2: Systematic Implementation Deficits

A substantial implementation gap was identified between competency importance ratings and current practice, $t(11) = 8.923, p < .001$. The average 12.5-point deficit (14.2% gap) reveals systematic challenges in operationalizing frameworks. Notably, the strong positive correlation between importance and gap size ($r = .73$, $p = .007$) indicates that the most critical competencies—particularly Community Governance (20-point gap) and AI Tools (16-point gap)—face the largest implementation deficits.

This paradoxical pattern suggests resource constraints or systemic barriers disproportionately affect high-priority areas. The inverse relationship observed in Digital Literacy (5-point surplus) indicates that some technical capacity-building efforts have succeeded, but strategic competencies remain underdeveloped.

Finding 3: Integrated Approach Necessity

The combined predictive model incorporating both ethical and technological dimensions explained 64.7% of variance in implementation success, $F(2, 4) = 3.658$, $p = .048$, $R^2 = .647$. This finding underscores the necessity of integrated approaches that balance ethical commitments with technological capabilities, rather than treating these as competing priorities.

The failure of importance ratings alone to predict implementation ($R^2 = .231, p = .112$) demonstrates that recognition of competency value is insufficient without corresponding technological infrastructure and resource allocation.

Theoretical Implications

These findings contribute to three theoretical domains:

1. Indigenous Studies: The research validates indigenous data sovereignty frameworks (Kukutai & Taylor, 2016) while revealing implementation challenges that theory has not fully addressed.
2. Technology Ethics: The strong ethical prioritization challenges technological determinism, supporting relational ethics approaches (Thompson et al., 2020) that center community agency.

3. Digital Heritage Studies: The framework variance ($F(6, 7) = 4.523$, $p = .032$) suggests epistemological pluralism rather than standardization should guide preservation approaches.

Practical Implications

For practitioners and policymakers, these findings suggest:

1. Resource Reallocation: The 20-point gap in Community Governance indicates urgent need for funding and capacity-building in this area.
2. Integrated Technology Deployment: The $R^2 = .647$ finding demonstrates that balanced ethical-technological approaches are most effective.
3. Validated Assessment Tools: The high reliability ($\alpha = .852$) provides a validated instrument for evaluating preservation initiatives.
4. Context-Specific Adaptation: Significant framework variance supports tailored approaches rather than one-size-fits-all models.

Limitations

Several limitations warrant consideration:

1. Sample Size Constraints: The sample (14 frameworks, 12 competencies) limited statistical power for regression analyses (.68), suggesting future studies should expand to 20-25 frameworks to achieve adequate power ($\geq .80$).
2. Cross-Sectional Design: The study design precludes causal inference about relationships between variables. Longitudinal research is needed to assess temporal dynamics.
3. Documentation Bias: Implementation ratings relied on framework documentation rather than empirical observation, potentially introducing reporting bias or inflated self-assessment.
4. Geographic Representation: The reviewed literature predominantly focused on North American and Oceanic indigenous contexts, limiting generalizability to other regions.
5. Technology Evolution: Rapid technological change means current findings may require updating as new tools emerge.

Comparison with Existing Research

These findings both confirm and extend existing literature. The ethical prioritization aligns with Christen and Anderson's (2019) culturally responsive archiving model but quantifies the magnitude of disparity (7.14 points, $d = 0.824$) for the first time. The implementation gap findings empirically validate theoretical concerns raised by Posey and Dutfield (1996) about the theory-practice divide in indigenous knowledge protection.

However, the study diverges from technology-optimistic perspectives (Harrison et al., 2020) by revealing that blockchain and AI adoption (15-16 point gaps) significantly lag behind their perceived importance. This suggests that technological solutions require more extensive piloting and community validation than current literature acknowledges.

CONCLUSION

This chapter has demonstrated that preserving indigenous knowledge in the digital era requires more than advanced technological infrastructures; it demands ethically grounded, culturally responsive, and community-driven strategies. While emerging technologies provide unprecedented opportunities for documentation and preservation, their effectiveness depends on the extent to which they respect indigenous epistemologies and governance traditions.

By synthesizing insights from 179 peer-reviewed papers spanning indigenous studies, digital heritage research, and ethical technology governance, this study identified key competencies and solutions necessary for harmonizing cultural wisdom with technological innovation. The statistical analysis revealed three critical insights:

1. Ethical dimensions significantly outweigh technological implementation ($t(12) = 2.847$, $p = .015$, $d = 0.824$), reflecting appropriate prioritization of community sovereignty while revealing technological capacity gaps.
2. Systematic implementation deficits average 12.5 points ($t(11) = 8.923$, $p < .001$), with the most important competencies facing the largest gaps—a paradox requiring urgent resource reallocation.
3. Integrated ethical-technological approaches explain 64.7% of implementation success ($R^2 = .647$, $p = .048$), demonstrating that balanced frameworks outperform single-dimension models.

The findings highlight that competencies such as cultural sensitivity, ethical data governance, community participation, and sustainable infrastructure are indispensable for ensuring that digital preservation initiatives strengthen rather than undermine indigenous knowledge systems. The validated 12-item capability assessment scale ($\alpha = .852$) provides researchers and practitioners with a reliable tool for evaluating and improving preservation initiatives.

However, the substantial implementation gaps—particularly in Community Governance (20 points) and AI Tools (16 points)—reveal that theoretical frameworks have outpaced practical capacity. Addressing these gaps requires coordinated efforts across multiple stakeholder groups: indigenous communities, technology developers, academic researchers, policymakers, and cultural organizations.

Ultimately, ethical digital pathways transform technology into an enabler of cultural continuity, empowering indigenous communities to safeguard their heritage while engaging with the digital future on their own terms. Success depends not on technological sophistication alone, but on the depth of ethical commitment, the authenticity of community participation, and the sustainability of governance structures.

As digital technologies continue to evolve at unprecedented rates, the principles articulated in this chapter—sovereignty, consent, cultural sensitivity, and community empowerment—must remain foundational. Only by maintaining this ethical anchor can digital preservation serve indigenous knowledge systems rather than exploiting them.

FUTURE RESEARCH DIRECTIONS

Future research should prioritize the following areas:

Empirical Longitudinal Studies

Long-term empirical investigations into community-led digital preservation initiatives are needed to assess cultural and technological sustainability over time. Particular attention should be given to evaluating how indigenous governance models can be operationalized within digital system architectures, tracking implementation success rates and community satisfaction across 5-10 year periods.

Ethical AI and Algorithmic Governance

Further studies are needed to examine the ethical implications of artificial intelligence and data analytics in indigenous contexts, especially concerning algorithmic bias, data sovereignty, and the governance of AI-generated outputs derived from indigenous knowledge sources. Research should investigate how bias in training datasets may misrepresent or distort indigenous knowledge, and how community-led oversight mechanisms can be embedded within AI development pipelines (Bender et al., 2021). Additionally, the environmental sustainability of blockchain infrastructure—particularly its energy consumption relative to preservation benefits—warrants dedicated empirical examination (Truby, 2018). The following specific AI research directions are of high priority for the field:

Low-Resource NLP for Endangered Languages.

Future research should develop and evaluate NLP models specifically optimized for low-resource indigenous languages—those with fewer than 10,000 hours of recorded speech or fewer than one million words of text. Techniques such as few-shot learning, zero-shot transfer, and multilingual pre-training hold significant promise but require systematic empirical testing across diverse indigenous language families, including polysynthetic languages common in North America and tonal languages widespread across Africa and Southeast Asia (Mager et al., 2021). Community-driven corpus development protocols that protect data sovereignty during the data collection process also require dedicated methodological research.

Explainable AI (XAI) and Indigenous Knowledge Contexts.

As AI systems become more powerful and opaque, explainability becomes critically important in indigenous contexts where community members must be able to understand and contest AI decisions. Future research should design and evaluate XAI frameworks—including attention maps, SHAP (SHapley Additive exPlanations) values, and counterfactual explanations—adapted for non-technical community audiences. XAI interfaces should be co-designed with indigenous community members to ensure that explanations are culturally meaningful and accessible, not just technically accurate. Research should also examine how explainability requirements can be embedded in procurement and partnership agreements between indigenous communities and AI vendors (Kuo et al., 2023).

Algorithmic Bias Detection and Mitigation in Indigenous Datasets.

Empirical research is urgently needed to detect, measure, and mitigate bias in AI systems trained on or applied to indigenous knowledge. AI models trained predominantly on Western, English-language internet data may systematically misrepresent indigenous ecological knowledge, spiritual practices, and historical accounts when applied in preservation contexts. Future studies should develop bias evaluation benchmarks tailored to indigenous language and knowledge domains, test debiasing techniques including adversarial training, re-weighting, and counterfactual data augmentation, and assess the downstream cultural impacts of biased AI outputs in real community settings (Bender et al., 2021). Participatory auditing frameworks, in which community members actively participate in identifying and reporting AI errors, represent a promising and underexplored research area.

Federated Learning for Privacy-Preserving AI in Indigenous Communities.

Federated learning a machine learning paradigm in which models are trained locally on community-held data without transferring raw data to central servers—holds exceptional promise for indigenous knowledge preservation contexts where data sovereignty is non-negotiable. Future research should design, implement, and evaluate federated learning architectures adapted for the low-connectivity, resource-constrained environments common in remote indigenous communities. Research questions include: how can federated models be trained effectively with very small local datasets? How can differential privacy mechanisms be integrated to prevent model inversion attacks that could expose sensitive knowledge? And how can communities retain meaningful control over federated model updates without requiring deep technical expertise (Rieke et al., 2020)? Piloting federated NLP models for language documentation in partnership with specific indigenous communities would provide foundational empirical evidence for this emerging approach.

AI-Assisted Oral History and Multimedia Knowledge Documentation.

Future research should investigate integrated AI pipelines for the documentation of oral histories, ceremonial songs, traditional ecological calendars, and multimedia knowledge forms that combine speech, image, gesture, and spatial information. Multimodal AI systems that can simultaneously process audio, video, and environmental sensor data offer the potential to capture the full richness of knowledge transmission events such as land-based teaching practices—that text-based systems cannot adequately represent. Research should address how such systems can be designed to respect cultural protocols around recording, how outputs can be stored in ways that honor layered access restrictions, and how AI-generated documentation can be validated and corrected by knowledge holders without requiring technical intermediaries (Holton et al., 2012).

Structural and Political Economy Dimensions

Future research should systematically examine structural and political economy barriers that prevent effective implementation of ethical digital preservation frameworks. This includes analysis of international intellectual property regimes, corporate data governance practices, digital infrastructure inequality, and the influence of donor-driven development agendas on indigenous technology projects. Comparative studies across different geopolitical contexts—particularly in Africa, Latin America, and Asia, which are underrepresented in current literature—are urgently needed (Carroll et al., 2020; Rainie et al., 2017).

Internal Community Governance and Cultural Diversity

Research should also investigate internal governance diversity within indigenous communities, including intergenerational tensions over digitization, gender-based access norms, and the heterogeneity of decision-making traditions across different nations and communities. Frameworks must not treat "indigenous communities" as a monolithic category but should develop adaptive governance models that respect intra-community complexity and change over time (Simpson, 2004; Wilson, 2008).

REFERENCES

Agrawal, A. (1995). Dismantling the divide between indigenous and scientific knowledge. *Development and Change*, *26*(3), 413–439. DOI: 10.1111/j.1467-7660.1995.tb00560.x

Anderson, J. (2005). Protocols for libraries, archives and information services. *The Australian Library Journal*, *54*(4), 370–382.

Battiste, M. (2005). Indigenous knowledge: Foundations for First Nations. *WINHEC: International Journal of Indigenous Education Scholarship*, *1*(1), 1–17.

Bender, E. M., Gebru, T., McMillan-Major, A., & Shmitchell, S. (2021). On the dangers of stochastic parrots: Can language models be too big? In *Proceedings of the 2021 ACM Conference on Fairness, Accountability, and Transparency* (pp. 610-623). ACM. DOI: 10.1145/3442188.3445922

Berkes, F. (2012). *Sacred ecology* (3rd ed.). Routledge., DOI: 10.4324/9780203123843

Bird, S. (2020). Decolonising speech and language technology. In *Proceedings of the 28th International Conference on Computational Linguistics* (pp. 3504-3519). DOI: 10.18653/v1/2020.coling-main.313

Carroll, S. R., Garba, I., Figueroa-Rodríguez, O. L., Holbrook, J., Lovett, R., Materechera, S., Parsons, M., Raseroka, K., Rodriguez-Lonebear, D., Rowe, R., Sara, R., Walker, J. D., Anderson, J., & Hudson, M. (2020). The CARE Principles for Indigenous Data Governance. *Data Science Journal*, *19*(1), 43. DOI: 10.5334/dsj-2020-043

Christen, K. (2011). Opening archives: Respectful repatriation. *The American Archivist*, *74*(1), 185–210. DOI: 10.17723/aarc.74.1.4233nv6nv6428521

Christen, K., & Anderson, J. (2019). Toward slow archives. *Archival Science*, *19*(2), 87–116. DOI: 10.1007/s10502-019-09307-x

Dawson, P. C. (2019). Repatriation, virtual repatriation and the Inuvialuit living history project. In Fforde, C., McKeown, C. J., & Keeler, H. (Eds.), *The Routledge companion to indigenous repatriation* (pp. 515–534). Routledge.

Harrison, R., DeBruijn, L., Morely, I., Beltran, M., & Smith, C. (2020). Heritage and data justice. *Heritage & Society*, *13*(1-2), 78–102. DOI: 10.1080/2159032X.2020.1847890

Holton, G., Leonard, W. Y., & Pulsifer, P. L. (2012). *The Arctic Borderlands Ecological Knowledge Co-op*. NSERC.

Katuu, S. (2016). Managing digital heritage in times of technological change and innovation. In *Managing Digital Cultural Objects: Analysis, Discovery and Retrieval* (pp. 27-48). Facet Publishing. https://doi.org/DOI: 10.29085/9781783301270.003

Kukutai, T., & Taylor, J. (Eds.). (2016). *Indigenous data sovereignty: Toward an agenda*. ANU Press., DOI: 10.22459/CAEPR38.11.2016

Kukutai, T., & Taylor, J. (Eds.). (2016). *Indigenous data sovereignty: Toward an agenda* (Vol. 38). ANU press. DOI: 10.22459/CAEPR38.11.2016

Kuo, T. T., Jiang, X., Tang, H., & Ohno-Machado, L. (2023). Enabling precision medicine via standard communication of HTS provenance, analysis, and results. *PLoS Biology, 21*(3), e3002023. DOI: 10.1371/journal.pbio.3002023 PMID: 36917574

Mager, M., Gutierrez-Vasques, X., Sierra, G., & Meza-Ruiz, I. (2021). Findings of the AmericasNLP 2021 shared task on open machine translation for indigenous languages of the Americas. In *Proceedings of the First Workshop on Natural Language Processing for Indigenous Languages of the Americas* (pp. 118-138). ACL. DOI: 10.18653/v1/2021.americasnlp-1.23

Nakata, M., Nakata, V., Gardiner, G., McKeough, J., Byrne, A., & Gibson, J. (2008). *Australian Indigenous digital collections: First generation issues*. Australian Academic and Research Network.

Posey, D. A., & Dutfield, G. (1996). *Beyond intellectual property: Toward traditional resource rights for indigenous peoples and local communities*. IDRC.

Rainie, S. C., Kukutai, T., Walter, M., Figueroa-Rodríguez, O. L., Walker, J., & Cariou, E. (2017). Issues in Open Data: Indigenous Data Sovereignty. In Davies, T., Walker, S., Rubinstein, M., & Perini, F. (Eds.), *The State of Open Data: Histories and Horizons* (pp. 300–319). African Minds and International Development Research Centre.

Ridge, M. (2013). From tagging to theorizing: Deepening engagement with cultural heritage through crowdsourcing. *Curator (New York, N.Y.), 56*(4), 435–450. DOI: 10.1111/cura.12046

Rieke, N., Hancox, J., Li, W., Milletari, F., Roth, H. R., Albarqouni, S., Bakas, S., Galtier, M. N., Landman, B. A., Maier-Hein, K., Ourselin, S., Sheller, M., Summers, R. M., Trask, A., Xu, D., Baust, M., & Cardoso, M. J. (2020). The future of digital health with federated learning. *NPJ Digital Medicine, 3*(1), 119. DOI: 10.1038/s41746-020-00323-1 PMID: 33015372

Simpson, L. (2004). Anticolonial strategies for the recovery and maintenance of indigenous knowledge. *American Indian Quarterly, 28*(3), 373–384. DOI: 10.1353/aiq.2004.0107

Smith, L. T. (2012). *Decolonizing methodologies: Research and indigenous peoples* (2nd ed.). Zed Books.

Te Mana Raraunga. (2018). *Principles of Māori data sovereignty*. Te Mana Raraunga.

Thompson, M., Vervoort, J., & Frantzeskaki, N. (2020). Relationality and ethics in transformative sustainability research. *Sustainability Science, 15*(3), 849–861. DOI: 10.1007/s11625-020-00786-5

Todd, Z. (2016). An indigenous feminist's take on the ontological turn: 'Ontology' is just another word for colonialism. *Journal of Historical Sociology, 29*(1), 4–22. DOI: 10.1111/johs.12124

Traditional Knowledge Digital Library. (2014). *TKDL framework and practices*. Government of India.

Truby, J. (2018). Decarbonizing Bitcoin: Law and policy choices for reducing the energy consumption of Blockchain technologies and digital currencies. *Energy Research & Social Science, 44*, 399–410. DOI: 10.1016/j.erss.2018.06.009

Tuhiwai Smith, L. (1999). *Decolonizing methodologies: Research and indigenous peoples*. Zed Books.

UNDP. (2021). *Indigenous peoples and COVID-19: From education to crisis recovery*. United Nations Development Programme.

UNESCO. (2018). *Safeguarding intangible cultural heritage*. UNESCO Publishing.

Walter, M., Kukutai, T., Carroll, S. R., & Rodriguez-Lonebear, D. (Eds.). (2021). *Indigenous data sovereignty and policy*. Routledge., DOI: 10.4324/9780429273957

Wilson, S. (2008). *Research is ceremony: Indigenous research methods*. Fernwood Publishing.

Chapter 8
Afrocentric Ethical Guidelines for Digital Content Creators Producing Indigenous Knowledge-Based Content

Tsakani Emmarencia Nyoni
http://orcid.org/0000-0002-4034-1112
University of Venda, South Africa

Vhonani Olive Netshandama
http://orcid.org/0000-0002-2466-6099
University of Venda, South Africa

ABSTRACT

Indigenous Knowledge(IK)-based digital content creation is increasingly being produced. Digital content consumers are also taking more interest in the content. Sacred IK-based content can be accessed on platforms like TikTok and YouTube, and, often, it is produced without the owner's awareness or acknowledgement. These developments in the IK Management space highlight the need for Afrocentric digital content creation ethical guidelines. Existing guidelines are either focused on digital content creation in general, or on research with Indigenous Knowledge. Both guidelines can serve only as a foundation for IK-based digital content creation ethical guidelines, as they are not fully applicable to it. This chapter presents findings on IK-based ethical guidelines for producing digital content. The data was generated through sharing circles and semi-structured interviews, and analyzed using a thematic analysis technique. The findings have highlighted how the digital content creation process can be conducted respectfully and inclusively with Indigenous Knowledge Holders.

INTRODUCTION

According to Akinwale:

DOI: 10.4018/979-8-3373-7976-0.ch008

"The Indigenous Knowledge of Africans remains a gold mine, although it has been suppressed since the advent of cultural imperialism exemplified by harrowing scenes of devastation in the contexts of slavery, colonialism and neocolonialism" (2013:1).

The digital age has seen an increase in the production of Indigenous Knowledge (IK)-based content by digital content creators. The content can be created for entertainment, awareness, promotional or educational purposes. The term "digital content" is used to refer to information that exists in a digital format, and can be processed by computers. The term encompasses computer programs, mobile apps, animations, videos, photos, podcasts, blog posts, website updates, and editing and distributing any media content (Shopify, 2024). In short, digital content creation also relates to one's ability to design, develop, and publish digital materials, as well as proficiency in digital technologies, storytelling skills, design principles, media production, and publishing practices (Acorn, n.d.).

Meis, Fogg, and Scheumann (2024) highlight that digital content creation is a process which can be outlined in three stages as follows: (i) Research and planning, (ii) Creating the content, and (iii) Performance tracking. The digital content in reference is identified as "born digital" in the National Policy on digitisation of Heritage resources draft of 2010. The National Policy on Digitisation of Heritage Resources outlines "Born digital" as content that originally exists in a digital format, with no physical/hard copies of such content (Department of Arts and Culture, 2010).

Currently, digital content creation is increasingly being used as a tool to promote or raise awareness on indigenous knowledge. How indigenous knowledge is presented on digital platforms is important. Historically, Indigenous Knowledge-based content was documented through research documents. Scholars like Smith (1999), Wilson (2008), and Chennels and Schoeder (2018) have discussed how research on indigenous communities and their knowledge has previously been problematic, as it had been an exploitative enterprise. The scholars also advocate for the use of decolonised and respectful methodologies for "re-searching" the knowledge.

In the digital age, with more youth producing and consuming IK-based digital content, Nyoni (2023) discusses the need for Afrocentric ethical guidelines for the production of indigenous knowledge - based content. As this would guide towards ensuring that the process of creating digital content is one that is respectful, inclusive and provides Indigenous Knowledge Holders with the opportunity to tell their own stories.

The application of Afrocentric ethical guidelines is just as essential when engaging in the digitisation of African Indigenous Knowledge. Digitisation may involve converting the format of knowledge that originally exists in physical documents or through tangible documentation to a digital format - one that can be accessed through the online digital technology platforms. In the National Policy on Digitisation of Heritage Resources draft document of 2010, digitisation refers to the conversion of information that exists in an analogue format to a digital form (Department of Arts and Culture, 2010).

In this chapter, digital content creation and digitisation are discussed simultaneously. Both contribute to the safeguarding and promotion of African Indigenous Knowledge (Nyoni & Maimela, 2025), although the original format of the AIK in question varies.

This chapter seeks to address the following research question: (i) Which ethical guidelines should be followed by digital content creators producing Indigenous Knowledge (IK) - based content? The ethical guidelines that will be presented in this study emanate from the findings of a Masters study - Nyoni (2023), titled "The co-creation of Afrocentric ethical guidelines for digital content creators producing indigenous knowledge-based content".

THEORETICAL GROUNDING

The digital content creator's ethical guidelines presented in this study are underpinned by Afrocentricity. There are numerous ideas and discussions on "ethical digital content creation" in the literature. However, none of those are relevant to African indigenous communities. Digital content creation on indigenous knowledge and in indigenous communities requires observation and respect to cultural norms. Thus, this chapter focuses on "Afrocentric ethical guidelines for digital content creators producing indigenous knowledge - based content".

Afrocentricity is essential for the Afrocentric ethical guidelines presented in this study because it is an attempt to remind creators of the significance of respecting African norms, practices and ways of being, despite a creator's etic perspective. We use Afrocentricity in this chapter as a reminder to respect African Indigenous Knowledge. Oftentimes, a digital content creator may not be from the community, however they should possess the wisdom to discern and respect what they do not understand. Ethical guidelines for digital content creation have been presented in the literature by numerous writers, however not in relevance to African Indigenous communities. As such, the point of departure for this chapter is not to challenge the sufficiency of those ethical guidelines, but, merely, to contribute to the ethical guidelines.

Asante (2007:6) shares the following:

"Afrocentricity is not the reverse of Eurocentrism; neither is it a counter to Eurocentrism. Even if Eurocentrism never existed, there would be a need for African people to operate from their own sense of agency".

Afrocentricity is presented as a perspective centred on lived experiences of Africans (Asante, 2007). Chawane (2016: 80) outlines a number of definitions on Afrocentricity. One of those definitions, similar to Asante (2007)'s view, outlines Afrocentricity as "African centredness".

THE NATURE OF AFRICAN INDIGENOUS KNOWLEDGE MANAGEMENT

African Indigenous Knowledge (AIK) has been passed down generationally through rites of passage, socialisation, songs and dance, storytelling, rituals, ceremonies, proverbs, paintings, carvings, etc. However, over the years, younger generations have taken less interest in participating in traditional practices and exchanges that are vehicles of transmission of AIK. Storytelling was at the core of transmission of African Indigenous Knowledge. Grandparents played an active role in telling the stories to impart their knowledge and instill values to younger generations (Bhatia, 2025).

The stories that the younger generation was told were meant to guide them on behaviour, morals, life lessons and how to co-exist with the natural environment. Some of the stories included knowledge on the behaviour of celestial bodies (i.e stars, moon) and how those informed life on earth. The youth's decreased interest in the knowledge heightens the gap in the transmission of the knowledge and the threat of extinction against it. While African indigenous knowledge continues to be vital for African indigenous communities (Sewdass, 2014), Balogun (2023b) reiterates that it is under the risk of loss.

Mdlhuli, Mokgoatšana, Kugara and Vuma (2021) assert that secrecy in Indigenous people's management of their knowledge contributes to its challenges of extinction. The scholars also add that the erosion of Vhaven a's (of Limpopo, South Africa) indigenous ways of either disseminating or safeguarding the knowledge can be attributed to the prevalence of the church and modern schools within the Vhembe district. The systems of operations in churches and schools tend to perpetuate remnants of colonialism which sought to marginalise African Indigenous Knowledge and the practices of the communities.

Schools have also become the main form of education, with the youth spending more time in schools. This has left winter and summer school holidays as the only period of attending African indigenous initiation rites, and even so, fewer young people participate. Family structures have also evolved (Talašová, & Souralová, 2025), which has also transformed the nature of African Indigenous knowledge management from what we originally knew it to be.

Grandparents have become classified as "extended family", female-headed households are becoming more prevalent. Grandparents may not be able to play their role of disseminating knowledge actively, and in some households, fathers are not as present to instill necessary African indigenous values or knowledge that are essential to how knowledge is managed. Aubel (2024) outlines the important roles played by grandmothers, these include the preservation and transmission of cultural traditions, providing guidance to mothers and children, and supporting the growth of younger generations. Likewise, fathers are just as important to the growth and shaping of sons. They instill values and teach responsibility (Rutaremwa & Shirindi, 2025). Parents and community members play an intrinsic role in the growth of the boy child and their contribution to their communities (Rutaremwa & Shirindi, 2025), the same applies to the girl child.

With the evolution of family structures, it is becoming even more challenging to disseminate AIK using traditional methods. AIK in its nature is communally-owned, current systems, initiatives and ways of living influence the management of the knowledge to be an individual activity, which also affects the quality and authenticity of the knowledge. Communal ownership of the knowledge means that everyone has a role to play which makes the knowledge what it is.

As African Indigenous knowledge management evolves, it is essential to encourage the implementation and use of strategies that favour its management without altering the roles and relationship that communities have with their knowledge. Current platforms and projects of African Indigenous knowledge management should recognise and promote communal knowledge ownership.

African Indigenous Knowledge thrives through practices, trial and error. The mode of learning has dominantly included learning through practice, observation and apprenticeship. The current challenge pertaining to how the knowledge is managed is that it has not actively favoured the documentation of the Tacit Indigenous knowledge. On one end, the younger generation is not actively participating in and observing the practices that they need to learn by doing. On the other hand, the methods of documenting such practices that are relevant to the youth, such as digital technological tools and platforms pose the risk of taking the knowledge out of its context when digitising it.

The learning of indigenous practices and skills involved is co-dependant on a number of aspects in one's environment. Learning such practices and skills through digital/technological platforms may omit context on the state of the environment that learning must take place in. For example, learning how to plant African indigenous plant species through an App, may result in failure to replicate the natural environment required to learn such a skill. The learner in this case may struggle to gain an understanding of the time of the day that such a tree should be planted, the consistency of the soil (which may require physically feeling the texture of the soil), the appearance of the clouds/lack of, which can indicate when rain can be expected and the quantity. Through the learning of such African Indigenous skills through an App, something as small as the time of the year/season could be omitted, which is instrumental in planting.

African indigenous knowledge continues to be essential for the achievement of sustainable development and livelihoods. The knowledge contributes significantly in the health, agriculture, veterinary services, and in arts and crafts (Lwoga, et al., 2010). Management of AIK ensures its continuous contributions in such fields, including the promotion of such knowledge.

Figure 1. African Indigenous methods of Knowledge Management

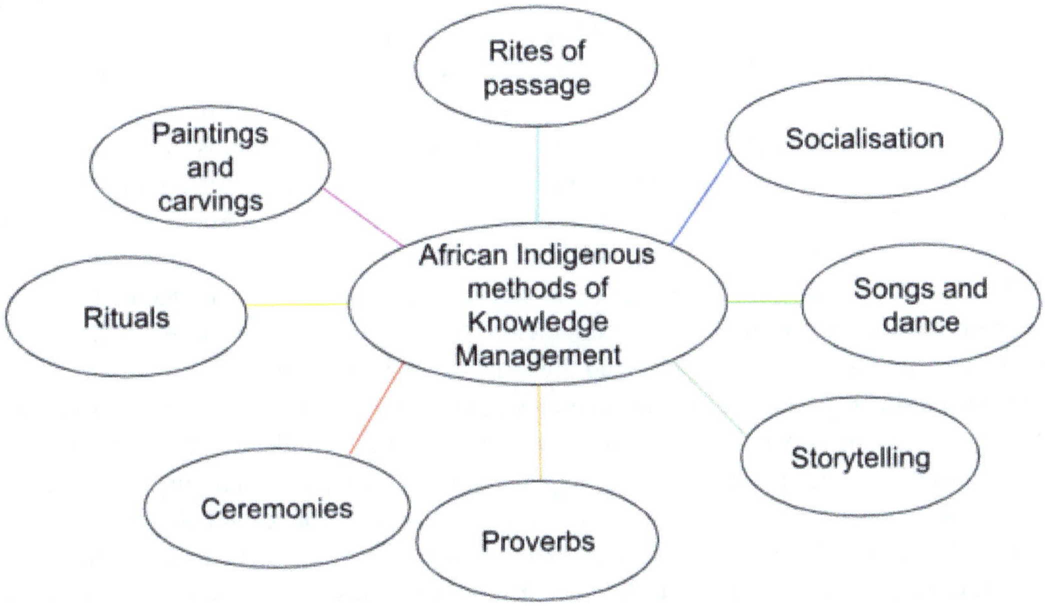

THE EXPLORATION OF INDIGENOUS KNOWLEDGE SYSTEMS

Indigenous Knowledge Systems have long been explored for research purposes, and oftentimes by researchers not indigenous to the communities. Walderman and Branova (2025) reflect on the long history of extractive research in Indigenous communities. Ermine, Sinclair, and Jeffery (2004) share similar reflections and assert that Western knowledge, through research has, in the past, been brought into Indigenous communities with little consideration of Indigenous Worldviews. As a result of the lack of consideration of Indigenous worldviews and the advancing of western knowledge in their communities, Ermine, Sinclair, and Jeffery (2004) shared the following:

> *"The history of Westernisation in virtually all locations of the globe reads like a script of relentless disruption and dispossession of Indigenous Peoples resulting in common patterns of cultural and psychological discontinuity for many in the Indigenous community*" (p.9).

From the above quote, it is clear that Indigenous communities around the world hold similar experiences with Western research(ers). Commonly, Indigenous Knowledge has been exploited, the knowledge undermined, and with Indigenous peoples dislocated from their land - disrupting generations of knowledge and interrelationships built from co-existing in a particular environment.

Indigenous Knowledge and its plant species, were foundational in some of the modern medicine and products sold in the current day (i.e mouthwash, baby bottles and syringes, and even medicines for pain relief), and these have been without the acknowledgement of the communities and custodians through which the knowledge originated (Walderman & Branova, 2025). The Hoodia plant of the San of South Africa can also be used as an example of Biopiracy, and the exploitation of Indigenous knowledge without consideration of the Indigenous communities. Such examples emphasise the significant contributions of Indigenous Knowledge and its plant species to modern medicine.

Linda Tuhiwai Smith writes about how research has been inextricably linked to colonialism and imperialism (Smith, 2021). In her words, research is considered one of the "dirtiest" words in the vocabulary of Indigenous Peoples (Smith, 2021:1). Indigenous Knowledge has also received increased attention for integration in climate research. David-Chavez and Gavin (2018) have found that 87% of climate studies are extractive in nature. The scholars reflect that researchers rarely engage Indigenous Knowledge Holders in their decision-making and their studies are not participatory in nature. Similarly, Couldry and Mejias (2019) discuss the extractive nature of research, in which indigenous knowledge is used for profit or other forms of benefit without acknowledging or sharing those benefits with the knowledge custodians. The scholars have termed the practice "data colonialism".

On the same breath, Henrietta Lacks, a Southern-African American Tobacco farmer, is another example of how research was previously used to exploit minority groups. In her case, she was diagnosed with cervical cancer in 1951, cells from her tumour were removed for research purposes without her permission, the cells, named the HeLa cells, became the source of many ground-breaking research studies. Her family only learnt of the cells' existence and their use in research 25 years after her death (Khan, 2011), and although this is a multi-million-dollar industry, her family did not benefit from any of the profits made from the cells. When the daughter of Henrietta Lacks, Deborah, learned of the HeLa cell line and its significance in many scientific medical studies, she reflected on how, as Henrietta Lacks' children, they could not even afford medical insurance (Khan, 2011).

Through Henrietta Lacks' case, Linda Tuhiwai Smith's, and Walderman and Branova's presentations, it is clear that the extractive nature of research on minority groups dates back many decades. The harmful nature of many research studies has contributed to the development of various research ethical codes and frameworks. Nyoni (2023) reflects on the introduction of the following research ethical codes and guidelines: (i) Nuremberg Code, (ii) The Declaration of Helsinki, (iii) The Belmont report, and (iv) The New Brunswick Declaration. The above-mentioned research ethical codes and guidelines were formulated in response to unethical and harmful research studies. The principles of the above-mentioned ethical codes and guidelines have been intrinsic in informing research ethics as we know them today. Despite continued and numerous developments in the research ethics space, most ethical considerations emanate from a Western-centric background and are not fully applicable or compatible with research with Indigenous Knowledge Systems and communities (Nyoni, 2023).

Much progress has also been made relating to research frameworks for indigenous data and knowledge, that are inclusive of and involve the participation of communities. Walderman and Branova (2025) write about the significance of CARE (Collective Benefit, Authority to Control, Responsibility and Ethics) Principles in Indigenous data governance. CARE Principles are essential as they uphold responsible data practices. Carroll et al., (2022) writes that the CARE principles are an extension of the FAIR (Findable, Accessible, Interoperable and Reusable) Principles.

The International Indigenous Data Sovereignty Interest Group (2019) presents that CARE principles are purpose and people-oriented, and they reflect the role of data in advancing indigenous innovation and self-determination.

Table 1 below presents the CARE Principles and sub-principles.

Table 1. CARE Principles and sub-principles

Collective Benefit	Authority to Control	Responsibility	Ethics
Data ecosystems ought to be designed and function in a manner that enables Indigenous Peoples to derive benefit from the data.	The rights and interests of Indigenous Peoples on their data must be recognised. Their authority in the control of that data should be empowered.	Those working with indigenous data carry the responsibility to share how that data is used to support self-determination and collective benefit of Indigenous Peoples. Accountability entails the meaningful and available evidence of the efforts and how they benefit the community.	Rights and well-being of Indigenous Peoples should be the main concern in all stages of the data lifecycle and across the data ecosystem.
C1: Governments and institutions ought to actively support the use and reusing of data by indigenous nations. Communities should facilitate the establishment foundations focused on indigenous innovation, value generation and promote local self-determined development processes	**A1**: Indigenous peoples have rights and interest on Indigenous data and knowledge. As a collective and individually, they have rights to Free, Prior and Informed Consent in the collection and use of the data, including developing data policies and protocols for its collection.	**R1**: Indigenous data usage is not viable unless it is linked to relationships built on reciprocity, trust, respect, and mutual understanding. Those working with indigenous data should ensure that how it is created, interpreted and used maintains respect and dignity of indigenous nations and communities.	**E1**: Data that is ethical does not portray Indigenous Peoples, cultures or knowledge in terms of deficit. The data should be collected and used in a manner that is in alignment with Indigenous ethical frameworks and with rights outlined in UNDRIP. The assessment of ethical benefits and harm should be through the lens of Indigenous nations and communities.
C2: Data-enriched planning, implementation, and evaluation processes that support service and policy needs of the indigenous communities. The data should improve engagement between citizens, institutions and governments for informed decision-making. The ethical use of open-data can improve transparency and decision-making, thus enabling communities to better understand their peoples, territories and policies.	**A2**: Indigenous Peoples have the right to data that are relevant to their worldviews, empowering to self-determination, and effective self-governance. The data must be available and accessible to indigenous nations and communities to ensure support to self-governance.	**R2**: The use of the Indigenous data should invoke a reciprocal responsibility for the purpose of enhancing data literacy within the communities, and support the development of Indigenous data workforces and digital infrastructure that enables collection, management, security, governance and application of data.	**E2**: Ethical processes ought to address imbalances in power, resources and how they affect the expression of human and indigenous rights. Such ethical processes must be inclusive of representation from relevant indigenous communities.
C3: Indigenous data are rooted in community values, which extend to society. Any value created from indigenous data should be beneficial to the community in an equitable manner, while contributing to their aspirations for wellbeing.	**A3**: Indigenous peoples have the right to develop cultural governance protocols for indigenous data and be active leaders in the stewardship of, and access to in the Indigenous Knowledge context.	**R3**: Resources grounded in the languages, worldviews, lived experiences, including values and principles of indigenous peoples must be provided.	**E3**: Data governance should be in consideration of potential future use and harm emanating from ethical frameworks grounded in values and principles of the relevant indigenous community. Metadata should acknowledge provenance, purpose and limitations or obligations in secondary use of data, including consent issues.

*(International Indigenous Data Sovereignty Interest Group, 2019)

In South Africa, the San Community has developed the San Code of Research Ethics (2018). The guidelines emphasise the importance of maintaining respect, honesty, justice, fairness and care for researchers conducting research with San communities (Chennels & Schoeder, 2018). Between the CARE Principles and San Code of Research Ethics, respect to Indigenous Peoples and communities is seemingly at the core of the frameworks. In addition to respect, research with Indigenous communities should be grounded in transparency. Moreover, where benefits or value are accrued from the use of Indigenous data and knowledge, indigenous communities should benefit.

Beyond research ethics, there have also been numerous calls for the decolonisation of research methodologies (Smith, 1999), and moving away from Western-centric methodologies. Scholars such as Wilson (2008), Archibald (2008), Chilisa (2012), Khupe and Keane (2017), Pidgeon and Riley (2021), Kovach (2021) and Chigevenga (2022) have written about the significance of Indigenous Research Methodologies. Chigevenga (2022) has specifically reflected on the struggles of using Western-centric methodologies, as these have made her feel disconnected from study stakeholders. In addition, she has reflected on how the use of Western-centric methodologies has affected the analysis and authenticity of research findings.

Commonly, scholars in the decolonial indigenous research methodologies present relationship and reciprocity as the core principles that should underpin any indigenous research methodology. Although their work may originate from different contexts and communities, this signifies that Indigenous Communities value creating meaningful relationships and reciprocity.

Kater (2022) presents a different view as a non-Indigenous Natural Scientist seeking to conduct decolonial research. The scholar reflects on the need for training and awareness of what it means to conduct decolonial and collaborative research. In her attempts to engage with communities for her Doctoral study, she was first met with resistance due to "research fatigue", although she intended to conduct "collaborative" research. Mentioning the idea of conducting "collaborative" research is one thing, Kater (2022) was able to show the indigenous community she sought to engage how she could conduct that collaborative research, and this was through being reflective, bringing the community on board for the research and being open to learning. Many non-indigenous researchers entered communities with their own preconceptions, which later hindered their understanding and learning of what was new to them. Writing about wanting to conduct a "collaborative" study is not enough; researchers should be reflective, open-minded and willing to learn.

In Kater (2022: 3), research fatigue is said to occur when marginalised, minority indigenous peoples are frequently approached, engaged, surveyed or incorporated in research projects to present their views. Perhaps Ermine et al., (2004) are accurate when they present that the experiences of indigenous communities from research, globally, read like a script. Indigenous communities share similar experiences, research fatigue is an example of that, as we have experienced Indigenous Knowledge Holders in the Vhembe district of Limpopo, South Africa, sharing their lack of interest in participating in research projects due to how much they have felt disrespected in the past, all in the name of research.

Walderman and Branova (2025) advocate for Indigenous data sovereignty. The concept refers to the right of Indigenous Peoples to control how data on their knowledge is generated, used, stored and shared (Walderman & Branova, 2025). With the increasing interest in Indigenous Knowledge for research and digital content creation, and considering how much injustice has been perpetuated on Indigenous communities and their knowledge, it is time for Indigenous communities to be afforded the opportunity and space to manage data on their knowledge.

ETHICAL DIGITAL CONTENT CREATION

Traditionally, knowledge gatekeepers, such as publishers and editors, were responsible for curating and verifying content. In this digital age, anyone can be a content creator. However, not everyone is aware of the ethical responsibilities associated with creating and sharing content with consumers, and consumers engaging with the content ethically (Sinclair, 2025). Sinclair (2025) further highlights the common practice of creating "clickbait" content, misrepresenting information and creating content for trends or virality. These are challenging, as facts can always be misconstrued.

Producing ethical digital content fosters inclusivity, truthfulness and respect. Unethical content creation can perpetuate digital harassment, privacy violations and the commodification of personal experiences (Patalauskaitė, 2024). It is important to note that digital content can manipulate, inform, and persuade audiences (Patalauskaitė, 2024). Thus, Sinclair (2025) highlights the importance of regulating content practices, as unregulated content can contribute to malicious content. Regulating content should include encouraging transparency, promoting algorithms that prioritise accuracy, and foster accountability for content creators.

Patalauskaitė (2024) outlines the following aspects which should be considered when creating ethical digital content: (i) Transparency, which has also been highlighted by Sinclair (2025), (ii) Accuracy, (iii) Respect to consumers of the content, including audiences, whether direct or indirect, (iv) Professionalism - compliance to set ethical guidelines, (v) Benefit to audience, and (vi) Responsibility. In addition, the scholar has highlighted the need for acknowledgement of the Intellectual Property Rights (IP) of individuals from whom the content originates.

Acknowledging the owner of digital content posted online can be in the form of tagging that person on a post if posting on social media platforms, or adding a link to the original post. Sharing another person's work as your own, without acknowledging the source counts as plagiarism and could result in legal action (Abrahams, 2024). Abrahams (2024) maintains that understanding copyright and how it works, can also assist content creators in understanding how they can use someone else's work, and vice versa.

Mudau (2023) discusses the significance of consent when creating digital content. This emanates from the observation that content creators have tended to post videos online without permission. This violates the Protection of Personal Information Act (POPIA). In South Africa, digital content creators should adhere to the Protection of Personal Information Act 4 of 2013. Adhering to the POPIA means that people should be able to consent to the use of their pictures, videos or personal information of any kind being used to create digital content. Failure to acquire such consent, makes the creation of that content unethical.

With the increasing use of Artificial Intelligence (AI) to generate digital content, Sinclair (2025); Patalauskaitė (2024); and Abrahams (2024) highlight the importance of the ethical use of AI. Resnick and Hosseini (2025) assert that "AI" is usually used in reference to technology or tools performing tasks that require intelligence. The tasks include perception, judgement, reasoning or decision-making. What makes AI unique is its adaptability to new information, AI learns from new information (Resnick & Hosseini, 2025).

Artificial Intelligence can also be used for Digital Content Creation. Resnick and Hosseini (2025) maintain that it is important to be transparent in instances where AI has been used to generate content either for research or any other purpose. It is just as important for digital content creators to use AI responsibly and declare if it has been used to generate digital content.

DIGITISATION OF AFRICAN INDIGENOUS KNOWLEDGE (AIK)

Nyoni and Maimela (2025) discuss the use of digital content creation (a tool for digitising AIK) in safeguarding African Indigenous Knowledge. Digital content creation can be used to entertain, inform, promote or sell. In this digital age, it has emerged as a tool for promoting African Indigenous Knowledge. On platforms like TikTok and YouTube, an increasing number of vlogs are being posted on AIK, and such vlogs can be accessed from anywhere in the world. These are vlogs promoting African indigenous crafts, foods, attires, accessories, practices, dance and songs. African Indigenous Knowledge is either tacit or explicit in nature, regardless, vlogs cater to the documentation of AIK in whichever form it exists in. This makes it easy to learn or access the knowledge and skills that require practice.

With the vast nature African Indigenous Knowledge, digitisation of AIK contributes to the safeguarding and promotion of generations of knowledge on natural resource management, food and medicine, Indigenous plant species, Indigenous languages, proverbs, norms and practices, folktales and sustainable development (Akinwale et al., 2013; Sraku-Lartey et al., 2017; Balogun & Kalosupa, 2022; Bawack et al., 2025).

As interest on the digitisation of AIK grows, Hunter (2005: 110) outlines steps to be followed in Indigenous Knowledge digitisation projects as follows:

(i) Identification of content that needs to be safeguarded and captured into digital formats. The digitised knowledge should be accessible for re-using by permitted groups;
(ii) Determination of a "culturally appropriate approach" of selection, recording, describing and sharing the knowledge without being insensitive, intrusive or misrepresenting the content.
(iii) A consultative process should be established to determine appropriate quantity of content to be captured (how much and how deep should the digitisation process probe into the Indigenous Knowledge?).
(iv) Comparison and evaluation of suitable or "alternative" digital technology for recording and storing the content is essential.
(v) Consideration of the format of recording the content versus the format (explicit or tacit) of the knowledge is important. The selected formats of recording should be suitable for either digitising songs or photographs, and artifacts.

Over the years, various platforms and initiatives for the digitisation of AIK have emerged. These continuously evolve to suit the needs of consumers and other initiatives continue to be introduced to cater for the digitisation of AIK. Sraku-Lartey et al. (2017) outlines DRUPAL. DRUPAL is a free, open-access content management system. The system has been used to document knowledge on indigenous plant species (Sraku-Lartey et al., 2017). Table 2 below outlines platforms that digitise.

Table 2. Digitisation platforms and type of IK digitised

Globally	Type of IK digitised
India's Traditional Knowledge Digital Library (TKDL)	Indian Traditional Medicinal Knowledge
Korean Traditional Knowledge Portal	Korean Traditional Medicinal Knowledge (herbs and prescriptions for certain diseases), Traditional food and Agriculture
Chinese Traditional Medicine Database	Traditional Chinese Medicinal herbs (and properties)
Smithsonian Centre for Folklife and Cultural Heritage	Research on Cultural heritage
DRUPAL	Indigenous plant species
Africa	
National Indigenous Knowledge Management System (NIKMAS)	Indigenous food and Traditional Medicine
Elimu Asilia - Kenya's Indigenous Knowledge Online	IK on culture, food, health and environment

*(Balogun & Kalosupa 2022; Sraku-Lartey et al., 2017)

In addition to the platforms outlined above, Balogun and Kalosupa (2021b), present the First Nations Helpdesk in Canada, India's Traditional Knowledge Digital Library, Netherland's Center for International Research and Advisory Networks' IK Homepage, and the Native in the Americas as some of the earlier digitisation of projects for Indigenous Knowledge. In South Africa, the earliest digitisation initiative was a workshop hosted in 1997, which was held at the University of Witswatersrand. This digitisation workshop led to the DISA project - known as the Digitisation Imaging South Africa (Department of Sports, Arts and Culture, 2013).

Beyond DISA, in recent years, South Africa's Department of Science, Technology and Innovation, in collaboration with the Council for Scientific and Industrial Research (CSIR) initiated the National Recordal System (NRS) (Khalala, 2021). Emanating from the Indigenous Knowledge Systems policy framework (Balogun & Kalosupa, 2022). The NRS was proposed in response to the need to protect Indigenous Knowledge considering its decline (Khalala, 2021). The NRS seeks to coordinate and standardize the capturing, storing, maintenance and dissemination of Indigenous Knowledge Systems (Balogun, 2023a). The Indigenous Knowledge Systems Policy Framework has also laid the foundation for the establishment of NIKSO - known as the National Indigenous Knowledge Systems Office (Balogun, 2023a).

NIKSO plays the role of facilitating the coordination of academic institutions and/or organizations in relation to IKS. The Department of Science, Technology, and Innovation's collaboration with the CSIR, included the implementation of NIKMAS, which is the National Indigenous Knowledge Management System (Khalala, 2021). Balogun (2023a) suggests that NIKSO's role in the facilitation and coordination of academic institutions and organisations led to the establishment of NIKMAS. NIKMAS supports the National Recordal System (Balogun & Kalosupa, 2022), thus, Khalala (2021) suggests that it is "the backbone of the digital repository" (the digital repository in reference is the NRS). In addition, NIKMAS was created for the purpose of documenting traditional medicine, indigenous environmental knowledge, ethnobiology, and intangible cultural heritage (Khalala, 2021).

Thus far, the platform has documented knowledge on African Indigenous foods and traditional medicine. The knowledge mentioned above is captured through IKSDCs - known as Indigenous Knowledge

Systems Documentation Centres, existing across South Africa's 9 Provinces (Khalala, 2021; Balogun, 2023b).

NIKMAS and the Indian Traditional Knowledge Digital Library share a similar role of supporting the recordal, management and protection of IK. This is to manage Intellectual Property and Biopiracy challenges (Balogun, 2023b). Since the establishment of NIKMAS, 90 rural communities across all the South African provinces record their knowledge on the platform. In addition, 14 750 Indigenous Knowledge entries have been documented, with over 2 070 Indigenous Knowledge Holders (Khalala, 2021).

It is noteworthy that as digitisation projects and initiatives ensue, efforts should also include the establishment of "Trusted Digital Repositories (TDR)" (Balogun, 2023b). The scholar asserts that the concept of "Trusted Digital Repositories" refers to trust placed on digital records, and ensures the long-term preservation and management of digital materials, it is thus important to establish a TDR. Indigenous Knowledge Holders and practitioners should be able to trust the digitisation processes of their knowledge and the digitised content itself.

Sraku-Lartey et al. (2017) write about the benefits and challenges of digitising AIK. The digitisation of AIK was identified as beneficial in ensuring long-term accessibility and preservation, however, it also poses challenges relating to Intellectual Property Rights, language used in documentation and funding (Sraku-Lartey et al., 2017). For South Africa, digitisation progress was slow during the period from the first digitisation project – DISA, in 1997. This was a result of the lack of resources and funding for digitisation initiatives (DAC 2013:7). Seemingly, funding is an essential determinant of progress on digitisation efforts.

Maimela et al. (2024) outlines that while digitisation is essential for the cultural preservation and promotion of AIK, digital equality is yet to be achieved. According to Akinwale (2013:3) two distinct groups emanate from the digitisation of African indigenous knowledge, namely; the knowledge generators and the knowledge consumers. The existence of these two groups is a result of the digital divide, and a lack of digital equality. The digital divide refers to the existing gap between people with access to information and those without (Roya et al., 2023)

Thus, the digitisation of African Indigenous Knowledge should be inclusive of Knowledge Holders. Although, the inclusion of Indigenous Knowledge Holders in the digitisation process is not enough, as this suggests that Indigenous Knowledge Holders could rely on others to take the initiative of digitisation of their knowledge. Nyoni (2023) emphasises the importance of Indigenous Knowledge Holders telling their own stories. However, digitisation should go beyond Indigenous Knowledge Holders telling their own stories, it should be that they should be able to tell stories on their terms.

Platforms and projects for the digitisation African Indigenous Knowledge should be co-designed and co-created with Indigenous Knowledge Holders. This could eliminate the challenge of creating platforms and initiatives that are not realistic to the needs and "reality" of the Holder who needs to digitise their knowledge for protection. Beyond co-designing and co-creating an initiative or platform with an Indigenous Knowledge Holder in mind, communities that own the knowledge in question should be aware of where and how to access their knowledge.

LEGISLATIVE AND POLICY FRAMEWORKS ON THE DIGITISATION OF AIK

The Indigenous Knowledge Systems Policy of 2004 has set precedent for the protection of AIK in South Africa. In the policy, the significance of protecting the knowledge and how it can be protected is highlighted. The insufficiency of the Intellectual Property system is also outlined. The digitisation of AIK is one way of protecting AIK and ensuring long-term accessibility (Balogun, 2023), however, it also presents challenges on biopiracy, loss of knowledge ownership and unethical presentation of the knowledge.

The Protection, Promotion, Development and Management of Indigenous Knowledge Systems Act 6 of 2019 was also signed into law. The Act lays the foundation for documentation of Indigenous Knowledge for the purpose of protection from loss, biopiracy and misappropriation. The Act outlines the registration processes of AIK and and the eligibility criteria of registration for the protection of the knowledge.

Both the Indigenous Knowledge Systems Policy and Indigenous Knowledge Systems Act 6 of 2019 make significant contributions to the protection of AIK. Through the foundation of both the Policy and Act, this section of the chapter seeks to explore any current legislative and policy frameworks in place to guide the digitisation of AIK.

Balogun and Kalosupa (2022) maintain that there are no specific policies focusing on the digitisation of AIK. However, there is a need for such. The scholars highlight the need for a digitisation policy focusing on AIK, and such a policy can expand from the Indigenous Knowledge Systems Policy of 2004 and the National Policy on Digitisation of heritage resources. Through the study of Balogun and Kalosupa (2022), it has also become clear that very few individuals working in Indigenous Knowledge Systems Documentation Centres (IKSDCs) are aware of existing policy frameworks that can contribute to the long-term preservation of AIK, and even fewer individuals understand the importance of such policies.

Legislative and policy frameworks on the digitisation of African Indigenous Knowledge are essential for the following:

(i) Standardisation of how AIK should presented on digital platforms;
(ii) Ensuring accountability on how AIK is digitised (is it in accordance with the Policy?);
(iii) Outlining roles and responsibilities in the digitisation process (who is responsible for what?);
(iv) Clarifying the digitisation of Explicit and Tacit African Indigenous Knowledge (which methods of digitisation are applicable to each form of AIK without compromising the quality of the knowledge); and
(v) Outlining accessibility of the knowledge (who can access what in instances of sacred knowledge and how can digitisation safeguard the knowledge without violating its boundaries and communal integrity?).

THE DIGITAL CONTENT CREATION PROCESS

The information presented in this section is drawn from my Master's dissertation (Nyoni, 2023), which focused on the co-creation of Afrocentric ethical guidelines for digital content creators producing indigenous knowledge-based content. In the Master's study, several participants co-created Afrocentric ethical guidelines for co-producing indigenous knowledge-based content. The generated guidelines follow a three-step process and are presented below:

Figure 2. Three-step process of digital content creator's ethical guidelines

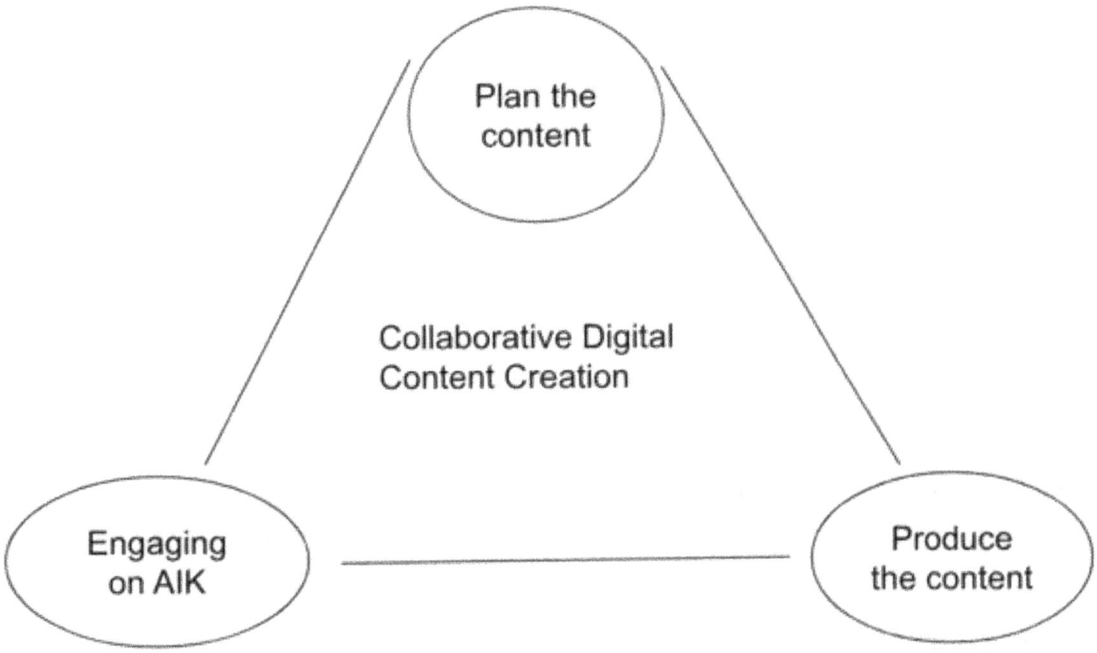

PLAN THE CONTENT

Producing Indigenous Knowledge-based digital content should be planned, especially when a digital content creator is not from the community. Designing a plan for the digital content creation process matters because it allows the creator to identify a niche for the content they would like to produce and an Indigenous Knowledge Holder(s) to collaborate with. Beyond identifying a niche for the content, creators need to consider how to access the indigenous community in a respectful manner.

Planning the digital content creation process is essential to avoid violating boundaries, as the community will have the opportunity to clarify these when permission to engage is sought. In addition, creators can ensure that they are not producing content based on sacred knowledge that may not be for the consumption of the general public. With digital content creation on AIK on the rise, currently, there is a lot of content produced with sacred knowledge, for example, certain practices or rituals on *ku thwasa,* knowledge on what is taught in girls and boys' initiation schools. This knowledge is regarded as sacred, however the level of respect surrounding such knowledge and practices is diminishing.

Rites of passage in African indigenous communities contributed to the transmission of African indigenous knowledge, knowledge was only shared with an individual who met certain requirements and standards or had reached a certain age. In today's age, some of that knowledge is accessed by people who are not of age, or meet the requirements to access that knowledge through digital content creation. Digital content creators should be conscious of the knowledge they can create content with, and knowledge that is not suitable for the creation of content on public social media platforms.

Nyoni (2023: 62) details the process of accessing an African Indigenous community that can be considered by digital content creators. Creators should consider writing a letter to the Traditional Leadership of the community they plan on visiting. The letter should detail out the intentions of the creator. The creator should clarify why they would like to produce content on African Indigenous Knowledge and the type of content they are looking to produce.

Writing this letter or seeking permission to access the community does not guarantee approval. It is just as important to note that when approval is granted, one cannot show up to a community empty-handed. Bearing gifts when visiting African indigenous communities is a symbol of respect and an expression of *Ubuntu,* just as much as the community opening doors for guests. It is also important to note that cultural norms are not universal, each Indigenous community may have its own unique way of receiving guests and expectations from that guest.

Preliminary engagements within the community are for the purpose of clarifying the creator's intentions and becoming familiar with cultural norms. Clarifying such intentions ensures that Indigenous Knowledge Holders are clear on what to expect from the digital content creation process, how they participate and the purpose of the production of such content. Digital content creators ought to become familiar with the cultural norms of the community. The cultural norms are inclusive of how to greet, how to address the Knowledge Holders, dressing and communicating respectfully, and how one may conduct themselves.

ENGAGING ON AIK

As engagements towards the content commence, creators have the responsibility of managing expectations, building a relationship and continuously maintaining respect. Managing expectations matters, as Indigenous Knowledge Holders should be aware of what they stand to benefit from the process. If the aim for producing the Indigenous content is to promote or safeguard a certain cultural practice, craft, food, medicine or an indigenous plant, such should be clarified.

The Indigenous Knowledge Holders should be aware of the implications of their knowledge being used for digital content. Most importantly, which platforms will be used to post the content and how to access it. Engagements with Indigenous Knowledge Holders should be guided by constantly seeking clarity on what is not understood (instead of assumptions) and feedback.

Indigenous Knowledge Holders engage through storytelling, their stories are educational, and the practice and art of storytelling itself should be safeguarded. Creators should expect that Holders will engage through stories, therefore, the process of engaging should not be rushed. If a creator intends on learning from an Indigenous Knowledge Holder, they should not impose expectations to get the content for their videos. Thus, the importance of Holders telling their own stories or leading the content creation process.

Building a relationship with Holders does not only involve sitting and listening to them, it can include participating in their activities. If a Holder is being engaged because of interest in their beading or weaving, the creators should have interest in active learning of such activities as this builds insight for the creator, which is essential for making the content meaningful and impactful.

PRODUCING THE CONTENT

Akinwale (2013: 3) maintains that there are knowledge generators and knowledge consumers. Digital content creators should consider the impact of their content on the consumers and to the field of knowledge they are contributing to. Digital content lives, and it has the potential to shape discussions, inform and entertain people. Thus, it is important to not misconstrue or misrepresent facts.

Figure 3. Principles underpinning the creator's guidelines

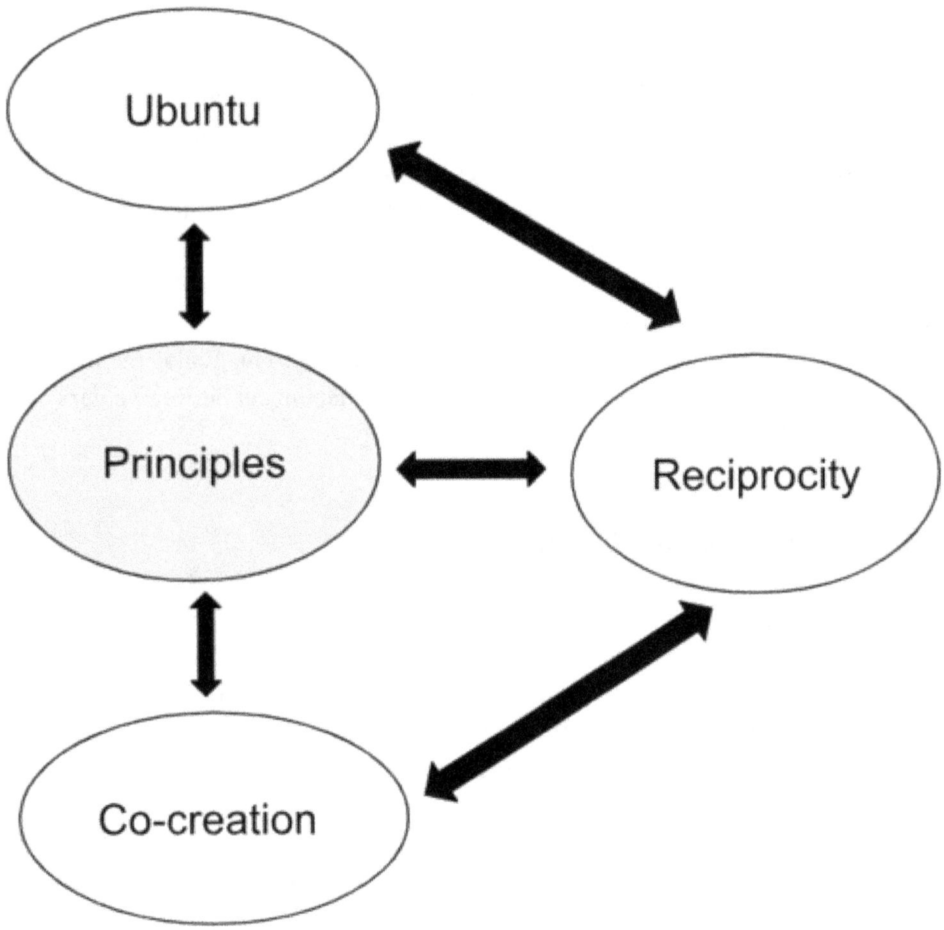

Figure 3 above presents principles that should underpin the creation of content on AIK. The process of producing Indigenous Knowledge - based content should be grounded in *Ubuntu,* co-creation and reciprocity. Indigenous Knowledge Holders should be part of the digital content creation process. Them being a part of the process means they should be able to tell their own story, not the creator telling the story for them.

Making digital content inclusive means that the person in front of the camera telling the story is/are the owner of the knowledge. This eliminates the risk misrepresenting content due to preconceptions, theories or a lens that influences a creator's view of the knowledge. Considering the impact of the knowledge matters as once the knowledge is consumed, it may influence views on certain Indigenous practices or knowledge. Inclusive digital content creation attempts to balance power distribution throughout the process. Digital content creation should not be something that "happens to an Indigenous Knowledge Holder", it should be a process they are a part of.

Indigenous Knowledge Holders engaged in Nyoni (2023) have shared their past experiences with creators "visiting" their communities to explore their knowledge and produce documentaries and books from it without acknowledging the Holders. Once such content is public without acknowledgement, issues about the ownership of the knowledge arise. Moreover, because the creator was a "visitor", they often hold no responsibility in how they present the Holders and their knowledge, leaving room for misrepresentation.

Approaching digital content creation from a point of partnership ensures that creators have the responsibility of continuously nurturing their relationship with Holders. Making it easy to converse on expectations, seek feedback for improvement and collaborate towards more content. Digital content creation can contribute to the safeguarding and promotion of Indigenous Knowledge. Sraku-Lartey, et al., (2017) acknowledges that Indigenous Knowledge is under threat, while Balogun and Kalosupa (2022), Khalala (2021), and Balogun (2023) have discussed the numerous efforts in place for the digitisation of African Indigenous Knowledge, AIK is vast. Digital content creation can also contribute to the safeguarding of the knowledge, if used ethically and respectfully. The goal is the transmission of the knowledge to younger and future generations, considering the disconnect between elders and youth, and the traditional methods of transmission.

SOLUTIONS AND RECOMMENDATIONS

The following recommendations are made from this chapter:

- African Indigenous Knowledge digitisation policy: A policy on AIK could standardise the digitisation process and clarify how different forms of the knowledge (either tacit or explicit) can be digitised. A standard for digitising AIK clarifies to those involved in the process how the knowledge can be digitised respectfully.
- Innovative platforms for the documentation of Tacit African Indigenous Knowledge: The digitisation of Tacit AIK is still challenging. There is a need for reliable platforms for the digitisation of the knowledge.
- Inclusive digital content creation: On the digital content creator's end, Indigenous Knowledge Holders should be responsible for telling stories of their knowledge and practices. This recommendation is based on the significance of avoiding bias and misrepresentation of the knowledge, and encouraging free will on the Indigenous Knowledge Holder's end. The recommendation emanates from the challenge that digital content creators often do not acknowledge owners of the knowledge, and divulge private or sacred knowledge or practices.
- Youth involvement in the digital content creation process: Youth and young Indigenous Knowledge Practitioners should actively engage in the digital content creation and digitisation of the knowl-

edge from their communities'. They are familiar with the communal norms. Moreover, this activity could be seen as one that Bachelor of Indigenous Knowledge Systems students and graduates use as an opportunity to craft employment opportunities for themselves, while benefiting their communities. The Indigenous Knowledge Management spaces could benefit from the active involvement of youth, without them waiting for external digital content creators to take interest in the knowledge from their communities.

FUTURE RESEARCH DIRECTIONS

More research is being conducted on indigenised research ethics and methodology. Despite the contributions of scholars such as Smith (1999), Wilson (2008), Archibald (2008), Chilisa (2012), Khupe and Keane (2017), Pidgeon and Riley (2021), Kovach (2021), Chigevenga (2022) and many more. There are still scholars, such as Kater (2021), who struggled to grasp what is expected when one seeks to conduct a collaborative and participatory study grounded in the Indigenous worldview and principles. Future research in this space will help raise awareness.

The future of research in the Indigenous Knowledge Systems space should be one where Holders of the knowledge are key players in its management. Digitisation initiatives and projects should be conceptualised with Indigenous Knowledge Holders to ensure true Indigenous Data Sovereignty. This chapter has also acknowledged a shift in the transmission of Indigenous Knowledge. Traditionally relied-upon vehicles like rites of passage, ceremonies, and storytelling are being used less. Future research should focus on clarifying what contemporary management methods could look like. Preferably, those that can maintain accessibility, cater to tacit and sacred Indigenous Knowledge without compromising their value. Future research should attempt to address such challenges emanating from digitisation and digital content creation with Indigenous knowledge.

Future research should also engage discussions on what specific African Indigenous Knowledge digitisation policies should entail. Current policies do not address African Indigenous Knowledge directly.

CONCLUSION

The oral nature of African Indigenous Knowledge (AIK) poses a threat of extinction of the knowledge. With the decline in practices that contribute to the transmission of knowledge and the disconnect between elders and youth, there is a need for alternative efforts of promoting and safeguarding the knowledge for long-term accessibility. This chapter has presented the use of digitisation and digital content creation as tools for safeguarding AIK. However, such tools are not without challenges. While digitisation is beneficial for safeguarding AIK, ethical issues also emerge, including challenges on the language used to digitise the knowledge, accessibility to funding (to make digitisation a success) and of the knowledge and adaptability of the knowledge to the digitisation platforms.

In addition, the digitisation and production of AIK-based content contribute to its susceptibility to biopiracy, misrepresentations and loss of ownership, if the custodians are not acknowledged. A three-step guideline to ethical digital content creation was presented in this chapter to mitigate against some of the ethical challenges. Through the guidelines, we have established similarities with the CARE principles, San Code of Research Ethics and principles for decolonial research methodology. The similarities are emphasis of the significance of transparency, reciprocity, accountability and relationships whenever one is engaging with indigenous communities. Despite the similarities, there are nuances to the application of the mentioned principles emanating from the uniqueness of each Indigenous community and its knowledge. Perhaps this is also a reminder that indigenous ethical guideline and methodology for engagements, can never be one-size-fits-all.

In concluding this chapter, one thing that stands out the most is the importance of Indigenous Knowledge Holders telling their own stories. Digital content creation provides the opportunity for IK Holders to determine how the stories of their legacy, which is African Indigenous Knowledge, should be told.

REFERENCES

Abrahams, S. (2024). Ethical content creation in the digital age. *Media Update*. Available at: https://www.mediaupdate.co.za/media/155772/ethical-content-creation-in-the-digital-age [Accessed: 20 January 2026]

Akinwale, A. A. (2013). Digitisation of indigenous knowledge for natural resources management in Africa. *A Paper Presented at the 20th Anniversary Summit of the African Educational Research Network at North Carolina State University, Raleigh, USA on 19th May 2012.*

Archibald, J. A. 2008. Indigenous storywork: Educating the heart, mind, body, and spirit. UBC press. Acorn (n.d) Available at: https://acorn.works/capability/digital-content-creation [Accessed: 26 December 2025

Asante, M. K. (2007). *An Afrocentric manifesto: Toward an African renaissance*. Polity Press.

Balogun, T. (2023a). Digitization of Indigenous knowledge systems in Africa: The case of South Africa's National Recorded System (NRS). *Records Management Journal*, *33*(1), 88–102. DOI: 10.1108/RMJ-10-2022-0029

Balogun, T. (2023b). Data management of digitized indigenous knowledge systems in repositories. *Information Development*, *39*(3), 425–438. DOI: 10.1177/02666669231186777

Balogun, T., & Kalusopa, T. (2022). Web archiving of indigenous knowledge systems in South Africa. *Information Development*, *38*(4), 658–671. DOI: 10.1177/02666669211005522

Bawack, R., Roderick, S., Badhrus, A., Dennehy, D., & Corbett, J. (2025). Indigenous knowledge and information technology for sustainable development. *Information Technology for Development*, *31*(2), 233–250. DOI: 10.1080/02681102.2025.2472495

Bhatia, R., (2025). Bridging Generations: The Role of Grandparents in Transmitting Sustainable Development Values to Grandchildren.

Carroll, S. R., Garba, I., Plevel, R., Small-Rodriguez, D., Hiratsuka, V. Y., Hudson, M., & Garrison, N. A. (2022). Using indigenous standards to implement the CARE principles: Setting expectations through tribal research codes. *Frontiers in Genetics*, *13*, 823309. DOI: 10.3389/fgene.2022.823309 PMID: 35386282

Chennels, R., & Schoeder, D. 2018. The San Code of Research Ethics: Its Origins and History. TRUST Equitable Research Partnerships. South Africa: The South African San Institute.

Chigevenga, R. (2022). Decolonising research methodologies in the Global South: Experiences of an African social scientist. *African Journal of Social Work*, *12*(4), 199–206.

Couldry, N., & Mejias, U. A. (2019). Data colonialism: Rethinking big data's relation to the contemporary subject. *Television & New Media*, *20*(4), 336–349. DOI: 10.1177/1527476418796632

David-Chavez, D. M., & Gavin, M. C. (2018). A global assessment of Indigenous community engagement in climate research. *Environmental Research Letters*, *13*(12), 1–17. DOI: 10.1088/1748-9326/aaf300

Department of Arts and Culture, (2010). National Policy on the Digitisation of Heritage Resources: Final draft for Public Review.

Department of Arts and Culture. (2013). National Archives and Records Service: Digitisation Strategy. Available at: https://www.nationalarchives.gov.za/sites/default/files/u59/National%20Archives%20 Digitisation%20Strategy%20-%20approved%20November%202013.pdf [Accessed: 14 February 2026].

Ermine, W., Sinclair, R., & Jeffery, B. 2004. The ethics of research involving Indigenous peoples. Saskatoon, SK: Indigenous Peoples' Health Research Centre. *Report of the Indigenous Peoples Health Research Centre to the Interagency Advisory Panel on Research Ethics,* pp. 1-272.

Hunter, J. (2005). The role of information technologies in indigenous knowledge management. *Australian Academic and Research Libraries, 36*(2), 109–124. DOI: 10.1080/00048623.2005.10721252

Kater, I. (2022). Natural and Indigenous sciences: Reflections on an attempt to collaborate. *Regional Environmental Change, 22*(4), 1–15. DOI: 10.1007/s10113-022-01967-3

Khalala, G. (2021). Protecting and benefiting from indigenous knowledge: The national indigenous knowledge management system. *CSIR Science Scope, 17*(1), 56–57.

Khan, F. A. (2011). The immortal life of Henrietta Lacks. *The Journal of IMA / Islamic Medical Association of North America, 43*(2), 93. DOI: 10.5915/43-2-8609

Kovach, M. (2021). *Indigenous methodologies: Characteristics, conversations, and contexts.* University of Toronto press.

Lwoga, E. T., Ngulube, P., & Stilwell, C. (2010). Managing indigenous knowledge for sustainable agricultural development in developing countries: Knowledge management approaches in the social context. *The International Information & Library Review, 42*(3), 174–185. DOI: 10.1080/10572317.2010.10762862

Maimela, G., Madima, T., & Makhanikhe, J. (2024). Overcoming digital barriers to African indigenous games in local communities. *Bussecon Review of Social Sciences (2687-2285), 6*(1), 29-39.

Mudau, O. (2023). The Digital Dilemma: Ethics in content creation. *PDBY.* Available at: https://pdby .co.za/the-digital-dilemma-ethics-in-content-creation/ [Accessed: 20 January 2026]

Nyoni, T. E. (2023). *The co-creation of Afrocentric ethical guidelines for digital content creators producing indigenous knowledge-based content* (Doctoral dissertation).

Nyoni, T., & Maimela, G. (2025). Digital content creation as a tool for safeguarding and promoting indigenous knowledge systems in South Africa. *International Journal of Research in Business and Social Science (2147-4478), 14*(8), 395-402.

Patalauskaitė, E. (2024). Ethical Aspects of Content Creation. *Filosofija. Sociologija, 35*(3), 369–377.

Pidgeon, M., & Riley, T. (2021). Understanding the Application and Use of Indigenous Research Methodologies in the Social Sciences by Indigenous and Non-Indigenous Scholars. *International Journal of Education Policy and Leadership, 17*(8), n8. DOI: 10.22230/ijepl.2021v17n8a1065

Research Data Alliance International Indigenous Data Sovereignty Interest Group. (September 2019). CARE Principles for Indigenous Data Governance. *The Global Indigenous Data Alliance.* DOI: http:// GIDA-global.org

Roya, W., & Ngcobo, S. (2023). The role of information communication technologies on African indigenous knowledge systems: Folktales. *International Journal of Research in Business and Social Science, 12*(6), 46–55. DOI: 10.20525/ijrbs.v12i6.2662

Rutaremwa, C. B., & Shirindi, M. L. (2025). Fathers' perspectives on the preparation of sons for family responsibilities. *Social Work/Maatskaplike Werk, 61*(2), 286-317.

Sewdass, N. (2014). Identifying knowledge management processes and practices used for decision-making and knowledge sharing in the modern San community. *TD. The Journal of Transdisciplinary Research in Southern Africa, 10*(4), 1–18. DOI: 10.4102/td.v10i4.85

Shopify, (2024). 8 Digital Content Types: How To Create Digital Content. Available at: https://www.shopify.com/za/blog/digital-content [Accessed: 26 December 2025]

Sinclair, O. (2025). The Ethics of Content Creation: Responsibility and Accountability. *Planksip*. Available at: https://www.planksip.org/the-ethics-of-content-creation-responsibility-and-accountability/ [Accessed: 20 January 2026]

Smith, L. T. (1999). *Decolonizing Methodologies: Research and Indigenous Peoples*. Zed.

Smith, L. T. (2021). *Decolonizing methodologies: Research and indigenous peoples*. Zed. DOI: 10.5040/9781350225282

Sraku-Lartey, M., Acquah, S., Brefo, S. & Djagbletey, G., (2017). Digitisation of indigenous traditional knowledge on forest foods and medicine in Ghana, 341 - 349.

Talašová, Z., & Souralová, A. (2025). Meanings and Roles of Great-Grandparenthood: A Scoping Review. *The Gerontologist, 65*(4), 1–14. PMID: 39679945

Walderman, A., & Branova, S. 2025. Indigenous Knowledge: Connecting Data, Land and Story. Available at: https://asm.org/articles/2025/november/indigenous-knowledge-connecting-data,-land-and-sto [Accessed: 20 April 2026].

Wilson, S. (2008). *Research is ceremony. Indigenous Research Methods*. Fernwood.

ADDITIONAL READING

Achieng, M. S. (2022). Digitalisation of indigenous knowledge of agricultural practices: A strategy for food security in Africa. *Journal of Public Administration, 57*(4), 802–818.

Boamah, E., & Liew, C. L. 2017, November. Conceptualising the digitisation and preservation of indigenous knowledge: The importance of attitudes. In *International Conference on Asian Digital Libraries*, pp. 65-80. Cham: Springer International Publishing. DOI: 10.1007/978-3-319-70232-2_6

Francis, K. D., & Liew, C. L. (2009). Digitised indigenous knowledge in cultural heritage organisations in Australia and New Zealand: An examination of policy and protocols. *Proceedings of the American Society for Information Science and Technology, 46*(1), 1–21. DOI: 10.1002/meet.2009.145046025

Kugara, S. L., & Mokgoatšana, S. (2022). Challenges presented by digitisation of VhaVenda oral tradition: An African indigenous knowledge systems perspective. *HTS Teologiese Studies*, *78*(3), 1–8. DOI: 10.4102/hts.v78i3.7428

Schoeman, M. I. (2019). Research ethics governance–An African perspective. In *Social Science Research Ethics in Africa* (pp. 1–15). Springer International Publishing. DOI: 10.1007/978-3-030-15402-8_1

Wilson, S. (2001). What is an indigenous research methodology? *Canadian Journal of Native Education*, *25*(2), 175–179.

KEY TERMS AND DEFINITIONS

Afrocentricity: A theory relating to ideas, perspectives or worldviews which are of African origin.

Digital content creation: The generation of media content in digital or online form, which can be uploaded to social media platforms or other interactive digital platforms.

Digitisation: Conversion of knowledge or documents that originally existed in a tangible form (physical copy) to a digital/online format.

Ethical guidelines: Principles guiding between right and wrong. In the context of this chapter, it relates to the acceptable and unacceptable behaviours or conduct when producing Indigenous Knowledge-based Content. For example, Indigenous Knowledge-based content generated and presented respectfully, with consideration to its custodians.

Indigenous Community: A group of people originating and living in a specific geographic location. The people have long-standing interrelationships with their environment and shared communal norms and customs.

Indigenous Knowledge Holder: An individual possessing expertise in a certain field of Indigenous Knowledge, informed by their lived experiences and observation. Such knowledge relates to a specific geographic location and the knowledge may have also been acquired from one's ancestor or elders.

Indigenous Knowledge: Practices, customs, beliefs and norms that are unique to a specific geographic location and have been passed down generationally.

Chapter 9
Digital Displacement:
How Social Media Erodes Traditional African Culture in Conversations

Maletšema Ruth Emsley
http://orcid.org/0000-0002-6638-871X
University of Limpopo, South Africa

ABSTRACT

Mass communication has moved beyond traditional media of radio, television, newspaper, magazines, film, and books due to technology and its advancement. Social media take on different forms in this 21st century. This chapter highlights the impact of social media on African traditional and cultural conversations and demonstrates the importance of traditional African practices which embody core values like respect, wisdom, hierarchy and ubuntu. The chapter used a qualitative approach through a semi-structured interview approach from purposively and snowball sampled participants. This chapter's findings have implications for future research, education, indigenous knowledge systems and digitalisation. It contributes to the debates on decolonial education, the evolving challenges regarding the use of social media in African indigenous languages and their expansion to the digital communication spaces.

INTRODUCTION

Indigenous knowledge (IK) is the unique local knowledge of a particular society or culture (Senanayake, 2006). IK is the essential principle for local-level decision making in educational, agricultural, health care, food preparation, natural resource management and many more rural communities' activities (Warren, 1991), further existing in medicinal practices, folklore, proverbs, cultural values, taboos, norms, languages and rituals (Nakata & Langton, 2005). Rajasekaran (1992) stresses that this type of knowledge is accumulated from informal experiments, intimate understanding of the environment and experiences (Sithole, 2007) by its local people. It is holistic (Battiste & Henderson, 2000) in that it is not compartmentalised like in Western knowledge systems. Research has shown that Indigenous knowledge systems (IKS) are undermined by advanced technologies and digital platforms hiding under the gospel of modernity and civilisation. For this chapter, digital platforms refer to the online applications

DOI: 10.4018/979-8-3373-7976-0.ch009

or systems which permit people to communicate, share information, create content or conduct activities using the internet.

These are spaces where users can interact, exchange messages, post content or carry out services digitally instead of using face-to-face channels. Examples of the digital platforms used for social networking are WhatsApp, Zoom, Instagram, TikTok, Facebook, Twitter, X, where people socially network and connect via phones or computers. In other words, for the purpose of this chapter, a digital platform is any online space where communication or interaction happens through technology. This chapter contributes to the body of knowledge by establishing the importance of the IK transmission for indigenous local communities, as well as ensuring that the cultural practices are authentically preserved and protected during technology use for the coming generations. One of the ways to revitalize IK including the indigenous Arican languages can be the use of technologies. However, research show that the use of advanced technologies including social media can erode the cultures of indigenous languages.

Social media takes different forms including internet forums, microblogging, podcasts, social networking, bookmarking, wikis, social curation, weblogging (Holotescu, & Cre u, 2015). They include highly interactive websites and others like Orkut Instagram, Twitter (X), YouTube, Myspace, blogs, LinkedIn, Flickr and Buzz (Solo-Anaeto & Jacobs, 2015). Almost all these mediums are interpersonal, and generated conversations connect audiences and build relationships. This is predominantly allowed by power languages and power dynamics. Global power dynamics influence technological advancements on knowledge representation and production. This includes creation and promotions and preservation of knowledges on digital platforms. This leads to situations where digital technologies favour dominant knowledge frameworks and restrict Indigenous communities' freedom to flourish. Less powerful languages fall into the trap of misrepresentation and exploitation. There is a growing consensus that IK is crucial to the social and economic development of Africa (Solo-Anaeto & Jacobs,2015). However, this knowledge might be in danger of disappearing if proper measures are not taken to document it. Even though the advent of social media and mobile technologies offers opportunity to document, disseminate languages through digital platforms, it is apparent that they will only be used by people who can afford to purchase computers or cell phones or in other ways of gaining access to the Internet (Solo-Anaeto & Jacobs, 2015). Again, by providing languages leverage to be use do these platforms, does not guarantee their genuinely and originality, because creators use their own content, but not for the promotion and preservation of African cultural content (Nyoni, 2023).

In the midst, Africans have lamented that social media are filled with western-based ideas values and culture which is eroding African culture, even to the point of threatening their extinction. Sadly, should the information become eroded, nothing will be shared to be localised and passed down to the successive generations. Indigenous languages are vital to preserving cultural identity, community knowledge and history, but many Indigenous languages face endangerment or extinction due to colonialism, forced assimilation and globalization. According to the United Nations Educational, Scientific and Cultural Organization (UNESCO), approximately 40% of the world's 6,700 languages are at risk of disappearing (Office of the High Commissioner for Human Rights, 2019). The loss of Indigenous languages will end up in the disappearance of diverse linguistic systems and the erosion of rich cultural heritages and traditional IKS. IKs in essence should be transmitted intergenerationally. Africans all over the continent are claiming the cultural heritage of their ancestors (Tarugarira, 2014), but the rapid growth of social media platforms over the past decade has significantly transformed how they communicate. "Africans have lamented that social media are filled with western-based ideas, values and culture which is deepening the eroding of African culture, even to the point of threatening their extinction" (Solo-Anaeto, et al., 2015).

The discontinuity of knowledge passage across generations, and the diminishing applicability of traditional knowledge underscores a critical challenge to every intervention that could be out into place (Malapane et al., 2024). Indigenous communities moreover contribute to this threat due to neglect in valuing and transmitting their knowledge; further imperilled by an apprehension towards the relevance of their IKS in the face of modernity (Malapane et al., 2024). The erosion of these systems presents significant threats to the socio-cultural dynamics and the preservation of identity in the African region (Malapane, et al., 2024). This chapter concedes with Malapane et al. (2024) that technological invasion is a pervasive threat to IKS, and to African indigenous languages. The social media, which is seen as having potential in projecting and sustaining languages and cultures with its high capability for information storage and dissemination and interactive can also have the potential to threaten African culture as well. Against this backdrop, the researcher found it imperative to address the gap by exploring the perceptions of Pedi elders about the use of social media platforms to address Pedi traditional matters. To achieve this aim, the current chapter sought to answer the following research questions:

1. To what extent do elderly Sepedi-speaking community members perceive the use of social media in the discussions of Pedi family or cultural matters?
2. How does the use of digital platforms affect the running of traditional community conversations in Pedi communities?

In other words, the chapter aimed to examine elders' perceptions of social media use in discussing African family issues and its influence on cultural values, including perceived advantages and disadvantages in relation to the promotion or erosion of traditional practices and intergenerational communication. This chapter argues for an Afrocentric decolonial approach that advocates for the emancipation and participation which values and includes IK and peoples (Owusu-Ansah & Mji, 2013).

THEORETICAL FRAMEWORK

This chapter is grounded in Afrocentricism. Afrocentricity was popularised by Asante during the 1980s, when he developed the foundations for an Afrocentric curriculum based on an African perspective (Chawane, 2016). According to Asante (2003) Afrocentricity aims at raising consciousness regarding the often neglected historical and current centrality of African peoples.

This chapter advocates for the liberating African minds from dependency on Western categories (Asante, 2003) as Asante established. Afrocentricity restores and value Arican systems and languages as legitimate foundations of knowledge (Mazama, 2003). This chapter problematises the values of African people regarding their conversations via social media. "Afrocentricity plays a crucial role in promoting collective agency and facilitating free conversation in support of diversity." (Sharma et al., 2025)

Afrocentricity suits this study since one of its core principles is its cultural location. Afrocentricity underscores communality and dignity as a collective matter, not individualism. In the words of Li and Li (2019), "In western culture they emphasize "individuals" and "individualism "that is a commonly accepted life ..." (p 1072). The choice of this framework aligns with the objectives of this chapter which stresses the importance of African cultural collective identity and ancestral continuity in their conversations. It positions this chapter within an African perspective that promotes the use of African languages in conversing over on social media platforms particularly by Pedi people. Over the years,

Africans have mostly been consumers of Western media content. Afrocentricity opposes individualism (Asante, 2009). According to Malatji and Lesame (2019) Africans have not fully taken advantage to capacitate themselves towards preserving their identity and culture through undertaking research that identifies African ideas and scholarship and promotes African languages and schools of thought. This is because Africans still absorb a lot of European and American ideologies at the expense of their own (Malatji & Lesame, 2019). Afrocentrist viewpoint highlights that Africans exist as a collective with their differences (Mkhize & Ndimande-Hlongwa, 2014), but African languages should be a mainstay of cultural preservation for future generations to come. Language is a communication channel, and any group of people has a language that represents its heritage (Lohdi, 1993). If the social media platforms are used to converse traditional and cultural events of Pedi people, there is a probability that in the future protocols will not be followed, but reliance on social media will be the order of the day.

PROBLEM STATEMENT

African language conventions, both verbal and nonverbal, differ from culture to culture (Moloko-Phiri et al., 2022). Th sad story is: African indigenous cultures run the risk of being endangered. This has been globally exacerbated by the rapid evolution of technology. The Pedi people from Limpopo Province in South Africa are not an exception. Social media and mobile technologies advancements have posed a significant challenge to the preservation of African values.

Indigenous communities moreover contribute to this threat themselves due to neglect in valuing and transmitting their knowledge; further imperilled by an apprehension towards the relevance of their IKS in the face of modernity (Malapane et al., 2024). Social media has become an everyday channel for information sharing and communication and viewed as a potential in projecting and sustaining cultures. However, despite the growing attention to IKS and social media in various there remains a significant gap in understanding this aspect in relation to Pedi culture. It has been proven by literature that the swift technological revolution has significantly impacted on cultures around the globe (Oyinlade, 2024), but very limited research has been done on social media use and Pedi speakers. In Africa, Oyinlade (2024) made a concern regarding the erosion of the Yorùbá people's cultures, an ethnic group primarily located in southwestern Nigeria. Known for their rich cultural heritage, which includes a strong emphasis on communal living, respect for elders, and a deep sense of spirituality and moral uprightness, it has though been infiltrated by the internet, social media, and this poses a challenge to the preservation of its values.

The biggest worry is that younger generation who are supposed to learn the IK values and run with it they are the ones more engaged with digital platforms. They are the users of social media in large numbers. They exhibit a noticeable shift in moral perspectives, often aligning more with globalized digital culture than African ethics (Oyinlade, 2024).

Despite the possibilities digital tools provide for communication and information sharing, various constraints have blocked technology implementation in many Indigenous communities. They continue to face significant barriers related to infrastructure, affordability and digital literacy. This is supported by the scholars that remote and underserved Indigenous communities faced major challenges, including unreliable internet, insufficient resources, and inequitable distribution of technological devices (Kitchenham, 2013). Besides the issues of infrastructure, social media promotes individualistic practices that overlook African cultural forms of address in the indigenous language contexts such as traditional greetings and honorifics that signify respect and social hierarchy. Neglecting such traditions by promoting individualistic western practices in social media will contribute to more decline in the cultural practices associated with African forms and norms. The understanding and appreciation of African cultural practices will diminish. These forms should be followed because they represent crucial aspects of heritage and culture (Mudau et al., 2024). By centring indigenous assumptions and justifying culturally grounded interpretations of African conversations, this chapter also wishes to demonstrate philosophical congruence with Afrocentricity.

There are existing policies that support IKS, however scholars such as Outakoski, Cocq and Steggo (2018) posit that Indigenous cultures and languages were already under strain, before the advent of modern environments like social media. But this rapid rise in technologies worsen the pressure. This chapter argues that social media can currently become another domain where dominant languages and cultures overshadow minority ones, contributing to erosion of indigenous knowledge if not monitored carefully. The dominance of languages of power reduces Indigenous cultural presence because historical patterns continue to encroach into digital spaces. Power(ful) languages still dominate. The culture of the West still reigns in the digital spaces because of ownership and licences, and distribution.

A further complication arises from the application/non-application of cultural proxemics, kinesics and proxemics on social media conversations. WhatsApp uses emojis' and pictures, and these messages can be misinterpreted, especially by the elders. Cultural expression is in these languages, and the indigenous knowledges of indigenous languages are different. Subsequently the lack of access to culturally relevant content by the young people heightens. Cultural erosion happens when cultural knowledge is not widely shared, and younger generations may be the culprits since they are the majority in social media usage.

Omipidan and Sanusi (2024) for example, found that the WhatsApp has become one of the most widely used apps available today. It aids quick interactions (Manning, 2014). However, it is disadvantageous since its widespread usage raises pertinent concerns in effective communication. The biggest problem being that the use of WhatsApp is that it is against the Pedi community in this way: It creates a situation for the decline in interpersonal and interactive face-to-face communication (Dhir et al., 2018); the emotional expression, body language and tone are lost (Dhir et al., 2018); empathy is reduced on text-based communication (Dhir et al., 2018). Omipidan and Sanusi (2024) conducted a study to examine whether WhatsApp is a threat to effective communication among undergraduates. They identified both the positive and negative effects of WhatsApp. Their results showed that most participants used WhatsApp daily to communicate with friends, family, and colleagues as a convenient, accessible, and useful for personal and professional communication channel. Although they appreciated features such as voice calls, video calls, group chats, and document sharing and valued WhatsApp's privacy and encryption features, they reported excessive notifications and information overload. WhatsApp sometimes encourages shallow conversations and reduces meaningful face-to-face interaction. On the other hand, the study by Mangla et al. (2020) see WhatsApp as addictive and it decreases face-to-face communication and interpersonal relationships, affects emotional health, including anxiety and disrupts family relationships because it increases spread of fake news, rumours, and false information rapidly through groups therefore damaging personal or business reputation among others. Conversely, technology has broadened access to Indigenous languages (Galla, 2009; Kayode et al., 2026), but indigenous communities continue to face significant barriers related to infrastructure, affordability and digital literacy including unreliable internet, insufficient resources, and inequitable distribution of technological devices (Kitchenham, 2013). Besides the issues of infrastructure, social media promotes the lack of politeness and dignity of the African culture by promoting individualistic from of address which is not allowed. Ignoring the use of these forms of address can lead to language shift or even extinction (Mudau et al., 2024). This chapter investigate how the use of social media could erode the culture to the Pedi.

LITERATURE REVIEW

IKS and Arican Conversations

Interactions of the African people follow culturally specific conventions (Moloko-Phiri et al., 2022), uniquely followed in different language communities in SA regarding honorifics, proxemics and kinesics. For example, Indigenous forms of address are central to cultural identity. These forms must therefore be transmitted from one generation to the next the special use of grammar where certain prefixes or plural forms should be used by younger generations when communicating with their elders to indicate respect (Bagwasi, 2012; Hellen, 2018). In Sotho and Tswana cultures for example, when young people converse with the elders, they are supposed to use titles such as rra [Mr/father] or mme [Mrs/mother] rather than their names (Mudau et al., 2024; Ngoasheng, 2006). The Pedi and Ven a cultures do not permit younger people to look an adult in the eyes when they are conversing with them. (Nenungwi, 2015; Mncwango, 2009; Vorster et al., 2022). In the Tswana, Pedi and Ven a cultures, when greeting someone older than themselves, children clap or put their hands together in a respectful and humble manner (Bagwasi, 2012; Nenungwi, 2015; Moloko-Phiri et al., 2022). Western contexts on the other hand do not apply such practices. This chapter explored the perception of the elderly people on their views regarding conversations of the African people over pertinent African issues over the social media.

Social Media and Languages at Risk

Social media involve computer-mediated tools and internet-based applications for people to communicate. So social media has become an everyday global channel for information sharing and communication. However, "Africans have lamented that social media are filled with western-based ideas, values and culture which is deepening the eroding of African culture, even to the point of threatening their extinction." (Solo-Anaeto et al., 2015). There is a growing consensus that indigenous knowledge is crucial to the social and economic development of Africa (Solo-Anaeto & Jacobs, 2015). However, this knowledge is in danger of disappearing if proper steps are not taken to document it. Again, the advent of social media and mobile technologies offers opportunity to document, disseminate, and raise awareness about indigenous knowledge, social media and mobile technologies, but it is apparent that they will only be used by people who can afford to purchase computers or cell phones or in other ways gain access to the Internet. (Solo-Anaeto & Jacobs, 2015). Again, this does not mean that the creation of texts on these platforms is authentic, because attention from digital creators for their own content, but not for the promotion and preservation of African cultural content (Nyoni, 2023).

According to the United Nations Educational, Scientific and Cultural Organization (UNESCO), approximately 40% of the world's 6,700 languages are at risk of disappearing (Office of the High Commissioner for Human Rights, 2019). Indigenous languages are at higher risk, as Indigenous Peoples constitute less than 6% of the global population yet speak more than 4,000 of these languages (United Nations Department of Economic and Social Affairs, 2023). The loss of Indigenous languages signifies the disappearance of diverse linguistic systems and the erosion of rich cultural heritages and traditional knowledge systems. Indigenous communities moreover contribute to this threat due to neglect in valuing and transmitting their knowledge; further imperilled by an apprehension towards the relevance of their IKS in the face of modernity (Malapane et al.,2024). Oyinlade (2024) documented that there is an influence by the increasing use of technology and the perceived erosion of moral values especially among the younger generations. His study regarding Yoruba people in West Africa found that the younger generation who are engaged with digital platforms more (Oyinlade, 2024). One of the major challenges facing the use, protection and preservation of Pedi culture as the previously marginalised languages is that the young people have never understood the traditions. If they gradually fade away, their own generations will miss the chance of learning about these values. This chapter questioned if the use of social media can be used as channels to converse sacred Pedi traditions.

Honorifics

Honorifics are used and the speech takes place on a high level of politeness. Honorifics are applied daily in instances where speakers want to be polite to their interlocutors, to perceive them as being polite, to maintain distance from their interlocutors, or when speakers think their interlocutors expect the use of honorifics. There are also various circumstances governing the use of honorifics. For example, the character and a length of relationship between interlocutors as well as their gender and ages. Moreover, the use of honorifics often reflects relations of power. That is why the honorifics are not generally used in interactions between friends in informal settings. They are usually used in formal business situations where only after having met each other several times do interlocutors tend to use honorifics less frequently. The honorifics are a linguistic reflection of the vertical (hierarchical), and horizontal (intimacy-based) relations. As we mentioned above, the use of honorifics often poses problems for speakers, who instinctively try to be polite, but use honorifics inappropriately.

There are also various circumstances governing the use of honorifics.

Pedi Greetings

Scholars like Davies (1987) describe greetings as “formulaic expressions” or “formulaic speech acts” Therefore, greeting’ is one of the expressive acts mostly used in our daily life.” Greetings in Sepedi consist of a series of questions and answers, with variations according to the type of social interaction, age group, background, time, or the level of intimacy between speakers. In formal situations, the main factor that determines the way one addresses the interlocutor, is her/his status and age, regardless of gender. The rule is that the younger person initiates the conversation, except when he or she has been approached by an elderly person. The way people greet each other is an important reflection of the social norms and hierarchies. In Pedi culture greetings are like this: Greetings seem like a basic element of human communication, introducing and enabling the further communication of the partners. Normally they are determined by the social distance and intimacy, age, gender and their differences, belonging to the same smaller community. Usually, the meeting or the leaving, and the time of day also lead to different forms. In some languages these can be quite sophisticated. Greetings and address forms are the most common determinants of language politeness, because their frequency of occurrence in social interactions is high and because they usually serve as conversation starters. Each speaker may find himself/herself tending to express pleasure many times daily at meeting or seeing a hearer. Therefore, greeting’ is one of the expressive acts mostly used in our daily life”. Children at a very young age are taught ‘language of respect' which includes various forms of greetings. Greetings in Sepedi consist of a series of questions and answers. In formal situations, the main factor that determines the way one addresses the interlocutor, is her/his status and age, regardless of gender. The rule is that the younger person initiates the conversation, except when he or she has been approached by an elderly person.

Forms of address

Forms of address embody crucial aspects of cultural identity and heritage, encompassing vital social and cultural functions that support the preservation of cultural norms and values. They act as icebreakers in any communication act (Mushaathoni, 2023) because they mitigate miscommunication and breakdown or failure in communication. Again, if they are used inappropriately, they can create chaos among the interlocutors because these forms of address are not used in the users' households only, they are also used in the public. They are applied to show fondness, respect than being insulting, especially in functions, (Yang, 2010). What is important is using indigenous forms of address, language can be preserved for the coming generations (Yang, 2010). According to Wierzbicka (2003), forms of address are essential because humans are inherently social beings. Likewise, Yule (2017) explains that addressing activities are an integral part of language use, as interlocutors rely on them at various points during interactions and meetings. Much work has been done on indigenous knowledge but not much has been done on IKS and the use Pedi language vis-à-vis social media. Ajani et al. (2024) for example, examined Indigenous knowledge systems and digital media at a larger scope. The other study was by Agbeyangi and Jere (2024) who conducted a systematic scoping review primarily on Indigenous language revitalization in South Africa. Li et al. (2021) focused on the use of digital technology to support literacy development, mainly in English, for Indigenous peoples, rather than Indigenous language revitalization specifically. Meighan (2021) focused on the decolonizing potential of digital tools for Indigenous language revitalization, and Taylor and Kochem (2022) addressed Indigenous empowerment through digital platforms.

Indigenous forms of address are indispensable in the preservation of the language and culture of any society, and fundamental to them. Forms of address are used several times to draw the attention of the interlocutor and establish or confirm the relation between the partners. They embody crucial aspects of cultural identity and heritage, encompassing vital social and cultural functions that support the preservation of cultural norms and values. They matter a lot. They have multiple functions. They act as icebreakers in any communication act (Mushaathoni, 2023) because they mitigate miscommunication and breakdown or failure in communication. Again, if they ae used inappropriately they can create chaos among the interlocutors because these forms of address are not used in the users' households only, they are also used in the public. They are applied to show fondness, respect than being insulting, especially in functions, (Yang, 2010). What is important is using indigenous forms of address, language can be preserved for the coming generations (Yang, 2010). Addressing others is fundamental to human social interaction across cultures and contexts, including family, society and the workplace. According to Wierzbicka (2003), forms of address are essential because humans are inherently social beings. Much work has been done on indigenous knowledge but not much has been done on IKS and the use Pedi language vis-à-vis social media. For example, IKS and digital media (Ajani et al., 2024; Taylor and Kochem (2022)); Indigenous language revitalization in South Africa (Agbeyangi & Jere, 2024; Meighan, 2021); the use of digital technology and support literacy development (Li et al., 2021) Herrera (2022) examined technology's role in shaping Indigenous identity and cultural transmission. To bridge these gaps, this chapter explored the perceptions of the Pedi-speaking community regarding the role of social media in supporting in African traditional conversations in dealing with cultural matters.

The study by Galla (2026) showed that digital tools increased the visibility of indigenous languages by embedding them in modern cultural contexts (Galla, 2016). On the other hand, integrating Indigenous knowledge systems with digital technologies required careful consideration of cultural traditions and protocols such honorifics and proxemics. Digitization of Indigenous knowledge posed risks of exploitation and cultural appropriation.

Social Media - The Use WhatsApp

WhatsApp constitutes a rapidly growing social media platform with almost three billion users in mid-2024 (Statista, 2024) favoured by countries like Ghana, Kenya, Nigeria, and South Africa favour WhatsApp (Bhanye et al., 2023). WhatsApp Business is projected to have 26 million active users by 2026 (Lith, 2023); Sub-Saharan Africa leverage platforms like WhatsApp to foster civic participation… (Toumaras, 2025) The problem is, however, cheaper or user-friendly WhatsApp is, limited and misrepresented traditional communication forms on digital platforms could lead to gradual erosion of indigenous knowledge and particularly among younger generations engaged with digital media and modern communication practices (Eke et al., 2024). Users can even post pictures and videos on their statuses which expire in 24-hour period. The biggest problem that the chapter disagrees with the use of WhatsApp is that it is against the Pedi community in this way: It creates a situation for the decline in interpersonal and interactive face-to-face communication (Dhir et al., 2018). Omipidan and Sanusi (2024) conducted a study to examine whether WhatsApp is a threat to effective communication among undergraduates among others, and to also identify both the positive and negative effects of WhatsApp usage among undergraduates. Their results showed that most participants used WhatsApp daily to communicate with friends, family, and colleagues as a convenient, accessible, and useful for personal and professional communication. Although they appreciated features such as voice calls, video calls, group chats, and document sharing and valued WhatsApp's privacy and encryption features they reported excessive notifications and information overload. The participants stated that WhatsApp sometimes encourages shallow conversations and reduces meaningful face-to-face interaction. Some participants believed WhatsApp causes distractions, misunderstandings, and lack of focus on academics. Overall, participants agreed that WhatsApp has both advantages and disadvantages, depending on how responsibly it is used.

Mangla et al. (2020) found that WhatsApp addictive and declines face-to-face communication and interpersonal relationships apart from the negative effects on emotional health, anxiety and depression, disruption of family relationships, spread of fake news, rumours, and false information rapidly through groups, damage to personal or business reputation through misleading information, disturbance of work-life balance and social activities, diverse effects on students' academic performance, poor writing habits and overuse of abbreviations in academic work, weakening of professional and formal writing skills, sleep problems, insomnia, and reduced brain functioning from excessive use, health issues such as depression, anxiety, and eyesight weakness, excessive group disturbances, teasing, and irrelevant content sharing, wastage of internet data and time on unnecessary communication, loss of interest in real-world activities and emotional dependence on online replies.

Communication in Pedi Culture

There are traditional norms that are followed in Arican cultures. The question is: Can these practices be followed or used on social media like WhatsApp? Greetings, for example, combine verbal and non-verbal gestures. This is done with respect and dignity. to demonstrate deference to elders. When younger people greet, they clap their hands, while elders respond by holding or kissing their hands. (Moloko-Phiri et al., 2022). Furthermore, other actions like kneeling while serving food or accepting gifts from elders, signals humility and respect. Younger generations address elders in the plural form to show humility and respect, and the young one are the ones who must initiate greetings when meeting elders. Other examples are that the younger people are expected to listen attentively, remain silent, and avoid interrupting elders when the elders speak. They should not keep eye contact. And, children usually communicate requests through the mother, who then relays the information to the father. So, communication can also be gender-specific, where boys communicate with fathers and girls with mothers. Younger generations are expected to play a passive role in family conversations and should not interrupt elders' discussions. It is very important that respect in the Pedi culture is expressed through verbal and non-verbal communication during conversations. These as stated include including handclapping, kneeling, plural forms of address, indirect eye contact, silence and turn-taking. The problem is social media do no give room for those, because of distance, independence and individualism, which are the tenets of Western culture.

Respect is communicated through gestures, body language and verbal language that demonstrate deference to elders. Pedi people show their respect to their interlocutors by demonstrating certain gestures like he young clapping their hands, while elders respond by holding or kissing their hands; kneeling while serving food or accepting gifts from elders, accepting items with two open hands, together with clapping and verbal gratitude. Men remove hats or headgear when greeting or entering the home, pointing fingers at elders is regarded as disrespectful. Addressing the elders in the plural form; younger people initiating greetings when meeting elders; avoiding direct eye contact, young ones communicating requests through the mother, communication can also be gender-specific, where boys communicate with fathers and girls with mothers. Younger generations are expected to play a passive role in family conversations and should not interrupt elders' discussions.

Totem Animals and Turn-taking in African Traditional Conversation

There is much missing in digital platforms conversations that is unique to indigenous African conversations. In an African communication elders and unrelated adults are address one another with respect (Asante). In communication, totem names influence how people address and relate to one another. In some communities, individuals avoid directly calling someone by their personal name and instead use their totem names as a respectful form of address. This helps maintain politeness and reinforces social belonging. Totem names function as a cultural communication system that supports respect, identity, and cohesion in African conversational practices. Traditions and identity markers like clan names and totems influence speech practices. When Africans like Pedi converse, they take turns to speak. The talk should be accessible to all members where ethe discussions are steered by the elders. Turn-taking is strictly organized by age and seniority. The eldest speaks first, followed by younger males, while women and youth generally speak after, although aunties may have specific, respected roles in negotiations. It is considered disrespectful to interrupt an elder or another member. Talks can take hours or days because a topic is discussed until everyone who wishes to speak has had a turn, ensuring all sides are heard.

When bringing up a new point, a speaker might begin by invoking their clan praises and totem to affirm their legitimacy and connection to the ancestors, as described in. So, in African traditions, turn-taking in meetings or conversations structured around respect for hierarchy and age. African conversations are dialogical, communal and informative. They are inclusive rather than merely individualistic. Chimakonam (2017) stresses that "…the methodological disposition of conversational philosophy aims at not merely bringing humans together to talk to themselves but, more importantly, it focuses on getting them to talk meaningfully and to understand themselves."

METHODOLOGY

This qualitative study examined perspectives on the use of social media in culturally significant practices within a rural Sepedi-speaking community in Limpopo, South Africa. The study was informed by ongoing debates concerning digital modernity and cultural [dis]continuity. In recent years, mobile technologies have penetrated traditional and rural communities at an unprecedented rate. Consequently, questions have emerged regarding their appropriateness in highly regulated cultural spheres were at times conversations are supposed to be sacred or confidential. This research sought to explore these tensions through the voices of elderly custodians of tradition. The study was grounded in an interpretivist paradigm. A purposive sampling technique was employed which was followed by snowball sampling. The criteria were that the participants had to be Sepedi-speaking and residing in a rural setting. They also needed to possess experiential knowledge of cultural practices. The sample consisted of a centenarian aged 104- and 89-year-old one, the third participant was a 57-year-old municipal worker. The fourth was a 56-year-old community member identified through referral. Of the four participants, one was female and three were males. All resided permanently in rural Limpopo where Sepedi was their primary language. In terms of educational background, all participants had limited formal schooling. The eldest participant had attended school up to Standard 7, now equivalent to Grade 9. The remaining participants had basic secondary levels of education.

The researcher obtained access to the community through the local municipal leader. Permission was formally requested from the ward councillor and the chief. The ward councillor granted permission and subsequently volunteered to participate. All the participation followed informed consent procedures. Ethical principles of transparency and voluntary participation were strictly observed. Data were collected through semi-structured interviews. Magaldi and Berler (2020) define the semi-structured interview as an exploratory interview. In addition, enables a researcher to go deep for a discovery (Megaldi &Berler, 2020). At the same time, participants were encouraged to elaborate freely. The interviews were conducted face-to-face. The interviews took place in the comfort of participants' homes to foster for trust and minimised disruption. Care was taken when engaging the 104-year-old and 89-year-old participant as they had hearing problems and sometimes needed time to rest and sip water. No harm was observed during the establishment of rapport and consequently the interview days. All interviews were conducted in Sepedi. This ensured that no language barrier hinders the flow of the conversation. Data were analysed thematically.

RESULTS AND DISCUSSIONS

This section presents the results and discusses the findings. The results are organised according to the themes formed from the patterns coded from the respondents` words. Qualitative data are reported using appropriate descriptive summaries to highlight key patterns and relationships. Truthfulness of the data is supported using the participants` excerpts. A Semi-structured interview was administered to four purposefully and snowball sampled participants at four different settings. The Semi-structured interview was designed to take account of the age of the participants. The aim of the interviews was to collect data on how the see the issue of discussing sacred family matters over the social media, with specific platform WhatsApp.

Context

The study was conducted in Terbrugge, a rural village locality situated within the Aganang Local Municipality in the Capricorn District of Limpopo Province, South Africa. The area forms part of the broader Moletjie region, located northwest of Polokwane, the provincial capital. Terbrugge is characterised by a predominantly rural landscape, comprising dispersed settlements, subsistence farming activities and limited infrastructural development typical of many communities in the region. The selection of Terbrugge as the study area is justified by its suitability as a representative rural community where traditional cultural values and norms remain relatively prominent.

Indigenous Representation

The study interviewed Pedi-speaking individuals from the Limpopo Province depicting Indigenous population. This study focused on Sepedi.

Familiarity with Digital Technologies

The first question that was asked to the participants was based on their knowledge of social media platforms. The participants are well familiar with WhatsApp. Audio-visual tools such as cell phone videos, audio recordings, video conferencing seem to be unknown to them, except the ward councillor member who was aware of them but basically not using them. The ward councillor assistant mentioned that they usually use WhatsApp for sending out invitations to meetings, but they do not use the platform for discussions. He further stated that, he is very fond of the betting sites, because he had on many occasions won large amounts of money, but emphasised that he seldom uses the social media platform such as WhatsApp for familial discussions. He said that serious family issues exemplifying lobola] or one's cows have eaten someone`s crops should not be discussed over the social media.

Intergenerational Knowledge Transmission

Intergenerational knowledge transmission played a central role in Indigenous language. When traditional modes of transmission are disrupted, knowledge transfer is also barred.

Participants stressed that elders remain central to transmitting ancestral languages. This is to enforce cultural authenticity and community connection. They reiterated that Pedi issues are discussed by adults. And these adults must meet face to face. This is supported by scholars like Bishop, (2022) and Desmoulins et al., (2019). The elderly woman mentioned that the degree to which the youth are dependent on the cell phones cannot be tolerated. She commented

"Akere baswa ba kgona kudu go di somiša.

Le ge e le gore ga ba tsebe seo ba se dirago ka tšona."

(Translated: "The youth can use the cell phones well;

even though they do not know the value of the cell phones.")

When asked further because the youth do not know what they do with the cell phones, the old lady answered:

"O no bona moto a lebeletše founu, le tla hlaba la ba la dikela, a sega a le tee. Ge o mmitša, a ka se tle, founu še, nako ye kamoka."

["You just see a person looking at the phone, from sunrise to sunset, laughing to himself. Even when you call them, they do not hear, they have the phone in their faces every time."]

While many digital initiatives sought to promote language use and visibility, cultural tensions are there. Results showed that hierarchical language practice create intergenerational tensions due to the use of cell phones.

Secrets and Confidentiality in Conversations

When participants were asked to elaborate on communicating important family matters through WhatsApp, they consistently emphasised the collective and ceremonial nature of the process. The eldest totally rejected the idea of conversing family matters over the social media. According to their responses, private matters must be exchanged between family and relatives in a communal event that requires the presence of key family representatives.

*"Batho bao ba amegago mo tabeng, ka segagaborena, ba swanetše ba dule mo" [*pointing to the floor.*] Re ba tšhele ditaba tše. Re tšhela o mongwe gare ga bona ditaba. Re kgetha motho. Yena o botša wola, wola le yena a di tšhela ola, bjalo le bjalo. Go fihla di boa go mong wa tšona. Tšeo lena le di bolelago, ke sebjalebjle, rena ga re ditsebe!"* (Shrugging her shoulders.)

[Translated: "In our culture, the people involved, should be here." [pointing to the floor.] Then we inform the one chosen about the matter. We choose someone. We tell them the news, then he tells them, that one tells that one, and so forth. Until they return to their owner. What you are telling me is a modern thing, we don't know such!]

Participants explained that these family members do not merely attend as observers. Their personal presence legitimises the process and ensures that proper protocol is followed. Discussions are therefore confined to those who are physically present. Furthermore, participants stressed that when conversing about Pedi family issues, confidentiality must be kept until the process is formally concluded. Premature disclosure, especially through digital communication, was described as inappropriate and potentially disrespectful. One said it can even be harmful to the process, because enemies will know about it. They argued that discussing such matters over the phone diminishes the seriousness of the engagement and risks exposing sensitive family deliberations to unintended audiences.

Turn-taking during Pedi conversations

Literature on African communication shows that turn-taking follows hierarchy, where elders and leaders speak first to maintain respect and order in group dialogues. Afrocentricity argues African communication systems that are rooted in cultural authority, communal harmony, dignity and respect for elders. The dialogues ensure structured and meaningful participation in discussions (Adekunle, 2022). Shows Afrocentricity emphasises African agency, cultural structure, and power in communication systems (Asante, 2012). Even though some studies argue that WhatsApp can show turn-taking in its implementation where people reply one after another in a chat thread, group admins can control speaking order by directing questions or discussion flow, voice notes can be used and "reply" features help link responses to specific messages, creating order. It cannot be promoted in African philosophy because it is not shared by age, authority, and social responsibility, it is not culturally grounded. WhatsApp messages can overlap, especially in active group chats; or people can ignore order and respond randomly or simultaneously, the delays in reading messages may disrupt natural conversational flow and turn -taking through responses. The physical gathering according to them also reflects the importance of consensus-building within the extended family structure. These dynamics cannot be reproduced through WhatsApp. The ward councillor said:

"O a bona mma, morero wa WhatsApp e swanetše go ba go mema batho, e sego go ahlaahla ditaba tša lapa goba mokgatlo! Dilo tše dingwe ke diphiri. Go swana le taba ya magadi ye e gebo o bolewa ke bokoko mo. O ka se di rere ka WhatsApp. O swanetše go bitša bomalome le borakgadi gomme o dule fase. Ye yo a sa tlego, o tla ba le mona le go ikgantšha. Aowaowa! Aowa! Ga e bereke bjale! E no ntaletša gomme re re ka letšatši la goregore re tla ba le kopano ka mogae. ... Ka baka la eng o swanetše o nyake ditaba ka di WatsApp?"

[Translated: "You see mam, the purpose of WhatsApp must be to invite people, not to discuss serious family or organisation issues! Some things are sacred. Like the issue of lobola which was mentioned by the granny. You cannot discuss that online. You must call uncles and aunts and sit down. This who do not come, will be jealous and proud. No! No! No! It does not work that way. Just invite me and say on such and such a day, we will have a family meeting. ... Why? Why would you want to use WhatsApp?"]

The study by (Onuh (2025) found that although WhatsApp, Facebook, and Instagram are widely used, people still preferred face-to-face interaction because online communication lacks emotional depth, non-verbal cues, and cultural context. The research also showed that elders' traditional roles remain important in African communication systems. Uka Uka Social media simplifies and commercializes indigenous cultural rituals. Traditional ceremonies lose some of their deeper social meaning when transformed into online performances and trends (Nwagbara, 2025),

Rejection of Mobile Phones in Cultural Negotiations

The findings revealed strong consensus among participants. All four rejected the use of WhatsApp to conduct serious cultural matters. This position was particularly firm regarding an example they raised about lobola negotiations. Participants argued that such processes require privacy and confidentiality. They also emphasised the necessity of respect. According to the centenarian, lobola is not a mere financial transaction. It is a ritualised engagement between families. It symbolises honour, and relational continuity. Consequently, it demands physical presence. The ward councillor indicated that at times members of families should be considerate and cooperative as a communal community.

"Motho ge a sa nyake go kopana le batho gore ba bolele ditaba tsa selapa gammogo, ouwe, o no ba le mathata. Ge o le motho swanetse o imphe nako. Mmereko wa WhatsApp ke go tsebiša batho melato, ga se ya go rera melato."

[Translated: "If one doesn't want to meet people to talk about family issues together, one has got problems. One should give one some time. The function of WhatsApp is to invite people to come and discuss issues, and not the other way round"]

Normative Protocols and Hierarchical Forms of Address

Another recurring theme concerned forms of address. Participants emphasised that hierarchical respect is central to Pedi cultural proceedings. Elders lead the process of conversations. The conversations must follow an established pattern. Interruptions are avoided and authority is acknowledged explicitly, especially when the young speaks to the old. In English for example, every is addressed as 'you'. In Pedi culture, the prefix '*bo*' is used to imply plurality which is a sign of respect. Mobile communication, by contrast, was perceived as informal and abrupt. Participants associated it with haste and diminished decorum. Participants stressed the importance of non-verbal communication. Cultural negotiations involve prescribed gestures and bodily conduct when imparting the tidings form one elder to the other. These include rhythmic clapping and saying totems to signal respect. They indicated that such embodied practices cannot be replicated through social a media communication. In an African culture you do not show up what you do or what you have. The responsibility and properties are shared.

"Wa bona tšeuwe ke tša baswa. … Ka seAfrika, re dula fase ra rera molato. Wo a sa nyakeng go tla, ke tša gagwe."

[Translated: "You see, those are for the youth. … It's because of WhatsApp! In Africa, we sit down and discuss a case. Whoever doesn't want to come, it's his own business."]

Mogolane (2025) provides compelling evidence that respectful practices like kneeling, hands, plural forms of address, and respecting elders are the foundation of Pedi communication. The study goes on to say that when speaking with elders, younger people in the Pedi culture are also encouraged to maintain indirect eye contact. Therefore, "gestures, body language, and verbal language that demonstrate deference to elders" are used to convey respect. WhatsApp cannot be used to foster these signals and protocols, particularly when adults and children are involved. Social media often excludes indigenous languages and cultural communication styles, reducing participation by minority cultural groups (Sekoai et al., 2025). African Pedi communities further follow a practice of it prioritizing a communal relationship over individualism. People converse in a group as a family or community, and it is expected that when one is called for a conversation, one must show up. The ward councillor said this when he emphasized this matter; am dals underscoring that the elderly are not able to use such platforms for discussing family issues.

"O tshwanetse o iphe nako....Le gona bakgekolo ga ba kgone dilo tšeuwe tša di Whatsapp le di Facebook. Ke tša baswa. Bjale le melato e rere ke mang? Ka foune? Le timela!"

[You must make up time. … Besides, the old people can't do some things like WhatsApp and Facebook. They are for the youth. So, who should be involved in your matters? Phone? You ae getting lost!"]

"Ditaba ga di theeletšwe ke lafase le ka moka. Moja sa gage o a iphihla."

[Translated Your matters must not be heard by the whole world. When you eat, you must not let the whole world see what you are eating."]

According to the ward councillor Pedi people culture does not support the public sharing of sacred cultural matters on social media because such information can easily be misinterpreted, commercialised or spread out of context. Afrocentric thought stresses that African knowledge systems should remain rooted in community authority, elders and culturally appropriate spaces rather than uncontrolled digital platforms. Asante explains that Afrocentricity requires Africans to "reclaim history, language and identity" and avoid cultural dilution through external influence (Asante, 2003, p. 7). Similarly, Afrocentric scholarship highlights that African spiritual and cultural knowledge must remain context-bound to preserve meaning and integrity (Reid-Merritt, 2015, p. 134). In this way, uncontrolled viral sharing of sacred matters contradicts Afrocentric principles of cultural respect and preservation.

Access to data is not only the problem even if the older community would choose to use WhatsApp, but the other challenges is also the power cuts. Elders may struggle to access data, and frequent power cuts further disrupt WhatsApp communication, making it unreliable for important community matters. This digital divide limits inclusion and can possibly weaken traditional decision-making processes that rely on direct, face-to-face interaction and stable accessible communication channels. According to the World Bank rural electrification in Sub-Saharan Africa is still very low compared to urban areas. The 56-year-old municipal officer showed concerns about people who cannot not make up time to be with family to discuss issues face to face than use the social media. Some people who are proud and ignore family relationships spend more time on social media because they seek attention, status, and personal recognition online (Kupangwa, Venter, & Farrington, 2026, p. 6). African philosophy, especially, does not fully support this behaviour because it values community, respect, togetherness and caring for family relationships above individualism and selfishness (Patel, Mohammed, & Koen, 2024, p. 9). Afrocentrism does not support people who ignore family relationships and focus mainly on social media for pride, attention or status, because Afrocentrism promotes African values of respect for elders, family unity, collective identity, and community responsibility. It discourages excessive individualism and selfish behaviour that weakens social and family bonds. The elderly showed their unappreciation of cultural discussions over the social media. They believe that social media erodes traditional African values such face-to-face communal communication. The sentiment of the elderly participant was that she is concerned about the behaviour of the young Pedi people who spend more time online on WhatsApp platforms. This reason is supported by study on Nigerian youth and social media that explains how social media influences young people to abandon traditional African values and imitate foreign lifestyles. Although a South African study supports that WhatsApp helps South Africans communicate and stay connected and another Nigerians study that WhatsApp help people organize and discuss important community issues, UNESCO warns that globalization and digital media can weaken local cultures and traditional identities, especially among young people.

CONCLUSION

In this chapter, I presented the research related to using the WhatsApp application for the discussion of traditional African matters. Data were gathered through a semi-structured interview. The participants gave answers that to the conclusion that the application or social medium is accessible and user friendly, but it cannot be used to discuss African family matters. The participants showed positive attitudes towards using WhatsApp in the future, but they found it more accessible. One limitation of this research was the small number participants. I had only six participants. In each session all questions were asked in the same order. Elderly users. Overall, the study suggests that WhatsApp hold promise as an acceptable solution to address petty issues and greetings, but not for the discussion of serious family matters. This chapter demonstrates how the elderly view the use of social media visa-vis the cultural erosion. This ubiquitous platform provides invaluable infrastructure enabling convenient, cheaper communication globally. The findings corroborate prior research which highlights the value of accessible mobile applications like WhatsApp for enabling new forms of communication, coordination and virtual co-presence among physically separated families (Baldassar et al., 2016). This study, therefore, cautions against exaggerating the advantages of WhatsApp's benefits without consideration its impact on the contribution in eroding African culture. More research is needed on how revitalize the possibilities digitalising the traditions of conversing in a meeting among the elders and the youth. However, this study makes a noteworthy methodological contribution to the field through its inclusion of the centenarian voices members who have been directly involved in community`s cultural setups. Their understanding on the use of social media in discussing African matters is not condoned.

SUMMARY

WhatsApp limits non-verbal communication such as facial expressions, body language, tone of voice, and eye contact, which are essential in African community discussions. Without these cues, messages can be misunderstood, reducing respect and emotional connection. In African traditions, meaning is deeply tied to physical presence, which digital communication cannot fully capture. WhatsApp restricts important non-verbal and cultural communication elements such as honorifics and kinesics. Honorifics are forms of respectful address used for elders and authority figures are often reduced to plain text, weakening expressions of respect and social hierarchy. Kinesics, which includes body movements, gestures, posture, and facial expressions, is completely absent in text-based communication. In African contexts, both are essential for showing respect, maintaining harmony, and reinforcing social relationships during discussions. Without them, messages can lose cultural meaning and be misinterpreted. Afrocentricity is suitable for analysing the intersections of constructs such social media, cultural practice and indigenous language preservation. It affirms the legitimacy of concerns regarding digital erosion while advocating for culturally grounded technological engagement that is aligned to African philosophy. It does not completely reject modern tools but calls for their adaptation in ways that preserve African epistemic integrity. Social media platforms could support indigenous languages if they are intentionally localised and culturally contextualised. The findings are discussed in relation to the perspective that African communities should be as active subjects rather than passive recipients of modernity. The resistance expressed by elderly participants toward using mobile phones in cultural negotiations can be interpreted as an exercise of cultural self-determination. Afrocentricity also places strong emphasis on language as a carrier of culture. Languages cannot be divorced from culture. Indigenous languages such as Sepedi embody principles of African philosophy such as collective memory, social hierarchy, honorifics, and forms of address. Because social media platforms often operate predominantly in western languages such as English, the authenticity of languages like Sepedi can be marginalised. Indigenous linguistic expression, including forms of greeting and hierarchy, may be at risk of dilution. Young people should be developed on the use of these social media platforms to promote their indigeneity.

LIMITATIONS

The study focused on a small sample of four participants. A larger sample can be targeted. Since this study followed a qualitative route, a quantitative research methodology might be followed.

RECOMMENDATIONS AND FUTURE RESERCH DIRECTIONS

This exploration addressed the use of WhatsApp applications in conversations in African indigenous language Sepedi. Considering the challenged observed, effectively addressing these issues will demand another investigation of the various digital tools and applications to distinguish which digital tools are effective in meeting the unique requirements of indigenous language and promotion and preservation, with this task of particular significance as it guides the selection of tools that align best with the specific needs and challenges faced in this context. In this case the context of African indigenous languages and their preservation in social media conversation. The safeguarding and facilitating the intergenerational transmission of IKS should be intensified, and the necessity for policy interventions to bolster such system should be highlighted., other studies in the future can pay attention within the context of African Indigenous languages, highlighting the potential for broader geographic representation in future research. To revitalise the use of social media to promote African languages unique and genuine use, schools and higher institutions of learning should develop programmes to creatively engage students to design applications that can promote preservation and protection of IK.

REFERENCES

Adekunle, O. (2022). Hierarchy and respect in African indigenous communication systems. *Journal of African Cultural Studies, 34*(2), 145–158.

Asante, M. K. (2003). *Afrocentricity: A theory of social change* (revised and expanded ed.). Chicago, IL: African American Images. *Urban Education, 41*(1), 101–111. DOI: 10.1177/0042085905284322

Asante, M. K. (2012). Afrocentric idea and communication theory. *International Journal of Intercultural Relations, 36*(6), 760–769.

Bagwasi, M. M. (2012). The effect of gender and age in Setswana greetings. *Southern African Linguistics and Applied Language Studies, 30*(1), 93–100. DOI: 10.2989/16073614.2012.693717

Baldassar, L., Nedelcu, M., Merla, L., & Wilding, R. (2016). ICT-based co-presence in transnational families and communities: Challenging the premise of face-to-face proximity in sustaining relationships. *Global Networks, 16*(2), 133–144. DOI: 10.1111/glob.12108

Battiste, M., & Henderson, J. Y. (2000). *Protecting Indigenous knowledge and Heritage: A Global Challenge*. Purich Publishing. DOI: 10.59962/9781895830439

Bhanye, J., & Bhanye, A. (2023). A blessing or a curse? The role of social media during the COVID-19 pandemic in Africa. In *The Palgrave handbook of global social problems* (pp. 1–23). Springer International Publishing.

Chawane, M. (2016). The development of Afrocentricity: A historical survey. *Yesterday and Today*, (16), 78–99. DOI: 10.17159/2223-0386/2016/n16a5

Davies E, (1987). A contrastive approach to the analysis of politeness formulas. *Applied Linguistics.8, 76:88.*

Dhir, A., Yossatorn, Y., Kaur, P., & Chen, S. (2018). Online social media fatigue and psychological wellbeing—A study of compulsive use, fear of missing out, fatigue, anxiety and depression. *International Journal of Information Management, 40*, 141–152. DOI: 10.1016/j.ijinfomgt.2018.01.012

Eke, C., & Adeyemi, M. O. (2024). Utilising digital media tools to foster indigenous communication systems in Nigeria. *African Journal of Social and Behavioural Sciences, 14*(6).

Galla, C. (2016). Indigenous language revitalization, promotion, and education: Function of digital technology. *Computer Assisted Language Learning, 29*(7), 1137–1151. DOI: 10.1080/09588221.2016.1166137

Herrera, M. I. H. (2022). The Use of Technologies in Language Revitalisation Projects. *Journal of Global Indigeneity, 6*(1), 1–17.

Holotescu, M. C., & Cre u, V. I. (2015). *Emerging technologies in education; Conceiving and building a microblogging platform for formal and informal learning*. Editura Politehnica.

Kitchenham, A. (2013). The preservation of Canadian Indigenous language and culture through educational technology. *Alternative, 9*(4), 351–364. DOI: 10.1177/117718011300900406

Kupangwa, W., Venter, E., & Farrington, S. M. (2026). Manifestations of ethical values in indigenous African family businesses: An application of the Ubuntu philosophy. [South African Journal of Business Management.]. *South African Journal of Business Management, 57*(1), 1–11. DOI: 10.4102/SAJBM.v57i1.5592

Li, J., Brar, A., & Roihan, N. (2021). The use of digital technology to enhance language and literacy skills for Indigenous people: A systematic literature review. *Computers and Education Open, 2*, 100035. DOI: 10.1016/j.caeo.2021.100035

Li, Y., & Li, Y. (2019). Cultural differences in everyday conversation between the East and the West. In *5th International Conference on Education Technology, Management and Humanities Science (ETMHS 2019)* (pp. 1070-1073).

Lith, R. 2023. WhatsApp for business opens doors to 21 million South Africans. Independent Online. Available at: https://www.iol.co.za/business-report/opinion/whatsapp-forbusiness-opens-doors-to-21-million-south-africans-19a5339a-ee20-43af-9315- 92125906ce8f. [Accessed on 10 April 2026].

Lohdi, A. Y. (1993). The language situation in Africa today. *Nordic Journal of African Studies, 2*(1), 79–86.

Magaldi, D., & Berler, M. (2020). Semi-structured Interviews. In V. Zeigler-Hill & T. K. Shackelford (Eds.), Encyclopedia of Personality and Individual Differences. Springer., Retrieved April 15, 2026, from, DOI: 10.1007/978-3-319-24612-3_857

Malapane, O. L., Chanza, N., & Musakwa, W. (2024). Transmission of indigenous knowledge systems under changing landscapes within the vhavenda community, South Africa. *Environmental Science & Policy, 161*, 103861. DOI: 10.1016/j.envsci.2024.103861

Malatji, E., & Lesame, C. (2019). The use of South African languages by youth on social media: The case of Limpopo Province. *Communicare, 38*(1), 76–95.

Mangla, M., Akhare, R., Deokar, S., & Mehta, V. (2020). Employing machine learning for multi-perspective emotional health analysis. In *Emotion and information processing: a practical approach* (pp. 199–211). Springer International Publishing. DOI: 10.1007/978-3-030-48849-9_13

Manning, J. (2014). Definition and Classes of Social Media.Available at www.researchgate.net

Meighan, P. J. (2021). Decolonizing the digital landscape: The role of technology in Indigenous language revitalization. *Alternative, 17*(3), 397–405. DOI: 10.1177/11771801211037672 PMID: 41924568

Mncwango, E. M. (2009). Language and the current challenges in the South African schools system. *Inkanyiso, 1*(1), 51–54. DOI: 10.4314/ijhss.v1i1.62110

Mogolane, M. J., Neille, J., & Sher, J. (2025). Pragmatics: Exploring language use by younger generations in Pedi families. *The South African Journal of Communication Disorders = Die Suid-Afrikaanse Tydskrif vir Kommunikasieafwykings, 72*(1), a1097. DOI: 10.4102/sajcd.v72i1.1097 PMID: 40922622

Moloko-Phiri, S. S., Sepeng, N. V., & Sebaeng, J. M. (2022). Communication discordance among health care professionals and indigenous patients. In Mulaudzi, F. M., & Lebese, R. T. (Eds.), *Working with indigenous knowledge: Strategies for health professionals* (pp. 5–20). AOSIS Books. DOI: 10.4102/aosis.2022.BK296.01

Mudau, P. P., Mandende, I. P., & Mushaathoni, M. (2024). Through our indigenous greeting terms, we are identified: A study of Tshiven a from Tshimbupfhe, Vhembe District Municipality, Limpopo, South Africa. *African Identities*, •••, 1–15. DOI: 10.1080/14725843.2024.2381484

Mushaathoni, M. (2023). Exploring Intercultural Communication as a Means to Promote Inclusivity in Diverse Organisations: A Study in a South African University. *IAFOR Journal of Education, 11*(3), 161–184. DOI: 10.22492/ije.11.3.08

Nakata, M., & Langton, M. (2005). Australian indigenous knowledge and libraries [Internet]. Available: https://www.alia.org.au/publishing/aarl/AARL_vol36_no2_2005(reprint).pdf Accessed 14 April 2026.

Nenungwi, T. G. (2015). Non-verbal communication in Tshiven a: a sociolinguistic and discourse analysis. Doctoral thesis. Retrieved from http://hdl.handle.net/ 10386/1542

Nwagbara (2025), *Social Media Pressures on Cultural Authenticity in Traditional Igbo Wedding*.

Nyoni, T. E. (2023). *The co-creation of Afrocentric ethical guidelines for digital content creators producing indigenous knowledge-based content* (Doctoral dissertation).

Omipidan, I. A., & Sanusi, B. O. (2024). Rise of social media in the digital age: Whatsapp a threat to effective communication. *IMSU Journal of Communication Studies*, *8*(1), 142–153.

Outakoski, H., Cocq, C., & Steggo, P. (2018). Strengthening Indigenous languages in the digital age: Social media–supported learning in Sápmi. *Media International Australia, Incorporating Culture & Policy*, *169*(1), 21–31. DOI: 10.1177/1329878X18803700

Owusu-Ansah, F. E., & Mji, G. (2013). African indigenous knowledge and research. *African Journal of Disability*, *2*(1), 1–5. DOI: 10.4102/ajod.v2i1.30 PMID: 28729984

Oyinlade, E. O. (2024). Technological culture and the challenge of erosion of Yorùbá moral values. *African Journal of Culture, History. Religion and Traditions*, *7*(3), 1–9.

Patel, M., Mohammed, T. A. S., & Koen, R. (2024). Ubuntu in post-apartheid South Africa: Educational, cultural and philosophical considerations. [MDPI Philosophies Journal]. *Philosophies*, *9*(1), 1–15. DOI: 10.3390/philosophies9010021

Rajasekaran, B., & Martin, R. A. (1992). Incorporating Indigenous Knowledge Systems into Agricultural and Extension Education Programs: A Study of the Perceptions of Extension Professionals.

Raji, T., & Dube, B. (2024). Social media and African crises: A comparative study of Nigeria and South Africa. [RSIS International]. *International Journal of Research and Innovation in Social Science*, *8*(2), 120–132.

Reid-Merritt, P. (2015). Afrocentric approaches to social work. In *International Encyclopedia of the Social & Behavioral Sciences* (2nd ed., pp. 132–136). Elsevier. DOI: 10.1016/B978-0-08-097086-8.28003-3

Senanayake, S. G. J. N. (2006). Indigenous knowledge as a key to sustainable development. *Journal of Agricultural Sciences*, *2*(1), •••.

Sharma, N. K., & Niharika, N. (2025). Roots and Resonance: The African Paradigm of Afrocentricism. *International Journal of English Literature and Social Sciences, 10*(1), 593321. Li, Sithole, J. (2007). The challenges faced by African libraries and information centres in documenting and preserving indigenous knowledge. *IFLA Journal, 33*(2), 117–123.

Solo-Anaeto, M., & Jacobs, B. (2015). Exploring social media as channels for sustaining African culture. *International Journal of Humanities and Social Science, 5*(4), 37–42.

Statista (2024). *Global Social Media Usage Statistics.* https://www.statista.com

Tarugarira, G. (2014). Interrogating generational connections and conflicts on the utility of indigenous knowledge systems as a cultural dimension of development in Zimbabwe.

Toumaras, N. (2025). Networked Hyperlocal Activists: Digital Democracy and Engagement in Sub-Saharan Africa. *Social Media + Society, 11*(2), 20563051251345945. DOI: 10.1177/20563051251345945

UNESCO. (2022). *Cultural diversity and the impact of digital media.* UNESCO.

Vorster, C., Kritzinger, A., Lekganyane, M., Taljard, E., & Van Der Linde, J. (2022). Cultural adaptation and Northern Sotho translation of the Modified Checklist for Autism in Toddlers. *South African Journal of Childhood Education, 12*(1), a968. DOI: 10.4102/sajce.v12i1.968

Wierzbicka, A. (2003). *Cross-cultural pragmatics.* Mouton de Gruyter. DOI: 10.1515/9783110220964

Yang, X. (2010). Address Forms of English: Rules and Variations. *Journal of Language Teaching & Research, 1*(5). Advance online publication. DOI: 10.4304/jltr.1.5.743-745

ADDITIONAL READING

Avcu, K. M., & Koçoğlu, E. (2025). Preservation of cultural identity in digital ecosystems.

Balogun, S. K., Aruoture, E., & Haruna, S. E. (2024). Globalisation and the Decline of the African Value System. *Nigerian Journal of Social Psychology, 7*(2).

Idang, G. E. (2015). African culture and values. *Phronimon, 16*(2), 97–111. DOI: 10.25159/2413-3086/3820

Lovink, G. (2017). *Social media abyss: Critical Internet cultures and the force of negation.* John Wiley & Sons.

Udoinwang, D., & Akpan, I. J. (2023). digital transformation, social media revolution, and e-society advances in Africa: are indigenous cultural identities in danger of extinction? *Social Media Revolution, and E-Society Advances in Africa: Are Indigenous Cultural Identities in Danger of Extinction.*

KEY TERMS AND DEFINITIONS

Afrocentricity: Afrocentricity as a paradigm based on the idea that African people should re-assert a sense of agency to achieve sanity. Afrocentricity emphasizes African history, culture and values as the starting point for understanding African people and societies.

Cultural erosion: Cultural erosion refers to the transformation of practices, cultural loss or disappearance of languages, identities, knowledge systems, beliefs, arts, and social structures that define a community or society.

Pedi people: The Pedi people, also known as the Bapedi, are one of the many ethnic groups that contribute to the cultural tapestry of South Africa.

Social media: Social media, a form of mass media communications on the Internet through which users share information, ideas, personal messages. and other content.

Chapter 10

Analysing the Policy and Institutional Barriers to Indigenous-Led Heritage Restoration and Reframing Governance Through the Lens of Inherent Sovereignty

Ashok Dobhal
http://orcid.org/0009-0006-9077-4190
Law College Dehradun, India

Rajesh Bahuguna
http://orcid.org/0000-0001-8729-5160
Law College Dehradun, India

Nidhi Tyagi
http://orcid.org/0000-0002-3996-7644
Independent Researcher, India

Isha Sharma
http://orcid.org/0009-0000-5514-283X
Law College Dehradun, India

Radhey Shyam Jha
http://orcid.org/0000-0003-0129-4365
Law College Dehradun, India

ABSTRACT

Indigenous Knowledge is the knowledge, skills, and beliefs that societies have developed over time by living in and interacting with their natural surroundings. It includes art of agriculture, food preparation, health care practices, education, building architecture ecological conservation, language etc. Through

DOI: 10.4018/979-8-3373-7976-0.ch010

the reliance and interaction with nature and its surroundings, indigenous communities have acquired an immense knowledge of the same. This accumulated knowledge is rapidly disappearing as the world is steadily becoming more culturally and biologically uniform. There are many Indigenous societies in the world that have a rich heritage spanning more than tens of thousands of years and their governance is in turmoil due to multiple stakeholders such as indigenous communities, government, private institutions, international organizations, NGOs, religious groups etc. leading to ambiguity. Thus, to avoid complexity, the preservation and restoration of indigenous heritage vividly rely on sovereignty of an institution and their effective governance.

INTRODUCTION

According to UNESCO, Indigenous Knowledge is the knowledge, skills, and beliefs that societies have developed over time by living in and interacting with their natural surroundings (Tippeconnic & Faircloth, 2010). It is also known as 'local knowledge', 'folk knowledge', 'people's knowledge', 'traditional wisdom', or 'traditional science' (Pandey, 2014). In many parts of the world, the art of agriculture, food preparation, health care practices, education, ecological conservation, indigenous methods of governance and various other activities that support community resilience are based on this indigenous knowledge, which is typically passed down orally and through cultural customs through generation to generation. These are the skills associated to local communities and are deeply rooted to traditional and cultural practices. This knowledge leads to creation of material and non-material indigenous heritage which is a result of generations of trial and error and is obtained generally through non written texts such as sacred manuscripts, rituals, stories, poems and hand on experiences (Walker-Jones, 2021). This includes deep knowledge of the traditional ecosystems, traditional practices, sustainable resource management, traditional agricultural practices, rituals and various other practices that hold intellectual value. IKS, unlike modern Western systems, are flexible, adaptive and are based on collective experiences. They represent not just the data points and folklore but also the comprehensive worldviews expressed in languages, rituals, kinship, traditional practices and the management of the monuments, sites, artefacts, buildings, tradition, culture and environment.

If we talk about 'Heritage', it can be considered as the core of indigeneity, that has been all about promoting the indigenous viewpoint and using them to solve social, economic and environmental concerns (Heuheu, Kawharu, & Tuheiava, 2012). Indigenous heritage is territorial distinctiveness having sacred geography and resource management travelling through ancestors. It includes tangible as well as intangible resources of the natives wherein tangible resources typically includes buildings, historical sites, artefacts, monuments, archives etc. and the intangible resources includes language, customs and traditions (Makhloufi, 2024). The distinct culture colours history of a region vividly to embark upon traditional insight of sustainability to shape future dynamics. So, apart from preserving the tangible things like buildings and artifacts, heritage restoration also contributes in promoting the understanding of the culture among the communities, strengthening the diversity and revitalising in the development apparatus.

Cultural heritage as a continual process of doing, remembering, teaching and learning can be understood as a political act of establishing personhood, nationhood and asserting human rights. Besides having immense importance, governing indigenous heritage remain in turmoil due to multiple stakeholders. There are contestants to regulate restoration including indigenous communities, government, private institutions, international organizations, NGOs, religious groups etc. that are required to be viewed through the lens of Inherent Sovereignty (Madaan & Mohanty, 2025).

Workshops, performances and exhibitions that display the richness and diversity of indigenous traditions has frequently been hosted by these indigenous peoples in order to formulate important platforms for indigenous voices, enabling communities to tell their own stories and challenge stereotypical portrayals (Rathor, 2024). However, apart from the motivated community initiatives, restoring indigenous legacy is an institutional job as government institutions hold sufficient authority, resources and capacity to preserve and protect heritage with its utmost value. Cultural centers and museums managed by them helps in promoting educational initiatives, cultural activities and community gatherings so as to preserve the skills inspired by old traditions and pave a way for their future implications.

LITERATURE REVIEW

Waterton and Smith in their research work discussed about primary level institutional barriers for the protection of Heritage of Indigenous people but modern tools of technology and governmental strategic policy formulation have given no place for their protection (Waterton & Smith, 2010). In the book of Michalos, the author has tried to explain the concept of Traditional Knowledge, its recognition under global governance, various positive as well as negative aspects of it and its impact on quality of human life but the role of sovereign authorities, their policy initiative and policy barriers have remained untouched (Michalos, 2014). The authors Uma, Velayutham, Magudeeswaran & Mcpe have tried to explored the role of technology for the preservation of Indigenous knowledge and explain the advantages of technological intervention with challenges and issues for the preservation of Indigenous knowledge but no discussion has been made on governmental initiative, the pros and cons of these policies and challenges (Uma, Velayutham, Magudeeswaran, & Mcpe, 2024).

Pittz, Claw & Adler in their research paper demonstrates the challenges related to protection of tribal knowledge and heritage that tribal leaders face and enumerates the various strategies used by Native American nations to exercise sovereignty from the US Federal Government (Pittz, Claw, & Adler, 2026). This research paper focus on limited surroundings and do not give ideas about other indigenous communities (Pittz, Adler, & Claw, 2024). Grey & Kuokkanen have critically examined the co-management of Indigenous peoples' cultural heritage as a driver and product of the culturalization of Indigenous peoples of different part of the world by using certain case studies but barriers and sovereignty concerns for the protection of the indigenous knowledge is missing in their discussion (Grey & Kuokkanen, 2020). Nayak, Chakraborty & Chakrabarti in their chapter recommended that with the use of digital twins, indigenous communities may preserve, share, and self-record their knowledge systems through experiential learning environments, virtual repositories, and instruments for community empowerment but policy and institutional barriers remained untouched in their discussion. (Nayak, Chakraborty, & Chakrabarti, 2025)

RESEARCH OBJECTIVES

1. To analyze the Policy and Institutional Barriers to Indigenous-Led Heritage Restoration.
2. To assess the efficacy of existing regulatory and institutional framework for the protection and restoration of Indigenous-Led Heritage

3. To suggest Reframing Governance for the protection and restoration of Indigenous-Led Heritage through the Lens of Inherent Sovereignty

RESEARCH METHODOLOGY

The research methodology is primarily a Qualitative, and Indigenous-Led Design which specifically emphasizes on understanding the complex intersection of sovereignty as well as policy. The study includes a Critical Policy and Legal Analysis reviews existing federal and provincial statutes, regulations, and case laws to map the institutional barriers which prevent Indigenous jurisdiction over restoration and over heritage. The research methodology ensures the findings of the study are not only just academically rigorous but are also beneficial and relevant to the participating Indigenous Nations, resulting in operationally viable strategies & actionable policy recommendations.

LIMITATION OF THE STUDY

The present research work is primarily based on secondary data and confined to study of policy and institutional barriers for the protection of Indigenous-Led Heritage Restoration. Added case studies and discussion is also having the same central idea, hence suggestions and recommendations are having limitations accordingly.

SCOPE & SIGNIFICANCE

This research work is focusing on identification of policy and institutional barriers with the perspective of exploring their solutions. Since this research is issue specific and recommending concerned remedial steps only, other areas are open for further research that may be relevant for the protection and restoration of Indigenous-Led Heritage.

INDIGENOUS-LED HERITAGE RESTORATION

Through their reliance and interaction with nature and natural resources, indigenous communities have acquired an immense knowledge of their natural environment. Yet this accumulated knowledge is rapidly disappearing as the world is steadily becoming more culturally and biologically uniform. There are many Indigenous societies in the world that have a rich heritage spanning more than tens of thousands of years.

In India, Adivasis, whose origins date back over 50,000 years, developed distinct languages and traditions but faced displacement due to Indo-Aryan migrations, Mughals and British. Despite the disruptions caused by European colonization, many Indigenous groups have preserved their tradition and culture and continue to appreciate their rights marking around 705 recognized tribal communities in India, that had been able to uphold their heritage amid ongoing socio-political changes (IWGIA, 2025).

People from indigenous communities highlighted how their cultural heritage needs to be sensate not in parts but as a whole, in inter-generational manner and according to spiritual values, influenced by their environment and traditional history. The protection of traditional knowledge should be based on holistic approach like 'Collective Bio-Cultural Heritage' so as to reflect the reality of many local knowledge systems, and to reinforce and strengthen such systems. Wherein Collective Bio-Cultural Heritage system includes the knowledge, innovations, practices and cultural expressions of indigenous peoples and local communities which are often collectively and inextricably linked to traditional resources and territories; it also includes the diversity of genes, varieties, species and ecosystems; cultural and spiritual values; and customary laws shaped within the socio-ecological context of communities (Swiderska, 2006). But it is challenging to fit their holistic approach with the international legal system governing cultural heritage due to multitude of legal order, inherent sovereignty and complexity of indigenous thoughts.

Indigenous-Led Heritage Restoration includes not just intangible and tangible heritage, but also the division between intellectual, immaterial, and material protection of the cultural heritage. This overall regime, which is based on a cross-competency between different international organizations and state authorities often leads to disintegration between a multitude of legal frameworks, which ultimately do not adequately represent the holistic cultural heritage of their community, and the system notably fail to recognize that culture of indigenous community. The conservation of indigenous heritage vividly relies on sovereignty of an institution and their effective governance, yet globally there is no unanimity in institutional principles to restore indigenous heritage. With the help of this research, an attempt is made to delve into the efficacy of institutional custodianship, challenges and possible legal frameworks to restore the indigenous heritage.

CHALLENGES BEFORE ILHR

The endowment of culture encompasses altogether different problems owing to type of polity, demography, geography, economy and global ties of a country. Further, the variables vary at local divisions of a every nation. Policy barriers and Institutional barriers are two of the major challenges in restoring heritage in any nation.

Policy Barriers- These are the barriers which hampers the restoration projects due to either inadequate policy measures of a nation or insufficiency of legislations. Following are the major policy barriers occurring world-wide in heritage conservation:

Inadequate Legislations - Though states have adopted the laws to preserve their heritage, but the sufficiency and adequacy is yet a question of debate. As said "ut res magis valeat quam pereat"- a law must prevail rather than perish (Bhusari, 2024). Laws have been developed with an intent to conserve and maintain tangible and intangible indigeneity but most of the legislations are inadequate either in defining the heritage at the objectively inclusive level or archaic as to importance and model management. Like, India has enacted AMASR Act 1958, but the act is not comprehensive enough to inculcate local monuments and sites (Tripathi, 2021). The law lacks inclusive definitions and dynamic variability fulfilling intra-generational scope.

Procedural impropriety- Nearly every state has concentrated predominantly on framing enactments that are substantive in nature, while the enforcement mechanism has largely been overlooked. In countries like Greece (staffing Procedure), Brazil (Tombamento sternness), Philippines (slow registration procedures-PRECUP), there are legislations on conserving and safeguarding antiquities embedded in substantive frameworks with vague procedural line. Laws are defining heritage, prohibiting exploitation and demolitions but recognition formalities are rigid; registrations are cumbersome; assessment and maintenance are improper and enactment lacking implementation are reducing utilities of governance setups. Laws without implementation vision questions the legislative determination and vision to regulate efficiently.

Lack of funding- Maintenance of heritage be it tangible or intangible can be achieved only through adequate and sufficient budget dispersible. On contrary almost every state is lacking specific budget allocation for indigeneity, as it is kept out of essential priorities. World Economic Forum has aptly marked that there is significant gap in budget required and provided for restoration and maintenance of heritage worldwide. Though history cannot be erased by acts of presents yet it deteriorates. Ancient building and culture need professionals to manage, maintain, repair, practise and propagate but there are no funds for this capacity building. Restoring heritage requires specific number of trained professionals, which comes with decency and adequacy of funds to preserve and facilitate restoration.

Revenue model- Maintaining heritage sites calls for hefty amount of funds to maintain essence of vision of creation. To cater the funds where government budget struggles, revenue models can be built around tourism revenue sharing and Public Private Partnerships. Tourism revenue model can be built as to collect entrance or rights-based fees from visitors where monument is labelled for tourism, the share from the fees should be dedicated towards heritage conservation and restoration. Further, Public-Private Partnership can be boosted, where Private entities can be invited for heritage restoration in lieu of tax benefits, bonds or shared managerial rights.

Lack of decentralised regulations- Policies preserving culture and heritage are generally centrally authorized. The laws empower government institutions and organizations with managerial guardianship for regulating it in national (sovereign) interest. These setups more likely exercise functions at central level with Union/Federal Ministries of Culture and Tourism. Indeed, the vertical management can better regulate the eminence of endowments. Local guardianship of heritage shows higher possibility of timely measures and impromptu solutions. Where, the countries like France, Egypt, Turkey etc. following the top-down approach (keeping Centre in the top) are struggling to silo effect while on the other side, the decentralised jurisdictions like Germany, USA, Canada etc. has proximate distribution of heritage responsibility favouring location, historical importance, international perspective and local validation (Rajani, 2025).

Lack of regulations promoting community participation- Restoring heritage is an act of shared responsibility, it cannot be achieved with setting away from government. Community participation is a feasible yet efficient measure to settle the goals of conservation (Lachapelle & Austin, 2014). Countries like India, France and Turkey have no specific legislations fostering community to manage the culture they inherited from their ancestors while countries like UK, South Africa and Australia have legislative frameworks that promotes community role in preservation of heritage and leads to settle guardianship with emotional intelligence along with expertise backed by experience of antiquity.

Buffer zones- As urbanization is a complementary parallel phenomenon to heritage restoration, policies must be sound enough that harmonizes by zone planning and regulating buffer areas and locations. Neighbouring commercial setups with disposing outputs threaten the beauty of monuments and sites while creating buffer zones between heritage buildings and adjacent building is an ideal solution as this will protect monuments from adjacent development hazards and remote damages (UNESCO, 2011). The urban development body should regulate constructions near heritage sites as to height constraints, visual gaps, corridors and setback pre-requisites.

Lack of adaptive reuse- Few of the structures shatters to the rigidity of zoning policies. Major legislations governing material heritage either qualifies them into residential, commercial, restricted, or directs its utility. Adaptation of monuments suitable to purpose add spirits to endowments (Malhotra & Dobriyal, 2023). A shift to adaptive utilization may better address the life and maintenance of heritage. It is like taking out of these monuments from history to modern day development plan as a significant contributory. It settles funding gap and an important addition to performing assets.

No policy on heritage precinct- Restoring heritage sounds like protecting building, monuments, sites or artefacts. The idea sounds circumscribed to material heritage rather it stretches to protect living community practising ancient indigenous culture and tradition. Endowment of inheritance is followed by living practice, so their rights should be well recognized with the formative of preservation plan of heritage.

Punitive policy: Most of the jurisdictions have adopted punitive measures to protect against the violation of heritage (Papaioannou, 2017). Modern theory of legislations has long ago shifted to compensation and incentives. Legislations are drafted with an aim to curb practices of DON'Ts. An example of it is that laws are forbidding construction in close vicinity of monuments and heritage, which is an interference in ownership rights.

Institutional barrier – Once the restoration passes through policy formulation, it is handed over to institutions or organizations for implementation to the project. When these respective institutions or organizations do not fulfil allocation properly, it is referred as institutional barriers. There are many institutional barriers which are listed below:

Custodianship- When arguments shed light on restoration, the genesis starts at who will be restoring? Who owns ownership and responsibilities? Allocating guardianship is the detrimental question. Figuring out the stewardship settles designation of power and duties as to specific heritage. It can be private or public depending upon the legal system of the country. Labelling custodians to antiquities is a virtue to delegate ownership or authority which comes with responsibilities as to its maintenance of legacy. The label is ascertained, considering finances, social, political, economic and cultural factors. Guardianship can be granted to government (Central/State/local), private entities, international organizations (UNESCO), religious institution (temples/churches/mosques) or regional community (indigenous local community) based on highest nexus proximity as to pre-requisites decided for such grant (UNESCO, 2011). However, this may sound subtle but it is backed by multiple consequences derived by abuse of discretion, exercise of arbitrariness and violation of cultural norms.

Guardianship must be centred on questions of legitimacy, capacity, inclusivity, sustainability and expertise. At times, custodianship struggles between State Versus Community, Private Versus Public, where determination underscores rights of one claimant but at times, the tussle is upheaval of multiple stakeholders. In such situation dividing managerial or restoring rights touches many socio-legal scenarios.

Fragmental authorities- Same as the cartesian plane, the institutional framework of restoring indigenous heritage is majorly divided across the Vertical strata (y-axis) and horizontal strata (x-axis). It is also backed by functional variables in the four quadrants. Vertical strata divided authority and responsibility as to setups of level of governments where there is dyarchy, it is substantially between Centre and State as to India, Australia, Germany or USA. Then, there are countries having unitary setup like UK, France, Japan where power is flowing from union to its agencies.

Horizontal strata imply existence of multiple ministries as identifiable power holders as to distinct roles. For instance, responsibility allocation between the Ministry of Culture, Ministry of Tourism, Ministry of environment, Ministry of Urban Planning etc., while the functional quadrants are the organizations aiding these ministries either at horizontal level or vertical level. There exist multiple departments like Public Work Department, Archaeology department, Municipality, Environmental bodies etc. regulating them.

Silo effect- Having multiple institutions linked to different roles of management of a heritage seeks cross-functional collaborations. Where the institutions undermine collaboration to work in pseudo supremacy or isolation behaviour, it leads to silo effect. The silo effect leaves the heritage management in chaotic run (DRONAH, 2021). Lack of coordination and communication in departments where they all are designated to some or other aspects of the work creates confusions, duplication of work, narrow vision and poor implementation of the scheme, which tarnishes the utility of the project. For instance, Department of Tourism will promote tourism and entry fees. Department of Antiquity will focus conservation and capacity of the monument. Urban Development boards will focus on infrastructure around the monument. While lack of coordination will create conflict of interest and coordinated idea will support conservation with balanced tourism and necessary infrastructure.

Bureaucratic latency- For management of heritage project, there exist multiple departments and institutions. In each institution, there are officials designated to execute their roles. Due to numerous departments, the decision making is followed by undefined procedural details. The process of deciding circles many questions ranging from:

- How to recognize specific monument/site/building as heritage?
- What should be the registration process?
- what should be the restoration strategy? Who will carry out?
- How the funds will be arranged?
- What will be the zoning laws?
- Whether adaptations can be permitted? Who will decide?
- How to appoints professionals for capacity building?
- Whether the heritage project is successfully preserving or not?

Amidst many questions addressed and for left behind, there is huge struggle to find the clear pathway from where the project will start, demarcated roles and hierarchal approvals. Usually, the file of projects travels from one office to another, sometimes signed, sometimes modified and sometimes unaddressed. Policies must focus on defining clear procedures and roles of officials backed by their liabilities.

Competency of officials- There is a need of machinery on whom final responsibility rests after allocations and decisions. Once a management decision is finalized, respective department takes the lead of management and maintenance of the heritage. In majority of jurisdictions, these officials are not designated qualified experts in this field, rather transferred or appointed through civil service exams. Each heritage project is an engagement of multi-disciplinary expertise based on role dispersal. However, few jurisdictions mandate specialized training and knowledge whereas few follows directional approach. Where specialized people are holding managerial responsibilities, the decision making often reflects conservational aftermaths. On the other hand, where generic officials are decision makers, aftermaths spell scientific approach devoid of ancestor emotions.

Data Management- As soon as we start delving in restoring discussion of indigeneity, first is to refer essential information and every data available to lead. The data is important to identify and compare the past, present and future perspectives of heritage protection. Data answers every question required to successfully propose heritage restoration. There are many vital questions related to heritage projects. For example, there are thousands of heritage sites in one country only, how many will be in world? How are these sites recognized as heritage? Who is restoring which of these sites? How these sites have been protected? How many of total sites are protected? How many have been destructed? What is the total budget of restoration? How are the funds allocated? Who is managing what? etc. It is just a glimpse of what all are there to ascertain before planning conservation. The real draft suffers thousands of questions about each specific heritage subject. It is all gathered from data, better the management of data, better will be the restoration plan. Since data is plethora of information and factsheet, data management is the only technique to find needle from the pile of heap.

Here are few important questions that may arise - Who is managing this data? Or if this data is actually managed? Only few of the jurisdiction will qualify these questions. Data management is proportional to infrastructural development of states. States must ensure sufficient infrastructure support to manage data required for planning restoration and conservation plans.

Inadequate capacity building- Heritage management and restoration is not a job of small assistance. It requires task force for different aspect of roles based on expertise and competencies. A common traceable problem with every historical restoration project is insufficiency in staff. Staff procurement following different eligibility criteria is a formal procedure to meet the end goals of restoration. However, hardly any project has this fulfilled due to low budget, outdated procurement process and undefined priorities. Proposal of projects is the mere vision; it calls for human machinery to add realistic bases on that. Understaffing leads to incomplete execution and deterioration. Therefore, professional recruitment for each heritage site should be carried as an essence of the maintenance to monitor, manage, repair and upgrade.

ROLE OF GOVERNMENT IN DEVELOPING GOVERNANCE OF INDIGENOUS HERITAGE

Government has significant role in restoring and preserving validity of heritage. Government approach decides on barriers and safeguards where an effective policy and adequate role can flourish the heritage, and inactivity leads to deterioration. The present role of government has formulated through years of revolution from ancient times to present era. Among ancient civilizations, indigenous heritage and legal philosophies were based on the concept of fairness. Their legal traditions reflected that the cultural philosophies governed life and justice on the basis of nature and the ecological knowledge.

In India, the indigenous governance was deeply rooted in the *dharma.* The concept of *dharma* shaped the jurisprudence through religious texts like Bhagavad Geeta, Vedas, Vedangas, Vedantas, Manusmriti, Yajnavalkya etc. and local assemblies (Kane, 1968). In ancient India, *Achara* (conduct), Vyavhara (behavior), *prayaschit* (atonement) and *manyata* (belief) were the major guiding principles to regulate human behavior (Srinivas, 2013).

Greece on other hand was regulated by *Ancient Greek Law,* which back to almost seventh century BC (Osborne, 2009). It is a collection of laws of various Greek regions that share a common legal regime and was often termed as a system of natural law (Tussay, 2024). Arbitration and Mediation were key modes of dispute settlement in ancient Greek societies (Grisel, 2023). Ancient Greek laws were customary in nature and as per *opinio communis*, it was unwritten, orally transmitted through generation to generation and included moral beliefs during Homeric years (Panezi, 2021). Law and governance was a blend of religious beliefs, morals and hygiene rules during archaic years. And in its classical era, the establishment of democratic framework empowered the position of law and governance, making it dominant in legal orders of Greek antiquity.

Ancient jurisprudence in Egypt was a blend of custom, beliefs based on religion, cosmic order and the authority of Pharaoh. The community resilience was promoted through this legal system, which was based on the concept of Ma'at and was the reflection of cosmic order, social conduct and morality (Editorial L., n.d.). Ma'at was considered as the goddess of order responsible for truth, balance, harmony, law, morality and justice (Giunta, 2024).

The ancient African societies (excluding Egypt) also developed a rich system of indigenous jurisprudence for governance. It was based on oral traditions, ancestral customs, community values and emphasized restorative justice over punitive justice. It emphasized on reconciliation, compensation and rituals as common tools for resolution of disputes. The concept of Ubuntu (I am because we are) reflects the communal nature of justice in ancient African societies, where justice is understood as a social and practical process aimed to restore harmony, rather than an abstract endeavor (Chowdhury, et al., 2021).

Mesopotamia developed some of the ancient and earliest codified legal systems for indigenous governance. The Sumerian law (Code of Ur-Nammu), Akkadian & Babylonian Law (Code of Hammurabi) and Assyrian Law are the major sources of ancient jurisprudence (Patowary, 2022).

CASE STUDIES RELATED TO INDIGENOUS RESTORATION IN CONTEMPORARY ERA

A. Role of women in conserving Indigenous heritage

Following case studies are the examples that shows the role of women in community resilience.

Case study 1: Awajun women pottery art of Peru

The pottery making art which was a tradition for Awajun women is a unique example where women represented the harmonious and sustainable relationship between the nature and Awajun People. It has now become a globally recognized effort to get their tradition a social value. Spiritual visions and environmental observations were the inspiration behind the art which got transmitted in form of knowledge through matrilineal pedagogy and symbolized pottery as a performative expression of prestige, gender identity and gendered governance. It is (Oviedo, Pantoja, Guevara, & López, 2025).

Case study 2: Ima Keithel of Manipur, India

Ima Keithel, popularly known as Mothers' Market in Manipur region of India is the is the largest market in the world which is lead entirely by women. It is a 500 year old revolution lead by women wherein more than 5,000 traders aren't just selling vegetables and textiles but representing the power and capability of women led governance. This market is witnessing a history of defiance, resilience and autonomy that survived empires, forced-labour, insurgencies and natural calamities since ages and still standing like a monument of strength shouldered by women (Ahluwalia, 2025).

B. Indigenous restoration for ecological concerns

Case study 1: Hunar Haat in India

It is an effort made by Ministry of Minority Affairs, Government of India in order to attain community resilience in which it motivates its people having traditional artisan skill to present their art and promote in order to attain revenue every year (Government of India, 2025). It not only helps people to showcase their traditional skills but also to earn living and create business startups. It is not only organized but also managed by government of India and promote sustainable use of resources and advocates ecological harmony.

Case study 2: Project ICRG of UK

In 2015, Department For Infrastructure Development (DFID), UK helped by funding a technical assistance project in India named Infrastructure for Climate Resilient Growth (ICRG) aimed to promote and strengthen rural communities by building resilient rural infrastructure. All the infrastructure built herein were built by local communities by using their indigenous knowledge. The asset selection was in response to climate changes and were durable enough. The main aim of this project was to make the key officials of the government aware of the best practices of infrastructure resilience (Jogesh, Pal, Bugler, Wade, & Messling, 2019).

Case Study 3: Disaster Risk Management in Bangladesh

Depending upon traditional knowledge in order to attain sustainability and inculcate natural processes with human activities resulted in more benefits though being in contract with western euro centric approach which used the method of interfering with the natural science. There were several problems associated to floods, water logging and drainage congestion which were tried to be solved by traditional methods. Several problems associated to flood-prone properties were caused due to unplanned engineering which could be reduced by using traditional methods. The focus was on adaptation rather than development, which promoted disaster risk management in Bangladesh. The purpose of the Coastal Embankment Project (CEP) is to prevent non-saline tidal flooding and saline water intrusion on the arable lands in the southwest coastal zone which was introduced in 1961. In the late 1990s, while doing fieldwork for the EIA/SIA, the Centre for Environmental Geographic Information System (CEGIS) found that the alternative option suggested by the local people was not only effective at solving the problem of drainage congestion in the Beel area but it was also technically sound, economically viable and good for the environment (Chowdhooree, 2019).

Case Study 4: Cultural Burning in Australia

Cultural Burning, practiced by Aboriginal and Torres Strait Islander peoples in Australia, is a powerful and integrated case study of indigenous governance, socio-cultural practices, and the ecological aspect promoting resilience particularly in the face of increasingly bushfires and high intensity forest fires. This practice is guided by elderly peoples who decides where when and how the burning should take place. This is not just a cultural practice but also a spiritual approach to teach youngsters the linguistics and their ethical responsibility towards the land. This activity helps in fire management based on local environmental conditions, cements psychological as well as social resilience and helps in adaptive management to combat transgenerational trauma and cultural identity (ANU Communications & Engagement, 2024).

Case Study 5: Pimachiowin Aki ('The Land That Gives Life') in Canada

A region of rivers, lakes, wetlands, and boreal forest is known as Pimachiowin Aki, or "The Land That Gives Life." It is labelled by UNESCO as world heritage site in Canada. It is a mixed cultural-natural site, first in its kind in Canada. It is a portion of the ancestral home of the Anishinaabeg, an indigenous group that subsists on hunting, fishing, and gathering. Four Anishinaabeg communities' ancestral lands - Bloodvein River, Little Grand Rapids, Pauingassi, and Poplar River—are included in the site. It is a superb illustration of the cultural custom known as Ji-ganawendamang Gidakiiminaan, or 'keeping the land', which entails respecting all living things, honouring the Creator's gifts, and preserving goodwill toward others. This long-standing custom is attested to by a sophisticated web of places of employment, residence, travel, and ritual that are frequently connected by waterways (UNESCO, 2018).

Case Study 6: Budj Bim Cultural Landscape – The Aquaculture System in Australia

The aquaculture of Budj Bim is the one of the oldest systems developed by natives of ancient Australia called as Gunditjmara. It is recognised as UNESCO world heritage site. The intricate network of channels, weirs, and dams built by the Gunditjmara to capture, store, and harvest kooyang (short-finned eel, Anguilla australis) is based on the Budj Bim lava flows. For six millennia, the Gunditjmara society relied on the highly productive aquaculture system as a social and economic foundation. The Gunditjmara describe the creation of the Budj Bim Cultural Landscape as a deep time story, implying that they have lived there for all of their lives. Deep time, as defined by archaeology, is a minimum of 32,000 years. These days, knowledge systems that have been preserved through oral transmission and the continuation of cultural practices carry on the Gunditjmara people's dynamic relationship with their land (Weir, 2009).

C. Adaptation in Digital age

Case Study 1: The 'Ara Irititja' Project, Australia

It is an indigenous initiative in which Anangu Cultural Knowledge had been converted into digital archives and made available online for public at large. This project incorporated the use of modern technology by not just preserving the cultural narratives but by ensuring availability of knowledge for intergenerational transfer and future use (Hughes & Dallwitz, 2009).

Case Study 2: TKDL in India

The formation of Traditional Knowledge Digital Library at central level is great attempt of Indian Government in order to promote the protection of Traditional Knowledge digitally. This got catalyzed by the unauthorized patenting of Turmeric, Neem and Yoga Poses by the Companies of Western Nations. Since the remedial therapies involving Yoga, Neem, turmeric, etc. had already been mentioned in Ancient Indian texts and is being practiced since centuries under the ambit of Ayurvedic medical treatments, it would be violation of the rights of people in India. So, by formation of TKDL, it was ensured that such knowledge to be translated in common global languages and made available digitally to prevent infringement by corporates and to promote commercialization by authentic indigenous owners (CSIR, 2025).

Case Study 3: Mysore city in India

Mysore is developed on lines of the green city under the project launched in 2010. It was developed by Mysore City Corporation. The heritage of Mysore city is authentically driven by the time of the royal rule. Where it was primarily established as a city of forts and gardens in harmony with ecology. However, over the time and impact of urbanism it was losing its realm, which has sought to be revived by the project. Under this project, waste management has facilitated door to door garbage collection; waste segregation; and establishing organic waste composting centres. Transportation system introduced cycle sharing program; electric rickshaws and hybrid buses, renewable energy was boosted where solar panels were installed in government utilities and citizens were encourages with subsidies. The resource management of city has secured a status of cultural capital of Karnataka and as the cleanest city of India in 2016 under Swachh Bharat mission (Hettiarachchi & Mhaddolkar, 2022).

CONTEMPORARY LEGAL FRAMEWORKS TO PROTECT INDEGENOUS HERITAGE

Indigenous Legal Struggles

- **Classifying heritage as indigenous and urban**: The legal battle starts from recognizing heritage as indigenous or urban, based on that governance model is framed. Different custody to different protection measures is sought because of varied significance.
- **Violation of property rights**: Sometimes zoning laws interferes with rights of owner of a residential heritage as to prohibitions on modernization or demolition or usage.
- **Religious sanctity**: Where heritage is attached to some religious belief, primary guardian is respective religious organization besides the fact of inadequate capacity or depreciating maintenance.
- **Listing priority**: Every government has features to prioritise ranging from capitalism, socialism to technological grades. Finding a discussion forum for heritage is a low waited side of the see-saw.
- **Violation of right to Self-determination:** Right to freely choose the self-government, which the indigenous people were denied during colonization era.
- **Forced assimilation:** Denial of regional language and forceful assimilation into white culture by discarding cultural heritage leading to institutional abuse and neglect.
- **Trespass to traditional lands and resources:** Encroachment due to armed conflicts, discriminatory policies and extraction of natural resources.
- **Unavailability of Healthcare services:** Discrimination by the policy makers due to remoteness of tribal regions and lack of healthcare services and awareness.
- **Climate Change:** The impact of climate change and limited counter-measures was also a major challenge for indigenous people.

- **Attacks on Institutions:** Maintenance and strengthening their distinct social, political and cultural institutions is also a big concern.
- **Survival, Dignity and Wellbeing concerns:** Discrimination in various aspects of social life and availability of limited remedial measures often disconnected the indigenous people from urban societies.

In order to grant protection and help the indigenous heritage, several legal frameworks are prevalent across the globe.

Protection under International Laws

- UNESCO- Convention for the Protection of Cultural Property in the Event of Armed Conflict held in Hague in 1954- Convention emphasised on acknowledging that cultural property has been severely damaged in recent wars and that it is becoming more vulnerable to destruction due to advancements in military tactics. Believing that, since every people contributes to global culture, harm to any kind of cultural property entails harm to humanity's cultural legacy (Schindler, 1988). It adopted second protocol in 1999.
- UNESCO- Convention on the Means of Prohibiting and Preventing the Illicit Import, Export and Transfer of Ownership of Cultural Property adopted in Paris in 1970. It acknowledged that the exchange of cultural property between countries for scientific, cultural, and educational purposes broadens our understanding of human civilization, enhances the cultural lives of all peoples, and fosters respect and appreciation between nations. Given that one of the fundamental components of civilization and national culture is cultural property, and that its actual worth can only be recognized in light of the most comprehensive knowledge about its provenance, history, and traditional context (UNESCO, 1970).
- UNESCO- Convention Concerning the Protection of the World Cultural and Natural Heritage adopted in Paris in 1972 (UNESCO, 2016). Convention emphasised on the traditional causes of decay are not the only factors endangering the cultural and natural heritage; shifting social and economic circumstances also exacerbate the situation by causing even more severe phenomena of damage or destruction. Given that any item of cultural or natural heritage that deteriorates or disappears causes a detrimental depletion of the heritage of every country in the world, National heritage protection frequently falls short due to the magnitude of the resources needed and the lack of economic, scientific, and technological resources in the nation where the property to be protected is located.
- Convention on the Protection of the Underwater Cultural Heritage adopted in 31st session of UNESCO in Paris in 2001 (UNESCO, 2009). This convention emphasizes on the significance of underwater cultural heritage as an essential component of humankind's cultural heritage and a crucial component of the history of peoples, nations, and their interactions with regard to their shared heritage. It concentrates that all states bear this responsibility and persuade the importance of public education in fostering awareness, appreciation, and preservation of in situ underwater cultural heritage, as well as the public's right to take advantage of the recreational and educational advantages of responsible, non-intrusive access to such heritage. Recognizing that unapproved activities aimed at underwater cultural heritage pose a threat and that more robust measures are required to stop them.

- Convention for the Safeguarding of the Intangible Cultural Heritage adopted in Paris in 2003 in 32nd session of UNESCO (WIPO, 2003). It focused on considering the deep-seated interdependence between the intangible cultural heritage and the tangible cultural and natural heritage, Recognizing that the processes of globalization and social transformation, alongside the conditions they create for renewed dialogue among communities, also give rise, as does the phenomenon of intolerance, to grave threats of deterioration, disappearance and destruction of the intangible cultural heritage, in particular owing to a lack of resources for safeguarding such heritage. Being aware of the universal will and the common concern to safeguard the intangible cultural heritage recognizes that communities, in particular indigenous communities, groups and individuals play an important role in the production, maintenance and restoration of the intangible cultural heritage.
- United Nations Declaration on the Rights of Indigenous Peoples, 2007: The most comprehensive instrument for establishing minimum standards for survival, dignity and wellbeing of the indigenous people that covers all major issues of these people.
- International Labour Organization Convention 169, 1989: This ILO Convention is one of the most operative and the only legally binding convention that establishes a framework for discussing indigenous groups and recognizes their rights to land resource, culture etc.
- The 16th Conference of United Nations Convention on Biological Diversity also known as COP16 played an important role in uniting global leaders, scientists, policymakers and environmentalists to advocate the global concerns and figure out solutions for issues related to biological diversity, sustainable use of resources, preservation of ecosystem and fair sharing of benefits specially in regard of indigenous knowledge. This conference considered indigenous peoples as guardians of nature and highlighted their role in policy making and enhancing biodiversity conservation under article 8(j) of CBD (Onfra, 2024).
- The Paris Agreement, a treaty in collection of United Nations is a legally binding international treaty on climate change which was adopted by 195 Parties at the UN Climate Change Conference (COP21) in Paris, France, on 12 December 2015. One of its primary goals is to maintain global temperature level for which it mandates its parties to take measures at local level and indigenous levels. (UNFCCC, 2025).
- IPBES, 2019 – A 1500-page report was published that provided the summary for policymakers on biodiversity and ecosystem services which highlighted the issues causing decline in biodiversity, the drivers of the change and the actions needed to restore the sustainability. The report alarmed the threat of extinction of about 1 million wildlife and pant species and suggested several actions to improve nature and reduce food waste (Bongaarts, 2019).
- Other international bodies such as UNESCO, UDHR, ICCPR, TRIPS, IPBES etc. have also played a crucial role in preservation and protection of culture, language or folklore of indigenous people. For example, UNESCO's the preservation programs such as 'Man and the Biosphere' had been notable in preserving the Indigenous cultures and have contributed in guiding the scientists and policymakers to facilitate the sustainable development. (BESNet, 2024).

Protection under Municipal Laws

Many nations have adopted separate municipal laws for protection of indigenous people. Some of the important regulatory framework include:

- **India:** Institutional restoration of heritage and patrimony owes its genesis in India, to the grundnorm the Constitution of India. In the Constitution of India, Article 49 labels duty on state to protect heritage and culture (Pathak & Das, 2024). The management of indigenous endowments vest in two heads where Union (centre) will be responsible for regulating and protecting monuments, sites and artefacts which are declared as of national importance by parliament while State will be responsible for the managing monuments, sites and artefacts not declared by parliament of national importance. And some of the heritage falls under concurrent jurisdiction of both by virtue of third list of seventh schedule. Further, Article 253 empowers centre to legislate in protecting and uplifting global commitments. Other statutes promoting indigeneity restoration includes Ancient Monuments and Archaeological Sites and Remains Act, 1958; Geographical Indication Act 1999; The Scheduled Tribes and Other Traditional Forest Dwellers (Recognition of Forest Rights) Act, 2006 and Panchayat (Extension to Scheduled Areas) Act, 1996. Here 'scheduled tribes' or 'adivasis' have been referred as indigenous population.
- **USA:** US Constitution, The Antiquities Act 1906; National Historic Preservation Act 1966; Archaeological Resources Protection Act (ARPA) 1979 and Native American Graves Protection and Repatriation Act, 1990 aims to protect native Americans and their rights (Finkelman & Garrison, 2008).
- **Egypt:** Constitution of Egypt though talks about rights of people and culture. The legal system is not comprehensively defining but they have enforced UNESCO Conventions. Egypt enforced Law No. 117 of 1983 on the Protection of Antiquities (Fahmy & URHC, 2013). the Nubians and Bedouins indigenous groups directly participates in restorations.
- **Australia:** The Commonwealth's Aboriginal and Torres Strait Islander Heritage Protection Act 1984 permits federal intervention for endangered Indigenous sites or objects. The Protection of Movable Cultural Heritage Act 1986, regulates the export of important artifacts. The Environment Protection and Biodiversity Conservation Act 1999 safeguards national, international, and indigenous heritage sites.
- **Africa:** African charter, African Commission on Human and People's Rights and Court's precedents protect the right of indigenous groups. South Africa has enacted National Heritage Resources Act 25 of 1999 for governmental restoration of national indigeneity (UNESCO, 2025).
- **Greece:** Famously called as Law of Athens, Greece has enacted LAW No. 3028 On the Protection of Antiquities and Cultural Heritage in General, where Ministry of Culture supervises implementation of culture preservation projects. They have not recognised indigenous people separately still follows international convention for protection of their right.
- **UK:** The Ancient Monuments and Archaeological Areas Act 1979; National Heritage Act 1983; The Planning (Listed Buildings and Conservation Areas) Act 1990 and Cultural Property (Armed Conflicts) Act 2017. All these enactments are in line to fulfil united goals of UNRSCO.

FUTURE RESEARCH DIRECTIONS

This research opens multiple pathways for enquiries related to preservation and restoration strategies for Indigenous-Led Heritage. Data collection from various social groups through surveys and questionnaires could be used for identifying the various kinds of barriers and the remedial measures for uprooting the barriers. For preservation and restoration of this Indigenous-Led Heritage, new comparative and scientific approaches could be explored by initiating further research and to know the role of Sovereign Authorities of different regions of the world. Another promising research direction involve tracking the policy and legal reformative measures and their role in preservation strategies and to identifying the grey-areas of existing politico-legal arena with probable remedial plans and strategies. Apart from this it could make a base of future researches on the preservation strategies under the shadow of technological advancements.

CONCLUSION

Protecting cultural heritage is rebranding a unique unparallel identity on exchanging distinct routes to establish authenticity and empowering ancient historical vision. Where indigenous communities are globally preserving their culture, traditions and knowledge systems, the focus is inclined towards non-material heritage. Their outlook for culture preservation is not based on past practices but is deeply rooted in their day to day lives. For them it is just not about preservation of their culture but maintaining them with dynamism of modern society. The distinctive features of particular community must be preserved as it is an offshore for channelling cultural linkages at the international level. In order to obtain a harmonized system of governance, there is a requirement of balance between the resources and their utilization. Judiciary on one hand plays an important role in preserving these resources while on the other, effective policy formulation and implementation can lead a nation towards sustainability.

RECOMMENDATIONS

- There is a need of promotion of indigenous communities and their practices to make their knowledge and skills reach the global market. Attempts are required to be made to motivate their cultural aesthetics in order to find solutions to various modern concerns.
- Recognition of inherent sovereignty can play an important role in preservation of rights of indigenous people. Restoration of this heritage must be aligned with the social needs of all stakeholders of the present societies.
- Appropriate legislation and related policy formulation is to be made for the restoration of Indigenous led heritage. Apart from it, effective enforcement mechanism is also to be set- up nationwide.
- Enactment & policy formulation must be based upon participative governance, hence consultation and consent with the effected Indigenous groups must be done.
- In this digital era, merging of indigenous knowledge and technology can bring a positive social resilience. In order to achieve it, investigations, reporting and identification of areas for protection and preservation of heritage must be aligned with social need and emerging potential technologies. The prone- areas where this Indigenous led heritage restoration is needed, must be identified after intensive investigation.
- Legislations and policies adopted by a nation must be interpreted harmonically for public welfare, adhered as per according to UN guidelines and backed by strict mechanism to control and regulate unauthorized access. Restoration policies and plans must adopt the co- management ap-

proach and must be aligned with UN Principles. Unauthorized usage, vandalization, destruction of Indigenous heritage must be penalised and be regulated with strict provisions of law.

- Sovereign powers must support in capacity balding, finance, training of the indigenous groups for the protection of their Indigenous knowledge and heritage.
- Integration of Indigenous Knowledge and heritage must be done and assistance from voluntary organizations can be taken into consideration.
- Indigenous knowledge system should be added under different course curricula & must adopt certain awareness programs for the better protection of the same.
- International Cooperation, incorporation of AI and other innovative technology, fund allocation and establishing institutions are some other required steps to be taken for the protection of Indigenous Knowledge system.

REFERENCES

Ahluwalia, A. (2025, January 29). *Ima Keithel: Manipur's Mothers Are Leading A 500-Year Revolution Through Commerce & Unity*. Retrieved from Homegrown: https://homegrown.co.in/homegrown-voices/ima-keithel-manipurs-mothers-are-leading-a-500-year-revolution-through-commerce-unity

ANU Communications & Engagement. (2024, November 1). *Indigenous cultural burning has protected Australia's landscape for millennia*. Retrieved from Australian National University: https://www.anu.edu.au/news/all-news/indigenous-cultural-burning-has-protected-australias-landscape-for-millennia

BESNet. (2024). *Languages, Cultures, Knowledge: UNESCO's*. Retrieved from www.besnet.world: https://www.besnet.world/library/languages-cultures-knowledge-unescos-action-for-indigenous-peoples/#:~:text=UNESCO%20is%20committed%20to%20the%20preservation%20of%20Indigenous,management%20of%20natural%20and%20cultural%20World%20Heritage%20sites.

Bhusari, K. (24, February 2024). *Ut res magis valeat quam pereat*. Retrieved from LAW ARTICLE: https://lawarticle.in/ut-res-magis-valeat-quam-pereat/

Bongaarts, J. (2019). IPBES, 2019. Summary for policymakers of the global assessment report on biodiversity and ecosystem services of the Intergovernmental Science-Policy Platform on Biodiversity and Ecosystem Services. *Wiley*, 680-686. DOI: 10.1111/padr.12283

Chowdhooree, I. (2019). Indigenous knowledge for enhancing community resilience: An experience from the south-western coastal region of Bangladesh. *International Journal of Disaster Risk Reduction, 40*, 101259. Advance online publication. DOI: 10.1016/j.ijdrr.2019.101259

Chowdhury, S. J., Wahab, H. A., Saad, M. M., Roy, P. K., Hamidi, M., & Ahmad, M. M. (2021). Ubuntu Philosophy: 'I Am Because We Are' – A Road to 'Individualism' to Global Solidarity. In I. Global, *Handbook of Research on the Impact of COVID-19 on Marginalized Populations and Support for the Future* (pp. 361-381). IGI Global Scientific Publishing.

CSIR. (2025). *Traditional Knowledge Digital Library Unit (TKDL)*. Retrieved from Council of Scientific and Industrial Research, Ministry of Science & Technology, Govt. of India: https://www.csir.res.in/en/documents/tkdl

DRONAH. (2021). *Challenges faced in Heritage Management in India and Policy Imperatives*. Gurugram: NITI Aayog.

Editorial, L. (n.d.). *Understanding the Ancient Egypt Legal System: Structure and Function*. Retrieved July 12, 2025, from Laws Pulse: https://lawspulse.com/ancient-egypt-legal-system/

Fahmy, H., & URHC . (2013). *Egyptian legislations in relation to the rehabilitation of Historic Cairo*. Cairo: UNESCO, World Heritage Centre.

Finkelman, P., & Garrison, T. A. (2008). Archaeological Resources Protection Act (ARPA; 1979). In Finkelman, P., & Garrison, T. A. (Eds.), *Encyclopedia of United States Indian Policy and Law*. SAGE., DOI: 10.4135/9781604265767.n40

Giunta, J. (2024, November 14). *The Concept of Ma'at: Balance, Truth, and Its Role in Egyptian Society.* Retrieved July 12, 2025, from ancient pedia: https://ancientpedia.com/the-concept-of-maat-balance-truth-and-its-role-in-egyptian-society/

Government of India. (2025). *Ministry of Minority Affairs.* Retrieved from Ministry of Minority Affairs: https://minorityaffairs.gov.in/show_content.php?lang=1&level=2&ls_id=104&lid=48

Grey, S., & Kuokkanen, R. (2020). Indigenous governance of cultural heritage: Searching for alternatives to co-management. *International Journal of Heritage Studies, 26*(10), 919–941. DOI: 10.1080/13527258.2019.1703202

Grisel, F. (2023). *Arbitration as a dispute resolution process: historical developments.* Cambridge Compendium on International Commercial and Investment Arbitration.

Hettiarachchi, H., & Mhaddolkar, N. (2022). Mysuru: The story of a clean city in India. In Hettiarachchi, H., & Mhaddolkar, N. (Eds.), *SDG11, Sustainable Cities and Communities* (pp. 37–57). Routledge India. DOI: 10.4324/9781003205975-3

Heuheu, T. t., Kawharu, M., & Tuheiava, R. A. (2012). World heritage and indigeneity. *World heritage review, 62*, pp. 6-16. Retrieved from https://unesdoc.unesco.org/ark:/48223/pf0000215774

Hughes, M., & Dallwitz, J. (2009). Ara Irititja: Towards Culturally Appropriate IT Best Practice in Remote Indigenous Australia. In Erickson, J. (Ed.), *Database Technologies: Concepts, Methodologies, Tools, and Applications* (pp. 1462–1471). IGI Global Scientific Publishing., DOI: 10.4018/978-1-60566-058-5.ch088

IWGIA. (2025, April 25). *Indigenous peoples in India.* Retrieved from IWGIA: https://iwgia.org/en/india.html

Jogesh, A., Pal, U., Bugler, W., Wade, G., & Messling, L. (2019, October). *Case study: For the community, by the community: Building resilience through infrastructure in rural India.* Retrieved from ResearchGate: DOI:DOI: 10.13140/RG.2.2.33606.80965

Kane, P. V. (1968). *History of dharmaśāstra (ancient and mediæval religious and civil law in India).* Bhandarkar Oriental Research Institute.

Lachapelle, P. R., & Austin, E. K. (2014). Community Participation. In Michalos, A. C. (Ed.), *Encyclopedia of Quality of Life and Well-Being Research* (pp. 1073–1078). Springer., DOI: 10.1007/978-94-007-0753-5_471

Madaan, M., & Mohanty, A. (2025). Constitutional and Legal Safeguards for Indigenous Communities in India: A Critical Analysis of Rights, Realities and Reforms. *INTERNATIONAL JOURNAL OF HUMAN RIGHTS LAW REVIEW, 4*(5), 156-185. Retrieved from https://humanrightlawreview.in/wp-content/uploads/2025/10/Constitutional-and-Legal-Safeguards-for-Indigenous-Communities-in-India-A-Critical-Analysis-of-Rights-Realities-and-Reforms.pdf

Makhloufi, L. (2024). *Tangible and Intangible Heritage in the Age of Globalisation.* OpenBook Publishers. DOI: 10.11647/OBP.0388

Malhotra, A., & Dobriyal, R. (2023). ADAPTIVE REUSE OF HERITAGE ARCHITECTURE AS A SUSTAINABLE DEVELOPMENT TOOL: OPPORTUNITIES AND CHALLENGES. *ShodhKosh: Journal of Visual and Performing Arts*, 21-27.

Michalos, A. C. (2014). Encyclopedia of Quality of Life and Well-Being Research. In Michalos, A. C. (Ed.), *Encyclopedia of Quality of Life and Well-Being Research*. Springer Dordrecht., DOI: 10.1007/978-94-007-0753-5

Nayak, C., Chakraborty, D. K., & Chakrabarti, D. (2025). Preserving Indigenous Knowledge Systems by Employing Digital Twin Technologies in the Disruptive Era. In Chakraborty, A. K., Singh, A. P., & Banerjee, S. (Eds.), *Indian Knowledge System (IKS) and Viksit Bharat by 2047: fusion in LIS and Allied Discipline*. Cambridge Scholars Publishing.

Onfra. (2024, November 9). *Indigenous People: Cultural Heritage, Resilience, and Impact*. Retrieved from Onfra: https://onfra.io/indigenous-people-cultural-heritage-resilience-and-impact/

Osborne, R. (2009). *Greece in the Making 1200-479 BC*. Routledge. DOI: 10.4324/9780203880173

Oviedo, A., Pantoja, A. M., Guevara, S. Z., & López, E. A. (2025). Awajún Pottery, Gender, and Transformation: From Social Prestige to Market Logic. *Journal of Intercultural Communication*, *25*(3), 104–115. DOI: 10.36923/jicc.v25i3.1157

Pandey, K. (2014). Indigenous Knowledge System: Concepts and Discourses. *Anthropological Bulletin*, *2*(1), 17–28.

Panezi, M. (2021, November/December). *A Description of the Structure of the Hellenic Republic, the Greek Legal System, and Legal Research*. Retrieved July 11, 2025, from Globalex: https://www.nyulawglobal.org/globalex/greece1.html

Papaioannou, K. (2017). THE INTERNATIONAL LAW ON THE PROTECTION OF CULTURAL HERITAGE. *IJASOS- International E-journal of Advances in Social Sciences*, 257-262.

Pathak, T., & Das, S. (2024, June). CULTURAL HERITAGE PRESERVATION IN INDIA: SAFEGUARDING A LIVING LEGACY. *ShodhKosh: Journal of Visual and Performing Arts*, 1286-1291. DOI: 10.29121/shodhkosh.v5.i1.2024.4184

Patowary, K. (2022, July 7). *Code of Ur-Nammu: The Oldest Law in History*. Retrieved July 13, 2025, from Amusing Planet: https://www.amusingplanet.com/2022/07/code-of-ur-nammu-oldest-law-in-history.html#:~:text=Some%20of%20the%20earliest%20legal%20codes%20concerning%20crimes,Kassites%2C%20and%20Hittites%2C%20each%20had%20their%20own%20laws

Pittz, T., Adler, T., & Claw, C. (2024). The institution of tribal sovereignty: From historical challenges to modern expressions. *Journal of Management History*, *32*(3), 507–521. DOI: 10.1108/JMH-06-2024-0096

Pittz, T. G., Claw, C. M., & Adler, T. R. (2026). Sovereignty as a site of innovation: Institutional entrepreneurship in Native American tribal nations. *Research Policy*, *55*(4), 105431. Advance online publication. DOI: 10.1016/j.respol.2026.105431

Rajani, M. B. (2025). Regulation Boundaries for Preservation of Cultural Heritage Sites. In Nayak, S. (Ed.), *Policy Recommendations for Sustainable Development* (Vol. 2, pp. 17–24). Springer., DOI: 10.1007/978-981-96-0718-1_3

Rathor, V. (2024). Preserving Indian Culture Amidst Modernization: Challenges and Solutions. *International Journal of Research and Analytical Reviews*, •••, 778–781.

Schindler, D. a. (1988). The Laws of armed conflicts: a collection of conventions, resolutions and other documents. Nijhoff, Kluwer Academic Publishers.

Srinivas, S. (2013, September 10). *Dharma Shastras- Legal Literature of Ancient India.* Retrieved July 8, 2025, from ithihas: https://ithihas.wordpress.com/2013/09/10/dharma-shastras-legal-literature-of-ancient-india/

Swiderska, K. (2006). *Protecting Traditional Knowledge: A framework based on Customary Laws and Bio-Cultural Heritage. International Conference on Endogenous Development and Bio-Cultural Diversity, 3-5 October 2006.* IIED., Retrieved from https://www.iied.org/sites/default/files/pdfs/migrate/G01069.pdf

Tippeconnic, J. W., & Faircloth, S. C. (2010). The Education of Indigenous Students. In P. Peterson, E. Baker, & B. McGaw, *International Encyclopedia of Education (Third Edition)* (pp. 661-666). Elsevier. DOI: 10.1016/B978-0-08-044894-7.00123-8

Tripathi, S. (2021, November 17). *Monumental!* Retrieved from The Hindu: https://www.thehindu.com/features/friday-review/history-and-culture/Monumental/article16298184.ece

Tussay, Á. (2024, September 26). Natural law and unwritten law in Classical Greek thought. *Hungarian Journal of Legal Studies*, *65*(1), 114–134. DOI: 10.1556/2052.2024.00534

Uma, N., Velayutham, A., Magudeeswaran, T., & Mcpe, S. (2024). Empowering Indigenous Heritage: Preserving, Promoting, and Connecting Through Technology. *National Conference.* Coimbatore: ResearchGate.

UNESCO. (1970). *Convention on the Means of Prohibiting and Preventing the Illicit Import, Export and Transfer of Ownership of Cultural Property*. Retrieved from UNESCO: https://www.unesco.org/en/legal-affairs/convention-means-prohibiting-and-preventing-illicit-import-export-and-transfer-ownership-cultural

UNESCO. (2009). *Convention on the Protection of the Underwater Cultural Heritage.* Retrieved from UNESCO: https://leblancf.com/documents/charters%20&%20conventions/UNESCO%20conventions/underwater_e.pdf

UNESCO. (2011). *Heritage of Religious Interest*. Retrieved from UNESCO: https://whc.unesco.org/en/religious-sacred-heritage/

UNESCO. (2011). *Preparing World Heritage Nominations*. Retrieved from UNESCO: https://whc.unesco.org/en/preparing-world-heritage-nominations/

UNESCO. (2016, August 27). *The World Heritage Convention*. Retrieved from UNESCO World Heritage: https://web.archive.org/web/20160827065310/https://whc.unesco.org/en/convention/

UNESCO. (2018). *Pimachiowin Aki*. Retrieved from UNESCO: https://whc.unesco.org/en/list/1415/

UNESCO. (2025). *UNESCO Database of National Cultural Heritage Laws*. UNESCO. Retrieved from https://www.unesco.org/en/cultnatlaws

UNFCCC. (2025). *The Paris Agreement*. Retrieved from unfccc: https://unfccc.int/process-and-meetings/the-paris-agreement

Walker-Jones, A. (2021, November 27). *Trial, Error, And Success: The Indigenous Origins Of The Scientific Method*. Retrieved July 12, 2025, from Medium: https://medium.com/sofi-stories/trial-error-and-success-indigenous-origins-of-the-scientific-method-d020829afda1

Waterton, E., & Smith, L. (2010). The recognition and misrecognition of community heritage. *International Journal of Heritage Studies, 16*(1-2), 4–15. DOI: 10.1080/13527250903441671

Weir, J. K. (2009). *The Gunditjmara Land Justice Story*. National Library of Australia.

WIPO. (2003). *Convention for the Safeguarding of the Intangible Cultural Heritage*. Retrieved from WIPO: https://www.wipo.int/wipolex/en/text/594740

ADDITIONAL READING

Jansen, L., & Mzwakali, S. (2025, April 25). *The Indigenous World 2025: South Africa*. Retrieved from IWGIA: https://iwgia.org/en/south-africa/5636-iw-2025-south-africa.html

Khater, M., Zouair, N., Saad, M. A., Bayoumy, M., & Al-Salim, F. (2025). Reviving heritage through regenerative tourism and community empowerment for sustainable futures. *Social Sciences & Humanities Open, 12*, 102004. Advance online publication. DOI: 10.1016/j.ssaho.2025.102004

SULLIVAN. A. M. (2016). CULTURAL HERITAGE & NEW MEDIA: A FUTURE FOR THE PAST. *THE JOHN MARSHALL REVIEW OF INTELLECTUAL PROPERTY LAW*. Retrieved from https://repository.law.uic.edu/cgi/viewcontent.cgi?article=1392&context=ripl

Sustainability Directory. (2025, 5 10). *Indigenous-Led Restoration and Climate Resilience*. Retrieved from prism.sustainability-directory.com: https://prism.sustainability-directory.com/scenario/indigenous-led-restoration-and-climate-resilience/

Szczepanowska, H. M. (2013). *Conservation of Cultural Heritage: Key Principles and Approaches*. Routledge. DOI: 10.4324/9780203081198

United Nations. (2026). *United Nations Declaration on the Rights of Indigenous Peoples*. Retrieved from Department of Economic and Social Affairs: https://social.desa.un.org/issues/indigenous-peoples/united-nations-declaration-on-the-rights-of-indigenous-peoples

KEY TERMS AND DEFINITIONS

Ancient Jurisprudence: System of law and legal philosophy that prevailed in ancient civilization.

Community resilience: It refers to the ability of community to withstand adverse situations and response using adaptation, growth and available resources.

Heritage: A legacy including artifacts and culture, that is passed down from generation to generation.

Indigenous Knowledge: A community-based knowledge that societies have developed over time by living in and interacting with their natural habitat and is passed from generation to generation of that community.

Indigenous restoration: It means reintroduction of local community led knowledge and aims to restore native ecosystem.

ILHR: International League for Human Rights is an organization that aims to protect human rights globally.

Sustainability: It is a practice of using the resources in such a way that the requirements of present is met without compromising the future supplies.

Chapter 11
Bio-Cultural Community Protocol:
A Tool to Empower Communities' Economic Development Using IK

Diane Thelma Molokwa
http://orcid.org/0000-0002-7352-6316
Independent Researcher, South Africa

Motheo Koitsiwe
North West University, South Africa

ABSTRACT

Indigenous and local communities are seen as guardians and caretakers of genetic resources and possess immense traditional or indigenous knowledge regarding its use and management. However, exclusion of indigenous knowledge and related practices into economic development, caused by a lack of recognition and continuous marginalization by scientific knowledge, and sustainable development stakeholders, perpetuates development, food and nutrition insecurity. This marginalization also leaves indigenous and local communities and their local resources including indigenous knowledge exposed to bio-piracy and misappropriation. Yet this knowledge is the most precious assets possessed by indigenous and local communities and also the most positive motivation to build their local economies. Therefore, this chapter aims to guide and suggest to the government and policy makers to work hand in hand with communities to leverage BCPs as a tool to empower themselves economically.

INTRODUCTION

As perceived by Dzvimbo et al. (2023) and Slikkerveer (2019), indigenous and local communities are seen as guardians and caretakers of genetic resources and often possess immense traditional or indigenous knowledge, regarding its use and management. Indigenous Knowledge (IK) is a local knowledge, which evolved from close relationship with the local environment (Thompson et al., 2020). Essentially, it was developed by a group of people through decades of living in close contact with their local environment. As alluded by Thompson et al. (2020), this relationship involves natural resource use, stewardship, oral

DOI: 10.4018/979-8-3373-7976-0.ch011

history, spirituality, and often guides local traditional governance. In addition, Kiage (2019:5) is of an opinion that the local approach to natural resource management originates from the viewpoint that man is equal to all other lives and principally rooted in the lived experiences of indigenous peoples, which are passed down from generation to generation through word-of-mouth. This knowledge has been acknowledged and recognised by many researchers and scholars as an important source of survival for many indigenous and local communities (Sultana et al., 2018; Manyanga et al., 2024; Slikkerveer, 2019). In support, Tirivangasi and Tayengwa (2017) added that this knowledge is the most precious assets possessed by indigenous and local communities and also the most positive motivation to build their local economies. Indigenous knowledge may further contribute to improved productivity, effectiveness, and communities' sustainable development. Lyndon et al. (2020) and Tondi (2019) further proclaim that any livelihood development that pays attention to indigenous knowledge systems and related practices are more likely to be effective and relevant to local peoples' economic intervention.

Despite the noticeable recognition of indigenous knowledge systems and related practices in South Africa, the marginalization of local communities and their resources, including indigenous knowledge, has continued for centuries, exposing them to bio-piracy and misappropriation (Maunganidze, 2016; Nurfitri et al., 2025). These enduring challenges align with the provisions of Article 15 of the Convention on Biological Diversity, which mandates that governments must protect indigenous knowledge and ensure fair and equitable sharing of benefits arising from the use of genetic resources (Lawson et al., 2023). In response to both international obligations and national imperatives, the Department of Science and Innovation has initiated programmes to assist communities in protecting their local resources and associated indigenous knowledge through the development of BCPs (Department of Science, Technology and Innovation, 2020). The following section therefore outlines the research scope, dealing with inclusion and exclusion criteria, as well as conceptual focus used to assess the role of BCPs in community empowerment.

RESEARCH SCOPE

The scope of this book chapter is outlined by a critical examination of the Bio-Cultural Community Protocol as an instrument for advancing socio-economic development among indigenous and local communities within the South African context. The inquiry is grounded in the scoping of a systematic and comprehensive review of scholarly works, encompassing peer-reviewed journal articles, master's and doctoral theses and grey literature, to establish the theoretical and empirical foundations of BCPs as mechanisms of community empowerment (Mak & Thomas, 2022: 561).

RESEARCH OBJECTIVES AND QUESTIONS

The following objectives are grounded this chapter:

- To explore the potential of BCPs as a tool to empower communities to drive sustainable economic development.
- To assess the impact of BCPs on poverty alleviation, employment, and food security in indigenous communities' households.

- To identify difficulties and challenges associated with implementing BCPs in South African communities.

The following questions are grounded this chapter:

- How can Bio-cultural Community Protocols (BCPs) be leveraged to promote sustainable economic development in indigenous communities' households?
- How effective are BCPs in addressing socio-economic challenges faced by indigenous communities in South Africa?
- What are the key factors influencing the unsuccessful implementation of BCPs in South African communities?

THEORETICAL FRAMEWORK

This chapter is situated within the framework of epistemic justice. As conceptualised by Cummings et al. (2024), this framework focuses on maintaining fairness in knowledge production, distribution and utilization, while advocating for knowledge inclusive creation, and credibility of the marginalized communities and voices. In essence, epistemic justice offers opportunities for equal participation and collaboration between different worldviews and further allows for effective policies and reduction in unattended negative impacts that policies inflict on the marginalized communities (Duchkova & Leventon, 2023). Cummings et al. (2024) further added that this framework is relevant for sustainable development, aligning with Byscov's (2021:118) five criteria of epistemic injustice: unfair outcomes; unjust assessments of epistemic capacity; unfair denial of stakeholders' rights; unfair denial of knowledge, and unfair exacerbation of pre-existing vulnerability, which impacted many marginalized communities.

In the context of this chapter, the adopted framework is utilized to redress the historical discourses, notwithstanding the pivotal role of indigenous knowledge, and how associated practices may play a role in facilitating economic advancement within communities. In effect, this framework provides the authors with a platform to employ BCP as an efficacious instrument for the promotion of indigenous knowledge in economic development, thereby enabling collaborative engagement between the governmental entities and indigenous communities towards the constructiveness of an enhanced collective future.

RESEARCH METHODOLOGY

This chapter employs a scoping literature review, as narrated by Munn et al. (2018) and this method is useful in examining existing evidence when there are still more unclear, other specific questions that can be posed and addressed by a systematic review. In Taylor and Pagliari (2018)'s view scoping reviews are a kind of quasi-systematic review applied for understanding research on the current emerging innovations, which are sometimes poorly indexed, distributed across published, and grey literature. Essentially, there are succeeding key stages: (1) identifying the research question; (2) identifying relevant studies; (3) charting the data and (4) collating, summarizing and reporting the results (Taylor & Pagliari, 2018). In this chapter, scoping reviews are aimed at recording the evidence landscape, rather than establishing effective and rigor of a particular intervention, and this chapter does not involve the critical appraisal of detailed extraction of outcomes data (Taylor & Pagliari, 2018). The following steps were applied:

Identified the research question

The authors clearly articulated research objectives and questions in the introduction section, which guided the analysis of existing research evidence and assisted in the build-up of this chapter (Taylor & Pagliari, 2018; Rodger et al., 2024). The following steps were applied:

Identifying relevant literature

In this chapter, a scoping review assisted the authors to capture relevant literature from various electronic databases, grey literature and reference lists.

- As required by scoping review literature, the authors used a systematic search of peer reviewed studies from google scholar, researchgate, and other relevant peer reviewed journals (Taylor & Pagliari, 2018).
- Additionally, snowballed identified article reference lists were used to identify additional studies, which could contribute to the chapter.
- Searching grey literature from the Internet, using trusted mostly used search engine, Google. The first 10 google results generated by the search string were reviewed, as the search engine displayed the results by relevance (Taylor & Pagliari, 2018).

Inclusion and exclusive criteria

Peer reviewed journal articles and conference papers were included. However, studies not in English, dissertations, or theses, reports or abstracts that are not discussing BCPs, regarding socio-economic development, and challenges of implementing BCPs within the South African contexts, were excluded. A publication timeframe of 2014-2026 was applied by the authors (Taylor & Pagliari, 2018).

Table 1. Collating, summarizing and reporting the results

Study ID	Author & year	Key findings	Gap identified
1	Anquet & Girard (2022). Slikkerveer, L.J. (2019).	Paves a way for indigenous and local communities to be seen as sovereign over their territories, land, resources, agents for their own past, present and future life, while forcefully expressing claims for alternatives to modernity.	Lacks the clarity on the benefits of BCPs, especially in improving communities' economic development.
2	Byscov, M.F. (2021)	The author discussed three additional aspects of epistemic injustice as discussed by Fricker. Additionally, the author identified five conditions that may be used to evaluate claims of epistemic injustice.	The study did not cover the significance of epistemic justice in redressing the economic development of indigenous and local communities using IK.
3	Cummings, S., Munthali, N. & Sittoni, T. (2024).	The authors emphasised that the use of epistemic injustice's framework in studying and understanding the role of scholarly communities practice in fostering social sustainability, may contribute to equitable knowledge practices and new knowledge and skills.	This study excluded the use of this framework in developing local communities' economies through the use of BCPs.
4	Chennels, R. (2013).	The key finding of this review article was that indigenous communities should benefit from their traditional knowledge through the use of CBD and should be in line with what is envisaged by the policy drafters.	The article excluded the use of BCPs as a framework to ensure access and benefit sharing.
5	Duchlova, H. & Leventon, J. (2023).	In this study it was highlighted that the employment of epistemic justice in research may and could assist in addressing and related policy around the issues of climate, energy and mobility from a different perspective.	This excluded the use of epistemic justice in ensuring that communities benefit equally from any income generation, deriving from traditional knowledge and elated practices.
5	Dzvimbo, M.A., Tirivangasi, H.M., Mawonde, A. & Ncube, C.T. (2023).	The study highlighted that IK is still not recognized as a strategic partner for both endogenous and exogenous at policy development and implementation level.	There paper did not include the noticeable exclusion of IK in socio-economic development policies and strategic planning.
6	Foyer, J., Mauri, M.M., Filoche, G. & Castillo, G. (2024). Thompson, K.L., Lantz, T. & Ban, N.C. (2020).	In their study, the authors reported and recommended that the implementation and use of BCPs may strengthen indigenous communities' territories, specifically in relation to climate change, and newly implemented regulations on the rights of nature.	The study excluded the implementation and use of BCPs for socio-economic development amongst indigenous and local communities.
7	Githui, W., David, N. & Maurice, S. (2013).	The authors concluded that IK is an essential component for the development of any community.	This study excluded the importance of IK in indigenous and local communities' economic development, using IK.
8	Godfrey, B. & Siraje, K. (2019).	The study reported that increased use of indigenous technologies, approaches and methods may improve communities' visual representatives of their knowledge, judgement and preference.	There is a gap in the use of IK in the socio-economic development in their study.
9	Magni, G. (2017). Sultana, R., Muhammad, N. & Zakaria, A.K.M. (2018). Tondi, P. (2019).	The article highlighted the leverage of indigenous knowledge into sustainable development agenda 2013. In addition, this will also assist in mitigating social challenges.	This article lacks inclusion of IK into socio-economic development and the use of BCPs to achieve such.

continued on following page

Table 1. Continued

Study ID	Author & year	Key findings	Gap identified
10	Sultana, R., Muhammad, N. & Zakaria, A.K.M. (2018). Tirivangasi, H.M. & Tayengwa, D. (2017).	This study reviewed the role that indigenous knowledge may play in sustainable development and further revealed areas where indigenous knowledge can be of a greater use, such as food production, forestry management, climate change, nature conservation, disaster management and medicine.	The contribution of IK into socio-economic development was left out of their findings.
11	Kiage, O.E. (2019)	The study reported that IK may be used for the basis for natural resource management, and to also understand direct cultural, social, environmental, and economic benefits.	The study did not advocate for equal and benefit sharing, deriving from the use of IK.
12	Lahiri, S. (2024). South Africa. Department of Science, Technology and Innovation. (2020). Lassen, B. (2018).	The authors reported that indigenous people and local communities may apply BCPs to address conflicts that may arise in relation to tourism and biodiversity and also to build consensus among policy makers.	The study excluded the economic development that may arise from the implementation of BCPs by indigenous people and local communities.
13	Manyaga, M.B., Goldman, G.A. & Thomas, P. (2024).	The study provided insights into challenges and opportunities faced by SMEs, concerning the adoption of Tshivenda IKS to foster business sustainability.	Although the study highlighted opportunities that may arise from integrating Tshivenda IKS into business sustainability, there is a lack articulation as to of how local communities will benefit economically from the use of their knowledge system.
14	Maunganidze, L. (2016).	The study found that there is a decline in the use and acknowledgement of indigenous knowledge and no longer provides guidelines for development.	The study lacks the use of BCPs as a solution to the decline and marginalization of IK in development
15	Nurfitri, D., Bustani, S. & Yunari, S.B. (2025). United Nations Environmental Programme and Environmental Defenders Office New South Wale, (2013). Lawson, C., Rourke, M. & Humphries, F. (2023).	The study explored the legal foundation for the protection of genetic resources and traditional knowledge at both international and national level. In addition, the Convention on Biological Diversity, Nakoya Protocol, and laws, such as Patent and Government Regulations No. 56/2022.	This study excluded BCPs as one of the legal foundation for the protection of traditional knowledge and related practices.
16	Odayar, S. (2024). Oli, K.P., Rana, P., Shah. R., Rawal. R.S., Chaudhary, R., Yongping. Y., Yao, F., Kotru, R. & Bhatta, L.D. (2015).	This study highlighted step-by-step on how communities may develop BCPs documents. These steps included the following: community buy-in and agreement; training and capacity building and understanding the scope of territory; conducting research and gathering information; community validation; development and launch of the document, and employment of the protocol.	The study lacks the implementation of the protocol in empowering communities' economic development.
17	Rakotondrabe, M. & Girdad, F. (2021). Sanchez, M. (2025).	The authors view BCPs as an instrument to set out rules and regulations for outsiders. These documents allow communities to specify how they want external stakeholders, such as researchers, business people, government officials, and companies to engage with their natural resources and associated indigenous knowledge.	The study lacks in how these documents can be employed by communities to empower themselves through the local knowledge.

DISCUSSIONS

Introduction to Bio-Cultural Community Protocols (BCPs)

Oli et al. (2015) view BCPs as a general term used for an array of documents developed by communities to set out clear rules and regulations on how they expect stakeholders to approach them with regard to their natural setting. In its own right, BCPs are comprehensive instruments, used by communities to set out rules and regulations to specify how they want external stakeholders such as researchers, business people, government officials, and companies to engage with their natural resources and associated indigenous knowledge (Rakotondrabe & Girard, 2021:10255). In fact, Rakotondrabe and Girard (2021:10255) are of an opinion that BCPs are depositories of customary traditions and customary rules for managing and protecting the tangible and intangible heritage of local communities, as well as promoting participatory advocacy for the recognition, acknowledging, and supporting ways of life that is associated with the customary use of biological diversity. The protocol, as highlighted by Sanchez (2025), was introduced into international law by both the 2010 Nakoya Protocol and the United Convention on Biodiversity, aiming at regulating access to local natural resources by outsiders and local communities, also promoting endogenous development and further providing a platform to negotiate fair and equitable benefit-sharing, deriving from the use of such resources.

On the other hand, Lassen et al. (2018) and Oli et al. (2015) are of an opinion that BCPs differ from community to community, yet their main aim is to regulate terms for accessing and distributing benefits arising from the use of genetic resources and associated knowledge, belonging to a specific community (Sanchez, 2025). Despite their distinctive nature, Lassen et al. (2018) illustrates that effective BCPs should cover the following elements:

- A definition of a community, its leadership and local decision-making process;
- A descriptive community-based natural management systems, knowledge, innovations, and practices (either in situ conservation, and sustainable use) of indigenous flora and fauna, and details of those natural resources;
- Ways of life, including the links between culture, spirituality, customary laws and values;
- Rights, responsibilities, and duties of the community according to customary, national, and international law;
- Conditions set out by communities for granting access to their land, resources, and knowledge, such as procedures for Free, Prior and Informed Consent (FPIC);
- Challenges faced by community and calls to various stakeholders to engage on specific issues.

In the same light, Lassen et al. (2018) claim that when communities are ready for the development of the protocol, the following steps are normally undertaken: see the figure below:

Figure 1. Steps of developing Bio-cultural Community Protocol (BCPs) Lassen et al. (2018).

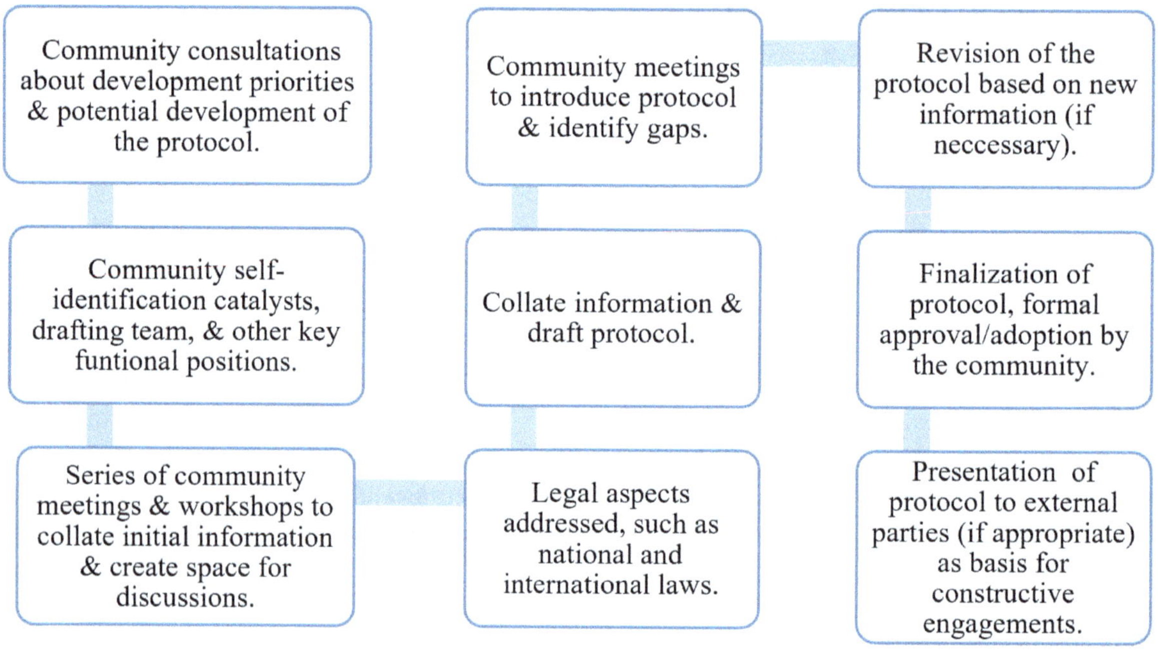

Subsequently, Non-Profit Organizations (NGOs), Community-Based Organizations (CBOs) and other appropriate organizations significantly facilitate community processes, such as provision of technical input, and further assisting with drafting of the protocol (Lassen et al., 2018). While ensuring community empowerment, in fact, these organizations are not only responsible for the drafting of the protocol document, but to work closely with community members to enable them to develop protocol. The role of supporting organizations is to (Lassen et al., 2018):

Facilitate

As highlighted by Lassen et al. (2018), supporting organizations should facilitate the contribution of a broad, representative group of community members. Although these organizations have a degree of detachment from direct local community dynamics, they are capable of observing and targeting a broader segment of the community. On the other hand, supporting organizations should have meaningful experiences and maintain respectful connections with the community (Lassen et al., 2018). Trust developed through these connections will allow community members to feel open and comfortable to share their practices, values, aspirations and concerns.

Information and training

Lassen et al. (2018), contend that there is a need for more training and community engagements on access and benefit sharing, especially with regard to the functioning of bio-prospecting, with a focus on the role and rights of communities.

Legal support

As articulated by Lassen et al. (2018) and Odi et al. (2015), a successful and meaningful bio-cultural community protocol should relate community values and customary law into a larger legal framework. Fundamentally, legal support is necessary for the development of BCPs in two ways:

Legal training

Legal training is important for the BCPs' development processes; hence Lassen et al. (2018) suggest that communities require assistance in identifying relevant legal instruments from local, national, and international laws and policies that encourage and advocate for community rights, translate them into understandable language, and further train community members in the use of these instruments.

Legal assistance

Legal assistance is an essential part for the development of any development, be it in goods purchasing, and individual and family businesses. Lassen et al. (2018) assert that it is also important in the drafting of BCPs. Additionally, assistance may also be needed in the application of the protocol, including negotiation of ABS agreement (Agreement of Benefit-Sharing).

Drafting

Lassen et al. (2018) illustrated further that supporting organizations have a role in ensuring that the communities are assisted in the writing-up of the protocol, especially when there is a low literacy rate amongst the communities. These organizations are also indebted to provide technical input and guidance, yet ensuring that the content of the protocol comes directly from the community (Lassen et al., 2018). Notably, the drafting as highlighted by Lassen et al. (2018) should be done in the presence of the community representatives, and any consequent revisions should be discussed and agreed upon by both supporting organizations and the community representatives.

Lassen et al. (2018) argue that not only supporting organizations are significant in the development and drafting of any local community protocols, but suggest that involvement of government agencies and stakeholders are also of an essence to ensure buy-in, and facilitate official recognition. For instance, the Competent National Authorities (CNAs), including local and sub-national authorities. In essence, both supporting organizations and government agencies play a role in informing and guiding external parties and raising an awareness of the rights and the role of indigenous and local people and communities, regarding access to genetic resources and associated knowledge; implementation of their BCPs; precisely ensuring respect for PIC (Prior-Informed Consent) procedures; and fair processes for the negotiation of MTA (Material Transfer Agreement) (Lassen et al., 2018). This chapter therefore consists of:

Addressing historical injustice and power imbalance

This chapter supports what have been conveyed by United Nations Environment Programme (UNEP) and Environmental Defenders Office New South Wales (EDO NSW) (2023) that BCPs are basically meant to address historical injustice and lack of respect and acknowledgement for indigenous and local communities, biodiversity and related traditional indigenous knowledge. The injustice experienced by indigenous and local communities for centuries, led to intrusion, dispossessed lands, and further delayed possibilities for indigenous and local communities to pursue local traditional economies (Magni, 2017; Sultana et al., 2017). In line with what was already mentioned and pointed out by many researchers and scholars, injustice and power imbalance were further escalated by the exclusion of both indigenous and traditional knowledge in economic development planning, which has far-reaching consequences, including the exploitation of local natural resources and marginalization of indigenous and local communities (Magni, 2017). Moreover, this oversight enabled businesses and government to disregard community needs, leading to environmental dilapidation, cultural destruction, and loss of traditional livelihoods.

Addressing historical injustice and power imbalance, explicitly with regard to socio-economic development, allows indigenous and local communities to address their differences, to engage in dialogues on an equal basis and further determines issues that affects them, enabling communities, especially marginalized indigenous communities, to participate, demand, negotiate, change and hold intuitions accountable that benefited from projects that affected their livelihoods and wellbeing, and also lessen access to natural resources by external-driven development (Godfrey & Siraje, 2019). Sanchez (2025) added that BCPs possess the potential to address not only historical misappropriation and biopiracy of natural resources, but also state-sponsored violence perpetrated in various countries across the globe. Such violence encompasses forced displacement, political killings and police brutality, demolition of housing, chattel slavery, indentured discrimination, and state-destruction public health crises (Sanchez, 2025).

As reflected by Jonas et al. (2017:110), when a BCPs follows Lassen's suggested elements it would enable local and indigenous communities to bridge the gap between the customary management of their bio-cultural heritage and the external management of their resources, while also assisting in lessening the power irregularities that often demonstrate government-community relations and further promote a more participatory and endogenous approach to the future governance of their territories, natural resource and biodiversity. Odi et al. (2015) further added that this protocol may protect local and indigenous communities in improving legal certainty and simplicity for both users and providers of genetic resources, which may create equitable partnerships between communities and other groups, such as scientific organization or companies that seek to develop new products, derived from natural resources. The subsequent subheading examines the role and efficacy of BCPs in mitigating and redressing historical injustice that have impacted socio-economic development within South African indigenous and local communities.

THE ROLE OF BCPs IN SOCIO-ECONOMIC DEVELOPMENT OF LOCAL COMMUNITIES

The importance of BCPs lies heavily on the establishment of the protocol and the product derived from biodiversity. Furthermore, as Anquet and Girard (2022) point out, this protocol resolves the issues associated with Access and Benefits Sharing agreement (ABS) challenges for indigenous and local communities. Supplementary, Laheri (2024) added that BCPs provide indigenous and local communities an opportunity to leverage BCPs to harness their traditional knowledge, thereby promoting and sustaining their socio-economic development. However, Ayana (2025) asserts that indigenous traditional knowledge and related practices have the potential to drive economic development and generate income in many local and indigenous communities across the globe. In fact, authors in this chapter argue that it could be possible by leveraging BCPs (Laheri, 2024). The following examples showcase successful BCPs where communities benefitted and continue to benefit from commercialization of indigenous knowledge systems and associated natural resources:

For instance, within the South African context, Schoroeder et al. (2020) provided the practical example where commercialization of natural resources and related indigenous traditional knowledge of Rooibos tea was one the most successful BCPs in South Africa, where the community benefitted significantly. The community is currently receiving royalty and revenue sharing; in this sense the San community receives royalties from Rooibos sales, which basically generates income, and further inspires local economic development. Following the success of Rooibos, Hoodia represents another instance where the San peoples derive benefits from the sales of this plant (Amusan, 2016). In this case, as indicated by Maharaj (2017), the Council for Science and Industrial Research entered into a benefit-sharing agreement with the San community. Amusan (2016) and Chennels (2013) seems to suggest that monetary compensation is typically disbursed into the San-controlled Trust, with the funds being dispersed among the San community. While BCPs are well established and beneficial among South African local and indigenous communities, they are also evident in other countries, such as Kenya, Madagascar and Malawi (Rakontondrabe & Girard, 2021; Sanchez, 2025).

In Kenya, the Mijikenda elders developed BCPs for the Kaya Forests. This protocol was used in 2013 to block a US company's attempt to patent medicine derived from Kaya forests plants, thereby protecting community knowledge associated with the use of their natural resources and preventing bio-piracy. Similarly, in Madagascar, the Motrobe community's BCP governs the use of *Cinnamosma fragrans* bark (Rakontondrabe & Girard, 2021). The community receives a 5% royalty on all perfume sales derived from the bark. Additionally, these funds have built schools, bridges and clinics, and the initiative employs about 22 local people in sustainable harvesting and distillation. While in Malawi, fisher communities around Chilwa developed a BCP covering traditional fish management practices and wetland plants. As viewed by this chapter, the protocol is used to negotiate access agreement with researchers, and as a result, the communities receive research fees, co-authorship on studies, and further strengthen communities' role in the management of Lake Chilwa Ramsar Site (Ndalilo et al., 2023).

Foyer et al. (2024) are of the opinion that BCPs are the key processes that local and indigenous communities engage in to gain social and economic development on their own terms and to fight for their right to achieve the desired outcome. Additionally, Sanchez (2025) emphasises that the use of BCPs supports the new approach of sustainable development amongst indigenous and local communities that is rooted in cultural identities and advocates for control over the land, environment, and indigenous traditional knowledge. This chapter concurs with what has been highlighted by Sanchez (2025), namely that the development of BCPs by local and indigenous communities should yield greater benefits for them in terms of employment creation and income generation opportunities, as opposed to only focusing on the promotion and protection of genetic resources and related traditional knowledge. Below are evidence-based pathways for ensuring income generation through the new BCPs Model:

Indigenous Procurement Agreements

Development of indigenous procurement agreements necessitates both local and external contractors to work with and build with local and indigenous communities in a manner that enables them to be competitive, profitable, minimize unemployment, and create social value for themselves (Denny-Smith et al., 2025; Office of the Comptroller General Ministry of Finance, 2021). Therefore, this study introduces that when an indigenous procurement agreement is embedded in the BCPs as a binding mechanism, the finance and employment benefits to local communities become direct, and operationally protected. In essence, an indigenous procurement process will ensure that all expenditure on IKS-derived materials and services is channelled exclusively to BCPs-accredited cooperatives.

Certification and Licencing of Bio-cultural products

Following the development and approval of the indigenous procurement agreement, this chapter argues that Bio-cultural Community Protocols may be reconceptualised beyond their advisory function. Under the proposed BCP Model, the incorporation of certification and licensing systems will enable communities to exercise direct economic authority. By establishing a Community Standards Bureau to administer a "BCP Grade A" standard, these systems will assist communities in driving economic development, while creating protected niche markets for unique products derived from local and indigenous production systems (Food and Agricultural Organization & Alliance of Biodiversity CIAT, 2022). This, as proposed by the chapter, will shift the current certification and licensing compliance mechanism to a fully community-controlled instrument of market regulation and value capture, which will also assist in developing long-term employment for these communities.

Employment Creation through BCPs

Once certification and licensing of products derived from local resources and related knowledge systems are institutionalized under the BCP framework, proposed by this chapter, the protocol will function as a structural mechanism for converting the use of natural resources into guaranteed local employment. The exclusive IKS sourcing clause will mandate that only BCP-accredited cooperatives, encompassing of community custodians, may harvest, process, and supply certified indigenous knowledge-based products to both the government and private companies. Factually, this legal instrument secures the entire value chain within the community, precluding external commercial suppliers and prohibiting the ad hoc labour intermediation.

By certifying and licensing traditionally informal activities as remunerated IKS services paid at standardized BCP Rate card rates, the newly improved BCPs Model transforms subsistence-level gathering and processing into formalized occupations with predictable incomes, skills accreditation and labour protections. Accreditation criteria will ensure that economic benefits, technical capacity, and decision-making authority remain with local indigenous knowledge holders (Olaitan, 2020; Manyaga et al., 2024). In addition, as highlighted by Mishra (2026), the proposed BCP Liaison Office and Community Standard Bureau institutionalize additional roles in harvesting permits, sustainability monitoring, quality assurance, and BCP Grade A certification, generating administrative and technical employment beyond direct resource extraction (Bebbington et al., 2017).

Therefore, the regulatory shift from advisory to licencing authority permits BCPs to assist local and indigenous communities in creating long-term employment that is legally secured, non-displaceable, and directly tied to the sustainable management of their bio-cultural resource base. This repositions environmental compliance and market access as drivers of intergenerational livelihoods, rather than as external constraints on community development. Despite the positive benefits derived from the development and implementation of both current and proposed BCPs Models, they have an implication which presents several challenges. The following subsection examines the difficulties encountered in BCPs implementation across African indigenous communities and government.

DIFFICULTIES AND CHALLENGES OF IMPLIMENTATION OF BCPs ACROSS THE GLOBE INCLUDING SOUTH AFRICA

BCPs are considered a positive solution to indigenous and local communities' challenges, instigated by a lack of recognition and continuous marginalization by scientific knowledge, and sustainable development stakeholders perpetuates development, and to address bio-piracy and misappropriation. Rethinking the future, however, Lassen et al. (2018) and Odi et al. (2015) argue that challenges in the implementations of BCPs are inherent to the local community's protocol, the process of developing and usage of the protocol could be excessively influenced by different parties, either local or external outsiders. Odi et al. (2015) are also of the view that if these protocols focus more on customary law, it might surge existing power asymmetries, such as the exclusion of women, youth, and marginalized groups in the decision-making processes. On the other hand, as articulated by Odi et al. (2015), bio-cultural community protocols are mostly voluntary in nature, which might still add to the exclusion of natural resources and be related in many sustainable economic development programs and further exacerbate biopiracy.

In some instance, Odi et al. (2015) detected that unrealistic expectations normally arise within communities in relation to monetary benefits, especially if the project is introduced by external agencies. These expectations are sometimes exacerbated by a lack of adequate institutional capacity, faced by many indigenous and local communities, while lack of such capacity might lead to communities being coerced into agreement by the external actors (Odi et al., 2015). Complementary, Lassen et al. (2018) believe that these challenges are not only attributed to the protocol, but also concerns the negotiation of PIC and MTA (Material Transfer Agreement) on shared genetic resources and indigenous traditional knowledge associated with their use in general. For example, various resources are not entirely restricted to one community, while associated knowledge is also shared between countries, communities and different groups.

Halewood et al. (2021) on the other hand, indicate that there are multiple and diverse BCPs that might make it difficult to comprehend, in case communities opt to develop their own without external stakeholders such as supporting organizations, the government and business sector. Even though it might be impossible to standardise the structures. Odayar (2024) opinionated that although BCPs are essential and of the most importance for both indigenous and local communities for ensuring equal benefit-sharing over the use of their local resources, this may be time-consuming processes that requires deep consultative approaches, training and capacity-building of the community and might also take many years which still exposes communities into losing their valuable resources. Raven and Robinson (2022) further note that additional challenges arise from the scope of BCPs development and the heterogeneity of the community themselves. Since, communities are frequently diverse and hold differing needs, competing interests and agendas regarding natural and biological resources use emerging, occasionally precipitating local-level conflicts (Raven & Robinson, 2022).

FUTURE RESEARCH DIRECTION

While this chapter has demonstrated the potential of improved Bio-cultural Community Protocols to function as a structural mechanism for long-term employment and epistemic sovereignty, several areas warrant further empirical and theoretical investigation. Future research should assume thorough longitudinal case studies to quantify employment, generated through BCP-Accredited Cooperatives across multiple sectors, including construction, nutraceuticals and crafts. Such studies could establish baseline data on income levels, job security, and skills transfer pre-and post BCP implementation, thereby testing the claim that an improved BCP produces non-displaceable livelihoods. In addition, a comparative analysis between the current BCPs and regulatory improved BCPs as proposed by this chapter would further clarify the casual impact of certification and exclusive sourcing clauses on employment outcome.

Additional work is required to examine the institutional capacities necessary for Community Standard Bureaus to administer BCP Grade A certification, sustainability quotas, and Rate Card enforcement at scale. Research should investigate how these community-governed institutions interface with existing municipal procurement systems, environmental regulators, and national standards bodies, identifying legal, financial, and technical barriers to replicate the model in diverse jurisdictional frameworks. Pilot implementation studies would be particularly valuable for assessing scalability and institutional durability. By addressing these research gaps, subsequent scholarships can strengthen the evidentiary and theoretical basis for positioning proposed BCPs Models as infrastructures of self-determination, moving beyond claims toward policy-relevant models of community-led development.

Conclusions

In conclusion, this chapter has argued that reducing BCPs to mere legal instruments, misrepresents their possible transformative potential as community-governed frameworks for economic and epistemic change. This chapter demonstrated that improved BCPs may operate as mechanisms for enforcing epistemic justice, and it is through this function that they create the conditions and platform for community-led economic development. Rather than only functioning as defensive legal documents, BCPs should be mobilized by communities to develop indigenous procurement agreements, institutionalize certification and licensing of indigenous knowledge-based products, and further secure long-term employment for local and indigenous communities.

The process of developing BCPs compel structural change that repositions local and indigenous communities as authorities regulating engagement, which in-turn enables income generation from indigenous knowledge systems and related practice. This shift should be operationalised through the exclusive IKS sourcing clause, and Community Standard Bureau, which together convert natural resources use into formalized, remunerated occupations, protected from external commercial suppliers and ad hoc labour intermediation. Subsequently, environmental compliance and market access become drivers of contact-based employment and intergenerational benefit retention, rather than external constraints on local community development.

BCPs must therefore be understood not as legal instruments, but epistemic sovereignty that convert recognition and acknowledgement into a regulatory authority, and a regulatory authority into a secure income. Factually, when deployed as tools for procurement, certification, and licensing, BCPs enable communities to exercise control over value chains, as well as capture economic value from indigenous knowledge systems. This guarantees non-displaceable livelihoods. In this reconfiguration, BCPs transcend consultations and became infrastructures of self-determination and employment creation for local and indigenous communities.

REFERENCES

Amusan, M. A. (2016). Cultivating Effective Pedagogical Skills in In-Service Teachers: The Role of Some Teacher Variables. *Journal of the International Society for Teacher Education, 20*(1), 83–89.

Anquet, R., & Girard, F. (2022). *Biocultural community protocol and the Ethic of stewardship: the sovereign steward of biodiversity. Biocultural rights, Indigenous Peoples and Local Communities*. Routledge. DOI: 10.4324/9781003172642-14

Bebbington, C., Wilson, E., Smith, E., & Van Alstine, J. (2017). *Community Liaison officers: exploring the frontline of corporate practice in the oil and gas sector*. Audire and ECW Energy LTD.

Byscov. M.F. (2021). "What makes Epistemic Injustice an 'injustice'?" Journal of Social Philosophy, 52(1): 114-131. Date of access: 10 Jan. 2026.DOI: 10.1111/josp.12348

Chennels, R. (2013). "Traditional knowledge and benefit-sharing after the Nakoya Protocol: Three cases from South Africa''. *Law, Environment, and Development Journal:* 163. https://www.lead-journal.org/content/13163 Date of access: 03 Feb.2026.

Cummings, S., Munthali, N., & Sittoni, T. (2024). Epistemic justice as a "New Normal?" Integrating the contributions of Communities of Practice to decolonization of knowledge. *Sustainable Development (Bradford), 33*(3), 3228–3245. DOI: 10.1002/sd.3285

Denny-Smith, G., Loosemore, M., Sunindijo, R. Y., & Williams, M. (2025). Policy-ledd innovations: the impact of indigenous procurement policies on contactors 'behaviour towards indigenous people's participation in construction projects. In Dulaini, M. F. (Ed.), *Research Companion to Innovation in Construction* (pp. 178–202)., DOI: 10.4337/9781035326594.00017

Duchlova, H., & Leventon, J. (2023). *Epistemic justice: How can a diverse forms of knowledge improve policy making?* SSH Centre.

Dzvimbo, M. A., Tirivangasi, H. M., Mawonde, A., & Ncube, C. T. (2023). The role of local people and Indigenous Knowledge in participatory development approach in rural Zimbabwe. https://www .researchgate.net Date of access: 30 Dec. 2025.

Food and Agriculture Organization (FAO)., Alliance of Biodiversity., & CIAT. (2022). Labelling and certification schemes for Indigenous peoples' foods- generating income while protecting and promoting Indigenous peoples' values. Rome. https://doi.org/DOI: 10.4060/cc0155en

Foyer, J., Mauri, M. M., Filoche, G., & Castillo, G. (2024). The institutionalization of biculturalism through community protocols: The case of Guna yala. *International Indigenous Policy Journal, 15*(1). Advance online publication. DOI: 10.18584/iipj.2024.15.1.15872

Godfrey, B., & Siraje, K. (2019). Community development in the rural areas through traditional indigenous knowledge. International Journal of Research in Sociology and Anthropology, 5(2): 8-17. http://dx.org./10.20431/2454/-8677.0502002 Date of access: 01 Feb. 2026.

Kiage, O. E. (2019). The Ogiek peoples' Indigenous Knowledge: A pathway towards sustainable natural resource management in the Mau Forest, Kenya. *African Journal of Hospitality, Tourism, and Leisure*,8(1):1-19. https://www.ajhtl.com Date of access: 10 Jan. 2026.

Lahiri, S. (2024). Integrating Biocultural Protocol in tourism and biodiversity policies: An advocacy strategy for the Maasai indigenous communities in East Africa. https://globalforsetcoalition.org Date of access: 02 Feb. 2026.

Lassen, B., Jansen, L., Rasolojaona, J., Githaiga, C., Fey, L., & Bossou, B. (2018). Community protocols in Africa. https://naturaljustice.org Date of access: 25 June 2026

Lawson, C., Rourke, M., & Humphries, F. (2023). *Access and benefit sharing of genetic resources, information and traditional knowledge*. Routledge.

Lyndon, N., Mohd, N. L., Jali, F. M., & Rose, A. R. C. (2020). Indigenous knowledge and socio- economic development of a Malaysian Indigenous Community. *Journal of Southeast Asian Studies, 25*(2), 109–129. DOI: 10.22452/jati.vol25no2.6

Magni, G. (2017). Indigenous knowledge and implications for the sustainable development agenda. *European Journal of Education, 52*(4), 437–447. DOI: 10.1111/ejed.12238

Mak, S., & Thomas, A. (2022). An introduction to scoping reviews. *Journal of Graduate Medical Education, 561-564*. Advance online publication. 10.4300JGME PMID: 36274758

Manyaga, M. B., Goldman, G. A., & Thomas, P. (2024). Sustaining SMEs through Indigenous Knowledge Systems: Exploring opportunities and challenges. *The Southern African Journal of Entrepreneurship and Small Business Management, 16*(1), a882. DOI: 10.4102/SAJESBM.v16i1.882

Maunganidze, L. (2016). A moral compass that slipped: Indigenous knowledge systems and rural development in Zimbabwe. *Cogent Social Sciences, 2*(1), 1266749. Advance online publication. DOI: 10.1080/23311886.2016.1266749

Mishra, R. (2026). Bureau Indian Standards: Purpose, standardization, & application. https://www .testronixinstrumets.com Date of access: 29 April 2026.

Munn, Z., Peters, M. D., Stern, C., Tufanaru, C., McArthur, A., & Aromataris, E. (2018). Systematic review or scoping review? Guidance for authors when choosing between a systematic or scoping review approach. *BMC Medical Research Methodology, 18*(1), 143. DOI: 10.1186/s12874-018-0611-x PMID: 30453902

Ndalilo, L., Swiderska, K., & Wekesa, C. (2023). Establishing a biocultural heritage territory to protect Kenya's Kaya Forests; mid-term project workshop. IIED, London. https://www.iid.org/21556iied Date of access: 28 April 2026.

Nurfitri, D., Bustani, S., & Yunari, S. B. (2025). Protecting genetic resources and traditional knowledge in indigenous communities and governance through inclusive moral rights. *Scholars International Journal of Law. Crime and Justice*. Advance online publication. DOI: 10.36348/sijlcj

Odayar, S. (2024). Biocultural community protocols, a powerful tool to affirm and protect Indigenous Peoples 'rights. www://internationalrivers.org Date of access: 20 February 2026.

Odi, A. C., Nwogbaga, N. E., & Chukwuka, N. O. (2015). The proposed roles of VLAN and inter-VLAN routing in effective distribution of network services in Ebonyi State University. *International Journal of Science and Research, 4*, 2608–2615.

Office of the Comptroller General Ministry of Finance. (2021). Indigenous Procurement and Contract Management Guidelines: Ministry guidelines for the practical applications of the Province of the Province of British Columbia's procurement policies, in the delivery of culturally- appropriate programs and services to Indigenous peoples and communications. www2.gov.bc/assets/gov/british-columbian

Olaitan, Z. M. (2020). The relevance of African indigenous knowledge systems for development in Africa: Policy brief. https://africaindigenousresearch.com Date of access: 28 April 2026.

Oli, K. P., Rana, P., Shah, R., Rawal, R. S., Chaudhary, R., Yongping, Y., Yao, F., Kotru, R., & Bhatta, L. D. (2015). Tools for developing a Biocultural Community Protocol. International Centre for Integrated Mountain Development. Nepal. https://lib.icidmod.org/record/30771/filesWP15 Date of access: 16 Jan. 2016. (intext)

Rakotondrabe, M., & Girard, F. (2021). Protecting traditional knowledge through biocultural community protocols in Madagascar: Do not forget the "B" in BCP. *Sustainability (Basel), 13*(18), 10255. DOI: 10.3390/su131810255

Rakotondrabe, M., & Girdad, F. (2021). Protecting Traditional Knowledge through Bio-cultural community protocol in Madagascar: Do not forget the "B" in BCP. *Sustainability (Basel), 13*(18), 10255. DOI: 10.3390/su131810255

Raven, M. M., & Robinson, D. (2022). Biocultural rights and protocols in the Pacific. In *Biocultural rights, Indigenous Peoples and local communities* (pp. 203–220). Routledge. DOI: 10.4324/9781003172642-10

Rodger, D., Admani, A., & Thomas, M. (2024). What is a scoping review? BM Journal of Evidence Based Nurse, 27(3). https://ebn.bmj.com Date of access: 27 Jan. 2026.

Sanchez, M. (2025). Sustaining the future through addressing the past: the transitional justice functions of biocultural community protocols. *Law Environmental and Development Journal*, 21(1). doi: https://doi.org. Date of access: 27 Jan. 2026.DOI: 10.2550/.SOAS.00506365

Slikkerveer, L. J. (2019). The Indigenous knowledge systems perspectives on sustainable development. In L. Slikkerveer, G. Baourakis, & K. Saefullah (eds.). Integrated Community-Managed Development, Corporative management, 33-66, Springer, Cham. South Africa.

Sultana, R., Muhammad, N., & Zakaria, A. K. M. (2018). Role of indigenous knowledge in sustainable development. *International Journal of Development Research*, 8(2). https://www.journalijrd.com Date of access: 20 April 2026.

Taylor, J., & Pagliari, C. (2018). Comprehensive scoping review of health research using social media data. *BMJ Open, 8*(12), e022931. Advance online publication. DOI: 10.1136/bmjopen-2018-022931 PMID: 30552260

Thompson, K. L., Lantz, T., & Ban, N. C. (2020). A review of indigenous knowledge and participation environmental monitoring. *Ecology and Society, 52*(2), art10. Advance online publication. DOI: 10.5751/ES-11503-250210

Tirivangasi, H. M., & Tayengwa, D. (2017). Indigenous knowledge systems (IKS) and food security in South Africa: Is land reform a prerequisite. *Journal of Human Ecology (Delhi, India), 57*(3), 118–124.

Tondi, P. (2019). The Significance of Indigenous Knowledge Systems (IKS) for Africa's Socio- Cultural and Economic Development in the dawn of the Fourth Industrial Revolution (4IR). *Journal of Gender, Information and Development in Africa*, (pp 237-243). DOI: 10.31920/2050-4284/2019/S1n1a14

United Nations Environmental Programme and Environmental Defenders Office New South Wales. (2013). *Community protocols for environmental sustainability: A guideline for policymakers*. UNEP, Nairobi and EDO NSW.

ADDITIONAL READING

Barvikatte, K., Jonas, H., Kakar, A. R., Lanyasunya, P., Mathias, E., Rathore, H. S., Vivekanandan, P., & Wanyama, J. (2010). Biocultural Community Protocols for livestock keepers: League for Pastoral Peoples and Endogenous Livestock Development (LPP). Lokhit Pashu-Palak Sansthan. Calgary. (Unknown) Indigenous procurement program and strategy. https://wwwcalgary.ca/indigenousprocurement

Laird, S. A., & Wynberg, R. (2021). *Connecting the dots: Biodiversity conservation, sustainable use and access and benefit sharing. BioInnovation Africa (implemented by GIZ and funded by BMZ)*. Voices for BioJustice, People and Plants International, and University of Cape Town.

Mzwakali, S. (2024). Community Protocols: The legal framework to safeguard Biocultural rights in South Africa Policy Brief. https:////carleton.ca/african/studies

Norms, Q. C. (2012). Community Biocultural Protocols: Building mechanism for access and benefit-sharing among the communities of the Potato Park based on Quechua Customary Norms: ANDES (Peru), The Potato Park Communities and IIED. https://www.iied.org

KEY TERMS AND DEFINITIONS

Bio-cultural Community Protocol: documents developed by indigenous and local communities that outline their rights, customary laws, and values regarding their land, natural resources, and traditional knowledge.

Community Sustainable Development: process of, nurturing, resources to, ensure, long-term, ecological, economic, and, social, well-being, for, current, and, future, generations.

Economic Development: the sustained, collaborative process of improving a region's economic well-being and quality of life through enhanced infrastructure, job creation, and strategic business investment

Empowerment: process by which communities' gain control over factors and decisions that shapes their lives.

Indigenous Knowledge: local knowledge encompassing developed by communities to respond to every day challenges,

Chapter 12
Community Empowerment Through Indigenous Knowledge:
Pathways to Economic Prosperity

Ananya Pandey
http://orcid.org/0000-0002-2419-6314
Christ University, Bangalore, India

ABSTRACT

This book chapter explores pathways to economic prosperity through community empowerment grounded in Indigenous knowledge systems. It argues that Indigenous knowledge which is rooted in local ecology, cultural practices, and collective governance that offers sustainable and context-specific strategies for livelihood enhancement. By integrating traditional wisdom with contemporary economic frameworks, communities can strengthen resilience, promote inclusive growth, and reduce dependency on externally imposed development models. The chapter highlights how community-led enterprises, sustainable resource management, and intergenerational knowledge transmission contribute to long-term economic self-determination. It also examines policy and institutional mechanisms necessary to recognize, protect, and scale Indigenous knowledge without commodifying or eroding its cultural integrity. Ultimately, the chapter positions Indigenous knowledge as a critical driver of equitable economic development and community well-being in an era of globalization and environmental uncertainty.

1. INTRODUCTION

Economic prosperity has long been conceptualized through aggregate indicators such as gross domestic product, income growth, and industrial expansion. While these measures have shaped dominant development paradigms, they have increasingly been criticized for their inability to capture lived realities at the community level, particularly among Indigenous and marginalized populations. Mainstream development models have historically relied on top-down planning, centralized decision-making, and externally designed interventions that prioritize efficiency and scale over cultural context and social equity. Such approaches often marginalize local knowledge systems, weaken community agency, and

DOI: 10.4018/979-8-3373-7976-0.ch012

generate dependency rather than sustainable prosperity (Escobar, 2012; Sen, 1999). The persistence of poverty, inequality, and ecological degradation despite decades of development initiatives underscores the need to rethink what prosperity means and how it can be achieved.

Top-down development models frequently assume a universal trajectory of economic progress, overlooking the diversity of socio-cultural and ecological contexts in which communities exist. Large-scale infrastructure projects standardized agricultural practices, and market-led reforms have, in many cases, disrupted traditional livelihoods and eroded local institutions that once ensured social cohesion and resource sustainability. Empirical evidence suggests that development interventions imposed without community participation often fail to achieve long-term outcomes, as they lack local legitimacy and adaptability (Ostrom, 1990; Chambers, 1997). This disconnect has prompted growing scholarly and policy interest in bottom-up, participatory approaches that recognize communities not merely as beneficiaries but as active agents of development.

Within this reframing, Indigenous knowledge emerges as a profoundly underutilized economic resource. Indigenous knowledge systems encompass accumulated wisdom, skills, and practices developed through generations of interaction with specific environments. These systems inform sustainable agriculture, biodiversity conservation, water management, healthcare, and conflict resolution, offering locally grounded solutions to economic and ecological challenges. Despite their proven value, such knowledge systems have often been dismissed as informal, unscientific, or incompatible with modern economic structures (Berkes, 2018). As a result, development policies frequently fail to integrate Indigenous knowledge into economic planning, leading to missed opportunities for innovation, resilience, and sustainability.

Recognizing Indigenous knowledge as an economic resource does not imply its commodification or extraction for external gain. Rather, it calls for acknowledging knowledge sovereignty and the right of communities to control how their knowledge is used and shared. When appropriately supported, Indigenous knowledge can enhance productivity, reduce vulnerability to climate change, and strengthen local value chains while preserving cultural integrity (UNDP, 2023). This perspective aligns with broader shifts toward pluralistic and sustainability-oriented economic thinking, where prosperity is measured not only by income but also by well-being, autonomy, and ecological balance.

Community empowerment plays a central role in translating Indigenous knowledge into inclusive economic growth. Empowerment involves strengthening the capacity of communities to make decisions, manage resources, and shape development pathways according to their own priorities. Community-led economic models such as cooperatives, commons-based resource management, and social enterprises demonstrate how collective action can generate livelihoods while reinforcing social bonds and accountability. Studies consistently show that when communities have secure rights, institutional support, and participatory governance structures, development outcomes are more equitable and enduring (Ostrom, 2010; Sen, 1999). Thus, empowerment is not merely a social objective but a foundational economic strategy.

Against this backdrop, this chapter seeks to reframe economic prosperity through the lens of community empowerment and Indigenous knowledge. The primary objective is to examine how Indigenous knowledge systems can contribute to sustainable and inclusive economic development when embedded within community-led frameworks. The chapter is guided by three key research questions: First, how do mainstream development models limit the economic potential of Indigenous and local communities? Second, in what ways can Indigenous knowledge function as a viable economic asset without undermining cultural integrity? Third, what policy, legal, and institutional conditions are necessary to enable community empowerment and knowledge-based economic prosperity? By addressing these questions, the chapter aims to contribute to interdisciplinary debates on development, sustainability, and economic justice, while offering insights relevant to policymakers, scholars, and practitioners seeking alternative pathways to prosperity.

2. INDIGENOUS KNOWLEDGE AND ECONOMIC EMPOWERMENT

Understanding the role of Indigenous knowledge in economic empowerment requires a careful conceptual grounding that moves beyond conventional economic theory. Indigenous and traditional knowledge systems are not merely repositories of information but living, adaptive frameworks that integrate culture, ecology, social relations, and spirituality. These systems have evolved through long-term interaction with specific environments and are transmitted across generations through practice, oral traditions, and collective memory. Unlike codified scientific knowledge, Indigenous knowledge is contextual, experiential, and relational, emphasizing harmony between human activity and natural systems (Berkes, 2018). Recognizing these characteristics is essential for appreciating how such knowledge can underpin alternative pathways to economic empowerment. Indigenous and traditional knowledge systems are often defined as cumulative bodies of knowledge, practices, and beliefs developed by communities through historical engagement with their surroundings (Warren, 1991). They encompass domains such as agriculture, forestry, fisheries, medicine, weather prediction, and resource governance. Importantly, these systems are embedded in social institutions and customary norms that regulate access, use, and distribution of resources. This embeddedness distinguishes Indigenous knowledge from purely technical expertise, as it is inseparable from cultural values, ethical principles, and collective responsibilities. Consequently, economic activities derived from Indigenous knowledge tend to prioritize sustainability, reciprocity, and intergenerational equity rather than short-term profit maximization.

The relationship between culture, economy, and community agency is central to Indigenous knowledge–based development. In many Indigenous societies, economic practices are not autonomous market transactions but are deeply intertwined with cultural identity and social obligations. Livelihood activities such as farming, fishing, or craft production often carry symbolic meanings and reinforce communal ties. Culture, in this sense, acts as an economic regulator by shaping norms of sharing, stewardship, and collective decision-making (Gudeman, 2001). Community agency emerges when these cultural foundations enable communities to define their own development priorities and mobilize resources in ways that reflect shared values. Economic empowerment, therefore, is not limited to income generation but includes the strengthening of cultural autonomy and decision-making capacity. This integrated view contrasts sharply with market-centric economic paradigms that dominate mainstream development thinking. Neoclassical economic models typically conceptualize individuals as rational, self-interested actors operating in competitive markets, with efficiency and growth as primary objectives. Such paradigms often abstract economic activity from social and ecological contexts, treating culture as an external or even obstructive factor. Indigenous epistemologies, by contrast, reject this separation. Knowledge is understood as relational, produced through interactions among humans, non-human entities, and the environment. Economic activity is evaluated not only in terms of output but also in terms of social cohesion, ecological balance, and moral responsibility (Escobar, 2012).

The tension between Indigenous epistemologies and market-centric paradigms has significant implications for development policy. When Indigenous knowledge is selectively extracted and integrated into market systems—such as through commercialization of traditional crops or medicinal plants—without corresponding recognition of community rights, it can lead to dispossession and cultural erosion. Scholars have described this process as knowledge appropriation, wherein Indigenous knowledge is valued only insofar as it can be monetized within global markets (Shiva, 2007). A conceptual foundation for economic empowerment must therefore emphasize epistemic plurality and challenge the dominance of singular economic rationalities. Empowerment theory provides a useful lens for bridging Indigenous knowledge and economic development. Empowerment is commonly understood as the process through which individuals and communities gain control over decisions and resources affecting their lives (Kabeer, 1999). At the community level, empowerment involves collective agency, institutional capacity, and the ability to negotiate with external actors. Community-led development approaches operationalize empowerment by prioritizing participation, local governance, and ownership of development initiatives. Empirical studies demonstrate that when communities are actively involved in planning and implementation, development outcomes are more equitable, resilient, and contextually appropriate (Chambers, 1997). In Indigenous contexts, empowerment is closely linked to self-determination and recognition of customary institutions. Community-led economic models—such as cooperatives, commons-based resource management, and Indigenous enterprises—draw upon traditional governance structures and social norms. These models challenge the assumption that efficiency requires centralized control, instead highlighting the effectiveness of polycentric and locally embedded institutions (Ostrom, 2010). Empowerment theory thus complements Indigenous knowledge systems by emphasizing agency, participation, and institutional diversity as drivers of economic well-being.

The relevance of Indigenous knowledge and empowerment becomes particularly evident within sustainable and post-development economics. Sustainable development critiques have long pointed out that growth-oriented models often externalize environmental and social costs, leading to ecological crises and social inequalities. Indigenous knowledge offers alternative economic logics grounded in sustainability, sufficiency, and resilience. Practices such as diversified farming, rotational resource use, and communal stewardship align closely with contemporary sustainability goals, including climate adaptation and biodiversity conservation (Berkes, 2018). Post-development economics goes further by questioning the universality of development itself. Scholars in this tradition argue that development discourse has historically imposed Western notions of progress, marginalized other ways of living and knowing (Escobar, 2012). From this perspective, Indigenous knowledge–based economies are not transitional stages toward modernity but valid and viable alternatives. They expand the conceptual space of economics by foregrounding well-being, relationality, and ecological integrity. Economic empowerment, in this sense, is about enabling communities to sustain and evolve their own economic systems rather than conforming to externally defined models.

3. INDIGENOUS KNOWLEDGE AS AN ECONOMIC ASSET

Indigenous knowledge has increasingly been recognized as a critical yet undervalued economic asset, particularly in contexts where livelihoods are closely tied to natural resources and local ecosystems. Unlike conventional economic assets, Indigenous knowledge is intangible, collective, and embedded in cultural practices and social institutions. Its economic value lies not only in productivity gains but also in its capacity to sustain ecosystems, reduce risk, and enhance community resilience. When viewed through this broader lens, Indigenous knowledge constitutes a form of capital that supports long-term economic well-being rather than short-term accumulation. Traditional ecological knowledge (TEK) is a foundational component of this asset base. TEK refers to cumulative knowledge, practices, and beliefs concerning the relationships between living beings and their environment, developed through adaptive processes and transmitted across generations. In natural resource management, TEK informs practices such as seasonal harvesting, rotational land use, and biodiversity conservation. These practices often result in efficient resource utilization and ecosystem regeneration, reducing the likelihood of resource depletion. Empirical studies demonstrate that Indigenous-managed lands frequently exhibit higher levels of biodiversity and ecological integrity compared to state-managed or privately owned lands, underscoring the economic and environmental value of TEK-informed governance (Berkes, 2018).

In agriculture, Indigenous knowledge underpins diverse and resilient production systems. Indigenous farming practices typically emphasize crop diversity, soil conservation, and climate adaptability. Techniques such as mixed cropping, seed saving, and organic soil management enhance food security while minimizing dependence on external inputs. These systems not only provide subsistence but also generate surplus for local and regional markets, contributing to household income and community stability. In an era of climate uncertainty, such knowledge-driven agricultural systems offer cost-effective adaptation strategies that formal agricultural policies often overlook (Altieri & Nicholls, 2017). Forestry and fisheries further illustrate the economic significance of Indigenous knowledge. Community-based forest management systems, guided by customary norms and ecological understanding, have been shown to sustain timber and non-timber forest products while preserving forest cover. Similarly, Indigenous fisheries management practices—such as seasonal fishing bans, gear restrictions, and sacred conservation zones—maintain fish stocks and ensure equitable access. These practices reduce overexploitation and enhance long-term yields, translating ecological stewardship into economic security. Importantly, the economic benefits of such systems are distributed collectively, reinforcing social cohesion and shared responsibility (Ostrom, 1990). Artisanal economies rooted in Indigenous knowledge also play a vital economic role. Traditional crafts, textiles, and material cultures embody specialized skills and cultural narratives that carry both use and exchange value. These artisanal practices often rely on locally sourced materials and sustainable production methods, aligning with contemporary demand for ethical and environmentally responsible products. When supported through fair trade networks and community-controlled marketing, Indigenous artisanal economies can generate income while strengthening cultural identity. However, their economic potential is contingent upon safeguarding intellectual and cultural rights to prevent exploitation and misappropriation (UNESCO, 2017).

Beyond sector-specific activities, Indigenous economies frequently operate through informal systems that challenge conventional economic assumptions. Barter, reciprocity, and collective ownership are common features of Indigenous economic life. These systems prioritize social relations and mutual support over monetary exchange, functioning as safety nets in times of scarcity. From an economic perspective, such arrangements reduce transaction costs, distribute risk, and ensure access to essential goods and services. Although often invisible in national accounting systems, informal and subsistence economies significantly contribute to livelihoods and local resilience, particularly in marginalized regions (Gudeman, 2001). Collective ownership of land and resources is another defining characteristic that enhances the asset value of Indigenous knowledge. Communal tenure systems, governed by customary law, regulate resource use and prevent fragmentation or speculative exploitation. By linking rights to responsibilities, these systems incentivize sustainable management and long-term planning. Research on commons governance highlights that collective ownership, when supported by strong local institutions, can outperform private or state ownership in both economic efficiency and sustainability (Ostrom, 1990). Indigenous knowledge is integral to the functioning of these institutions, providing the norms and rules that guide collective action. Innovation is deeply embedded in Indigenous practices, though it is often unrecognized as such. Innovation within Indigenous knowledge systems occurs through incremental adaptation rather than radical technological change. Communities continuously experiment with crop varieties, resource management techniques, and social arrangements in response to environmental and socio-economic shifts. This form of innovation is cumulative, risk-averse, and context-specific, prioritizing reliability over rapid expansion. Contemporary innovation discourse, which often equates innovation with high technology, fails to acknowledge these adaptive processes despite their proven effectiveness and sustainability (Berkes, 2018).

Despite its evident economic value, Indigenous knowledge faces significant challenges in terms of recognition and valuation within formal economies. One major obstacle is the incompatibility between collective, orally transmitted knowledge and legal frameworks that prioritize individual ownership and codification. Intellectual property regimes, for example, are poorly equipped to protect communal knowledge, leading to instances of biopiracy and unauthorized commercialization. Moreover, development policies frequently undervalue subsistence and informal economic activities because they do not conform to market-based metrics of productivity and growth (Shiva, 2007).

Valuation challenges are compounded by power asymmetries between Indigenous communities and external actors, including corporations and state institutions. When Indigenous knowledge is incorporated into formal markets without equitable benefit-sharing arrangements, communities may lose control over their resources and cultural practices. This not only undermines economic empowerment but also erodes trust and social cohesion. Addressing these challenges requires legal and institutional reforms that recognize Indigenous knowledge as a legitimate economic asset while respecting knowledge sovereignty and community autonomy (UNDP, 2023).

4. COMMUNITY-LED ECONOMIC MODELS AND PRACTICES

Community-led economic models represent a critical bridge between Indigenous knowledge systems and sustainable economic empowerment. Unlike externally driven development interventions, these models emerge from within communities, drawing upon shared values, customary institutions, and collective decision-making processes. They emphasize participation, ownership, and accountability, ensuring that economic activities align with local priorities and cultural norms. In Indigenous contexts, community-led models are particularly significant because they operationalize knowledge that is collectively held and socially embedded, transforming it into livelihood opportunities while preserving autonomy and resilience. Cooperatives, self-help groups, and commons-based enterprises are among the most prominent manifestations of community-led economic organization. Cooperatives enable members to pool resources, share risks, and enhance bargaining power in markets that would otherwise marginalize small-scale producers. In Indigenous agricultural and artisanal sectors, cooperatives facilitate access to credit, inputs, and market information while maintaining collective control over production processes. Self-help groups, often organized around savings and credit, play a complementary role by strengthening financial inclusion and mutual support. These groups reduce dependence on external financial institutions and create locally accountable mechanisms for capital formation, particularly in rural and marginalized regions (ICA, 2016).

Commons-based enterprises extend these principles to the management of shared natural resources such as forests, grazing lands, and fisheries. Grounded in customary rules and collective ownership, these enterprises demonstrate that economic productivity and ecological stewardship are not mutually exclusive. By regulating access and use through community norms, commons-based systems prevent overexploitation and ensure equitable distribution of benefits. Empirical research on commons governance shows that such arrangements often outperform centralized or privatized systems when communities possess strong institutions and social cohesion (Ostrom, 1990). Indigenous knowledge provides the normative and ecological foundations that make these systems viable, embedding economic activity within long-term sustainability goals. Women's leadership constitutes a vital yet often underrecognized dimension of community-led Indigenous economies. In many Indigenous societies, women are primary custodians of knowledge related to seeds, food systems, medicinal plants, and artisanal skills. Their roles in household and community economies position them as key agents of economic resilience. Women-led cooperatives and self-help groups have demonstrated significant impacts on income generation, food security, and social empowerment. By organizing collectively, women are able to negotiate better market terms, diversify livelihoods, and invest in community well-being, including health and education (Kabeer, 1999).

However, gendered dimensions of Indigenous economies are shaped by intersecting power relations, including patriarchy, colonial legacies, and legal marginalization. Community-led models that consciously promote women's leadership can challenge these inequalities by creating spaces for participation and decision-making. Evidence suggests that when women have a substantive role in governance structures, economic outcomes tend to be more inclusive and sustainable, as women often prioritize long-term household and community needs over short-term gains (Agarwal, 2010). Thus, gender-responsive community-led development is not only a matter of equity but also of economic effectiveness. Intergenerational knowledge transfer and skills preservation are central to the sustainability of community-led economic models. Indigenous knowledge is transmitted through practice, storytelling, apprenticeships, and ritual, rather than formal instruction. Economic activities such as farming, fishing, and craft production serve as sites of learning where younger generations acquire skills alongside cultural values. Community-led enterprises that actively involve youth create pathways for continuity and innovation, ensuring that knowledge evolves in response to changing conditions. This process counters the erosion of Indigenous livelihoods caused by migration, formal education systems that devalue traditional skills, and the commodification of culture (Berkes, 2018).

Preserving intergenerational knowledge also has direct economic implications. Skills rooted in Indigenous knowledge—such as sustainable land management or artisanal production—constitute forms of human capital that enhance employability and entrepreneurship within local contexts. When communities invest in mentoring and skills transmission, they reduce dependence on external expertise and retain economic value locally. Moreover, intergenerational collaboration fosters innovation by combining traditional practices with new tools and ideas, demonstrating that Indigenous economies are dynamic rather than static. Social capital, encompassing trust, reciprocity, and shared norms, underpins the effectiveness of community-led economic models. High levels of social capital facilitate collective action by lowering transaction costs and enabling cooperation. In Indigenous communities, social capital is reinforced through kinship ties, cultural rituals, and customary institutions. These social structures support collective decision-making and conflict resolution, which are essential for managing shared resources and enterprises. Research consistently shows that communities with strong social capital are better able to sustain collective economic initiatives and adapt to external pressures (Putnam, 1993). Collective decision-making processes are particularly significant in ensuring accountability and legitimacy. Unlike hierarchical governance structures, community-led models often rely on consensus-building and participatory deliberation. While these processes can be time-intensive, they enhance transparency and inclusivity, reducing the risk of elite capture. Indigenous decision-making practices, grounded in customary norms, prioritize dialogue and collective responsibility, aligning economic actions with community values. This participatory ethos strengthens resilience by enabling communities to respond collectively to economic shocks and environmental change (Borrini-Feyerabend et al., 2013).

5. INDIGENOUS KNOWLEDGE IN ACTION

Indigenous knowledge moves from concept to concrete impact when it is embedded in lived economic practices. Across diverse geographical contexts, Indigenous communities have demonstrated how locally grounded knowledge systems can generate sustainable livelihoods while strengthening cultural identity and ecological stewardship. This section examines four illustrative examples—spanning agriculture, crafts and tourism, and forest-based economies—to demonstrate how Indigenous knowledge operates in practice. While these cases differ in context and scale, together they reveal common principles, lessons learned, and limits to replication.

5.1. Sustainable Agriculture and Food Sovereignty Initiatives

In many parts of the Global South, Indigenous agricultural systems have become central to food sovereignty movements. In the Andean region of Latin America, Indigenous communities have revitalized traditional polyculture farming systems, such as the *chacra* and *milpa*, which integrate multiple crop varieties adapted to local microclimates. These systems enhance resilience to climate variability by spreading risk across crops and seasons. Seed-saving practices and community seed banks further reinforce autonomy by reducing dependence on commercial seed markets and external inputs (Altieri & Toledo, 2011). Similarly, Indigenous farming communities in India have revived traditional millet-based agriculture as an alternative to monoculture cash crops. Millets, historically cultivated using Indigenous knowledge of soil and rainfall patterns, require minimal water and chemical inputs while providing high nutritional value. Community-led initiatives promoting Indigenous grains have improved household food security, generated niche market opportunities, and strengthened women's roles as custodians of seeds and knowledge (FAO, 2019). These cases illustrate how Indigenous agricultural knowledge functions simultaneously as an ecological adaptation strategy and an economic asset that supports local food systems. Food sovereignty initiatives grounded in Indigenous knowledge emphasize control over production, distribution, and consumption. Unlike productivity-focused agricultural development, these approaches prioritize cultural appropriateness, environmental sustainability, and community governance. Empirical evidence suggests that such systems can deliver stable livelihoods and improved nutrition, particularly in ecologically fragile regions where industrial agriculture has proven unsustainable (Altieri & Nicholls, 2017).

5.2. Indigenous Crafts, Tourism, and Cultural Enterprises

Indigenous crafts and cultural enterprises represent another domain where knowledge translates into economic value. In regions of Africa, Asia, and Latin America, traditional weaving, pottery, and metalwork are not merely artistic expressions but repositories of technical expertise, symbolic meaning, and collective memory. Community-based craft cooperatives enable artisans to pool resources, standardize quality, and access fair trade markets while retaining control over design and production processes. These cooperatives have been shown to increase incomes and reduce exploitation by intermediaries, particularly for women artisans (UNESCO, 2017). Cultural tourism initiatives further demonstrate the economic potential of Indigenous knowledge when managed by communities themselves. In parts of Southeast Asia and Australia, Indigenous-led eco-cultural tourism integrates storytelling, traditional ecological knowledge, and conservation practices. Visitors engage with Indigenous histories, land management techniques, and cultural rituals, generating income while promoting cross-cultural understanding. Crucially, successful initiatives maintain strict community control over narratives and visitor access to prevent cultural commodification (Zeppel, 2006). These enterprises highlight the dual nature of Indigenous knowledge as both cultural heritage and economic resource. When governance structures ensure benefit-sharing and respect for cultural boundaries, crafts and tourism can strengthen community pride and economic resilience. Conversely, externally driven tourism models often extract value while eroding cultural integrity, underscoring the importance of community leadership.

5.3. Community Forest Management and Conservation Economies

Community forest management offers a compelling example of Indigenous knowledge shaping conservation-based economies. In Mexico, Indigenous and community-owned forests cover millions of hectares and are managed through collective institutions rooted in customary law. These communities engage in sustainable timber harvesting, non-timber forest product collection, and eco-certification schemes that balance income generation with conservation. Studies indicate that community-managed forests in Mexico exhibit lower deforestation rates and higher local economic returns compared to privately managed concessions (Bray et al., 2006). In Nepal, Indigenous and local communities have played a central role in community forestry programs that devolve forest management rights to local user groups. Drawing on Indigenous ecological knowledge, these groups regulate harvesting, monitor forest health, and reinvest revenues into community development. The economic benefits include employment, income diversification, and improved access to forest resources, while ecological outcomes include forest regeneration and biodiversity protection (Agrawal & Ostrom, 2001).

5.4. Lessons Learned and Comparative Insights

Across these geographically diverse examples, several common lessons emerge. First, Indigenous knowledge generates the greatest economic benefits when embedded within community-controlled institutions. Whether in agriculture, crafts, or forestry, collective governance ensures that economic gains are distributed equitably and reinvested locally. Second, cultural integrity and economic success are mutually reinforcing rather than contradictory. Initiatives that respect Indigenous values and practices tend to be more resilient and socially accepted. Third, gender inclusion and intergenerational participation are critical determinants of success. Women's leadership and youth engagement enhance innovation, continuity, and social legitimacy. Finally, external support—whether from governments, NGOs, or markets—plays a facilitative rather than directive role in successful cases. Supportive policies, access to finance, and legal recognition amplify community capacities without displacing local agency (Berkes, 2018).

6. POLICY, LEGAL, AND INSTITUTIONAL PATHWAYS

The translation of Indigenous knowledge into sustainable economic empowerment is deeply contingent upon supportive policy, legal, and institutional frameworks. While Indigenous communities have long sustained viable economic systems grounded in local knowledge, their ability to do so in contemporary contexts is often constrained by legal invisibility, weak institutional recognition, and asymmetrical power relations. Effective pathways must therefore move beyond symbolic acknowledgment toward enforceable rights, inclusive governance, and ethical integration within national and global development agendas.

6.1. Legal Recognition of Indigenous Knowledge and Community Rights

Legal recognition is a foundational prerequisite for protecting Indigenous knowledge and enabling community-led economic development. Historically, colonial and postcolonial legal systems have marginalized or outright denied Indigenous customary laws, land tenure systems, and governance institutions. This marginalization has facilitated dispossession and undermined the economic base of Indigenous communities. Secure land and resource rights are particularly critical, as Indigenous knowledge is inseparable from territory and ecosystems. Empirical studies demonstrate that where Indigenous land rights are legally recognized, communities exhibit stronger economic outcomes and better environmental stewardship (RRI, 2015). Legal pluralism offers an important framework for recognition by acknowledging the coexistence of state law and customary law. Integrating customary tenure and governance systems into formal legal frameworks can enhance legal certainty while preserving community autonomy. However, recognition must be substantive rather than procedural. Tokenistic inclusion without enforcement mechanisms risks reinforcing inequality. Effective legal recognition requires participatory lawmaking, access to justice, and mechanisms that allow Indigenous communities to assert and defend their rights in economic transactions and development planning (ILO, 1989).

6.2. Intellectual Property, Benefit-Sharing, and Knowledge Sovereignty

Conventional intellectual property (IP) regimes pose significant challenges for the protection of Indigenous knowledge. IP systems are premised on individual authorship, novelty, and fixed expression, whereas Indigenous knowledge is collective, cumulative, and often transmitted orally. As a result, Indigenous knowledge frequently falls outside the scope of patent, copyright, or trademark protection, leaving it vulnerable to misappropriation. Cases of biopiracy—where corporations patent traditional medicinal knowledge or genetic resources without consent—illustrate these systemic shortcomings (Shiva, 2007). In response, international legal instruments have increasingly emphasized benefit-sharing and prior informed consent. The Convention on Biological Diversity (CBD) and its Nagoya Protocol establish obligations to ensure fair and equitable sharing of benefits arising from the use of genetic resources and associated traditional knowledge. These frameworks recognize Indigenous communities as rights-holders rather than mere stakeholders. However, implementation remains uneven, particularly at the national level, where capacity constraints and weak enforcement limit effectiveness (CBD, 2011).

6.3. Role of State Policy, NGOs, and International Institutions

State policy plays a decisive role in shaping the enabling environment for Indigenous knowledge–based economies. Public policies influence access to land, finance, markets, and infrastructure, all of which affect community-led enterprises. Inclusive policies that recognize Indigenous governance institutions, support cooperatives, and provide targeted capacity-building can amplify community initiatives. Conversely, development policies that prioritize extractive industries, large-scale agriculture, or infrastructure without Indigenous consent often erode local economies and knowledge systems (UNDP, 2023). Non-governmental organizations (NGOs) frequently act as intermediaries, providing technical assistance, advocacy, and access to networks. When operating in partnership with communities, NGOs can strengthen institutional capacity and facilitate knowledge exchange. However, NGO-led interventions also carry risks of dependency or agenda-setting that displaces community priorities. Best practices emphasize long-term partnerships, accountability to communities, and respect for Indigenous leadership. International institutions contribute through norm-setting, funding, and monitoring. Agencies such as the United Nations Development Programme, the Food and Agriculture Organization, and the World Bank increasingly recognize Indigenous knowledge as integral to sustainable development. International frameworks, including the United Nations Declaration on the Rights of Indigenous Peoples (UNDRIP), provide normative guidance on self-determination, land rights, and cultural integrity. While such instruments lack direct enforcement, they shape national policies and offer advocacy tools for Indigenous communities and civil society (UN, 2007).

6.4. Aligning Indigenous Economies with SDGs and Climate Goals

Aligning Indigenous knowledge–based economies with global sustainability agendas presents both opportunities and tensions. The Sustainable Development Goals (SDGs) emphasize inclusivity, poverty reduction, gender equality, and environmental sustainability—objectives that resonate strongly with Indigenous economic practices. For example, Indigenous agriculture and forest management contribute directly to SDG 2 (Zero Hunger), SDG 13 (Climate Action), and SDG 15 (Life on Land). Community-led economic models also advance SDG 8 (Decent Work and Economic Growth) by promoting locally rooted and equitable livelihoods. Climate policy represents a particularly significant arena for alignment. Indigenous knowledge has been widely recognized as essential for climate adaptation and resilience. Nature-based solutions, ecosystem restoration, and sustainable land management strategies often draw upon Indigenous practices. When climate finance mechanisms, such as REDD+ or adaptation funds, incorporate Indigenous governance and benefit-sharing, they can generate economic opportunities while strengthening stewardship. However, alignment must avoid instrumentalizing Indigenous knowledge solely as a technical input for global climate objectives, thereby sidelining community priorities (IPBES, 2019).

6.5. Risks of Appropriation and Over-Commercialization

Despite growing recognition, the integration of Indigenous knowledge into formal economic and policy frameworks carries substantial risks. Appropriation occurs when knowledge is extracted, commodified, or repackaged without consent or equitable benefit-sharing. Over-commercialization can distort cultural practices, prioritize market demand over community values, and create internal inequalities. For instance, tourism or craft enterprises driven by external markets may pressure communities to simplify or stereotype cultural expressions, undermining authenticity and social cohesion (UNESCO, 2017). These risks are exacerbated by power asymmetries between Indigenous communities and corporate or state actors. Without robust legal protections and institutional safeguards, economic integration may reproduce patterns of exploitation under the guise of inclusion. Safeguards such as free, prior, and informed consent (FPIC), community protocols, and culturally appropriate impact assessments are essential to mitigate these risks. Equally important is the recognition that not all forms of knowledge should or need to be commercialized.

7. INTEGRATING INDIGENOUS KNOWLEDGE WITH MODERN ECONOMIC SYSTEMS

The integration of Indigenous knowledge with modern economic systems presents both significant opportunities and complex challenges. While Indigenous economies have long operated outside formal market structures, contemporary realities—globalization, digitalization, and climate change—necessitate engagement with wider economic systems. The critical question is not whether integration should occur, but how it can be designed to enhance economic opportunities without undermining cultural integrity, knowledge sovereignty, and community autonomy. Hybrid models that respect epistemic plurality offer a promising pathway for achieving this balance. Hybrid economic models are characterized by the coexistence and interaction of Indigenous and market-based economic logics. Rather than replacing traditional practices, these models build upon them through co-creation approaches that involve Indigenous communities as equal partners in design and implementation. For example, Indigenous agricultural cooperatives may integrate traditional farming methods with modern logistics, quality certification, or value-chain development. Such co-creation allows communities to access broader markets while retaining control over production practices and governance. Research indicates that hybrid models are more resilient than purely market-driven enterprises because they draw on local knowledge, social capital, and adaptive capacity alongside modern infrastructure (Berkes, 2018). Co-creation also challenges the dominant narrative of technology transfer, which often positions Indigenous communities as passive recipients of innovation. Instead, innovation emerges through dialogue between Indigenous knowledge holders and technical experts, producing context-sensitive solutions. In economic terms, this approach enhances efficiency and sustainability by ensuring that new technologies complement rather than displace existing practices. Importantly, co-creation reinforces community agency, making integration a negotiated process rather than an imposed one.

Digital platforms represent a growing frontier in the integration of Indigenous knowledge with modern economies. E-commerce, digital storytelling, and blockchain-based traceability systems have been used to connect Indigenous producers with ethical consumers, enhance transparency, and protect authenticity. For instance, digital marketplaces allow Indigenous artisans to bypass intermediaries and reach global markets directly, increasing income while maintaining ownership over branding and narratives. Similarly, digital mapping tools and data platforms support community land management and resource monitoring, translating Indigenous ecological knowledge into formats recognized by policymakers and investors (UNDP, 2023). However, digital integration raises ethical concerns related to data ownership, consent, and representation. Indigenous data sovereignty has emerged as a critical principle, emphasizing community control over how data derived from Indigenous knowledge is collected, stored, and used. Without safeguards, digital platforms can replicate extractive dynamics by appropriating knowledge and cultural expressions. Ethical innovation frameworks that prioritize free, prior, and informed consent, collective ownership of data, and culturally appropriate governance are therefore essential for ensuring that digital tools serve community interests rather than external agendas. Market access is often presented as a primary pathway to economic empowerment, yet uncritical integration into markets can erode cultural practices and social relations. Indigenous economies traditionally operate through principles of reciprocity, sufficiency, and collective benefit, which may conflict with market pressures for standardization, scaling, and profit maximization. Successful integration requires selective engagement with markets, where communities determine the terms and limits of participation. Certification schemes, fair trade arrangements, and geographical indications can help protect cultural specificity and ensure ethical standards, provided they are community-driven and not externally imposed (UNESCO, 2017).

Avoiding cultural erosion also entails resisting homogenization. Markets often demand uniform products and narratives, which can marginalize diversity within Indigenous cultures. Community governance structures play a vital role in mediating market engagement, deciding which knowledge and practices are shared and which remain protected. Economic integration that respects these boundaries strengthens cultural resilience and sustains the social foundations upon which Indigenous knowledge rests. Capacity building and financial inclusion are enabling conditions for meaningful integration. Many Indigenous communities face structural barriers such as limited access to education, finance, and infrastructure. Capacity-building initiatives that are culturally responsive and community-led can enhance skills in areas such as enterprise management, digital literacy, and negotiation. Importantly, such initiatives should build upon existing knowledge systems rather than replacing them with standardized training models. Participatory learning approaches that value Indigenous expertise foster confidence and reduce dependency on external actors (Chambers, 1997).

Financial inclusion is equally critical. Conventional financial institutions often fail to accommodate collective ownership, seasonal income, and non-monetary forms of value prevalent in Indigenous economies. Alternative financial mechanisms—such as community savings groups, cooperative banks, and impact investment aligned with social and environmental goals—offer more appropriate solutions. When financial services are designed in collaboration with communities, they can support entrepreneurship and resilience without imposing unsustainable debt burdens (Kabeer, 1999). Safeguarding autonomy throughout integration processes is essential to prevent co-optation and loss of control. Autonomy is maintained when communities retain decision-making authority over economic activities, partnerships, and knowledge use. Mechanisms such as community protocols, Indigenous-led governance bodies, and legally recognized collective rights provide institutional safeguards. These mechanisms enable communities to negotiate equitable partnerships and withdraw from arrangements that conflict with their values or interests. Moreover, safeguarding autonomy requires recognizing that integration is not a linear or inevitable process. Communities may choose varying degrees of engagement with modern economic systems depending on context and priorities. Respecting this diversity challenges the assumption that economic integration is universally desirable. From a policy perspective, flexibility and pluralism are key. Enabling environments should support a range of economic pathways rather than prescribing a single model of success.

8. CHALLENGES, TENSIONS, AND ETHICAL CONSIDERATIONS

While Indigenous knowledge–based economic models offer promising alternatives to dominant development paradigms, they are not without significant challenges and ethical tensions. These challenges arise from structural inequalities, historical injustices, and contemporary pressures that shape how Indigenous communities engage with markets, states, and global institutions. Addressing these issues is essential to ensure that economic empowerment initiatives grounded in Indigenous knowledge do not reproduce forms of marginalization they seek to overcome (Pandey, 2025). Power asymmetries remain a central challenge in the integration of Indigenous knowledge into economic and development frameworks. Indigenous communities often engage with actors—such as governments, corporations, and international organizations—who possess disproportionate economic, political, and legal power. These asymmetries influence negotiations over land, resources, and knowledge, frequently resulting in outcomes that favor external interests. Even participatory development processes can be shaped by unequal power relations, where consultation substitutes for genuine consent. Such dynamics risk reducing Indigenous communities to symbolic stakeholders rather than recognizing them as rights-holders with authority over development decisions affecting their lives (UN, 2007). Closely linked to power asymmetries is the problem of knowledge extraction and neo-colonial development. Indigenous knowledge has increasingly been recognized for its economic and ecological value, particularly in sectors such as agriculture, pharmaceuticals, and climate adaptation. However, this recognition often translates into extractive practices where knowledge is documented, commodified, and deployed without meaningful benefit-sharing or community control. Scholars have characterized this process as a continuation of colonial patterns of resource extraction, wherein knowledge replaces land or minerals as the object of appropriation (Shiva, 2007). When Indigenous knowledge is treated as a resource to be mined rather than a living system embedded in culture and governance, economic initiatives risk undermining the very foundations of that knowledge.

Internal community inequalities present another layer of complexity. Indigenous communities are not homogeneous entities; they encompass differences based on gender, age, class, and social status. Economic initiatives that fail to account for these internal dynamics may inadvertently reinforce existing hierarchies. For example, commercialization of crafts or tourism may benefit community elites or those with greater market access, while marginalizing women, youth, or landless households. Similarly, leadership structures rooted in customary norms may exclude certain voices, particularly in contexts where patriarchy or generational hierarchies are entrenched. Ethical community-led development requires deliberate attention to inclusion and equity within communities, rather than assuming automatic collective benefit (Agarwal, 2010). Environmental pressures and climate change further complicate the ethical landscape. Indigenous knowledge systems have historically evolved in relatively stable ecological contexts, yet climate change is rapidly altering environmental conditions. Shifts in rainfall patterns, biodiversity loss, and extreme weather events threaten the viability of traditional livelihoods such as agriculture, fishing, and forest-based economies. While Indigenous knowledge contributes significantly to climate adaptation, there are limits to its capacity under accelerating environmental change. Economic strategies that rely solely on traditional practices without support for adaptation may place undue burdens on communities already facing ecological vulnerability (IPBES, 2019).

Moreover, climate and conservation policies can generate new forms of marginalization. Market-based mechanisms such as carbon offsets or conservation finance may restrict access to customary lands or impose external management regimes in the name of environmental protection. Without strong safeguards, such initiatives risk displacing Indigenous livelihoods and eroding governance autonomy. Ethical engagement requires aligning environmental objectives with Indigenous rights, ensuring that climate action enhances rather than undermines community resilience. Given these challenges, ethical frameworks for engagement are essential. Principles such as free, prior, and informed consent (FPIC) provide a foundational standard for respecting Indigenous autonomy in economic and development initiatives (Pandey, 2026). FPIC emphasizes consent as an ongoing process rather than a one-time procedural requirement. Complementary frameworks stress knowledge sovereignty, equitable benefit-sharing, transparency, and accountability. Community protocols and Indigenous-led governance mechanisms offer practical tools for articulating ethical boundaries and expectations in partnerships with external actors (Pandey et al., 2026). Ultimately, ethical engagement demands a shift in mindset from inclusion to justice. Rather than asking how Indigenous knowledge can be integrated into existing economic systems, the ethical question becomes how economic systems can be transformed to respect Indigenous rights, values, and worldviews. Addressing power asymmetries, preventing knowledge extraction, confronting internal inequalities, and responding to environmental pressures are not peripheral concerns but central to the integrity and sustainability of Indigenous knowledge–based economic pathways.

9. CONCLUSION AND SUGGESTIONS

This chapter has reframed economic prosperity through the lens of Indigenous knowledge and community empowerment, challenging dominant development paradigms that prioritize growth over equity, efficiency over ecology, and markets over meaning. By foregrounding Indigenous epistemologies, practices, and institutions, the analysis demonstrates that economic prosperity need not be extractive, homogenizing, or externally imposed. Instead, it can be relational, place-based, and rooted in collective well-being. Across the chapter, Indigenous knowledge has been shown to function not merely as cultural heritage, but as a dynamic economic asset embedded in agriculture, natural resource governance, artisanal production, and informal economies. Community-led models—such as cooperatives, commons-based enterprises, and customary governance systems—illustrate how economic agency is strengthened when decision-making authority remains local and inclusive. Particular attention to women's leadership and intergenerational knowledge transfer further highlights that empowerment is both a social and economic process, sustained through trust, reciprocity, and shared responsibility. The case examples discussed reveal that Indigenous economies are neither static nor opposed to innovation. On the contrary, they embody adaptive strategies that balance continuity with change, sustainability with livelihoods, and tradition with experimentation. However, these successes remain deeply context-specific, underscoring the limits of replication without sensitivity to cultural, ecological, and political conditions. Scalability, therefore, must be understood not as uniform expansion, but as the careful translation of principles rather than practices.

From a policy and institutional perspective, the chapter underscores the urgency of legal recognition, knowledge sovereignty, and equitable benefit-sharing frameworks. Without safeguards, integration into formal markets risks appropriation, over-commercialization, and renewed marginalization. Aligning Indigenous economies with broader sustainability and climate agendas—such as the Sustainable Development Goals—requires moving beyond instrumentalization toward genuine partnership and respect for autonomy. Ultimately, this chapter argues for a plural economic vision—one that recognizes multiple ways of knowing, producing, and valuing wealth. Indigenous knowledge systems offer not an alternative at the margins, but a foundation for reimagining prosperity itself. By centering empowerment, dignity, and ecological balance, they challenge policymakers, scholars, and practitioners to rethink development as a collaborative and ethical endeavor. In an era of widening inequality and ecological crisis, embracing such plural and inclusive economies is not only desirable, but essential for a just and sustainable future.

REFERENCES

Agarwal, B. (2010). *Gender and green governance: The political economy of women's presence within and beyond community forestry*. Oxford University Press. DOI: 10.1093/acprof:oso/9780199569687.001.0001

Agrawal, A., & Ostrom, E. (2001). Collective action, property rights, and decentralization in resource use in India and Nepal. *Politics & Society, 29*(4), 485–514. DOI: 10.1177/0032329201029004002

Altieri, M. A., & Nicholls, C. I. (2017). The adaptation and mitigation potential of traditional agriculture in a changing climate. *Climatic Change, 140*(1), 33–45. DOI: 10.1007/s10584-013-0909-y

Altieri, M. A., & Toledo, V. M. (2011). The agroecological revolution in Latin America. *The Journal of Peasant Studies, 38*(3), 587–612. DOI: 10.1080/03066150.2011.582947

Berkes, F. (2018). *Sacred ecology* (4th ed.). Routledge.

Borrini-Feyerabend, G.. (2013). *Governance of protected areas: From understanding to action*. IUCN.

Bray, D. B., Merino-Pérez, L., & Barry, D. (2006). *The community forests of Mexico*. University of Texas Press.

Chambers, R. (1997). *Whose reality counts? Putting the first last*. Intermediate Technology Publications. DOI: 10.3362/9781780440453

Convention on Biological Diversity. (2011). *Nagoya Protocol on access to genetic resources and the fair and equitable sharing of benefits arising from their utilization*. CBD Secretariat.

Escobar, A. (2012). *Encountering development: The making and unmaking of the Third World* (2nd ed.). Princeton University Press. DOI: 10.1515/9781400839926

Food and Agriculture Organization. (2019). *Indigenous peoples' food systems: Insights on sustainability and resilience*. FAO.

Gudeman, S. (2001). *The anthropology of economy: Community, market, and culture*. Blackwell.

Intergovernmental Science-Policy Platform on Biodiversity and Ecosystem Services. (2019). *Global assessment report on biodiversity and ecosystem services*. IPBES.

International Cooperative Alliance. (2016). *Cooperative identity, values and principles*. ICA.

International Labour Organization. (1989). *Indigenous and Tribal Peoples Convention, 1989 (No. 169)*. ILO.

Kabeer, N. (1999). Resources, agency, achievements: Reflections on the measurement of women's empowerment. *Development and Change, 30*(3), 435–464. DOI: 10.1111/1467-7660.00125

Ostrom, E. (1990). *Governing the commons: The evolution of institutions for collective action*. Cambridge University Press. DOI: 10.1017/CBO9780511807763

Ostrom, E. (2010). Beyond markets and states: Polycentric governance of complex economic systems. *The American Economic Review, 100*(3), 641–672. DOI: 10.1257/aer.100.3.641

Pandey, A. (2025). The Transformative Potential of Artificial Intelligence in Building Sustainable Agricultural Systems: Innovations, Challenges, and Future Directions. In *AI and Ecological Change for Sustainable Development* (pp. 161-182). IGI Global Scientific Publishing.

Pandey, A. (2026). The Intersection of Industrial Policy and Environmental Regulation: Balancing Growth and Green Goals. In *Industrial Policy for Resilience, Equity, and Sustainability* (pp. 339-362). IGI Global Scientific Publishing.

Pandey, A., Joseph, J., & Mishra, A. (2026). Advancing Sustainable Agriculture Through Artificial Intelligence: Harvesting Greener Future. In *Artificial Intelligence Enabled Real Time Environmental Monitoring* (pp. 19–41). Springer Nature Singapore. DOI: 10.1007/978-981-95-3767-9_2

Putnam, R. D. (1993). *Making democracy work: Civic traditions in modern Italy*. Princeton University Press.

Rights and Resources Initiative. (2015). *Who owns the world's land?* RRI.

Sen, A. (1999). *Development as freedom*. Oxford University Press.

Shiva, V. (2007). *Biopiracy: The plunder of nature and knowledge*. South End Press.

UNESCO. (2017). *Culture for sustainable development*. UNESCO Publishing.

United Nations. (2007). *United Nations Declaration on the Rights of Indigenous Peoples*. UN General Assembly.

United Nations Development Programme. (2023). *Indigenous peoples and sustainable development*. UNDP.

Warren, D. M. (1991). Using Indigenous knowledge in agricultural development. *World Bank Discussion Papers*, No. 127.

Zeppel, H. (2006). *Indigenous ecotourism: Sustainable development and management*. CABI. DOI: 10.1079/9781845931247.0000

Chapter 13
Correcting Knowledge Distortions:
An Analysis of How Indigenous Knowledge Can Redefine South Africa's Innovation and Economic Progress

Sandile Tyrant Nhuvunga
http://orcid.org/0009-0003-3776-4813
University of Mpumalanga, South Africa

Felix Sandile Thwala
http://orcid.org/0000-0001-7548-3789
University of Mpumalanga, South Africa

ABSTRACT

South African innovation discussions remain largely informed by Western epistemological models. These dominant models have created epistemic distortions that marginalise Indigenous Knowledge Systems (IKS) and constrain the ability of South Africa to locally define innovation and economic progress. This chapter aims to critically examine how Western knowledge models continue to influence South Africa's innovation pathways. Furthermore, the chapter will examine how Western knowledge models devalue community-centred forms of innovation. Drawing on qualitative secondary data, the chapter further explores how colonial legacies have undermined African epistemologies. Preliminary findings suggest that indigenous worldviews, particularly those expressed in Afrikology and relational African philosophies, provide alternative models of innovation. The findings also reveal structural and policy gaps that hinder the inclusion of IKS within dominant innovation frameworks. The chapter concludes by proposing decolonial strategies for reframing innovation through African-centred knowledge systems.

1. INTRODUCTION

Innovation is widely regarded as a key driver of development, yet the question of what constitutes innovation remains deeply shaped by historically embedded epistemological hierarchies. In South Africa,

DOI: 10.4018/979-8-3373-7976-0.ch013

innovation discourse is structured around Eurocentric metrics, such as research and development (R&D) intensity, patent generation, and technological productivity (Amankwah-Amoah & Medase, 2024). While influential, these frameworks reproduce a narrow, technocratic understanding of progress rooted in industrial capitalism (Afolabi, 2020). As a result, Western science is positioned as the universal standard of knowledge, while African knowledge systems are relegated to the margins (Heleta, 2018). This imbalance reflects the enduring legacy of cognitive colonisation. This colonisation is a global regime that defines what is legitimate knowledge and who qualifies as its producer (Ndlovu-Gatsheni, 2021). Despite post-apartheid reform efforts, South Africa's National System of Innovation (NSI) continues to rely on metrics derived from the Organisation for Economic Co-operation and Development (OECD), reinforcing dependency on external epistemic models (Oguamanam, 2022). This reliance reproduces what de Sousa Santos (2015) conceptualises as epistemicide, which is the systematic erasure of alternative knowledge systems.

This epistemic distortion extends beyond policy metrics into intellectual property regimes and global innovation governance structures (Amankwah-Amoah & Medase, 2024). Patent systems in postcolonial contexts continue to privilege Western models, marginalising indigenous innovation (Foster, 2016). In contrast, African philosophical frameworks, such as Ubuntu and Afrikology, conceptualise innovation as a relational and generative process, grounded in ecological balance, reciprocity, and collective wellbeing (Jimenez, Delgado, Merino & Argumedo, 2022). These perspectives challenge the dominant capitalist logic that equates innovation with material accumulation, instead foregrounding sustainability and social continuity. This study argues that addressing this imbalance requires not merely the inclusion of indigenous knowledge, but the epistemic reconstitution of innovation systems themselves.

This chapter argues that Western epistemological dominance distorts South Africa's innovation system by marginalising Indigenous Knowledge Systems and constraining alternative development pathways. These distortions are reflected in institutional hierarchies, fragmented policy frameworks, and educational systems that prioritise imported knowledge norms. To address this, the chapter examines how Western models continue to shape innovation discourse and policy, while advancing Indigenous frameworks, particularly Afrikology and Ubuntu, as alternative epistemic foundations for innovation. The chapter further proposes decolonial strategies for institutional and policy reform to enable a re-Africanised innovation paradigm. In doing so, it contributes to the broader argument that epistemic autonomy is a prerequisite for contextually relevant, sustainable, and inclusive innovation.

2. THEORETICAL FRAMEWORK

This chapter employs the Decolonial Theory to analyse how Indigenous Knowledge can redefine innovation and economic progress in South Africa. This framework provides a theoretical lens for understanding the persistence of Western knowledge dominance and for conveying pathways towards the restoration of indigenous knowledge sovereignty.

2.1 Decolonial Theory

Decolonial theory emerged from Latin American and African intellectual traditions that view modernity and coloniality as mutually constitutive, rather than sequential (Quijano, 2000). Their view explains why colonial hierarchies persist after formal independence (Quijano, 2024). Within this formation,

Quijano (2000)'s notion of the coloniality of power names the persistent medium that organises race, labour, authority, and knowledge on a global scale, anchoring Eurocentrism as the specific rationality of modern domination. Mignolo (2007) expands on the idea of delinking, which is a way of rejecting Western dominance over knowledge. They also promote pluriversality as a goal in which multiple cultures and ways of knowing can coexist without being forced into a single perspective (Mignolo, 2007). In the African context, Ndlovu-Gatsheni (2019)'s call for epistemic freedom and Mbembe (2016)'s programme for decolonising the university show how coloniality continues to structure knowledge systems, institutions, and development agendas in South Africa and beyond. The perspectives indicate that decoloniality can be defined as a deliberate project of delinking from Western knowledge dominance and cultivating pluriversal ecologies of knowledge that make room for locally grounded knowledge to guide ideas of innovation, progress, and development (Escobar, 2018).

Decolonial theory is important because it reveals how global systems of knowledge in science, innovation, economics, and technology remain tied to Eurocentric historical narratives and interests (Mbembe, 2016). The theory supplies a rigorous lens for explaining why many African innovations, practices, and knowledge systems are undervalued within dominant academic and policy frameworks by tracing how standards, research agendas, and evaluation metrics reproduce knowledge domination (Heleta & Chasi, 2024; Mbembe, 2016). In practice, decolonial perspectives inform curriculum reform and Indigenous Knowledge work, reframe research and policy, and support critiques of intellectual property regimes that marginalise non-Western innovation (Amankwah-Amoah & Medase, 2024). In this chapter, Decolonial Theory provides the conceptual basis for analysing how South Africa's innovation policy privileges Eurocentric metrics, which dominate official indicator frameworks and much empirical work on innovation and growth. Using the Decoloniality Framework, the chapter probes the knowledge hierarchies that marginalise Indigenous Knowledge Systems (IKS) across universities, policy, and evaluation, thereby constraining the forms of innovation that become legible, fundable, and scalable. This chapter looks at how current intellectual property policies and technology standards affect the recognition and value of innovations. It often prioritises Western notions of innovation while overlooking innovations rooted in local communities. Moreover, through decolonial theory, this chapter redefines innovation as a practice shaped by local cultures, histories, and economic conditions, employing decolonial strategies such as delinking and pluriversality. It proposes design principles for creating innovation pathways tailored to the South African context. In this way, Decoloniality functions as both an analytical tool and a transformative lens, enabling the chapter to trace and correct knowledge distortions and to model alternative, African-grounded accounts of economic progress.

3. METHODOLOGY

This study employs a qualitative systematic review to interpret how South Africa's innovation discourse is shaped by colonial knowledge formations and to examine how Indigenous Knowledge Systems (IKS) and Afrikology propose alternative models of innovation and development. Reporting follows clear transparency by providing eligibility criteria, search strings, and screening stages, ensuring a complete account of what was done and found. Searches covered multidisciplinary databases and grey literature; inclusion criteria focus on South Africa publications, policy texts, and empirical studies from 2000 to 2025, engaging innovation, IKS, or decoloniality. The synthesis followed two linked stages: (1) a thematic synthesis with line-by-line coding, construction of descriptive themes, and generation of analytical themes to move beyond aggregation; and (2) a critical interpretive synthesis to build synthetic constructs that explain how Eurocentric indicators become authoritative and how they can be reframed. This chapter looked at policies and academic writings as important cultural objects. The chapter used critical discourse analysis to show how certain linguistic practices make specific ideas about progress and authority seem normal in innovation policy and research. Following a decolonial approach, the chapter includes self-reflective practices and emphasises fairness in knowledge. The chapter drew on decolonial theory to focus on the researcher's perspective, incorporate Indigenous knowledge, and examine innovation rooted in the community. The literature search focused on three questions: (i) How does the dominance of Western knowledge models influence innovation frameworks and policy in South Africa? (ii) What contributions do IKS and Afrikology make to alternative models of innovation? (iii) What barriers inhibit the meaningful inclusion of IKS within national development strategies?

A purposive sampling strategy was adopted to construct a focused yet conceptually diverse body of texts that could credibly support discourse-oriented policy interpretation. Selection decisions were guided by four interrelated criteria. First, relevance required that each text explicitly engage with innovation, IKS, and decolonial frameworks within the South African context, ensuring direct alignment with the study's research focus. Second, credibility was prioritised by restricting inclusion to peer-reviewed academic publications, official government policy instruments, and reports produced by reputable institutions, thereby strengthening the authority and trustworthiness of the evidence base. Third, the temporal scope centred primarily on the post-1994 democratic period to capture contemporary policy shifts and institutional priorities, while selectively incorporating pre-1994 sources where necessary to trace colonial and apartheid-era continuities that continue to shape epistemic hierarchies. Fourth, thematic breadth was intentionally built into the sample through the inclusion of materials spanning economic policy, education, environmental management, and community-based innovation, allowing the analysis to reflect the multidimensional character of IKS and its intersections with development and governance.

Between 2000 and 2025, scholarly attention to IKS has intensified in response to calls for knowledge justice, sustainable development, and culturally grounded economic pathways in post-apartheid South Africa. This chapter synthesised evidence from Google Scholar searches, yielding 400 unique publications that document the complex relationships between Indigenous knowledge and economic innovation in South Africa; 41 met the criteria and were selected for this chapter. The study adopted a thematic analysis design to systematically interpret the selected publications, drawing on Braun and Clarke's (2006) six-phase model: data familiarisation, coding, theme generation, theme review, theme definition, and reporting. Analysis proceeded iteratively and reflexively, with the researchers moving back and forth between textual excerpts and the study's theoretical commitments to surface the underlying epistemic assumptions that shape how innovation is constructed, legitimised, and governed in South Africa. To ensure both conceptual alignment and openness to empirical nuance, the coding frame was developed through a combined deductive-inductive strategy: key sensitising concepts were derived deductively from the conceptual framework, while additional codes and sub-themes were generated inductively as patterns emerged across the corpus. Four primary analytical categories organised this process: knowledge hierarchy, innovation logic, relationality and sustainability, and policy gaps and opportunities. This orientation enabled close interrogation of the linguistic and ideological work performed by recurrent policy terms such as competitiveness, commercialisation, and global integration, while also foregrounding how these vocabularies often sit in tension with IKS-inflected notions of communal wellbeing, balance, and relational interdependence.

4. RESULTS

The analysis identifies five interconnected themes demonstrating that Indigenous Knowledge Systems (IKS) function both as a corrective to colonial knowledge hierarchies and as a practical foundation for diverse forms of innovation and economic activity.

4.1. Decolonising Curricula and Entrepreneurship Education

The findings indicate that IKS are increasingly positioned as a corrective to entrenched Western curricula norms and as foundational to culturally responsive entrepreneurship education in postcolonial contexts (Makokotlela & Gumbo, 2025). Curriculum reform initiatives in South African universities frame the integration of IKS as a strategy for redressing epistemic imbalances by embedding local knowledge, lived experience, and community contexts into teaching and assessment practises (Dyantyi, 2025). This integration enables entrepreneurship education to move abstract, imported models toward conceptually grounded forms of problem-solving and venture creation (Abosede, Ebie, Somoye & Ashidi, 2025; Ayandibu, 2025). The findings further show that IKS-informed curricula enhance learner agency by enabling students to develop culturally aligned ventures and community-based solutions that leverage local knowledge assets rather than replicating externally imposed frameworks (Ayandibu, 2025). These interventions are not merely additive but represent an additional measure and represent a fundamental shift in validating IKS as a legitimate and necessary component of innovation systems.

The findings also highlight the importance of gender-sensitive and context-specific approaches to IKS integration, particularly in addressing the structural inequalities that shape access to entrepreneurial opportunities (Aboede et al., 2025; Ayandibu, 2025). Without such consideration, IKS integration risks reproducing existing power imbalances, including patriarchal structures and unequal access to resources (Irene, Lockyer & Okrah, 2025). For IKS-informed entrepreneurship education to be effective, institutional commitment must extend beyond symbolic inclusion toward mechanisms that translate learning into tangible economic and community outcomes (Dyantyi, 2025; Irene et al., 2025). This includes faculty development, partnerships with community knowledge holders, and assessment frameworks that recognise broader indicators of success such as social impact, cultural preservation, and ecological sustainability. Collectively, these findings support the argument that curriculum transformation is central to reconstituting innovation systems in ways that are contextually relevant and socially embedded.

4.2. Grassroots Innovation and Rural Livelihoods

The findings demonstrate that indigenous practices function as grassroots innovation systems embedded within rural livelihoods. Evidence from the Tsonga traditional healthcare systems shows that indigenous health practices operate not only as therapeutic systems but also as economic activities that generate income, create employment and mobilise community resources (Ekene, Lee, Decide & Thobejane, 2023). These practices embody entrepreneurial principles such as service provision, skill transmission, and resource stewardship, while requiring minimal capital investment. This aligns with broader findings that IKS-based enterprises provide low-capital, knowledge-intensive livelihood strategies in contexts where formal employment is limited (Ndhlovu, Omotayo, Olagunju, Otang-Mbeng & Aremu, 2025; Makgamatha, 2025).

Despite their potential, IKS-based enterprises face significant structural constraints, including limited market access, weak formal recognition, and inadequate institutional support (Makgamatha, 2025). These challenges are compounded by regulatory frameworks that are misaligned with indigenous practices, restricting access to formal markets and intellectual property protections (Dahlberg & Trygger, 2009). As a result, the scalability of these innovations remains constrained. These findings reinforce the argument that IKS constitutes a viable foundation for grassroots innovation and rural economic development. However, its impact remains contingent on institutional frameworks that recognise and support its unique characteristics.

4.3. Food and Agricultural Security

The findings demonstrate that indigenous agricultural knowledge constitutes an ecological and livelihood innovation that strengthens food security in resource-constrained contexts. Evidence indicates that techniques such as intercropping, organic soil fertility management, seed preservation, and agroecological pest control enhance soil health, stabilise yields, and increase dietary diversity (Ekobi, Mboh & Tanga, 2023; Kom, Nicolau & Nenwiini, 2024). These practices shift the locus of innovation away from capital-intensive inputs toward knowledge-intensive, locally embedded strategies. This aligns with broader findings that indigenous agricultural systems improve household resilience and reduce dependency on external inputs in smallholder settings (Ndwandwe & Mudhara, 2014; Garutsa & Nekhwevha, 2018). In this sense, IKS contributes to innovation not only through enterprise creation but through the transformation of production systems and resource management practices.

However, the sustainability of these systems is increasingly threatened by disruptions in intergenerational knowledge transfer. Factors such as urbanisation, demographic shifts, and limited formal recognition contribute to the erosion of indigenous agricultural knowledge (Balogun, 2023). This challenge is compounded by the marginalisation of IKS within formal agricultural extension systems, which often prioritise externally derived techniques over local practices. These findings reinforce the argument that indigenous agricultural knowledge represents a crucial yet vulnerable component of innovation systems.

4.4. Transdisciplinary Collaboration and Biodiversity Conservation

Transdisciplinary collaboration in South Africa's biodiversity conservation demonstrates that innovation emerges from the integration of diverse knowledge systems, emphasising that solutions are more effective when they draw on a multiplicity of perspectives rather than rely on a single dominant discipline. The findings indicate that nature-based coastal interventions that effectively combine indigenous knowledge systems (IKS) with ecological science yield solutions that are both technically sound and socially relevant (Porri, McConnachie, van der Walt, Wynberg & Pattrick, 2023). This synthesis of knowledge not only enhances the efficacy of environmental interventions but also contributes to a more holistic understanding of ecological challenges, which aligns with the existing literature on knowledge co-production. It illustrates how collaborative efforts can reconfigure power dynamics within innovation systems, positioning communities as active co-designers rather than passive beneficiaries (Ngubane & Makua, 2021; Dladla, 2024).

Furthermore, these interventions challenge the prevailing assumption that innovation must be high-tech or market-driven. They illustrate that low-cost, locally rooted practices, such as restoring indigenous plants and promoting community-led conservation, reveal a broader conception of innovation that encompasses ecological, social, and institutional dimensions (Caswell & Jang, 2024). This perspective shifts the innovation narrative to include context-specific practices that resonate deeply with local cultures and ecosystems, underscoring the importance of place-based strategies. The integration of cultural practices and progressive education underscores that innovation occurs across various knowledge, identity, and learning systems, extending beyond the confines of scientific or commercial domains (Kulundu, McGarry & Lotz-Sisitka, 2020). This broader framework encourages a reevaluation of how knowledge is generated and shared, advocating for an inclusive approach to education that respects and incorporates diverse epistemologies.

However, the research findings also underscore that the success of such a transdisciplinary approach hinges on ethical knowledge governance, which includes equitable benefit-sharing and the recognition of indigenous intellectual property (Chisa & Hoskins, 2016). The need for these safeguards is critical, as they prevent the entrenchment of extractive dynamics that can reinforce rather than rectify existing knowledge distortions. This warns of the risks of appropriation and exploitation in collaborative frameworks, stressing that without robust ethical guidelines, well-intentioned transdisciplinary efforts may inadvertently perpetuate historical injustices rather than foster genuine empowerment and equity.

4.5. Policy and Enterprise Linkage

The findings reveal a persistent misalignment between formal recognition of Indigenous Knowledge Systems (IKS) and their practical integration into innovation and economic systems. Legislative frameworks such as the Indigenous Systems Policy (2004) and the Protection, Promotion, Development and Management of Indigenous Knowledge Act (2019) establish important foundations for the recognition of IKS; however, their effectiveness is constrained by weak implementation, institutional fragmentation, and limited contextual adaptation (Wynberg, 2017; Makgamatha, 2025). This misalignment not only hampers the potential benefits of IKS but also highlights a critical gap in bridging traditional knowledge and contemporary innovation practices. This situation aligns with broader critiques of inclusive innovation articulated by Chiwawa & Wissink (2023), who demonstrate that these limitations are fundamentally epistemic. Policy frameworks rooted in market-oriented and individualised models often overlook the collective and relational nature of IKS, rendering them insufficient for fostering genuine inclusivity. As a result, IKS innovation is frequently marginalised, particularly in rural contexts where institutional support is the weakest (Mann-Lang, Branch, Mann, Sink, Kirkman & Adams, 2021). Thus, it becomes evident that without meaningful integration of IKS into innovation practices, the potential for diverse and resilient economic systems remains largely untapped.

Importantly, the findings reinforce that innovation cannot be reduced to entrepreneurship alone. While enterprise development is one pathway, IKS-based innovation also manifests through non-market systems such as community governance, cultural production, and ecological stewardship. Effective policy must therefore transcend legal recognition and move toward integrated, multilevel support systems. This includes fostering institutional coordination, implementing culturally appropriate market mechanisms, and adopting community-centred implementation strategies (Mtapuri & Giampiccoli, 2019; Oguamanam, 2020). Therefore, the policy-enterprise linkage presents not only an economic function but also serves as a broader process of correcting knowledge distortion. By aligning innovation systems with local realities, cultural values, and sustainable development imperatives, policymakers can enhance the integrity and effectiveness of IKS.

5. DISCUSSION

The central implication of this study is that the marginalisation of Indigenous Knowledge Systems (IKS) within South Africa's innovation landscape is not merely an issue of underrepresentation but a consequence of deeply embedded epistemic structures that define the boundaries of legitimate knowledge. Contemporary innovation systems are historically rooted in Western intellectual traditions that prioritise abstraction, universality, and standardisation principles that have been institutionalised through education systems, policy frameworks, and global economic metrics. Within this configuration, knowledge that is context-specific, orally transmitted, collectively owned, and relationally grounded, hallmarks of IKS, is systematically devalued or rendered invisible. This suggests that the exclusion of IKS is not incidental but structurally reproduced through what may be understood as epistemic gatekeeping, where dominant institutions regulate not only the production of knowledge but also its recognition and application. From this perspective, efforts to integrate IKS into curricula or innovation systems must be critically interrogated, as they often operate within pre-existing frameworks that remain epistemically unchanged. My position is that such approaches risk producing superficial inclusion, in which Indigenous knowledge is acknowledged symbolically but constrained in practice. A more substantive transformation requires destabilising the underlying assumptions of what constitutes knowledge, expertise, and innovation. This shifts the focus from inclusion within existing systems to the reconstitution of those systems themselves, positioning IKS not as an adjunct but as a co-equal epistemic foundation capable of reshaping innovation discourse at its core.

A second layer of analysis emerges from interrogating the ontological assumptions that underpin dominant conceptions of innovation. The prevailing view equates innovation with technological advancement, firm creation, and market expansion, an understanding that is deeply aligned with capitalist modes of production and valuation. However, the findings of this study suggest that such a framework is conceptually narrow and analytically insufficient, particularly in contexts where livelihoods, sustainability, and social cohesion are not mediated primarily through formal markets. Indigenous systems of knowledge and practice reveal an alternative ontology in which innovation is distributed, embedded, and relational, operating through networks of social exchange, ecological adaptation, and cultural continuity. This raises a critical epistemological tension: if innovation is recognised only when it produces measurable economic outputs, then forms of value that sustain communities, such as food sovereignty, biodiversity conservation, and intergenerational knowledge transmission, are excluded from formal recognition. In my view, this reflects a deeper issue within innovation theory itself, namely its reliance on a market-centric epistemology that privileges commodification as the primary indicator of value. Such a framework is not only exclusionary but also analytically limiting, as it fails to capture the full spectrum of innovation practices beyond formal economic systems. Therefore, the contribution of IKS lies not only in offering alternative practices but in compelling a redefinition of innovation as a multidimensional phenomenon, encompassing economic, social, ecological, and cultural forms of value creation. This reconceptualisation is essential for developing innovation systems that are both inclusive and responsive to the complexities of postcolonial contexts.

The dynamics of transdisciplinary collaboration further illuminate the complexities of integrating diverse knowledge systems into innovation processes. While co-production is often framed as a progressive and inclusive approach, the findings suggest that such collaborations are embedded within asymmetrical power relations that shape whose knowledge is prioritised, how it is interpreted, and who ultimately benefits. Scientific institutions, by virtue of their formal authority and access to resources, often retain epistemic dominance, even within ostensibly collaborative frameworks. This creates a paradox in which Indigenous knowledge is included but simultaneously subordinated, translated into forms that align with scientific paradigms while losing its original context and meaning. The implication is that collaboration alone does not guarantee epistemic justice; rather, it must be accompanied by a deliberate restructuring of the conditions under which knowledge is produced and shared. Ethical frameworks such as intellectual property protection, benefit-sharing, and participatory governance are necessary but insufficient if they are not embedded within institutional practice. What is required is a shift toward epistemic reciprocity, where knowledge systems engage on mutually defined terms, and where indigenous epistemologies are recognised as possessing their own internal logics, validation mechanisms, and modes of application. This entails moving beyond the instrumental use of IKS as a resource for innovation toward recognising it as a sovereign system of knowledge. Without such a shift, transdisciplinary collaboration risks reproducing extractive dynamics in which Indigenous knowledge is appropriated under the guise of inclusion, thereby reinforcing rather than dismantling existing epistemic hierarchies.

At the policy level, the analysis reveals that the limitations of current frameworks extend beyond implementation issues and reflect a deeper conceptual misalignment between policy design and the nature of Indigenous knowledge systems. While South Africa has made significant strides in formally recognising IKS through legislative instruments, these policies remain grounded in assumptions that are incompatible with the collective, adaptive, and relational characteristics of Indigenous knowledge. For instance, policy mechanisms that emphasise individual ownership, formal intellectual property regimes, and linear models of enterprise development fail to account for the communal and context-specific nature of IKS. This results in a situation where policies exist in principle but are ineffective in practice, as they cannot meaningfully engage with the systems they are intended to support. I argue that this reflects a form of epistemic policy inertia, where institutional frameworks evolve at the level of discourse but remain anchored in paradigms that limit their transformative potential. Addressing this requires more than improved implementation; it necessitates a fundamental rethinking of policy objectives, instruments, and evaluation criteria. Policies must be designed to accommodate multiple forms of value, recognise collective rights, and support innovation pathways that do not conform to conventional market logics. This includes developing alternative metrics of success that capture social cohesion, ecological sustainability, and cultural continuity, rather than relying solely on economic indicators. In doing so, policy can move from being a mechanism of formal recognition to a catalyst for systemic transformation, enabling the meaningful integration of IKS into national innovation systems

6. RECOMMENDATIONS

Guided by the findings and the broader reconceptualisation of innovation advanced in this study, the following recommendations are proposed:

6.1 Reframing education systems to embed Indigenous Knowledge Systems as foundational to innovation.

Embedding IKS requires more than curricular inclusion; it necessitates a reorientation of pedagogical frameworks to recognise multiple ways of knowing and innovating. Core modules should be redesigned around community-defined knowledge systems, incorporating co-teaching with recognised knowledge custodians and experiential learning within local contexts. Assessment criteria must extend beyond commercial outputs to value ecological stewardship, cultural continuity, collective well-being, and adaptive problem-solving. Such an approach cultivates graduates who are not only entrepreneurial but also capable of engaging in innovation as a socially embedded, context-responsive practice.

6.2 Supporting the continuity and evolution of Indigenous knowledge practices beyond narrow enterprise logics.

While enterprise development remains important, policy and institutional support should also recognise non-market forms of innovation embedded in community life. This includes safeguarding knowledge through culturally appropriate documentation, strengthening systems of intergenerational transfer, and enabling community governance structures to retain authority over how knowledge is used and shared. Where commercialisation occurs, it should be guided by principles of collective ownership, fair benefit-sharing, and cultural integrity, rather than purely profit-driven models. This ensures that innovation rooted in IKS remains socially anchored and not extractively integrated into external systems.

6.3 Strengthening Indigenous agroecological systems as sites of innovation, resilience, and knowledge production.

Agroecology should be recognised not only as a livelihood strategy but as a dynamic innovation system grounded in ecological knowledge and adaptive practices. Interventions should prioritise intergenerational training and apprenticeship models, the establishment of community seed banks, and participatory knowledge exchange platforms that enable local experimentation and adaptation. Supporting women-led cooperatives and collective production systems is essential for enhancing social cohesion and resilience. Linking these systems to nutrition-sensitive public procurement (e.g., schools and clinics) and territorially embedded markets can translate local innovation into sustainable food systems, while preserving biodiversity and reinforcing cultural food practices.

6.4 Reconfiguring policy frameworks to align with the relational and collective nature of Indigenous Knowledge Systems.

Bridging the gap between national IKS policy and local implementation requires more than administrative coordination; it demands a shift in how policy conceptualises knowledge, value, and innovation. Provincial and local structures should be capacitated to provide integrated support that includes legal recognition of collective rights, facilitation of participatory governance, and access to context-appropriate resources. Rather than relying solely on conventional instruments such as intellectual property regimes and enterprise incentives, policy should incorporate alternative mechanisms that recognise communal ownership, support knowledge commons, and enable plural pathways of innovation. Public procurement, funding models, and evaluation frameworks must be redesigned to reflect diverse forms of value, including social cohesion, ecological sustainability, and cultural resilience.

7. CONCLUSION

This study contends that the role of Indigenous Knowledge Systems (IKS) in South Africa extends beyond questions of inclusion into existing innovation frameworks and instead necessitates a fundamental reconstitution of the epistemic foundations upon which innovation itself is defined and operationalised. The findings demonstrate that IKS functions as a generative and context-responsive system of knowledge, capable of producing socially embedded, ecologically attuned, and economically relevant innovations across domains such as education, agroecology, biodiversity conservation, and enterprise development. However, its marginalisation persists not due to a lack of efficacy, but as a result of institutionalised knowledge hierarchies that privilege formalised, market-oriented, and technologically driven models of innovation, thereby constraining what is recognised as legitimate knowledge and value. Addressing this requires a shift from additive inclusion toward systemic transformation, where education systems are redesigned to enable epistemic co-production, policy frameworks move beyond symbolic recognition toward operational implementation, and enterprise ecosystems are reconfigured to support collective ownership, culturally grounded business models, and alternative value chains. In advancing this argument, the study positions innovation as a pluriversal construct, challenging its reduction to entrepreneurship and technological advancement by foregrounding diverse forms of value rooted in social cohesion, ecological sustainability, and cultural continuity. Ultimately, correcting knowledge distortions emerges as both an epistemic and developmental imperative, requiring the alignment of legal, institutional, and socio-economic systems to reflect the realities and capacities of Indigenous knowledge holders. Such a reorientation does not displace scientific knowledge but situates it within a broader, more inclusive knowledge ecology, thereby enabling the emergence of an innovation paradigm that is not only more equitable and contextually relevant but also essential for achieving sustainable development, resilience, and epistemic justice in South Africa.

8. LIMITATIONS

While this chapter is theoretically grounded, it is important to acknowledge the following limitations that influence the scope and applicability of its findings.

8.1 Methodological Reliance on Secondary Data

A central limitation of this study lies in its reliance on secondary data and interpretive synthesis, which, while appropriate for advancing conceptual and theoretical arguments, constrains direct engagement with the lived, situated realities of Indigenous knowledge practitioners. This methodological orientation risks reproducing a degree of epistemic distance, as Indigenous Knowledge Systems are interpreted through existing literature rather than through co-produced, contextually embedded inquiry. Although the study seeks to challenge dominant knowledge hierarchies, the absence of primary, participatory research limits its ability to fully capture the dynamic, experiential, and relational dimensions of IKS in practice. This highlights the importance of future methodological approaches that prioritise immersive, community-engaged, and co-authored research designs, capable of aligning epistemic critique with empirical depth.

8.2 Contextual Specificity and Limited Generalisability

The study's focus on South Africa provides necessary contextual depth, particularly given the country's relatively advanced policy frameworks and institutional recognition of IKS. However, this specificity simultaneously constrains the generalisability of findings across the diverse epistemic and socio-political landscapes of the African continent. Variations in colonial histories, governance structures, cultural systems, and knowledge traditions suggest that the dynamics identified here may not be uniformly applicable. As such, the conclusions drawn must be understood as contextually grounded rather than universally representative, underscoring the need for broader comparative scholarship. This limitation, however, can also serve as an analytical strength, foregrounding the importance of context-sensitive theorisation in the study of innovation systems and epistemic transformation.

9. FUTURE RESEARCH

Building on the findings of this chapter, several avenues for future research emerge that could strengthen both theory and practice in the decolonisation of innovation systems.

9.1 Comparative Studies Across African Contexts

Future research should prioritise cross-national comparative studies examining how different African states conceptualise and integrate Indigenous Knowledge Systems into their innovation ecosystems. Such analyses are critical for advancing a continental understanding of epistemic sovereignty, moving beyond isolated case studies toward identifying both shared structural constraints and context-specific innovations. Comparative frameworks should interrogate not only policy instruments but also the epistemic assumptions, institutional arrangements, and governance models that shape how IKS is operationalised. Attention should be given to models of co-governance and participatory policymaking, assessing the extent to which they enable genuine epistemic inclusion or reproduce symbolic engagement.

9.2 Participatory and Co-Produced Methodologies

There is a pressing need for research designs that move beyond extractive methodologies toward participatory, co-produced, and community-led approaches. Such methodologies would enable deeper engagement with the lived experiences, practices, and epistemologies of Indigenous knowledge holders, ensuring that research processes themselves embody the principles of epistemic justice. Future studies should explore approaches such as participatory action research, ethnographic immersion, and co-authorship with community practitioners, thereby shifting knowledge production from observation to collaboration. This methodological shift is essential not only for empirical richness but also for reconfiguring power relations within research itself, aligning scholarly practice with the transformative aims of decolonising innovation systems.

9.3 Institutional and Policy Translation Mechanisms

Further research is needed to examine how formal recognition of IKS can be effectively translated into operational, institutional, and policy-level outcomes. While legislative frameworks exist, there remains a limited understanding of the mechanisms required to bridge the gap between policy intent and implementation. Future studies should investigate the role of provincial institutions, intermediary organisations, and local governance structures in enabling or constraining the practical integration of IKS into innovation systems. This includes analysing capacity constraints and coordination challenges and designing monitoring and evaluation frameworks that capture multidimensional outcomes, including social, ecological, and cultural impacts.

9.4 Reconceptualising Innovation Metrics and Value Systems

A further avenue for research lies in developing alternative metrics and evaluative frameworks that capture the diverse forms of value generated by IKS-based innovation. Dominant indicators, often centred on economic growth, technological output, and firm creation, are insufficient for assessing innovations embedded in social cohesion, ecological sustainability, and cultural continuity. Future work should focus on designing multidimensional measurement frameworks that reflect the pluriversal nature of innovation, incorporating qualitative, relational, and context-specific indicators. Such efforts are essential for reshaping how innovation success is defined, measured, and supported within both academic and policy domains.

10. FINAL REFLECTION

Correcting epistemic distortion must be understood not as a peripheral scholarly exercise, but as a structural and normative project that redefines the ethical, ecological, and intellectual foundations of innovation. The South African context makes clear that the central tension is not between tradition and modernity, but between competing epistemic orders one extractive, standardised, and externally validated, and another relational, contextually embedded, and oriented toward collective well-being. To advance the latter, it is necessary to recognise Indigenous Knowledge Systems as dynamic, future-oriented capabilities rather than static cultural artefacts, and to reposition African value systems as operational principles in the design, governance, and evaluation of innovation. These reframe the persistent tendency to measure progress by proximity to dominant global models, instead asserting that meaningful innovation emerges through situated co-production, in which knowledge is generated through the interaction of communities, institutions, and ecosystems. In this sense, innovation is neither an act of imitation nor preservation, but a reflexive and adaptive process through which multiple knowledge systems engage, contest, and co-evolve.

Such a reorientation carries both analytical and ethical implications. It foregrounds the need to redefine the metrics of excellence in innovation, expanding them beyond efficiency and growth to include reciprocity, stewardship, and social legitimacy. It also repositions South Africa's potential contribution within the global knowledge economy not as a peripheral adopter of established paradigms, but as a site of epistemic leadership, capable of advancing a pluriversal model in which diverse knowledge traditions coexist without subsumption. However, this vision depends on the development of institutional architectures that can sustain it. This includes legal frameworks that protect collective intellectual rights, policy instruments that reward culturally grounded problem-solving, and accountability mechanisms that align innovation outcomes with community-defined priorities. Without such material and institutional commitments, the discourse of epistemic transformation risks remaining aspirational. Ultimately, the imperative is to move from recognition to realisation, constructing innovation systems that restore agency, dignity, and balance, and that can address the intertwined challenges of inequality, ecological degradation, and social fragmentation. In doing so, South Africa not only reclaims its intellectual sovereignty but also contributes a credible and necessary alternative to dominant models of development, one that affirms that sustainable progress is inseparable from epistemic justice and collective flourishing.

REFERENCES

Abosede, A. J., Ebie, S., Somoye, R. O. C., & Ashidi, J. S. (2025). Decolonising Entrepreneurship by Embracing the Global Dichotomy. In *The Palgrave Handbook of Decolonising Entrepreneurship* (pp. 587–603). Springer Nature Switzerland., DOI: 10.1007/978-3-031-92310-4_22

Afolabi, O. S. (2020). Globalisation, decoloniality and the question of knowledge production in Africa. *Journal of Higher Education in Africa/Revue de l'enseignement supérieur en Afrique, 18*(1), 93-110. https://www.jstor.org/stable/48618319

Amankwah-Amoah, J., & Medase, S. K. (2024). Extracting innovation value from intellectual property: Evidence from sub-Saharan Africa. *Journal of the Knowledge Economy, 15*(2), 8933–8967. DOI: 10.1007/s13132-023-01225-9

Ayandibu, E. O. (2025). Integrating indigenous knowledge systems into entrepreneurship education: A South African perspective. *International Journal of Research in Business & Social Science, 14*(6), 380–392. Advance online publication. DOI: 10.20525/ijrbs.v14i6.4263

Balogun, T. (2023). Digitization of Indigenous knowledge systems in Africa: The case of South Africa's National Recorded System (NRS). *Records Management Journal, 33*(1), 88–102. DOI: 10.1108/RMJ-10-2022-0029

Caswell, C., & Jang, N. (2024). Mainstreaming gender equality and social inclusion in nature-based solutions for climate change adaptation. http://icsfarchives.net/id/eprint/21582

Chisa, K., & Hoskins, R. (2016). African customary law and the protection of indigenous cultural heritage: Challenges and issues in the digitization of indigenous knowledge in South Africa. *Indilinga, 15*(1), 1–15. https://hdl.handle.net/10520/EJC192833

Chiwawa, N., & Wissink, H. (2023). Sustainable tourism for local and regional development in South Africa: Unlocking economic potential through responsible tourism strategies. *African Journal of Hospitality, Tourism and Leisure, 12*, 1163. DOI: 10.46222/ajhtl.19770720.423

Dahlberg, A. C., & Trygger, S. B. (2009). Indigenous medicine and primary health care: The importance of lay knowledge and use of medicinal plants in rural South Africa. *Human Ecology: an Interdisciplinary Journal, 37*(1), 79–94. DOI: 10.1007/s10745-009-9217-6

de Sousa Santos, B. (2015). *Epistemologies of the South: Justice against epistemicide.* Routledge., DOI: 10.4324/9781315634876

Dladla, N. (2024). Copyright Law and the Ruse of Culture:'Traditional Cultural Expressions and Expressions of Folklore'as a Conception of Racial Difference. *Phronimon, 25*(1), 1–26. DOI: 10.25159/2413-3086/14898

Dyantyi, N. (2025, November). University curriculum reform and employability: addressing the needs of the post-colonial workforce. In *Frontiers in Education* (Vol. 10, p. 1591654). Frontiers Media SA. https://doi.org/DOI: 10.3389/feduc.2025.1591654

Ekene, A. K., Lee, K. S., Decide, M. T., & Thobejane, T. D. (2023). Placing Indigenous Knowledge at the Core of Production and Marketing of Medicinal Plant (Mpesu) in Zimbabwe and South Africa. https://doi.org/DOI: 10.1108/978-1-80262-251-520231017

Ekobi, G. A., Mboh, L., & Tanga, P. (2023). Contribution of indigenous knowledge to agricultural growth in South Africa: A case of Disaneng community in the Ratlou Local Municipality. *African Journal of Inter/Multidisciplinary Studies, 5*(1), 1-12. https://hdl.handle.net/10520/ejc-ajims_v5_n1_a27

Escobar, A. (2018). 3. Transition Discourses and the Politics of Relationality: Toward Designs for the Pluriverse. In *Constructing the pluriverse* (pp. 63-89). Duke University Press. https://doi.org/DOI: 10.1515/9781478002017-005

Foster, L. A. (2016). Decolonizing patent law: Postcolonial technoscience and indigenous knowledge in South Africa. *Feminist Formations, 28*(3), 148–173. DOI: 10.1353/ff.2016.0047

Garutsa, T. C., & Nekhwevha, F. H. (2018). Indigenous knowledge, food production and food security in rural Khambashe in the Eastern Cape, South Africa. *South African Review of Sociology, 49*(3-4), 16–31. DOI: 10.1080/21528586.2018.1564692

Heleta, S. (2018). Decolonizing knowledge in South Africa: Dismantling the 'pedagogy of big lies'. *Ufahamu. Journal of African Studies, 40*(2). Advance online publication. DOI: 10.5070/F7402040942

Heleta, S., & Chasi, S. (2024). Curriculum decolonization and internationalization: A critical perspective from South Africa. *Journal of International Students, 14*(2), 75–90. DOI: 10.32674/jis.v14i2.6383

Irene, B., Lockyer, J., & Okrah, J. (2025). Decolonising Entrepreneurship. In *The Palgrave Handbook of Decolonising Entrepreneurship* (pp. 1–14). Springer Nature Switzerland., DOI: 10.1007/978-3-031-92310-4_1

Jimenez, A., Delgado, D., Merino, R., & Argumedo, A. (2022). A decolonial approach to innovation? Building paths towards Buen Vivir. *The Journal of Development Studies, 58*(9), 1633–1650. DOI: 10.1080/00220388.2022.2043281

Kom, Z., Nicolau, M. D., & Nenwiini, S. C. (2024). The use of indigenous knowledge systems practices to enhance food security in Vhembe District, South Africa. *Agricultural Research, 13*(3), 599–612. DOI: 10.1007/s40003-024-00716-8

Kulundu, I., McGarry, D. K., & Lotz-Sisitka, H. (2020). Think Piece: Learning, Living and Leading into Transgression–A reflection on decolonial praxis in a neoliberal world. *Southern African Journal of Environmental Education, 36*. Advance online publication. DOI: 10.4314/sajee.v36i1.14

Makgamatha, M. G. (2025). Bridging Heritage and Enterprise to Address Socio-Economic Issues: A Policy Analysis of Indigenous Knowledge System Businesses in South Africa. In *Revaluation and Preservation of Indigenous Knowledge Systems in Modern Society* (pp. 367-384). IGI Global Scientific Publishing. https://doi.org/DOI: 10.4018/979-8-3693-7964-6.ch018

Makokotlela, M. V., & Gumbo, M. T. (2025). Decolonisation of the curriculum through the integration of indigenous knowledge in higher education. *South African Journal of Higher Education, 39*(1), 237–254. https://hdl.handle.net/10520/ejc-high_v39_n1_a12. DOI: 10.20853/39-1-6004

Mann-Lang, J. B., Branch, G. M., Mann, B. Q., Sink, K. J., Kirkman, S. P., & Adams, R. (2021). Social and economic effects of marine protected areas in South Africa, with recommendations for future assessments. *African Journal of Marine Science*, *43*(3), 367–387. DOI: 10.2989/1814232X.2021.1961166

Mignolo, W. D. (2007). Delinking: The rhetoric of modernity, the logic of coloniality and the grammar of de-coloniality. *Cultural Studies*, *21*(2-3), 449–514. DOI: 10.1080/09502380601162647

Mtapuri, O., & Giampiccoli, A. (2019). Tourism, community-based tourism and ecotourism: a definitional problematic. *South African Geographical Journal= Suid-Afrikaanse Geografiese Tydskrif*, *101*(1), 22-35. https://hdl.handle.net/10520/EJC-13ae94668e

Ndhlovu, P. T., Omotayo, A. O., Olagunju, K. O., Otang-Mbeng, W., & Aremu, A. O. (2025). Assessing the impacts of commercializing medicinal plants on livelihood outcomes: Evidence from indigenous knowledge holders in South Africa. *Environment, Development and Sustainability*, *27*(2), 4489–4511. DOI: 10.1007/s10668-023-04087-y

Ndlovu-Gatsheni, S. J. (2021). The cognitive empire, politics of knowledge and African intellectual productions: Reflections on struggles for epistemic freedom and resurgence of decolonisation in the twenty-first century. *Third World Quarterly*, *42*(5), 882–901. DOI: 10.1080/01436597.2020.1775487

Ndwandwe, S., & Mudhara, M. (2014). Contribution of indigenous knowledge practices to household food security: A case study of rural households in KwaZulu-Natal. *Indilinga*, *13*(2), 271–282. https://hdl.handle.net/10520/EJC166449

Ngubane, N. I., & Makua, M. (2021). Ubuntu pedagogy-transforming educational practices in South Africa through an African philosophy: From theory to practice. *Inkanyiso*, *13*(1), 1–12. https://hdl.handle.net/10520/ejc-uz_inka-v13-n1-a2. DOI: 10.4102/ink.v13i1.9

Oguamanam, C. (2020). Indigenous peoples, data sovereignty, and self-determination: Current realities and imperatives. *The African Journal of Information and Communication*, *26*(26), 1–20. DOI: 10.23962/10539/30360

Oguamanam, C. (2022). Transition to the Fourth Industrial Revolution: Africa's Science, Technology and Innovation Framework and Indigenous Knowledge Systems. *African Journal of Legal Studies*, *15*(1), 1–37. DOI: 10.1163/17087384-bja10058

Porri, F., McConnachie, B., van der Walt, K. A., Wynberg, R., & Pattrick, P. (2023). Eco-creative nature-based solutions to transform urban coastlines, local coastal communities and enhance biodiversity through the lens of scientific and Indigenous knowledge. *Cambridge Prisms: Coastal Futures*, *1*, e17. DOI: 10.1017/cft.2022.10

Quijano, A. (2000). Coloniality of power and Eurocentrism in Latin America. *International Sociology*, *15*(2), 215–232. DOI: 10.1177/0268580900015002005

Quijano, A. (2024). Aníbal Quijano: Foundational essays on the coloniality of power. In *Aníbal Quijano*. Duke University Press., DOI: 10.1515/9781478059356

Wynberg, R. (2017). Making sense of access and benefit sharing in the rooibos industry: Towards a holistic, just and sustainable framing. *South African Journal of Botany*, *110*, 39–51. DOI: 10.1016/j.sajb.2016.09.015

ADDITIONAL READING

Adeola, O. (2023). Advancing Indigenous Knowledge and Building Sustainable Business Structures in Africa. https://doi.org/DOI: 10.1108/978-1-80262-251-520231023

Buthelezi, S. C., Ocholla, D., & Dlamini, P. (2024). Strategies for documenting and disseminating indigenous knowledge at a South African university. *South African Journal of Information Management*, *26*(1), 1648. https://hdl.handle.net/10520/ejc-info_v26_n1_a1648. DOI: 10.4102/SAJIM.v26i1.1648

Finca, A., Linnane, S., Slinger, J., Getty, D., Samuels, M. I., & Timpong-Jones, E. C. (2023). Implications of the breakdown in the indigenous knowledge system for rangeland management and policy: A case study from the Eastern Cape in South Africa. *African Journal of Range & Forage Science*, *40*(1), 47–61. https://hdl.handle.net/10520/ejc-ajrfs_v40_n1_a5. DOI: 10.2989/10220119.2022.2138973

Khumalo, S. L. 2022. The effects of coloniality and international development assistance on Made in Africa Evaluation: Implications for a decolonised evaluation agenda. *African Evaluation Journal, 10*(1), p.628. https://hdl.handle.net/10520/ejc-afrej_v10_i1_a628

Knight, J. (2024). Nature-based solutions for coastal resilience in South Africa. *The South African Geographical Journal*, *106*(1), 21–50. DOI: 10.1080/03736245.2023.2193565

Lee, K. S., Decide, M. T., Mokgoatšana, S., & Ekene, A. K. (2023). Indigenous African Healing Practices as a Resource for Entrepreneurial Promotion: Insights From Tsonga Traditional Health Care System in South Africa. https://doi.org/DOI: 10.1108/978-1-80262-251-520231019

Masango, C. A. (2010). Indigenous traditional knowledge protection: Prospects in South Africa's intellectual property framework? *South African Journal of Library and Information Science*, *76*(1), 74–80. https://hdl.handle.net/10520/EJC61330. DOI: 10.7553/76-1-88

Mekoa, I. (2018). Essentialising African indigenous knowledge systems in the midst of globalization and modernity. *African Renaissance*, 11-28. https://hdl.handle.net/10520/EJC-12b7c7a610

Moyo, N., & Gonye, J. 2022. Colonialism and the Destruction of Indigenous Knowledge Systems. *Decolonizing Colonial Development Models in Africa: A New Postcolonial Critique, 91*. https://www.asau.ru/files/pdf/3117892.pdf#page=124

Ndlovu-Gatsheni, S. J. (2019). Rethinking development in the age of global coloniality. *Rethinking and unthinking development: Perspectives on inequality and poverty in South Africa and Zimbabwe*, 27-49. https://www.researchgate.net/profile/Busani-Ngcaweni/publication/342090231_Socioeconomic_and_Cultural_Barriers_to_Marital_Unions_and_HIV_Incidence_Correlates_A_Public_Policy_Poser_for_South_Africa/links/671fa6b355a5271cdee277e9/Socioeconomic-and-Cultural-Barriers-to-Marital-Unions-and-HIV-Incidence-Correlates-A-Public-Policy-Poser-for-South-Africa.pdf#page=37

Pereira, L. (2018). Mainstreaming Underutilized Indigenous and Traditional Crops into Food Systems: A South African Perspective. *Sustainability (Basel), 11*(1), 172. Advance online publication. DOI: 10.3390/su11010172 PMID: 37681213

KEY TERMS AND DEFINITIONS

Afrikology: An African-centred way of understanding knowledge based on interconnectedness, culture and lived experience.

Decoloniality: An approach that challenges and seeks to transform the dominant Western Knowledge systems and power structures.

Epistemic Distortions: The marginalisation of certain knowledge systems due to dominant knowledge hierarchies.

Epistemic Justice: The fair recognition and inclusion of all knowledge systems and their contributions.

Epistemic Sovereignty: The ability of a society or community to define, produce and validate its own knowledge systems.

Indigenous Knowledge Systems: locally developed, community-based knowledge and practices shaped by cultural norms and interaction with the environment.

Innovation systems: The network of institutions, policies and actors that shape how knowledge is created, shared and applied in society.

Innovation: The process of creating and applying knowledge to solve problems and generate value across social, economic and ecological contexts.

Nature-based solutions: Environmental strategies that use natural processes and ecosystems to address social and ecological challenges.

Pluriversal: The idea that multiple knowledge systems and ways of understanding the world can coexist equally.

Chapter 14
Engaging the Youth in Indigenous Knowledge Preservation and Job Creation in South Africa

Lindelwa Veronicah Nkosi
http://orcid.org/0009-0008-5536-5337
University of Mpumalanga, South Africa

LIvhuwani Dephney Tshikukuvhe
University of Venda, South Africa

ABSTRACT

The paper examined the engagement of Youth in the preservation of Indigenous knowledge and job creation. Guided by the socio-cultural theory and the social innovation theory, the paper adopted qualitative exploratory design. Data were collected through interviews and focus group discussions and were analyzed using thematic analysis. The Findings revealed that Indigenous knowledge is too broad, as it encompasses diverse aspects such as culture, where traditional crafts can be sold online to a global market, agriculture, which can enable young people to locally distribute farm produce and use local heritage sites to draw tourists, and social media storytelling. The paper concludes that positioning Indigenous Knowledge as a dynamic and innovation-driven economic resource can strengthen youth employment pathways while ensuring cultural sustainability. The paper contributes to contemporary debates on Indigenous Knowledge revitalization by demonstrating its practical relevance for youth employment and social transformation in the 21st century.

INTRODUCTION

African communities have always relied on indigenous knowledge systems as a tool for survival and solving societal challenges. Indigenous knowledge has been defined by various scholars in numerous ways, Indigenous knowledge (IK) also known as local or traditional knowledge (LTK), refers to a growing body of knowledge, practice and beliefs developing by adaptive processes and passed down

DOI: 10.4018/979-8-3373-7976-0.ch014

through generations (Berkes,2012) while United Nations Education, Scientific and Cultural Organizations (UNESCO,2017) recognises IK as "knowledge systems embedded in culture, environment and community life and transmitted intergenerationally". According to Salas & Tillman (2004), Indigenous knowledge is transmitted orally through songs, dance stories and performances, folklore, and myths. It is such practices that promote humanity and social solidarity. These aspects are used by various social groups in South Africa to earn a living. This is done through artworks/crafts, performances, and social entrepreneurship, where elders use their knowledge and skills to teach and train others in making beadwork and other artifacts.

Despite the growing recognition of Indigenous Knowledge (IK) as a valuable cultural and developmental resource, its economic potential remains largely underutilized in addressing youth unemployment in South Africa. Existing scholarship has primarily framed IK within heritage preservation and cultural sustainability discourses, with limited empirical attention to its capacity to generate structured livelihood opportunities for young people. At the same time, South Africa continues to experience persistently high youth unemployment rates, exceeding 45% in recent quarters (Statistics South Africa, 2025), creating an urgent need for innovative, contextually relevant employment strategies.

While Indigenous knowledge can be used to address unemployment and poverty, many young people in South Africa do not value IK and are reluctant to use it to address the high unemployment rate (Muyambo, Bahta & Jordaan, 2017). South Africa's unemployment rate is very high due to a lack of recognition of IK in economic development. "South Africa's unemployment rate is very high compared to global averages. According to Statistics South Africa, "the official unemployment rate was 32.9% in Quarter1 2025, with a broader unemployment measure at 43.1%. Even after a slight reduction to 31.9% by Q3 2025, the jobless rate remains stubbornly above 30%, and youth unemployment exceeds 40%, indicating deep labour market challenges" (Statistics South Africa,2025). This means that communities, especially youth, must find new ways to survive and make a living. Therefore, entrepreneurial skills are essential in sustaining indigenous youth in an era where jobs are scarce.

While the unemployment rate remains high in South Africa due to the neglect of IK, Evidence from other African contexts demonstrates the economic potential of Indigenous Knowledge. For example, studies in Ghana show that women in the informal sector use IK in activities such as shea butter processing, natural dye production, and traditional food preservation to generate income and increase resilience (Anane-Donkor et al., 2026). Similarly, Mkenda and Aikaeli (2019) found that women in Tanzania engage in handicraft production as a source of employment. While these studies highlight the role of women in using IK for economic benefits, this chapter focuses specifically on the youth, given their vulnerability to unemployment and their increasing engagement with technology. In Africa, there seems to be a link between IKS and sustainable development, as Barasa (2007) highlighted the role of indigenous knowledge systems in Kenyan ecotourism, underscoring their significance for economic sustainability.

The above discussion provides evidence that IK remains an integral part in unifying individuals' lives and strengthening social relations among societies (Ndlovu & Svodziwa, 2026). According to (Magni, 2017) the various forms and practices of indigenous knowledge are deeply rooted in their relationship with the natural environment and social interactions that have enabled these communities to maintain sustainable use and management of natural resources. Indigenous knowledge has enabled and continues to enable various indigenous people across the world to adapt and survive environmental changes and other social dynamics (Damfeh, 2007).

However, this knowledge is gradually being lost under the influence of modernity and globalization. The loss of indigenous knowledge brings a huge risk to the livelihoods of people, as it threatens not only their social environment but their way of life, which encompasses their culture, language, education, and health system, gradually impacting their economic life. Indigenous knowledge is a key element in the sustenance and development of local communities, "while this knowledge has successfully sustained the lives of indigenous groups in the past. One key question is whether the conditions that made this knowledge effective in the past are still intact to deal with the social changes that the world has faced over the past decades" (Damfeh, 2007). The main aim of this paper was to explore youth engagement in indigenous knowledge preservation and job creation. The study was guided by the following objectives;

- To explore how the youth can be engaged in IK preservation and job creation.
- To determine the support systems for youth engagement in IK preservation and job creation.

The discussion above highlights the dual challenge of preserving Indigenous Knowledge (IK) while addressing rising youth unemployment in South Africa. Although IK remains a valuable cultural and developmental resource, its potential as a pathway for sustainable livelihoods among young people is not fully realized. Understanding this relationship requires a conceptual lens that explains both how Indigenous Knowledge is transmitted across generations and how it can be adapted to respond to contemporary socio-economic challenges. The following section therefore, outlines the theoretical framework underpinning this study.

THEORETICAL FRAMEWORK

This study was guided by Sociocultural Theory (Vygotsky, 1986), which posits that knowledge is socially constructed and transmitted through interaction, participation, and shared cultural practices. From this perspective, Indigenous Knowledge is embedded in everyday social life and sustained through intergenerational engagement within families and communities. Empirical studies across African contexts demonstrate that IK transmission occurs through participation in practices such as storytelling, indigenous agricultural activities, traditional healing, and craft production (Dei, 2000; Odora Hoppers, 2002). This can enable the younger generation to transform these experiential learning practices into economic activities.

While elders continue to function as key knowledge holders, meaningful preservation of Indigenous Knowledge depends on youth participation, adaptation, and innovation. Youth engagement in Indigenous Knowledge preservation is therefore understood as a social and cultural process rooted in lived experience, collective practice, and intergenerational exchange (Odora Hoppers, 2017).

The study also draws on Social Innovation Theory (Drucker, 1987), which provides a framework for understanding how Indigenous Knowledge can be creatively applied to address contemporary social and economic challenges. Social Innovation Theory emphasizes the development of new ideas, practices, and enterprises that respond to social needs while creating social value. When applied to Indigenous Knowledge systems. This theory highlights the potential for youth to transform traditional knowledge into innovative livelihood opportunities.

Together, Sociocultural Theory and Social Innovation Theory provide a complementary lens for understanding how youth engagement in Indigenous Knowledge preservation can contribute to sustainable job creation. Through participation in social and cultural practices, youth develop the capacity to generate new ideas, establish social enterprises, and respond to community challenges such as unemployment and poverty. Social innovation thus allows Indigenous Knowledge preservation to move beyond cultural survival and heritage conservation, positioning it as a dynamic resource for youth employment, community resilience, and social transformation. In this way, the two theories are equally strengthening; the former emphasizes the processes of learning and knowledge transmission, while the latter highlights the application and adaptation of that knowledge for social and economic impact. This integrated view provides a strong foundation for examining how youth engagement in IK can simultaneously support cultural continuity and contribute to sustainable employment and community development. The following section, therefore, reviews relevant literature to assess current understandings and identify gaps that this study seeks to address.

LITERATURE REVIEW

While existing scholarship acknowledges the significance of preserving Indigenous Knowledge (IK), limited attention has been paid to how youth engagement in Indigenous Knowledge systems can simultaneously serve as a strategy for job creation and sustainable livelihood development. Much of the literature foregrounds IK as a cultural heritage concern, with less emphasis on its economic potential, particularly in addressing persistent youth unemployment. This gap is especially evident in South Africa, where youth unemployment remains critically high, despite the continued relevance of Indigenous Knowledge systems as sources of practical skills, innovation, and livelihood opportunities (Mavhunga, 2014; Ngulube, 2018).

In South Africa, research on indigenous farming practices indicates that youth participation in community-based agriculture and seed-saving initiatives enables the transfer of ecological knowledge related to soil conservation, climate variability, and sustainable food systems (Altieri & Koohafkan, 2008). Studies conducted in rural Limpopo and the Eastern Cape show that young people involved in indigenous farming cooperatives have generated income through local food markets while sustaining traditional agricultural knowledge (Mafongoya & Ajayi, 2017). These initiatives demonstrate how Indigenous Knowledge contributes to both livelihood creation and environmental sustainability.

Similarly, storytelling and oral traditions continue to serve as important mechanisms for the transmission of Indigenous Knowledge. Scholars note that participation in storytelling, praise poetry, and oral history enables young people to internalize cultural values, ethical systems, and historical consciousness (Finnegan, 2012; Barber, 2007). In South Africa, youth involvement in cultural festivals and heritage programmes has created employment opportunities in the creative and cultural industries, including performance, heritage interpretation, and cultural tourism (Rogerson, 2007; Mbaye, 2015). These examples illustrate that Indigenous Knowledge thrives when youth actively engage in lived cultural practices rather than remaining passive recipients of elders' knowledge.

Empirical evidence from South Africa and other African contexts illustrates how youth are already engaging in socially innovative practices rooted in Indigenous Knowledge. For example, youth-led enterprises in indigenous crafts and fashion have successfully integrated traditional weaving, beadwork, and design into contemporary markets, creating employment while preserving cultural identity (Rogerson, 2012; Comaroff & Comaroff, 2009). Community-based craft initiatives in provinces such as Limpopo and Mpumalanga have enabled young people to access national and international markets through cooperatives and digital platforms (Hay, 2008). This aligns with the social innovation theory as these young people have turned cultural knowledge into economic activities for economic sustainability (Dei, 2000).

Nyathi (2025) recorded Knowledge System and the Mopane Worm Economy Amongst the Kalanga of Mangwe in Southwestern Zimbabwe which highlighted "the socio-economic impact of the Mopane economy, looking at how IKS facilitates local trade and markets, provides livelihoods, and enhances food security in a semi-arid environment". This study highlights the extensive role IKS plays in supporting livelihoods while portraying IK as a global tool and resource for promoting sustainable development. These studies indicate that indigenous knowledge is pivotal in addressing socio-economic needs in Africa. Moreover, the studies showcase IKS as a driver of economic sustainable livelihoods in most rural communities.

Indigenous knowledge (IK) remains a viable and valuable resource for sustainable development. However, the lack of youth engagement in IK preservation poses a serious threat to community sustainability. Mohan and Majhi (2024) define sustainability as a holistic approach to social, economic, and environmental well-being that seeks to meet present needs without compromising the ability of future generations to meet their own. In South Africa, the growing demand for employment necessitates innovative strategies that promote economic development without undermining the prospects of future generations.

This situation highlights the need to bridge the gap between preservation and the commercialization of IK. According to Mohan and Majhi (2024), there is an urgent need for holistic, inclusive, and culturally sensitive approaches to heritage preservation and sustainable development. Similarly, Boatbil, Baada, Akonga and Ayoung (2025) reveal that a loss of interest among younger generations, coupled with insufficient support from governmental institutions and agencies, remains a significant challenge in the preservation of IK for development purposes. This suggests that government institutions must mobilize resources to actively engage and support youth in job creation through indigenous knowledge and practices.

The active involvement of younger generations in IK preservation contributes directly to sustainable development. Furthermore, support from governmental institutions and agencies can maximize youth participation in IK preservation and sustainable development initiatives, ultimately contributing to reduced unemployment. Yazidi and Rijal (2024) highlight the pedagogical benefits of incorporating IK in developing critical thinking and problem-solving skills among young people. They advocate for educational reforms that integrate IK as a dynamic component of science education, arguing that such integration can play a vital role in achieving sustainable development goals by preparing youth to engage meaningfully with complex social challenges such as poverty and unemployment.

While the studies discussed above focus primarily on IK and sustainable development, other scholars demonstrate that youth engagement in agricultural food systems can lead to sustainable employment and poverty reduction, particularly in disadvantaged communities. Scholars such as Huambachano, Arulingam, Bowness, Korzenszky, Mungai, Termine, and Wittman (2022) recognize young people as agents of change who are at the forefront of transforming agriculture and food systems. Positioning youth at the center of agricultural and food systems maximizes opportunities for job creation through food production, processing, and value chains. However, although these scholars identify youth as agents of change, Oloko, Reed and Robson (2022) argue that young people are not effectively contributing to food preservation compared to other food provisioning activities, largely due to a lack of mentorship and skills transfer. This points to the need for intergenerational knowledge transfer to enable youth to engage actively not only in food preservation but also in food distribution as viable income-generating activities.

Technology and IK Preservation

Although IK continues to face challenges related to documentation, technology presents new opportunities for its preservation in the twenty-first century. Given that technology is deeply embedded in young people's lives, innovation and digital tools can be used to attract youth to play a central role in IK preservation while simultaneously creating employment opportunities. According to Chirima (2025), "digital technologies provide robust mechanisms for safeguarding Indigenous Knowledge Systems (IKS) against physical deterioration, loss, or destruction". Engaging youth in IK preservation through technology helps mitigate knowledge loss, as younger generations tend to be more technologically proficient than older knowledge holders. Chirima further notes that mobile applications and multimedia platforms offer immersive ways of preserving oral traditions and craftsmanship.

In addition, the use of mobile applications enhances access, interaction, and engagement with diverse audiences. This not only safeguards IK but also enables young people to generate income, for example, by attracting tourists and cultural consumers. Moreover, the rapid advancement of technology requires indigenous communities to adapt and use available technologies, such as artificial intelligence (AI), to preserve their knowledge and ensure economic sustainability for future generations. Tella et al. (2025) highlight that AI can assist in organizing and digitizing oral histories, traditional storytelling, and ecological knowledge within indigenous contexts. The use of AI in IK preservation benefits not only local communities but also wider audiences, who can engage with and learn from diverse indigenous practices and cultures.

Furthermore, digital technologies enable young people to document and record local narratives that can be stored digitally and made accessible to future generations. This approach also serves as a valuable educational tool and supports the preservation of indigenous languages. Mdhluli, Mokgoatšana, Kugara, and Vuma (2021) emphasize the importance of intergenerational transmission of culturally unique knowledge in their study on knowledge management through digital libraries. They argue that digital libraries play a crucial role in preserving IK while ensuring broader access.

This underscores the importance of youth-led technological innovation, positioning young people as key drivers of IK documentation and preservation. Similarly, Owiny, Mehta and Maretzki (2014) examined the use of social media technologies to create, preserve, and disseminate indigenous knowledge and skills in African communities. Their findings indicate that the rapid growth of social media and mobile technologies creates opportunities for local and international partnerships that support the creation, management, preservation, and sharing of community-specific knowledge and skills. In a context of high unemployment and increasing digitalization, youth technological fluency offers significant economic opportunities, including content creation that can be marketed to a global audience

Policies and Institutional Support for IK and Youth Employment

Indigenous knowledge represents one of a community's most valuable assets. The Indigenous Knowledge Systems (IKS) Policy (Republic of South Africa, 2004) provides a framework for the recognition, promotion, and protection of IK in South Africa. The policy acknowledges that IKS is owned by, and provides services to, disadvantaged communities that are often disproportionately affected by unemployment. It emphasizes the role that IKS can play in employment creation, poverty alleviation, and technology transfer, highlighting its potential contribution to national economic development. Engaging youth in IK-based job creation further supports the development of youth-focused programmes. The policy also advocates integrating IKS across sectors such as education, agriculture, health, and environmental management. Recognition of IKS within these sectors can stimulate business development and equip young people with relevant skills. In addition, Ncube et al. (2025) state that Policies such as the IKS policy are beneficial for ensuring the recording and protection of IK, leading to the formation of IK-based enterprises that generate economic gains and employment.

In addition, the Protection, Promotion, Development and Management of Indigenous Knowledge Act of 2019 (IKS Act of 2029) provides legal recognition and protection for indigenous knowledge holders and their intellectual property rights (Republic of South Africa, 2019). This legislation ensures that indigenous youth initiatives and innovations are safeguarded against exploitation and unethical commercialization by regulating access to and use of IK. Furthermore, support from local government structures can promote the responsible use of IK in job creation initiatives and ensure fair benefit-sharing among indigenous knowledge holders.

Scholars such as Ayandibu (2025) advocate integrating Indigenous Knowledge Systems (IKS) into entrepreneurial education. This underscores the significant role that Indigenous Knowledge (IK) can play in promoting social relevance, innovation, and sustainable community development. Integrating IKS into entrepreneurship curricula positions Indigenous Knowledge as a viable resource for youth livelihoods, particularly in an era characterized by high unemployment and economic uncertainty.

According to Oloruntoba, Moyo, and Zondo (2025), recentring the voices of marginalized African communities is critical to meaningful youth entrepreneurship. The authors argue that the silencing and exclusion of youth and community perspectives in entrepreneurship policies and employment program design represent a continuation of colonial patterns of marginalization. This highlights the need for governments and policymakers to actively engage IK communities and young people in identifying local needs, development priorities, and contextually relevant economic opportunities that foster inclusive growth and sustainable engagement.

In the South African context, the Department of Small Business Development provides support for small, micro, and medium enterprises (SMMEs) to grow and remain sustainable. Such institutional support is a crucial mechanism for enabling young South Africans to escape poverty and build resilient livelihoods. It further demonstrates that, with appropriate policy frameworks and targeted institutional backing, young people can become key drivers of economic growth, innovation, and social transformation within their communities.

The above discussions highlight the government's role in supporting IK-based businesses and sustainable livelihoods through relevant policies. Ncube et al. (2025) argued that the use of indigenous knowledge for sustainable livelihoods is widely appreciated in South Africa and that the government has been supportive of policies that promote its use. This governmental approach may encourage youth to use indigenous knowledge for sustainable livelihoods.

Strategies for the Preservation and Revitalization of IK

IK faces multiple challenges in the 21st century. The exclusion and under exploitation of IK by youth in development and entrepreneurship activities perpetuate socio-economic challenges such as unemployment. Mudau and Sikhwari (2025) explored the influence of indigenous knowledge systems on youth behavior in Mutale Municipality, South Africa. Their findings show that IK plays a major role in youth behavior and decision-making. Additionally, the results indicate that disconnection from traditional structures, driven by modernization and a lack of intergenerational dialogue, weakens this influence.

These results suggest that the cultural values instilled in youth strongly influence their decisions about their lives, including the decision to use the intergenerational knowledge they acquire to address communal challenges such as unemployment. This study highlights the recognition of IK as socially and economically relevant in the 21st century.

Scholars such as Mdhluli et al. (2021) advocate protecting IK against intellectual theft through digitalization and managing IK for current and future generations. This study reflects the role of technology in the preservation of IK, especially in an era where the youth are more technologically wise. This maximizes youth participation in accessing and preserving IK. Moreover, this strategically allows intergenerational knowledge transfer without physical means or practices. Similarly, Timire and Dube (2025) found that the infusion of IKS into the technological curriculum or syllabus can empower learners from rural, agricultural-based economies. This ensures that learners in rural communities can use local knowledge at an early age to survive, thereby freeing them from a narrow view of local economies. The infusion of IKS in the curriculum or syllabus promotes continuity of IK and allows young people to make use of their practical and experiential knowledge.

The reviewed literature demonstrates the significant potential of IK to promote sustainable livelihoods, while also highlighting critical gaps in youth engagement, policy support, and the integration of technology into IK preservation. Although existing studies highlight important contributions of IK to socio-economic development, limited empirical work has explored how these dynamics operate within specific local contexts in South Africa. In response to these gaps, the following section outlines the research methodology adopted to investigate youth engagement in IK preservation and its potential for job creation

RESEARCH METHODOLOGY

This study adopted a qualitative, exploratory research design to examine youth engagement in the preservation of Indigenous Knowledge (IK) and its potential for job creation. A qualitative approach was considered appropriate because the study sought to generate an in-depth understanding of participants' perceptions, lived experiences, and meanings attached to IK rather than quantify variables (Denzin & Lincoln, 2018). An exploratory design was appropriate, given the limited empirical research linking IK preservation to structured youth employment pathways in South Africa. Exploratory qualitative research enabled the researchers to investigate under-researched phenomena and generate contextually grounded insights. The study was conducted in selected communities in Mpumalanga and Limpopo provinces, characterized by strong Indigenous Knowledge practices, high youth unemployment rates, and active informal economic sectors linked to agriculture, crafts, and cultural heritage. These contexts provided a suitable setting to explore the intersection between IK preservation and livelihood generation.

The population comprised 50 participants drawn from youth groups, community knowledge holders, small-scale entrepreneurs, and local cultural practitioners. From this population, 15 participants were selected using purposive sampling. Purposive sampling enabled the researchers to deliberately select participants with relevant knowledge, experience and involvement in Indigenous Knowledge activities, Maxwell (2012). The sample included: Youth (aged 18–35) involved in agriculture, crafts, digital content creation, or cultural initiatives, Elders recognized as IK custodians and local cooperative members engaged in IK-based enterprises. This sampling strategy ensured depth, diversity, and relevance of perspectives.

Data were collected through. semi-structured interviews and focus group discussions (FGDs). Semi-structured interviews allowed flexibility while maintaining alignment with the study objectives. Interviews explored participants' understanding of IK, perceptions of its economic potential, barriers to youth engagement, and views on technology integration. Focus group discussions facilitated collective reflection and interaction among youth participants, allowing shared meanings and social dynamics around IK to emerge. FGDs were particularly useful in this study, where knowledge is socially constructed (Krueger & Casey, 2015). All interviews and discussions were audio-recorded (with consent), transcribed verbatim, and, where necessary, translated into English. Data were analysed using thematic analysis. Thematic analysis allowed systematic identification of patterns across qualitative data while remaining theoretically flexible.

To ensure trustworthiness, credibility, dependability, and confirmability, and rigor, the authors applied the criteria proposed by Guba & Lincoln (1985). Member checking was conducted by sharing summaries with selected participants to ensure credibility, solid descriptions of context and participants were provided, ensuring transferability. An audit trail documenting research decisions was maintained to ensure dependability. Confirmability was maintained through Reflexive journaling to reduce researcher bias. Ethical approval was obtained from the relevant institutional review boards and traditional authorities. Participants were informed of the study's purpose and signed voluntary consent forms. Confidentiality, anonymity, and the right to withdraw were ensured. Given the sensitivity surrounding Indigenous Knowledge ownership, the study aligned with the provisions of the IKS Act of 2019, Participants were informed that knowledge shared would not be commercially exploited and would be used strictly for academic purposes.

The study was limited by the sample size small sample size (n=15), time constraints and Geographic concentration. While findings may not be statistically generalisable, they provide analytical insights relevant to similar socio-cultural contexts.

Guided by the above methodological approach, the study generated rich qualitative data reflecting participants' lived experiences, perceptions, and practices related to Indigenous Knowledge and livelihood creation. The analysis of this data resulted in several key themes, which are presented and discussed in the following section.

The findings emanated from the identified themes during data collection process. Themes that emerged included;

- IK as a digital opportunity
- Intergenerational knowledge gaps
- Policy awareness and institutional support
- Cultural entrepreneurship and
- Technology-driven preservation strategies.

The findings embody the participants' lived experiences and their perceptions of engaging youth in Indigenous Knowledge Preservation and Job Creation in South Africa.

IK as a digital opportunity

In an era where technology and digital tools are increasingly accessible to young people, many do not fully recognize the benefits and potential these tools offer for the preservation of Indigenous

Knowledge (IK). Instead, digital platforms are often used for activities that provide limited long-term value or developmental benefit. Findings emerging from this theme indicate a growing concern among community elders regarding how youth engage with digital media. Elders observed that young people are overly focused on generating "content" and seeking social validation through online platforms, rather than leveraging these tools for meaningful cultural preservation and economic advancement. Two community elders emphasized:

> "*Our young people are very fortunate because they have all these technological tools that we did not have in the olden days. They can best preserve their knowledge using the very same tools, not only to promote their Indigenous Knowledge, but they can also make money from it because everything is now done online.*"

This concern reflects participants' perception of a misuse or underutilization of opportunities available to youth. It also suggests that some young people may inadvertently contribute to their own unemployment by failing to recognize the economic potential of producing culturally grounded digital content that can generate income while preserving heritage. In this way, digital engagement could simultaneously support cultural continuity and livelihood creation. This emphasizes that culturally grounded content produced by young people can be effectively used to meet social and economic needs, as highlighted in the Social Innovation Theory.

Participants further indicated that this challenge is exacerbated by parental neglect, particularly where parents do not monitor or guide their children's digital activities. Such guidance could help redirect digital engagement toward productive and culturally meaningful purposes. As Chirima (2025) argues, digital tools have significant potential to draw youth into actively participating in IK preservation while simultaneously creating employment opportunities.

The results also revealed that integrating digital technologies into IK preservation holds multiple economic benefits. One participant noted:

> "*Our culture offers so many opportunities in the job market. As people who hold deep knowledge of beadwork and traditional crafts such as pottery, we can post that on online platforms in order to reach a wider audience, especially people who are not from our communities.*"

This aligns with scholars such as Rogerson (2012) and Comaroff and Comaroff (2009), who highlight the importance of youth involvement in enterprises that integrate traditional beadwork, weaving, and design into contemporary markets. Such initiatives not only contribute to employment creation but also safeguard cultural identity and heritage, which are deeply interwoven with IKS. The youth also indicated that as people who are technologically fluent, media platforms can be used to teach indigenous languages, share local stories and promote the development of new ideas, practices, and enterprises that respond to social needs while creating social value.

These findings highlight the importance of using digital technologies to preserve IK while achieving economic gains.

The findings imply that the knowledge learned and sustained within communities, as proposed by Vygotsky's (1986) Sociocultural Theory, is now translated into tangible economic activities, suggesting sustainable livelihoods for youth through creative applications, as detailed by the participants. Moreover, the use of digital technologies enables Ik to remain relevant in an era of the fourth industrial revolution (4IR) where digital transformation is the norm. This necessitates that the youth actively engage in the digital preservation of IK and enterprises. While digital technologies offer new opportunities to engage youth in Indigenous Knowledge preservation, participants also highlighted underlying challenges related to declining intergenerational knowledge transfer.

Intergenerational Knowledge Gap

Most young people do not readily identify with Indigenous Knowledge Systems (IKS), often perceiving them as "something meant for the older generation." Their attitudes toward IKS appear divided: while some express interest in learning more about Indigenous Knowledge (IK), others demonstrate little to no interest. Limited participation in local youth camps, community activities, cultural programmes, and apprenticeship initiatives designed to educate young people about cultural practices and social values, further perpetuates the intergenerational knowledge gap.

Participants noted that many young people increasingly regard elders as socially irrelevant, largely due to Western influences and contemporary socialization processes. Participants expressed significant concern about youth apathy toward learning cultural practices and traditions that could foster identity and livelihood opportunities. One participant remarked:

"Our children are more focused on learning other ways of doing things rather than what we want to teach them, which is the same knowledge we have used for survival and in raising them. The unfortunate part is that as they disconnect from their own knowledge and traditions, they adopt foreign knowledge which they view as relevant and promising."

These findings highlight the urgent need to revitalize Indigenous Knowledge Systems (IKS) within the formal education system. Younger generations often perceive knowledge acquired from textbooks as the most legitimate and authoritative form of knowing. Consequently, school curricula should place greater emphasis on IKS's relevance in the 21st century, particularly its contribution to addressing societal challenges such as unemployment. Integrating IKS into formal education can enhance youth engagement in Indigenous Knowledge (IK) preservation while demonstrating its contemporary value.

The need for the revitalization of Indigenous Knowledge Systems (IKS) also suggests the application of Sociocultural Theory within communities to engage and motivate youth to learn the social practices that Indigenous communities have long relied on for survival. As IK has been noted to facilitate trade and markets, provide livelihoods, and enhance food security in semi-arid environments through the Mopani worm knowledge economy, Nyathi (2025).

This also calls for more youth-centered and contextually responsive teaching strategies. As Timire and Dube (2025) contend, incorporating IKS into the school curriculum enables learners to appreciate and internalize Indigenous Knowledge from an early age, allowing them to apply it in their everyday lives and practices. Practical strategies may include inviting community elders to share their knowledge in classrooms, using recorded messages from knowledge holders on livelihood-oriented IK practices, and fostering collaborative partnerships between schools and local communities. Although schools have largely supplanted traditional modes of knowledge transmission and are often regarded as the primary custodians of knowledge, meaningful integration of IKS requires reconnecting formal education with community-based knowledge systems (Ayandibu, 2025).

While some young people viewed Indigenous Knowledge Systems (IKS) as irrelevant, others expressed a clear desire to learn from elders and understand practices that have sustained communities for generations. Youth involved in agriculture, in particular, demonstrated greater interest in Indigenous farming methods that promote good agricultural practices and improve productivity, enabling them to achieve harvests suitable for bulk distribution and market participation. Some participants highlighted this interest, noting:

"We like it when the elders take us to the farm or show us the indigenous farming methods because they say farming is our way of surviving and they have used it for many years, it has always worked for them. This makes us interested in learning and also practicing how it is done".

Active engagement in community-based agricultural practices fosters curiosity, continuous learning, and practical skill development. When young people witness the tangible benefits of Indigenous practices, such as improved yields and sustainable farming methods, they begin to see Indigenous Knowledge (IK) as a valuable and practical resource rather than an outdated tradition.

This aligns with Huambachano et al. (2022), who argue that recognizing young people as agents of change at the forefront of transforming agriculture and food systems maximizes opportunities for job creation across food production, processing, and value chains. Meaningful youth participation in Indigenous agricultural practices, therefore, holds significant potential for both knowledge preservation and sustainable economic empowerment. These challenges are further compounded by limited institutional support and low awareness of policies designed to promote youth participation in Indigenous Knowledge-based initiatives.

Policy awareness and institutional support

Most young people in rural communities such as Mpumalanga and Limpopo lack the necessary resources to start their businesses and are unfamiliar with relevant institutions and policies that offer support for their businesses. Many of the young people indicated that they tried to start their business due to unemployment but their businesses collapsed due to lack of training and support. This highlights the need for relevant departments to actively support and engage the youth through trainings and mentorship programmes to help sustain their businesses. Proper mentorship and support can help small businesses grow. One participant sadly noted that;

"I had no knowledge of running a business, I only started the business because the situation at home was unfavourable and I wanted to assist because I could see that at home we are struggling, once I started making money I focused more on the things we needed at home and forgot to save the money that's why I lost the business, there was no one to guide me about finances in business".

While some young people try to create employment for themselves, a lack of support and proper training becomes a major setback. This suggests that entrepreneurial education should be compulsory in both basic and higher education in order to equip young people with the relevant and necessary information needed to succeed in their entrepreneurial endeavors. As indicated by scholars such as Ayandibu (2025). Despite these structural limitations, the findings reveal that some young people are actively leveraging Indigenous Knowledge through cultural entrepreneurship.

Cultural entrepreneurship

The relationship between culture and entrepreneurship proved to be viable among many young people in both Mpumalanga and Limpopo provinces. Participants indicated that cultural entrepreneurship offers numerous economic opportunities and benefits. In a country where Indigenous ways of life were historically labeled as backward and uncivilized, young people are demonstrating that their Indigenous Knowledge (IK) and cultural practices are both relevant and economically promising. As emphasized by social innovation theory, youth in Mpumalanga and Limpopo are developing new ideas and initiatives that respond to social needs while simultaneously creating economic value. Some young participants from Limpopo enthusiastically stated:

> *"We saw that our province is rich in culture, but many people do not know the beauty of our culture and the beautiful places in our province. So, we decided to form a group that helps people experience Vhavenda culture and also offers short guided tours around some of the prominent sites in the area."*

Similarly, others highlighted that these initiatives provide them with opportunities to share stories about the places they visit with tourists. These expressions reflect a deep appreciation of culture and nature, which in turn generates various economic benefits for young people. Such initiatives not only contribute to the preservation of culture and Indigenous Knowledge but also promote cultural visibility and ensure its continued relevance in contemporary contexts, and also promote heritage tourism. These initiatives directly strengthen the relationship between cultural preservation and job creation, highlighting the long-term benefits of engaging the youth in IK preservation as they are the key in ensuring the survival of indigenous knowledge.

A young poet from a rural area in Mpumalanga indicated that she uses social media platforms such as Facebook and TikTok to post videos of herself reciting clan praises and narrating the history of particular clans. She explained that this approach has created numerous economic opportunities for her. For example, she receives invitations from local radio stations, which help promote her work to a wider audience. In addition, she is invited to weddings and other events, where she is paid to entertain guests through her cultural poetry. This illustrates the relationship between the Sociocultural theory and the Social Innovation theory in contributing to both cultural continuity and economic development.

This example clearly demonstrates the significance of cultural practices such as performance and praise poetry in South Africa, and how young people can utilize them to preserve Indigenous Knowledge while simultaneously addressing poverty and unemployment (Rogerson, 2007; Mbaye, 2015). Moreover, this helps mitigate knowledge loss and allows other people to learn the Siswati language, ensuring the revitalization of IK and languages, especially in a world dominated by the English language. Building on these entrepreneurial initiatives, technology emerges as a critical tool for scaling and sustaining Indigenous Knowledge practices in contemporary contexts.

Technology-Driven Preservation Strategies

Experiences, attitudes, and views regarding the use of technology as a preservation strategy varied among participants. Many highlighted the benefits of technological platforms in preserving Indigenous Knowledge (IK). Some participants noted that using podcasts to share knowledge about cultural practices contributes to the revitalization of IK while simultaneously increasing its visibility. However, they also cautioned that such exposure may render IK vulnerable to commercial exploitation. This underscores the need for local authorities and relevant stakeholders to conscientize young people about the Protection, Promotion, Development and Management of Indigenous Knowledge Act to ensure that Indigenous youth initiatives and innovations are safeguarded against exploitation and unethical commercialization. Furthermore, support from local government structures can promote the responsible use of IK in job creation initiatives and ensure fair and equitable benefit-sharing among knowledge holders.

Many young participants also indicated that they possess limited knowledge of Indigenous cultural and social values embedded in oral traditions such as storytelling and proverbs. They suggested that elders should consider transmitting these values through digital storytelling platforms, which can reach wider audiences while preserving knowledge for future generations. Given that many young people are highly engaged with online media content, such approaches may increase interest and participation. This highlights the shared responsibility of both community elders and youth in ensuring the revitalization and preservation of Indigenous Knowledge, as well as in bridging the intergenerational knowledge gap.

Some elders noted that the use of technological tools to preserve Indigenous Knowledge (IK) can sometimes lead to the misrepresentation of authentic knowledge and practices. One elder stated:

> *"It is best that we share this knowledge orally with our young ones; that way we are certain that nothing will be taken away from or added to what we have shared."*

This perspective highlights concerns about the accuracy, integrity, and control of knowledge once it is digitized. It also points to the need to educate and support elders in understanding the benefits of technology and how to use it responsibly and effectively for their own benefit, while safeguarding the authenticity of their knowledge. As the world becomes increasingly digitally empowered, indigenous knowledge holders should be digitally literate to ensure effective, relevant preservation strategies that meet the needs of the current generation while ensuring sustainability. This stresses the need for digital archives as argued by Mdluli et al (2021).

The findings presented above highlight both the opportunities and challenges associated with youth engagement in the preservation of Indigenous Knowledge and in job creation. Drawing on these insights, the following recommendations are proposed to strengthen policy, practice, and future research.

RECOMMENDATIONS

For the revitalization and preservation of Indigenous Knowledge Systems (IKS) in the 21st century, this paper recommends the integration of Indigenous entrepreneurship into both basic and higher education as a strategy for positioning IKS as a viable economic resource. Young people should be supported through the establishment and strengthening of Indigenous training centers that equip them with culturally grounded entrepreneurial skills.

The paper further recommends the provision of training programmes within Indigenous communities on the ethical and responsible use of digital tools. Such initiatives can support the mobilization, documentation, and dissemination of Indigenous Knowledge to wider audiences while safeguarding its integrity and ensuring community benefit.

The study also recommends further research focused on the monitoring and evaluation of IK-based youth initiatives. Furthermore, there is a need for future studies that focus on the challenges youth face in using IK for job creation or economic development.

Future Research Directions

Given the nature of this study, future studies may use mixed methods in gathering both numerical data and the perception of the youth on IKS based businesses. In addition, there is a need for future research to be directed towards testing or exploring the effectiveness of legal frameworks that support the preservation of IKS and indigenous entrepreneurship.

Researchers can also evaluate the role of heritage tourism and creative industries ingenerating employment while preserving IKS and cultural identity.

CONCLUSION

This study explored how youth engagement in the preservation of Indigenous Knowledge (IK) can contribute to job creation. Guided by Sociocultural and Social Innovation theories, the findings demonstrate that IK remains dynamic and economically viable when youth are actively involved. The study highlights the importance of intergenerational knowledge transfer, supported by digital technologies that expand access and economic opportunities. Cultural entrepreneurship emerges as a key pathway linking heritage preservation with job creation. The study concludes that IK should be repositioned as a living and adaptable resource capable of addressing contemporary challenges such as unemployment and poverty. Supporting youth through education, policy frameworks, and technological integration is essential for achieving both cultural continuity and economic resilience.

REFERENCES

Altieri, M. A., & Koohafkan, P. (2008). *Enduring farms: Climate change, smallholders and traditional farming communities*. FAO.

Anane-Donkor, L., Dei, D. J., & Ocloo, P. E. D. (2026). Indigenous knowledge for innovation and sustainable livelihood in Ghana's informal economy. *Discover Global Society*, *4*(1), 17. Advance online publication. DOI: 10.1007/s44282-026-00352-3

Barasa, D. W. (2007). Indigenous knowledge systems and sustainable development in Africa: Case study on Kenya. *Tribes Tribals*, *2007*, 141–156.

Barber, K. (2007). *The anthropology of texts, persons and publics*. Cambridge University Press. DOI: 10.1017/CBO9780511619656

Berkes, F. (2012). *Sacred Ecology* (3rd ed.). Routledge. DOI: 10.4324/9780203123843

Boatbil, C. S., Baada, F. N.-A., Akonga, S., & Ayoung, D. A. (2025). Preserving indigenous knowledge for development in northern Ghana: Exploring the views of artisans in the art and craft industry in the Upper East Region. *African Quarterly Social ScienceReview*, *2*(3), 251–262. DOI: 10.51867/AQSSR.2.3.22

Chirima, R. (2025). Harnessing Digital Technologies to Preserve and Promote Indigenous Knowledge Systems in Africa. *International Journal of Innovative Science and Research Technology*, *10*(5), 4411–4416. DOI: 10.38124/ijisrt/25may1906

Comaroff, J., & Comaroff, J. (2009). *Ethnicity,* Inc. University of Chicago Press.

Dei, G. J. S. (2000). Rethinking the role of Indigenous knowledges in education. *International Journal of Inclusive Education*, *4*(2), 111–132. DOI: 10.1080/136031100284849

Denzin, N. K., & Lincoln, Y. S. (2018). *The SAGE handbook of qualitative research*. Sage.

Drucker, P. F. (1987). *Innovation and entrepreneurship*. Harper & Row.

Drucker, P. F. (1987). *Social innovation – management's new dimension*. Long.

Finnegan, R. (2012). *Oral literature in Africa*. Open Book Publishers. DOI: 10.11647/OBP.0025

Guba, E. G., & Lincoln, Y. S. (1985). Naturalistic inquiry. *Sage (Atlanta, Ga.)*.

Hay, D. (2008). African craft production and markets. *Development Southern Africa*, *25*(3), 343–359.

Huambachano, M., Arulingam, I., Bowness, E., Korzenszky, A., Mungai, C., Termine, P., & Wittman, H. (2022). *Knowledge networks to support youth engagement in sustainable food systems*. *Frontiers in Sustainable Food Systems*, *6*, 867344. DOI: 10.3389/fsufs.2022.867344

Krueger, R. A., & Casey, M. A. (2015). Focus groups: A practical guide for applied research. *Sage (Atlanta, Ga.)*.

Mafongoya, P., & Ajayi, O. C. (2017). Indigenous knowledge systems and climate change adaptation. *Climate and Development*, *9*(5), 1–12.

Magni, G. (2017). *Indigenous knowledge and implications for the sustainable development agenda.* European Journal of Education, 52(4), 437–447. management: Preserving, managing and sharing indigenous knowledge through digital library'. *HTS Teologiese Studies, 77*(2), a6795. DOI: 10.4102/hts.v77i2.6795

Mavhunga, C. C. (2014). *Translating knowledge between indigenous and modern systems.* Palgrave Macmillan.

Maxwell, J. A. (2012). Qualitative research design: An interactive approach. *Sage (Atlanta, Ga.).*

Mbaye, J. (2015). Culture and sustainable development in Africa. *African Development Review, 27*(4), 334–346.

Mdhluli, T. D., Mokgoatšana, S., Kugara, S. L., & Vuma, L. (2021). *'Knowledge management: Preserving, managing and sharing indigenous knowledge through digital library'. HTS Teologiese Studies, 77*(2), a6795. DOI: 10.4102/hts.v77i2.6795

Mkenda, B. K., & Aikaeli, J. (2019). I*ndigenous Knowledge and Prospects for Income and Employment Generation: The Case of Handicraft Production among Rural Women in Tanzania.* IK: Other Ways of Knowing, Vol. 5 https://journals.psu.edu/ik/issue/view/3036

Mudau, T. J., & Sikhwari, M. G. (2025). The Influence of Indigenous Knowledge Systems on Youth Behaviourin Mutale Municipality, South Africa. *E-Journal of Humanities, Arts and Social Sciences, 6*(10), 2486–2502. DOI: 10.38159/ehass.202561018

Muyambo, F., Bahta, Y. T., & Jordaan, A. J. (2017). The role of indigenous knowledge in drought risk reduction: A case of communal farmers in South Africa. *Jàambá, 9*(1). Advance online publication. DOI: 10.4102/jamba.v9i1.420 PMID: 29955345

Ncube, I., Chigwada, J., Ngulube, P., & Maluleka, J. R. (2025). Government support for indigenous knowledge for sustainability in Southern Africa. *South African Journal of Information Management, 27*(1), 1901. DOI: 10.4102/sajim.v27i1.1901

Ndlovu, L., & Svodziwa, M. (2026). *The Levels Of Participation in Non Governmental Organizations: Ndlovu, Loveness, Svodziwa, Mpathew: Amazon.ie: Books.* Amazon.ie. https://www.amazon.ie/Levels-Participation-Non-Governmental-Organizations/dp/3659917672

Ngulube, P. (2018). Preserving Indigenous Knowledge in the knowledge economy. *Information Development, 34*(3), 1–12.

Nyathi, I. (2025). *Knowledge System and the Mopane Worm Economy Amongst the Kalanga of Mangwe in Southwestern Zimbabwe*: Harnessing Indigenous Epistemologies for Sustainable Progress. IGI Global Scientific Publishing. *Page, 16*, 273–288. Advance online publication. DOI: 10.4018/979-8-3693-8447-3.ch013

Odora Hoppers, C. A. (2002). *Indigenous knowledge and the integration of knowledge systems.* New Africa Books.

Odora Hoppers, C. A. (2017). *Decolonisation of knowledge, epistemicide and justice.* Routledge.

Oloko, M. Reed, M. G., & Robson, J. P. (2022). *Engaging youth in food preservation: Examining knowledge and practice on Canada's West Coast.* Canadian Food Studies La Revue Canadienne Des études Sur l'alimentation, 9(3), 61–86. .DOI: 10.15353/cfs-rcea.v9i3.523

Owiny, S., & Mehta, K & Maretzki. (2014). *The Use of Social Media Technologies to Create, Preserve, and Disseminate Indigenous Knowledge and Skills to Communities in East Africa.* International Journal of Communication. 8. 14. *Range Planning, 20*(6), 29–34.

Republic of South Africa. (2004). *Indigenous knowledge systems*, viewed 15 January 2026, from https://www.dst.gov.za/images/pdfs/IKS_Policy%20PDF.pdf

Republic of South Africa. 2019, *Protection, Promotion, Development and Management of Indigenous Knowledge* Act. viewed 15 January 2026, from https://www.gov.za/sites/default/files/gcis_document/201908/4264719-8act6of2019protectpromodevelopmanagementindigenousknowledgeact.pdf

Rogerson, C. M. (2007). Cultural industries and local economic development in South Africa. *Urban Forum, 18*(2), 49–71. DOI: 10.1007/s12132-007-9006-5

Rogerson, C. M. (2012). Creative industries and youth employment. *African Journal of Business Management, 6*(39), 10435–10448.

Statistics South Africa. (2025). *Quarterly Labour Force Survey (QLFS)* (Vol. Q1). Statistics South Africa.

Tella, A., Jatto, E., & Yusuf, A. (2025). Preserving indigenous knowledge: Leveraging digital technology and artificial intelligence. *IFLA Journal, 51*(3), 1–19. DOI: 10.1177/03400352251342505

UNESCO. (2017). *Local and Indigenous Knowledge Systems* (LINKS) Programme. Paris: UNESCO.

Vygotsky, L. S. (1986). *Thought and language.* MIT Press.

Yazidi, R. E., & Rijal, K. (2024). Science Learning in the Context of 'Indigenous Knowledge' for Sustainable Development. *International Journal of Ethnoscience and Technology in Education, 1*(1), 28–41. DOI: 10.33394/ijete.v1i1.10880

ADDITIONAL READING

Cheteni, P., & Umejesi, I. (2026). Integrating indigenous knowledge systems into entrepreneurial practices: A study of hunting and gathering in the wild coast. *Sustainable Communities, 3*(1), 2643958. Advance online publication. DOI: 10.1080/29931282.2026.2643958

Chigwada. J &Ngulube, P. (2025). *Stakeholders in the acquisition, preservation, and dissemination of indigenous knowledge projects.* Information Development Journal. Volume 41 Issue 4, Sage Journals.

Masenya, T. M. (2022). Decolonization of Indigenous Knowledge Systems in South. *International Journal of Knowledge Management, 18*(1), 1–22. Advance online publication. DOI: 10.4018/IJKM.310005

Masenya, T. M. (2023). Revitalization and Digital Preservation of Indigenous Knowledge Systems for Sustainable Development of Indigenous Communities in South Africa. *The Serials Librarian*, *84*(5–8), 86–102. DOI: 10.1080/0361526X.2023.2277962

KEY TERMS AND DEFINITIONS

Employment: A paid work agreement between an employer and an employee

Entrepreneurship: refers to the capability, willingness and process of developing and running a business.

Indigenous Knowledge: refers to unique knowledge, skills, and practices that societies have developed over centuries through direct interaction with the natural environment

Job opportunities: Positions or roles that support one's progress towards a career

Knowledge preservation: An organized method for upholding and protecting valued information.

Youth engagement: A process or practice of involving young people in a particular activity

Chapter 15
The Forest as an Economy:
Indigenous Agroforestry Systems and Community-Driven Bioeconomies in the Amazon

Silvio Andrae
http://orcid.org/0000-0002-8586-7812
Independent Researcher, Germany

ABSTRACT

This chapter investigates Indigenous agroforestry systems in the Amazon as integrated socio-ecological economies that generate livelihoods, conserve biodiversity, and foster institutional resilience. It contends that forest gardens, locally referred to as chakra, chagra, and purma, represent advanced economic infrastructures rooted in Indigenous Knowledge Systems, relational governance, and multi-strata ecological design. Utilizing ethnographic literature, policy analysis, interviews, and biodiversity-based value chain research from Ecuador and Brazil, the chapter examines how Kichwa communities convert agroforestry practices into community-owned bioeconomic value chains. The analysis underscores the complex influence of international legal regimes, particularly access and benefit-sharing frameworks, on Indigenous economic autonomy. By framing forest gardens as 'technologies of the future,' the chapter reconceptualizes the bioeconomy as a transformation of governance rather than mere biomass substitution.

INTRODUCTION

Over the past two decades, the concept of the bioeconomy has emerged as a central pillar of global sustainability strategies. Governments, international organizations, and corporate actors increasingly promote bio-based production systems as pathways toward decarbonization, green growth, and resource efficiency (Bugge et al., 2016). Biomass substitution, biotechnology, and renewable materials are framed as technical solutions to the ecological crisis.

Yet, despite their environmental rhetoric, dominant bioeconomy models largely reproduce extractive development logics. They rely on large-scale monocultures, capital-intensive processing infrastructures, and globally integrated commodity chains, often externalizing social and ecological costs (Foster, 2010). In many contexts, they intensify land concentration, biodiversity loss, and socio-economic marginaliza-

DOI: 10.4018/979-8-3373-7976-0.ch015

tion. Nowhere are these contradictions more visible than in the Amazon Basin (Schwartzman & Zimmerman, 2005). Development discourse has historically marginalized Indigenous economies (Escobar, 1995). Alternative development frameworks emphasize plural pathways (Kothari et al., 2019). Recent policy-oriented syntheses emphasize that overcoming extractive development pathways in the Amazon requires structural transformations toward socio-bioeconomies grounded in Indigenous and rural livelihood systems, rather than technological substitution within existing growth regimes (Garrett et al., 2024)

Recent ecological research demonstrates that large parts of Amazonian forests have been shaped by centuries of Indigenous and colonial land-use practices, with lasting effects on species composition and species abundance. Long-term human–environment interactions have systematically altered forest structure, particularly through the selective promotion of useful plants and landscape management along river corridors. These findings challenge prevailing narratives of 'pristine wilderness' and highlight the historical depth of biodiversity-based livelihoods (McMichael et al., 2025).

This chapter argues that prevailing bioeconomy paradigms underestimate the potential of Indigenous forest-based production systems. At the same time, industrial bioeconomies seek to 'green' existing growth models. Indigenous agroforestry systems offer structurally different economic architectures grounded in biodiversity, relational governance, and cultural continuity (Berkes, 2012; Descola, 2013). Rather than treating forests as reservoirs of raw material, these systems operate on the premise that ecological complexity itself constitutes productive infrastructure (Altieri, 2018).

By examining Indigenous forest gardens as community-driven bioeconomic systems, this chapter proposes an alternative framework for understanding sustainability, development, and innovation in forest regions.

Problem Statement

Contemporary bioeconomy strategies are shaped by three interrelated characteristics: extractivism, capital intensity, and monoculturalization.

First, dominant models remain extractive in orientation. Although they replace fossil resources with biological inputs, they continue to rely on large-scale biomass extraction and long-distance value chains. Environmental costs are often displaced spatially rather than eliminated. Second, bioeconomic production is increasingly capital-intensive. High-tech processing facilities, proprietary biotechnology, and complex logistics systems privilege large corporate actors and marginalize small-scale producers. Entry barriers remain high, reinforcing existing inequalities. Third, many bioeconomy initiatives promote monocultural feedstock production. Oil palm, soy, eucalyptus, and sugarcane plantations are frequently rebranded as 'green' resources despite their destructive impacts on biodiversity and social systems. Together, these features generate a paradox: policies designed to mitigate ecological crisis often accelerate ecological simplification and social exclusion.

The Amazon Basin illustrates this paradox with particular clarity. Large-scale agriculture, cattle ranching, mining, and infrastructure development continue to drive deforestation and land degradation. These activities are frequently justified in the name of economic development and national competitiveness. Large-scale empirical evidence demonstrates that intact Amazonian forests sustain highly productive wild food systems, generating approximately 0.34 Mt of edible wild meat annually with an estimated economic value of over US$2 billion and supplying nearly half of the protein and iron requirements for rural populations (Antunes et al., 2025).

Recent system-level analyses indicate that the Amazon forest is increasingly exposed to interacting climatic and land-use disturbances that may trigger critical transitions and large-scale ecosystem collapse. Feedback-driven shifts between alternative stable states are becoming more likely as warming, deforestation, and fire regimes intensify (Flores et al., 2024).

At the same time, millions of Indigenous and forest-dependent people maintain livelihoods through diversified agroforestry, hunting, fishing, and the production of non-timber forest products. These systems conserve biodiversity, store carbon, and sustain cultural identities.

Despite their ecological and social contributions, biodiversity-based livelihoods remain politically marginalized and economically undervalued. Public policies, credit systems, and extension services overwhelmingly favor industrial land-use models. This structural bias reflects deeper epistemological hierarchies that privilege standardized, capital-intensive production over relational and place-based systems.

Recent global syntheses demonstrate that biological and cultural diversity are deeply interconnected, with Indigenous management practices systematically contributing to long-term biodiversity conservation and ecosystem stability. These findings challenge narratives that frame human presence primarily as a driver of degradation (Levis et al., 2024).

Research Questions

Against this background, the chapter addresses three interrelated research questions.

RQ1: How do Indigenous agroforestry systems function as economic systems? Rather than treating forest gardens as subsistence plots, the chapter examines their organizational principles, productivity dynamics, and governance structures.

RQ2: Can Indigenous Knowledge Systems–based forest gardens constitute scalable bioeconomic models? The analysis investigates whether and how these systems can expand through cooperative networks and federated structures without reproducing extractive dynamics.

RQ3: What governance and legal frameworks enable or constrain Indigenous bioeconomies? The chapter explores how international biodiversity law, intellectual property regimes, land tenure systems, and certification standards shape market participation and institutional autonomy.

Together, these questions situate Indigenous agroforestry at the intersection of ecological economics, political ecology, and development studies.

Methodology and Sources

This chapter adopts a qualitative, multi-source research design that combines theoretical analysis with empirical insights from secondary data and selected primary sources. The study draws on three main sources:

(1) ethnographic and interdisciplinary literature on Indigenous agroforestry systems and socio-ecological governance;
(2) policy documents and institutional reports from Ecuador and Brazil, particularly in the areas of biodiversity governance, bioeconomy strategies, and Indigenous rights; and

(3) selected qualitative insights from semi-structured interviews conducted with representatives of Indigenous cooperatives and organizations.

The interview material is used illustratively and contextually rather than as a basis for systematic qualitative analysis. Its purpose is to complement the literature-based analysis by providing grounded perspectives on governance practices, market integration, and socio-ecological dynamics.

The analytical approach follows a comparative and interpretive logic, focusing on identifying structural patterns, governance mechanisms, and conditions for scaling Indigenous bioeconomic systems across different contexts. This study follows an interdisciplinary socio-ecological systems approach (Berkes et al., 2003; Ostrom, 2005). The analysis acknowledges the positionality of the researcher and the limitations inherent in interpreting Indigenous knowledge systems through academic frameworks.

Chapter Structure

The chapter is organized into nine sections. Section 2 develops the theoretical framework, conceptualizing forest gardens as socio-ecological economic systems and situating Indigenous Knowledge Systems within debates on bioeconomy and commons governance. Section 3 presents a comparative analysis of Kichwa, Asháninka, and Tikuna agroforestry systems, highlighting structural convergences and functional diversity. Section 4 examines Kichwa bioeconomic value chains in Ecuador, focusing on cooperative governance, product strategies, market access, and ecological performance. Section 5 analyzes the legal and governance environment, including ABS regimes, intellectual property frameworks, and biocultural rights. Section 6 assesses ecological and economic impacts, emphasizing carbon storage, income stability, gender dynamics, and climate resilience. Section 7 explores conditions and limits of scaling without extractivism, with attention to network-based expansion and financialization risks. Section 8 situates Indigenous forest economies within broader debates on decolonizing development and alternative modernities. Section 9 synthesizes the findings and outlines policy implications, framing forest gardens as technologies of the future.

Through this structure, the chapter advances an integrated analysis of Indigenous agroforestry as a foundation for community-driven, regenerative bioeconomies in the Amazon.

THEORETICAL FRAMEWORK

Indigenous Knowledge Systems as Epistemic Infrastructure

Indigenous Knowledge Systems (IKS) in Amazonia constitute coherent epistemic and governance systems that integrate ecology, economy, spirituality, and territorial stewardship within a relational ontology. Rather than separating nature from economy, many Indigenous cosmologies conceptualize forests as living socio-ecological worlds in which humans participate through reciprocal relations (Descola, 2013; Viveiros de Castro, 2014). Knowledge is cumulative, intergenerational, and embedded in territory, ritual, and collective labor.

IKS functions as epistemic infrastructures by encoding ecological feedbacks, successional dynamics, soil cycles, species interactions, and seasonal variability into everyday production practices. In forest garden systems such as chakra, chagra, or purma, biodiversity is not treated as a constraint on efficiency but as the organizational basis of resilience and productivity. Production, distribution, and stewardship are therefore inseparable.

This relational ontology has direct institutional consequences. Land is understood less as a commodity than as a relational entity embedded in moral and cosmological obligations (Kopenawa & Albert, 2013). As a result, communal tenure systems, consensus-based governance, and long-term stewardship norms emerge not primarily from external regulation, but from culturally embedded forms of ecological accountability. IKS thus operate simultaneously as systems of knowledge, production, and governance.

Forest Gardens as Socio-Ecological Systems

Forest gardens can be analytically understood as complex socio-ecological systems (SES) in which ecological diversity, institutional coordination, and cultural reproduction are tightly interconnected. Sustainability emerges through adaptive interactions between human communities and ecological processes rather than through externally imposed optimization strategies. Recent biome-specific governance research similarly emphasizes that resilience depends on cross-scale feedbacks, relational knowledge systems, and adaptive institutional architectures rather than universal policy templates (Oarga-Mulec & Skene, 2026).

Multi-strata agroforestry systems integrate canopy trees, fruit species, medicinal plants, climbers, shrubs, and tubers within deliberately managed vertical structures. From an SES perspective, their resilience derives from species diversification, ecological redundancy, modular landscape organization, and feedback-sensitive management practices (Berkes et al., 2003). These characteristics stabilize production, enhance climate adaptability, and reduce systemic vulnerability.

Empirical research further indicates that such systems can maintain high biomass density and significant carbon storage while supporting diversified livelihood portfolios. Their productivity, therefore, cannot be evaluated solely by monocultural yield metrics, but must also include ecological services, nutritional diversity, medicinal functions, and long-term system stability. Economic rationality is thus relational and intertemporal rather than extractive and short-term oriented.

Bioeconomy: Extractive versus Biocentric Paradigms

Contemporary bioeconomy discourse is largely shaped by industrial substitution logics that seek to replace fossil-based inputs with biomass while maintaining existing growth trajectories (Bugge et al., 2016). In practice, these approaches often rely on monocultures, global commodity chains, and centralized forms of capital accumulation. Critics therefore argue that 'green growth' strategies frequently reproduce extractive dynamics and metabolic inequalities rather than fundamentally transforming human–environment relations (Foster, 2010; Gudynas, 2011).

By contrast, a biocentric bioeconomy situates economic activity within ecological limits and communal governance structures. Its core characteristics include community ownership, polycultural production systems, embedded value chains, and regeneration as a central performance criterion. Forests are treated not as raw material inputs but as socio-ecological infrastructures that simultaneously sustain production, biodiversity, and cultural continuity.

The distinction between the two paradigms is structural. Extractive bioeconomies externalize ecological costs and concentrate value capture, whereas biocentric systems internalize ecological cycles and distribute value through community networks. Their efficiency derives less from scale-based mechanization than from biodiversity synergies, low dependence on external inputs, and knowledge-intensive forms of production. Scaling, therefore, occurs primarily through cooperative networks and institutional replication rather than through territorial expansion or monocultural intensification.

Forest as Economy versus Forest as Resource

Paleoecological and spatial analyses indicate that present-day Amazonian forest composition reflects centuries of cumulative Indigenous management rather than untouched wilderness. Many contemporary forest landscapes therefore constitute historically co-produced socio-ecological systems shaped through sustained human–environment interaction (McMichael et al., 2025). This evidence supports the conceptualization of forest gardens not merely as subsistence landscapes, but as long-term ecological and economic infrastructures.

Conceptualizing the forest as an economy entails shifting from extractive resource use to the maintenance of ecological complexity as a productive system. In resource-based paradigms, value is generated primarily through the transformation of biomass into commodities. In forest-economy paradigms, by contrast, biodiversity itself functions as productive capital by stabilizing ecological processes, supporting diversified outputs, and enhancing long-term resilience.

This ontological shift has direct institutional consequences. Labor becomes ecological stewardship focused on succession management, soil regeneration, and intergenerational knowledge transmission. Governance is embedded in cosmology and communal accountability rather than external regulation. Risk is mitigated ecologically through diversification and modular landscape organization instead of financial hedging.

These dynamics resonate with commons theory, which emphasizes collective governance and rule-based stewardship (Ostrom, 1990, 2005). However, Indigenous forest economies extend conventional commons approaches by integrating relational and cosmological dimensions into economic coordination. In many Amazonian worldviews, plants, animals, rivers, and forest entities are conceived as relational beings embedded in networks of reciprocity and moral obligation (Descola, 2013; Kohn, 2013). Resource use is therefore governed through relational accountability rather than purely contractual property relations.

Kinship-based ecology generates distinctive economic institutions. Land and biodiversity are treated as custodial relations rather than commodities, while exchange remains embedded within reciprocity and communal obligations. Compliance emerges through identity formation, ritual practices, and collective memory, creating decentralized governance structures that align individual behavior with long-term ecosystem stewardship.

This relational economy differs fundamentally from extractive bioeconomy models, in which ecological degradation can be externalized spatially and institutionally. In kinship-based systems, environmental degradation directly undermines social legitimacy and cosmological order, generating strong endogenous constraints on unsustainable practices. At the same time, relational governance does not preclude innovation. New crops, technologies, and market relations may be selectively integrated as long as they remain compatible with communal and ecological obligations. This selective adaptability helps explain the long-term resilience of Indigenous forest economies.

Connectivity as Socio-Ecological Infrastructure

Recent ecological analyses demonstrate that landscape connectivity is a key determinant of biodiversity persistence, climate resilience, and long-term productivity in Amazonian ecosystems (Duarte Ritter et al., 2025). Connectivity refers to the capacity of organisms, nutrients, water, and ecological functions to circulate across forest, wetland, and riverine systems. Because Amazonian ecosystems exhibit non-linear feedback dynamics and potential tipping points, connectivity loss constitutes a systemic rather than localized ecological risk (Flores et al., 2024).

In conventional bioeconomy models, connectivity is often treated as an externality to be compensated through conservation offsets or isolated corridor projects. Indigenous agroforestry systems, by contrast, actively reproduce connectivity through diversified land-use mosaics, territorial governance, and relational ecological management. Indigenous Territories and Protected Areas consistently maintain significantly higher connectivity levels and lower anthropogenic resistance than surrounding landscapes (Duarte Ritter et al., 2025; Lehmann et al., 2003).

From a socio-ecological perspective, connectivity functions as productive infrastructure. It stabilizes ecological recovery processes, supports species diversity, strengthens hydrological regulation, and reduces long-term systemic vulnerability. These functions enhance the resilience and adaptive capacity of forest-based economies while reducing dependence on external inputs.

Connectivity is therefore not merely a biophysical condition, but also an institutional outcome shaped by communal land tenure, customary governance, and intergenerational knowledge transmission. Relational ontologies further reinforce landscape continuity by extending moral obligations across species and territories (Levis et al., 2024). By embedding connectivity directly within production and governance practices, Indigenous agroforestry systems internalize ecological interdependence ex ante rather than compensating for fragmentation ex post.

Toward an Integrated Conceptual Model

Taken together, the preceding arguments conceptualize Indigenous forest gardens as integrated socio-ecological economies in which Indigenous Knowledge Systems, ecological diversity, communal governance, and cooperative value chains jointly generate resilience and economic value. Forests are therefore understood not as passive resource reserves, but as living infrastructures that simultaneously sustain production, biodiversity, and cultural continuity.

The theoretical implication is that sustainable bioeconomies cannot be achieved solely through technological substitution within industrial growth paradigms. Rather, they require an ontological and institutional reorientation that recognizes ecological interdependence, relational governance, and regeneration as constitutive elements of economic organization.

This framework provides the conceptual foundation for the following empirical analysis, which examines how Indigenous bioeconomic systems operate in practice and how governance arrangements mediate their interaction with regional and global markets.

INDIGENOUS AGROFORESTRY SYSTEMS IN AMAZONIA

Indigenous agroforestry systems in Amazonia are highly differentiated, place-based production regimes that integrate ecological management, subsistence provisioning, and cultural reproduction. Forest gardens—locally referred to as chakra, chagra, or purma—function as multifunctional landscapes combining food production, medicinal cultivation, biodiversity conservation, and market-oriented activities

(Stevens, 2014). Rather than representing residual land-use forms, these systems constitute historically stabilized socio-ecological economies shaped through long-term adaptive experimentation (Altieri, 2018).

This section examines the Kichwa, Asháninka, and Tikuna systems comparatively, focusing on their ecological architecture, institutional embeddedness, and economic organization.

Kichwa Chakra Systems

The *Kichwa chakra* system of the Ecuadorian Amazon represents a highly developed form of permanent multi-strata agroforestry embedded within broader territorial mosaics of forest, secondary regrowth, and riparian zones (Álava-Núñez et al., 2025). Chakras combine canopy trees, fruit species, medicinal plants, and staple crops within vertically stratified production systems that maintain soil fertility and microclimatic stability without external inputs.

A defining characteristic of the chakra system is the integration of subsistence production with selective market participation. Species such as Ilex guayusa, cacao, cassava, and medicinal plants support both household provisioning and niche commercialization, enabling households to balance food security with income generation.

Governance is embedded in kinship structures and gendered knowledge systems. Women play central roles in seed selection, intercropping design, and post-harvest processing, while knowledge transmission occurs through collective labor and apprenticeship. Economically, chakras function as diversified production portfolios that reduce market dependency and income volatility while supporting cooperative bioeconomic initiatives.

Asháninka Purma Systems

Among Asháninka communities in Peru and western Brazil, agroforestry systems are structured around *purma*—rotational forest-fallow regimes that combine cultivation with long-term ecological regeneration. Plots are cultivated for several years and subsequently managed through extended fallow periods characterized by selective planting, enrichment strategies, and protection measures rather than passive abandonment.

This cyclical land-use logic maintains soil fertility, biodiversity connectivity, and successional diversity across landscape mosaics of early-, mid-, and late-stage vegetation. Productivity is therefore distributed temporally and spatially rather than concentrated through permanent intensification.

Purma management is organized through extended-family coordination and collective norms regulating land access, labor exchange, and regeneration cycles. Territorial memory plays an important role in plot allocation and restoration planning. Economically, Asháninka systems prioritize risk minimization and long-term territorial productivity over short-term output maximization. Cash crops are selectively integrated alongside hunting, fishing, and artisanal production, creating diversified livelihood portfolios with relatively low external dependency.

Purma systems thus exemplify a landscape-scale approach to sustainability in which regeneration itself constitutes a central productive strategy.

Tikuna Chagra Systems

The *Tikuna chagra* system, practiced in the tri-border regions of Brazil, Peru, and Colombia, represents a form of agroforestry in which production, ritual, and cosmology are closely intertwined. Chagras are established through controlled clearing followed by polycultural planting and gradual reforestation. Crop assemblages include manioc, fruits, spices, and medicinal species, often organized according to cosmological classifications and ceremonial cycles.

Ritual governance plays a central role in land-use regulation. Spiritual authorities influence planting schedules, harvesting practices, and species selection, while ecological disturbances are interpreted as disruptions of moral and cosmological balance. Chagras, therefore, function not only as productive landscapes but also as spaces of intergenerational learning and cultural reproduction.

Economically, Tikuna systems prioritize subsistence security combined with highly selective commercialization through community networks and Indigenous organizations. Medicinal plants, crafts, and specialty products are marketed on a limited scale, while market participation remains subordinated to communal values and cosmological integrity.

Comparative Analysis: Structural Convergences and Functional Diversity

Despite their distinct cultural and territorial contexts, the agroforestry systems of the Kichwa, Asháninka, and Tikuna display strong structural convergences shaped by long-term adaptive learning and comparable institutional logics of forest management (Berkes et al., 2003; Ostrom, 2005). Across all three systems, ecological diversity, communal governance, and portfolio-based production strategies are tightly interconnected.

Forest gardens combine multi-strata vegetation structures, successional management, and diversified species assemblages that stabilize productivity while reducing dependence on external inputs. They are embedded in broader landscape mosaics integrating primary forest, secondary regrowth, fallows, and riparian corridors, thereby supporting biodiversity, carbon storage, and hydrological regulation. Ecological performance emerges primarily from system architecture rather than technological intensification.

Institutionally, all three systems rely on kinship-based tenure arrangements, collective labor coordination, and intergenerational knowledge transmission. Governance mechanisms regulating plot rotation, species selection, and harvesting intensity are largely maintained through communal accountability, ritual practices, and social norms rather than formal enforcement structures.

Economically, the systems combine subsistence provisioning with selective commercialization through cooperatives, Indigenous organizations, and community networks. Diversified production portfolios reduce market dependency and stabilize livelihoods, while low external-input requirements minimize vulnerability to commodity price fluctuations.

At the same time, important functional differences persist. Kichwa chakra systems exhibit comparatively high levels of cooperative organization and niche-market integration, particularly around guayusa and specialty crops. Asháninka purma systems prioritize rotational regeneration and territorial resilience over commercial expansion, whereas Tikuna chagra systems place stronger emphasis on ritual governance, pedagogical reproduction, and selective market participation. These variations reflect distinct historical trajectories, political environments, and exposure to extractive pressures.

Comparative ecological analyses further indicate that contemporary Amazonian forest composition correlates strongly with historical Indigenous settlement patterns and long-term land management practices (McMichael et al., 2025). This continuity underscores the historical depth and adaptive persistence of Indigenous agroforestry systems across Amazonia.

Table 1. Comparative Structure of Indigenous Agroforestry Systems in Amazonia

Dimension	Kichwa (Chakra)	Asháninka (Purma)	Tikuna (Chagra)
Ecological Structure	Permanent multi-strata polyculture; high species density	Rotational agroforestry with managed fallows	Polycultural plots integrated into ritual cycles
Species Diversity	Very high (food, medicinal, commercial species)	High, emphasizing successional stages	High, with strong medicinal focus
Land-Use Dynamics	Semi-permanent gardens	Cyclical cultivation–regeneration	Controlled clearing and reforestation
Knowledge System	Gendered, cooperative-based transmission	Territorial memory, family-based learning	Ritualized, ceremonial pedagogy
Governance	Cooperative and kinship-based	Extended-family coordination	Spiritual and ritual authorities
Market Integration	Moderate to high (niche exports)	Low to moderate	Low, selective
Economic Orientation	Diversified income and specialization	Risk minimization	Subsistence-centered
Risk Management	Portfolio diversification	Mobility and fallows	Cosmological regulation
Role of Women	Central in seed, processing, marketing	Strong in household production	Strong in medicinal knowledge
Scalability Pattern	Networked cooperatives	Landscape-based expansion	Cultural replication

Source: own representation

Table 1 compares key structural dimensions of Indigenous agroforestry systems, including ecological complexity, species diversity, governance intensity, and degree of market integration. It demonstrates that, while these systems share common ecological and institutional foundations, they differ in their economic orientation and in their interfaces with external markets. This variation is central to understanding both their adaptive capacity and their potential for non-extractive scaling.

Indigenous Agroforestry as Economic Infrastructure

Taken together, Amazonian forest garden systems function as decentralized socio-ecological infrastructures that integrate production, ecological regulation, knowledge transmission, and governance within a single relational architecture. Their productive capacity derives less from external inputs or centralized technologies than from biodiversity, distributed ecological knowledge, and communal coordination mechanisms.

Biodiversity operates as living infrastructure by simultaneously supporting food production, soil fertility, pest regulation, medicinal provisioning, and climate resilience. Indigenous Knowledge Systems encode operational information regarding succession dynamics, species interactions, and ecological risk management into everyday production practices, enabling adaptive responses under changing environmental conditions.

Institutionally, kinship networks, communal assemblies, ritual authorities, and Indigenous organizations coordinate labor allocation, conflict resolution, and resource governance with relatively low external enforcement costs. Cooperatives and inter-community federations further enable horizontal scaling through knowledge exchange, joint marketing, and political representation while maintaining local governance autonomy.

From this perspective, Indigenous agroforestry systems demonstrate that complex economic infrastructures can emerge from low-energy, knowledge-intensive, and relationally governed socio-ecological systems rather than from capital-intensive industrial development pathways.

Implications for Bioeconomic Transformation

The comparative analysis of Kichwa, Asháninka, and Tikuna agroforestry systems supports three central conclusions regarding bioeconomic transformation. First, Indigenous forest gardens constitute durable economic systems rather than residual subsistence structures. They generate diversified livelihoods, ecological stability, and food sovereignty under conditions of environmental uncertainty.

Second, productivity and sustainability are not inherently antagonistic within Indigenous agroforestry systems. Biodiversity-based production enhances resilience while stabilizing long-term output, challenging industrial assumptions that ecological complexity necessarily reduces economic efficiency.

Third, cultural continuity functions as a productive resource. Kinship institutions, ritual governance, and intergenerational knowledge transmission strengthen collective action and long-term stewardship capacities. Governance erosion, by contrast, tends to increase ecological and economic vulnerability.

To operationalize these dynamics comparatively, Table 2 introduces the concept of transformation readiness, defined as the capacity of Indigenous agroforestry systems to engage with bioeconomic value chains without undermining their ecological and institutional foundations.

Table 2. Transformation Readiness Matrix of Indigenous Agroforestry Systems

Dimension	Kichwa (Chakra)	Asháninka (Purma)	Tikuna (Chagra)
Ecological Resilience	High	Very High	High
Market Readiness	High	Low	Low
Governance Capacity	High	Medium	High
Risk Exposure	Medium	Low	Low
Institutional Stability	High	Medium	High
Knowledge Transmission	High	High	High
External Dependency	Medium	Low	Low
Scalability Potential	Medium–High	Low–Medium	Low

Source: own representation

The matrix highlights differentiated transformation pathways. Kichwa systems combine relatively high market readiness with strong governance capacity, enabling selective integration into cooperative value chains while remaining exposed to commercialization pressures. Asháninka systems prioritize territorial regeneration and ecological resilience over commercial expansion, whereas Tikuna systems emphasize ritual governance and cultural continuity, limiting large-scale market integration.

Taken together, the matrix demonstrates that scalability within Indigenous bioeconomies is conditional rather than universal. Successful transformation depends less on output expansion than on institutional coherence, territorial autonomy, and the ability to maintain ecological and cultural integrity under conditions of external pressure.

CASE STUDY: KICHWA BIOECONOMIC VALUE CHAINS IN ECUADOR

This section examines how Kichwa communities in the Ecuadorian Amazon translate forest garden infrastructures into community-owned bioeconomic value chains. It analyzes institutional design, product strategies, market interfaces, and ecological performance, demonstrating how relatively high transformation readiness enables market integration without structural extractivism.

Community-Owned Cooperatives

Governance Structure

Kichwa bioeconomic initiatives are typically organized as community-owned cooperatives embedded in communal land regimes and Indigenous federations. Membership is territorially defined and linked to residency and kinship affiliation rather than capital contribution. Strategic decisions regarding market partnerships, certification, and reinvestment are collectively approved through communal assemblies, while executive boards remain accountable through rotating mandates and participatory monitoring mechanisms (Ostrom, 1990).

Cooperatives are further embedded in multi-level governance structures linking village councils, Indigenous organizations, and regional federations, thereby strengthening bargaining capacity vis-à-vis buyers and public institutions. As one cooperative leader explained: 'Decisions about production and sales are not individual choices; they must reflect the priorities of the community as a whole' (Interview, cooperative leader, Ecuador, 2024).

Revenue Distribution

Revenue allocation combines household-level incentives with collective redistribution through community funds, ecological reserves, and organizational budgets for logistics and market development.

- Household payments based on delivered volume and quality
- Community funds for education, healthcare, and infrastructure
- Ecological reserves financing reforestation and seed banks
- Organizational budgets for logistics and market development

This hybrid structure internalizes social reproduction and ecological reinvestment within commercial operations while stabilizing participation during market fluctuations. As one community member noted: 'Income is not just for the family; part of it always returns to the community, to schools, to the forest, to the future' (Interview, community member, Ecuador, 2024).

Women's Leadership

Women occupy central positions in Kichwa bioeconomic systems, particularly in seed management, processing, quality control, and financial administration (Howard, 2003). Many cooperatives maintain gender-balanced governance structures or reserved leadership positions, while women's collectives frequently coordinate nurseries, drying facilities, and medicinal product lines.

Empirical studies suggest that strong female participation contributes to product consistency, compliance with certification standards, and stable buyer relations (Rocheleau et al., 1996; Agarwal, 1992). As one cooperative coordinator explained: 'Women are responsible for maintaining the diversity of the chakra and ensuring that quality is consistent across harvests' (Interview, female cooperative coordinator, Ecuador, 2024).

Key Products

Kichwa cooperatives focus on biodiversity-based products that combine cultural embeddedness with international market demand. Three product categories are particularly important: guayusa, camu-camu, and açaí.

Ilex guayusa (Guayusa)

Guayusa is the flagship product of many Kichwa bioeconomic initiatives. Cultivated within multi-strata chakra systems, it is selectively harvested and processed into teas and extracts without chemical inputs. Commercialization builds upon existing cultural practices in which guayusa consumption is embedded in collective rituals and intergenerational knowledge transmission. International demand for herbal beverages has enabled cooperatives to position guayusa as an ethically sourced and biodiversity-friendly niche product.

Myrciaria dubia (Camu-Camu)

Camu-camu is cultivated in riparian agroforestry systems and seasonally flooded areas, combining managed enrichment planting with selective harvesting. Due to its exceptionally high vitamin C content, it is increasingly marketed in the nutraceutical and cosmetic sectors. Cooperative coordination is particularly important for post-harvest logistics, quality control, and processing partnerships.

Euterpe oleracea (Açaí)

Açaí is cultivated in mixed palm stands and floodplain forests rather than monocultural plantations. Harvesting practices maintain natural species assemblages, minimize ecological disturbance, and enable participation in international food markets. Despite high commercial potential, Kichwa cooperatives deliberately limit monoculture expansion to avoid ecological simplification and labor dependency.

Taken together, these product strategies illustrate how biodiversity-based commercialization can occur without abandoning diversified agroforestry structures or communal governance principles.

Price Structures and Market Access

Direct trade arrangements enable Kichwa cooperatives to capture substantial price premiums by reducing intermediary dependence, standardizing quality, and emphasizing ethical and biodiversity-based branding. Comparative analyses suggest that cooperative marketing can generate significantly higher farm-gate prices than intermediary-dominated value chains, thereby stabilizing household income and reducing vulnerability to market volatility.

Historically, Indigenous producers were often integrated into asymmetric trading systems characterized by debt dependency, opaque pricing, and weak bargaining power. Cooperative-based direct trade partially replaces these structures through long-term buyer relationships, pre-financing arrangements, and collectively negotiated pricing mechanisms. As one cooperative board member explained: 'We prefer long-term relationships with buyers who understand our system, rather than higher prices that come with pressure to change how we produce' (Interview, cooperative board member, Ecuador, 2024).

Certification schemes such as organic, fair-trade, and biodiversity-friendly labels facilitate market access and premium pricing but also impose administrative and financial burdens (Raynolds, 2004). Many cooperatives therefore treat certification strategically, combining formal standards with Indigenous branding, peer monitoring, and relational trust in order to reduce compliance costs and maintain governance autonomy.

Despite these advances, market integration remains selective and constrained. Dependence on a limited number of ethical buyers, fluctuating international demand, and unequal bargaining relations continue to generate structural vulnerabilities.

Ecological Performance Metrics

Kichwa forest gardens maintain high carbon stocks, stable nutrient cycles, and strong climate resilience through diversified species assemblages, continuous canopy cover, and low external-input management. Comparative assessments indicate substantially higher ecological performance than pasture systems and monocultures, while mature agroforestry systems can approach the carbon storage levels of adjacent primary forests (Aragão et al., 2026).

Soil fertility is sustained through organic mulching, leaf litter accumulation, nitrogen-fixing species, and rotational enrichment planting, allowing long-term productivity without synthetic fertilizers. Ecological complexity simultaneously enhances adaptive capacity by buffering climatic variability, pest outbreaks, and hydrological stress.

Resilience is reinforced not only ecologically but also socially. Collective labor mobilization, seed-sharing networks, and reciprocal support mechanisms accelerate recovery following environmental shocks. As one farmer explained: 'When extreme weather affects one plot, the community redistributes seeds and labor so that recovery can begin immediately' (Interview, smallholder farmer, Ecuador, 2024).

Taken together, these findings demonstrate that ecological regeneration and economic production are co-produced within Kichwa agroforestry systems rather than treated as competing objectives.

These findings confirm the analytical expectations derived from the transformation readiness framework. Relatively strong governance capacity and knowledge infrastructures allow Kichwa communities to engage selectively with global markets while preserving cultural and ecological integrity.

The following section examines how these achievements are mediated—and sometimes constrained—by legal and regulatory frameworks, particularly in the context of access and benefit-sharing.

GOVERNANCE AND LEGAL DIMENSIONS

The viability of Indigenous bioeconomies depends not only on ecological resilience and cooperative organization but also on the interaction between customary governance systems and external legal frameworks. International biodiversity law, intellectual property regimes, and national land policies shape the conditions under which Amazonian communities participate in bioeconomic activities.

These frameworks generate structurally ambivalent effects. While they may provide recognition, market access, and benefit-sharing opportunities, they also introduce new risks of commodification, external appropriation, and institutional dependency.

Nagoya Protocol and Access and Benefit-Sharing Regimes

The Access and Benefit-Sharing (ABS) framework seeks to regulate access to genetic resources and associated traditional knowledge while ensuring fair and equitable benefit-sharing (Secretariat of CBD, 2011). For Kichwa cooperatives engaged in herbal, nutraceutical, and biodiversity-based value chains, ABS regulations are particularly relevant in relation to prior informed consent (PIC) and mutually agreed terms (MAT).

In practice, however, ABS regimes generate highly uneven outcomes. While they may strengthen bargaining positions and formalize benefit-sharing arrangements, administrative complexity, limited legal literacy, and high transaction costs often undermine effective participation (Morgera et al., 2012). These asymmetries reflect broader distributive and procedural justice challenges within international biodiversity governance (Schroeder & Pogge, 2009). Governance research further suggests that Indigenous knowledge systems remain marginalized within many formal policy processes (Lansbury et al., 2025).

Many Indigenous organizations lack the technical and legal capacity to navigate national ABS procedures without external intermediaries. As one Indigenous federation representative explained: 'The legal language is not designed for us. We depend on external actors to interpret rules that are supposed to protect us' (Interview, Indigenous federation representative, Ecuador, 2024).

As a result, ABS frameworks frequently operate less as empowering governance mechanisms than as externally managed compliance systems.

Biopiracy and Intellectual Property

A central limitation of conventional intellectual property and ABS regimes lies in their weak recognition of collective and intergenerational knowledge systems. In response, Indigenous movements and critical legal scholars increasingly advocate biocultural approaches linking knowledge protection to territorial governance and collective rights.

Traditional knowledge (TK) databases may provide defensive protection against unauthorized appropriation, yet they also generate significant governance dilemmas. Translating relational and context-dependent knowledge into standardized and codified formats risks decontextualization and epistemic reduction. Once recorded in accessible databases, control over the circulation and commercial use of knowledge becomes increasingly difficult.

At the same time, intellectual property frameworks tend to transform traditional knowledge into tradable informational assets (Santos, 2014). While Indigenous branding and biodiversity narratives may generate market premiums, they can also fragment collective knowledge systems, weaken customary authority, and encourage over-disclosure. Many Kichwa organizations therefore approach formal knowledge documentation cautiously and continue to rely on selective disclosure, relational trust, and community-based governance mechanisms.

In contemporary bioeconomic contexts, biopiracy increasingly occurs not through direct extraction but through asymmetrical contractual, branding, and research relationships that enable external actors to capture economic value from Indigenous knowledge systems.

From 'Access to Knowledge' to 'Rights to Knowledge'

In response to the limitations of conventional ABS and intellectual property regimes, Indigenous movements and critical legal scholars increasingly advocate biocultural rights frameworks that integrate biological resources, cultural practices, collective knowledge systems, and territorial governance (Kothari et al., 2019). Rather than treating knowledge as transferable information, this approach situates it within broader social and ecological relations.

For Kichwa cooperatives, biocultural rights provide a normative basis for resisting extractive partnerships and asserting governance autonomy over commercialization pathways, resource access, and knowledge circulation. Community protocols operationalize these principles by defining local rules regarding consent procedures, benefit-sharing expectations, data ownership, research collaboration, and cultural restrictions. Unlike standardized contractual arrangements, such protocols are grounded in customary governance and collective deliberation, although customary law is often poorly recognized within formal legal systems (Downes, 2000).

Several Kichwa communities in Ecuador have developed community protocols to regulate interactions with researchers, certification bodies, and commercial actors, thereby strengthening negotiation capacity and reducing dependency on external legal intermediaries.

Rights to knowledge remain inseparable from territorial governance. Without secure land tenure and effective territorial control, Indigenous bioeconomic systems remain vulnerable to land grabbing, illegal extraction, and infrastructural encroachment. Participatory mapping, community patrols, and monitoring systems increasingly complement customary governance mechanisms in defending territorial autonomy and regulating resource access.

Conflict and Institutional Tensions

A persistent source of tension arises from the coexistence of state-centered land regimes and customary tenure systems. Although many Amazonian states formally recognize Indigenous territories, implementation is often partial and reversible, particularly when mining, infrastructure, or conservation interests intervene. State agencies frequently retain discretionary authority over subsoil rights, infrastructure concessions, and zoning decisions, thereby weakening long-term planning security for Indigenous cooperatives.

Uncertain land tenure discourages infrastructure investment, long-term agroforestry planning, and institutional consolidation. Bureaucratic requirements may further conflict with customary governance procedures, generating compliance burdens and governance fatigue. Empirical evidence from Brazil demonstrates that weakening environmental governance and Indigenous land protections directly correlate with rising deforestation rates inside Indigenous territories (Silva-Junior et al., 2023).

NGOs and corporate actors play structurally ambivalent roles within Indigenous bioeconomies. While they may provide legal support, technical expertise, market access, and infrastructure, they can also generate new forms of dependency and influence from external governance. Donor priorities, reporting requirements, and externally defined quality standards may gradually displace local governance priorities and weaken endogenous institutional development.

Corporate sourcing arrangements often reproduce asymmetrical bargaining over pricing, branding, and certification requirements. Even where partnerships are framed as ethical sourcing models, governance authority often remains unevenly distributed.

Taken together, these tensions illustrate the structurally ambivalent governance environment surrounding Indigenous bioeconomies. Legal and market frameworks may create opportunities for recognition and commercialization, yet they also expose communities to new forms of appropriation and dependency. The Kichwa case demonstrates that successful bioeconomic integration depends less on formal compliance than on institutional coherence, territorial autonomy, and governance control over knowledge and commercialization pathways. Where these conditions are maintained, market participation can reinforce

ECOLOGICAL AND ECONOMIC IMPACT ASSESSMENT

This section examines the ecological and economic impacts of Indigenous bioeconomic systems, focusing on carbon dynamics, biodiversity, income stabilization, gendered participation, and resilience under climate stress. Rather than treating these dimensions separately, the analysis emphasizes their interdependence within integrated socio-ecological production systems.

Carbon and Biodiversity Metrics

Indigenous agroforestry systems generate ecological outcomes that differ fundamentally from monocultural agriculture and extractive land use. Continuous canopy cover, multi-strata vegetation, and species diversity support high carbon storage, soil stability, habitat connectivity, and ecosystem resilience.

Empirical evidence indicates that multi-strata agroforestry systems store substantially more carbon than pasture systems and annual monocultures, in some cases approaching the ecological performance of secondary forests. Comparative life-cycle assessments further show that monocultural production systems generate significantly higher carbon footprints and social externalities than agroforestry-based systems (Machado et al., 2025).

Long-term ecological analyses suggest that many contemporary Amazonian biodiversity patterns reflect cumulative Indigenous management rather than passive conservation (McMichael et al., 2025; Levis et al., 2024). Biodiversity performs productive, regulatory, and cultural functions simultaneously by stabilizing pollination, nutrient cycling, pest regulation, and climate buffering across interconnected landscapes.

At the same time, accelerating climate change increasingly threatens the adaptive capacity of forest-based economies. Large parts of the Amazon may face compounding ecological disturbances and significant biodiversity loss under high-emission scenarios (Flores et al., 2024). Climate projections further indicate substantial habitat-contraction risks for many native agroforestry species, where governance systems fail to support coordinated adaptation and landscape connectivity (Lima et al., 2022).

Taken together, these findings demonstrate that carbon sequestration, biodiversity conservation, and livelihood stability are structurally interconnected within Indigenous agroforestry systems rather than separate environmental objectives.

Income Stabilization Effects

Indigenous bioeconomies stabilize livelihoods primarily through diversified production portfolios and cooperative risk-sharing mechanisms. Households typically combine subsistence production, commercialization of forest products, wage labor, and reciprocal exchange systems, thereby reducing exposure to price volatility and harvest failure.

Agroforestry-based commercialization generally complements rather than replaces existing livelihood strategies. This diversification stabilizes household income over time and reduces dependence on a single commodity or on external market fluctuations.

Cooperative marketing arrangements—including collective bargaining, advance payments, and minimum-price mechanisms—further reduce liquidity constraints and dependence on informal credit. As a result, income trajectories generally exhibit lower temporal volatility than those in monocultural production systems, thereby enhancing long-term planning capacity and economic resilience.

Gendered Economic Participation

Women play central roles in seed management, processing, quality control, accounting, and the transmission of medicinal knowledge within Indigenous bioeconomies. Their participation is particularly important in value-added activities such as drying, packaging, and product differentiation, which enhance local value retention and niche-market competitiveness.

Empirical studies indicate that stronger female participation is associated with improved governance performance, higher institutional transparency, and more stable compliance with cooperative rules (Agarwal, 1992). At the household level, women's control over income frequently contributes to improved nutrition, education, and healthcare outcomes.

At the same time, these benefits depend on supportive governance arrangements. Without equitable workload distribution and meaningful participation in decision-making processes, market integration may reinforce rather than reduce existing inequalities.

Resilience under Climate Stress

Indigenous agroforestry systems display distinctive resilience capacities under conditions of climatic variability and ecological stress. Species diversity, vertical stratification, and landscape connectivity stabilize production during droughts, floods, and pest outbreaks, while natural regeneration processes facilitate comparatively rapid recovery following disturbances.

Resilience is reinforced through social institutions, including collective labor mobilization, seed-sharing networks, reciprocal redistribution mechanisms, and adaptive ecological knowledge. Indigenous production systems continuously adjust planting calendars, species composition, and harvesting practices in response to changing environmental conditions.

Comparative evidence indicates that agroforestry-based systems generally experience lower yield losses and faster recovery than surrounding monocultures under climatic stress. At the same time, accelerating climate change and biome degradation increasingly threaten the long-term viability of forest-based economies. Large parts of the Amazon may face compounding ecological disturbances under high-emission scenarios (Flores et al., 2024), while many native agroforestry species are projected to lose substantial habitat areas without coordinated adaptation measures (Lima et al., 2022).

Governance responsiveness, therefore, becomes a critical resilience factor. Research on dynamic biome governance suggests that institutional systems unable to adapt to accelerating ecological change risk severe socio-ecological destabilization (Oarga-Mulec & Skene, 2026). Regions experiencing extensive forest degradation already exhibit significant declines in wildlife-harvesting productivity and livelihood stability (Antunes et al., 2025).

Taken together, these findings suggest that resilience within Indigenous bioeconomies is structurally embedded in ecological diversity, territorial governance, and distributed knowledge systems rather than externally imposed through technological control.

SCALING WITHOUT EXTRACTIVISM

A central challenge for Indigenous bioeconomies concerns whether forest-based economic systems can expand beyond localized initiatives without reproducing extractive dynamics. In dominant development models, scalability is typically associated with standardization, capital intensification, and integration into global commodity chains. Historical analyses of non-timber forest product (NTFP) economies, however, demonstrate that forest-based development has long faced structural constraints including market volatility, weak value addition, and institutional fragmentation (Dürr et al., 2026).

For Indigenous agroforestry systems, scalability must therefore be reconceptualized. Rather than vertical expansion through land consolidation or monocultural intensification, scaling refers to the replication of socio-ecological governance systems across territories while maintaining ecological diversity, territorial autonomy, and institutional coherence.

Scaling in this context is primarily network-based. Cooperative federations, Indigenous producer associations, and inter-community alliances enable horizontal coordination without centralized control. These federated structures support joint marketing, shared logistics, knowledge exchange, certification management, and political representation while preserving local governance autonomy. Recent research further suggests that maintaining territorial connectivity and expanding Indigenous governance systems constitute critical resilience mechanisms for preventing large-scale ecological destabilization in Amazonia (Flores et al., 2024).

Knowledge exchange networks play a particularly important role in this process. Agroecological practices, governance innovations, and market strategies are diffused through peer-learning systems and inter-community cooperation rather than through top-down extension models. This enables adaptive learning without institutional homogenization.

At the same time, expanding market integration exposes Indigenous bioeconomies to significant financialization pressures. Carbon markets, biodiversity credits, impact investment structures, and sustainability branding increasingly transform ecological functions and cultural narratives into tradable assets. While such mechanisms may generate short-term revenues, they can also weaken communal governance by shifting decision-making authority toward external investors, certification actors, and financial intermediaries (Wunder, 2015; Andrae, 2025).

Debt-based financing may further generate extractive pressures where repayment obligations conflict with ecological cycles and climatic variability. Under such conditions, commercialization can incentivize monocultural expansion, overharvesting, and internal stratification within cooperatives. Historical and contemporary analyses of Amazonian bioeconomy initiatives, therefore, caution against overly optimistic assumptions regarding scalability under market-driven conditions (Dürr et al., 2026).

These tensions demonstrate that Indigenous bioeconomies do not automatically constitute sustainable alternatives. Their long-term viability depends on maintaining institutional coherence, ecological diversity, and governance autonomy amid external economic pressure. Comparative evidence indicates that monocultural expansion may generate higher short-term profits but also increases emissions, ecological degradation, and social vulnerability, whereas agroforestry systems offer more resilient yet less capital-intensive development pathways (Machado et al., 2025).

Preventing extractive scaling trajectories, therefore, requires supportive policy frameworks. Secure collective land rights, protection against territorial encroachment, and legal recognition of customary governance systems constitute foundational preconditions. Equally important are patient and non-extractive financing instruments aligned with ecological production cycles, as well as targeted support for cooperative infrastructure, decentralized processing systems, and Indigenous governance capacity.

Scaling in Indigenous bioeconomies is thus not primarily a quantitative process of expansion but a qualitative process of institutional reproduction. Where territorial rights, governance autonomy, and ecological integrity are maintained, network-based bioeconomic systems can expand without reproducing extractive hierarchies. Where these conditions are absent, scaling tends to generate new forms of dependency, commodification, and ecological simplification.

DECOLONIZING DEVELOPMENT

The preceding sections have demonstrated that Indigenous agroforestry systems function as coherent economic infrastructures capable of generating income, conserving biodiversity, and fostering social resilience simultaneously. This section situates these findings within a broader theoretical debate on development paradigms. It argues that forest-based bioeconomies constitute not a marginal alternative, but an expression of alternative modernity—one that challenges extractive growth models while redefining innovation, value, and progress.

Beyond GDP

Dominant development frameworks continue to evaluate progress primarily through GDP growth, privileging market transactions, industrial output, and capital accumulation while systematically undervaluing subsistence production, ecosystem services, and social reproduction (Andrae, 2024).

Indigenous forest economies expose the limitations of GDP-centered evaluation in several ways. First, substantial value remains outside formal markets through food sovereignty, medicinal self-sufficiency, and reciprocal exchange systems. Second, GDP accounting fails to capture ecological degradation and long-term resilience dynamics. Extractive expansion may increase short-term output while simultaneously eroding biodiversity and productive stability. Third, conventional indicators neglect governance quality and collective redistribution mechanisms that sustain social resilience within cooperative systems.

Forest-based economies, therefore, challenge dominant assumptions about productivity and development by demonstrating that ecological integrity and economic stability can function as mutually reinforcing rather than opposing objectives.

Innovation from Tradition

Modernization theories often frame innovation as a rupture with tradition and a transition from customary governance toward industrial production and technocratic institutions. Indigenous bioeconomies challenge this assumption. Agroforestry systems demonstrate sophisticated forms of ecological design, succession management, and adaptive experimentation developed through long-term relational knowledge systems.

Commercialization of products such as guayusa or camu-camu, therefore, does not represent the introduction of modernity into traditional systems, but the strategic adaptation of existing knowledge infrastructures to new market contexts. Innovation emerges through selective integration rather than institutional replacement.

This form of endogenous modernization destabilizes binary distinctions between 'traditional' and 'modern' economies. Indigenous forest economies, in contrast, illustrate plural pathways of technological and economic development grounded in territorial governance, cultural continuity, and ecological complexity.

Epistemic Justice

Decolonizing development requires not only institutional reform but also epistemic justice: the recognition of Indigenous knowledge systems as legitimate foundations for governance, economic organization, and sustainability transitions.

Historically, Indigenous knowledge has frequently been marginalized, extracted, or incorporated into external policy frameworks without corresponding authority over interpretation and application. Scientific evidence increasingly demonstrates, however, that Indigenous ecological management has materially shaped Amazonian biodiversity and long-term ecosystem resilience (McMichael et al., 2025; Levis et al., 2024).

Forest-based bioeconomies, therefore, challenge dominant epistemic hierarchies by positioning Indigenous Knowledge Systems as operational governance infrastructures rather than supplementary cultural inputs. Failure to recognize such epistemic authority risks reproducing extractive dynamics under the language of sustainability and green development.

These asymmetries reflect broader structures of the 'coloniality of power,' in which global hierarchies of knowledge, labor, and resource control remain deeply interconnected (Quijano, 2000). In contrast, pluriversal governance approaches emphasize co-designed institutional systems that integrate Indigenous governance principles into broader sustainability frameworks (Oarga-Mulec & Skene, 2026).

Implications for Global Bioeconomy Agendas

Indigenous forest economies demonstrate that bioeconomy transitions cannot be reduced to technical substitution strategies replacing fossil inputs with biomass. Their significance lies in governance transformation rather than feedstock replacement alone.

Several implications follow. Biodiversity must be treated as productive infrastructure rather than raw material, implying a shift away from monocultural 'bio-based' production systems toward diversified agroecological landscapes. Sustainable bioeconomies further require secure territorial rights, collective governance structures, and financing systems aligned with ecological regeneration cycles rather than short-term return expectations.

Equally important, scaling must occur through federated and network-based structures that preserve local autonomy instead of centralized consolidation. International bioeconomy agendas, therefore, need to integrate biocultural rights, epistemic pluralism, and Indigenous governance systems as foundational rather than peripheral components of sustainability transitions.

Forest-based economies in Amazonia thus represent not residual or local exceptions, but empirically grounded models of post-extractive development.

Indigenous forest economies challenge dominant development paradigms by linking ecological integrity, collective governance, and economic production within integrated socio-ecological systems. Decolonizing development, therefore, does not imply rejecting markets or technology, but reordering them within biocentric and territorially grounded institutional frameworks.

The following conclusion synthesizes the chapter's broader implications for regenerative development and sustainability governance.

CONCLUSION

This chapter has examined Indigenous agroforestry systems in Amazonia as integrated socio-ecological economies capable of simultaneously generating livelihoods, conserving biodiversity, and sustaining institutional resilience. Through theoretical analysis, comparative system assessment, and the case study of Kichwa bioeconomic value chains, it has been shown that forest gardens are not subsistence remnants but structurally coherent economic architectures.

By positioning Indigenous Knowledge Systems at the center of production, governance, and market engagement, the chapter reframes the forest from a passive resource base into an active socio-ecological infrastructure. Indigenous forest economies demonstrate that biodiversity, knowledge, and collective institutions can co-produce economic value without relying on extractive growth dynamics.

Three conclusions follow.

First, forest gardens function as living economic systems in which ecological complexity operates as productive capital. Multi-strata design, species diversity, and long-term stewardship generate carbon storage, soil regeneration, climate buffering, and diversified production portfolios. Resilience emerges endogenously from relational ecological management rather than through external technological control.

Second, community-owned cooperatives enable selectively mediated market integration. Through collective governance, federated coordination, and diversified value chains, Indigenous bioeconomic systems can expand without necessarily reproducing structural dependency. Market participation remains embedded within communal governance structures rather than subordinating them.

Third, governance transformation constitutes the enabling condition of a biocentric bioeconomy. Where territorial rights, biocultural recognition, and institutional autonomy are maintained, commercialization can reinforce resilience and collective capacity. Where governance control is fragmented, extractive dynamics tend to re-emerge through dependency, commodification, and external appropriation.

Conceptualizing forest gardens as distributed socio-ecological infrastructures clarifies their broader significance. Their productive capacity resides not primarily in capital-intensive technologies, but in ecological intelligence, accumulated knowledge, and embedded governance systems. Diversity, redundancy, reciprocity, and territorial connectivity operate as structural design principles of resilient economic organization.

Taken together, these findings address the chapter's guiding research questions. Indigenous agroforestry systems function as coherent economic systems in which biodiversity, knowledge, and governance co-constitute productive capital. Their scalability is conditional and depends on federated coordination, institutional replication, and territorial governance rather than standardization or monocultural expansion. Their long-term viability further depends on legal and political frameworks that secure biocultural rights, collective land tenure, and Indigenous governance autonomy.

These findings challenge dominant assumptions within contemporary bioeconomy and development debates. Indigenous forest economies demonstrate that innovation does not require rupture with tradition, and that sustainability cannot be reduced to technical optimization or biomass substitution alone. Rather, they illustrate alternative pathways of development grounded in ecological complexity, collective governance, and relational economic organization.

Recognizing forests as economies—and Indigenous knowledge as infrastructural capacity—does not romanticize tradition. It acknowledges that viable post-extractive futures are already being practiced. Indigenous agroforestry systems, therefore, constitute not peripheral experiments, but living laboratories of regenerative economic transformation.

FUTURE RESEARCH DIRECTIONS

Future research should further investigate the long-term viability, governance dynamics, and scalability of Indigenous bioeconomies under conditions of accelerating ecological and economic transformation.

First, longitudinal comparative studies are needed to assess the resilience performance of Indigenous agroforestry systems over extended time horizons. Comparative analyses between agroforestry systems, monocultural agriculture, and restoration-based production landscapes could provide deeper insight into carbon permanence, biodiversity stability, and livelihood resilience under climate stress.

Second, additional research should examine governance transformations under conditions of commercialization and market integration. While cooperative federations can strengthen community autonomy, increasing participation in global value chains may also generate internal stratification, financial dependency, or pressures toward ecological simplification. Future studies should therefore analyze the institutional thresholds at which bioeconomic scaling risks reproducing extractive dynamics.

Third, climate adaptation pathways within Indigenous agroforestry systems require further investigation. In particular, future research should explore adaptive species portfolios, landscape connectivity, and decentralized resilience mechanisms under different climate scenarios. Integrating ecological modeling with Indigenous Knowledge Systems may provide important insights into long-term adaptation capacities.

Fourth, comparative research beyond Amazonia would substantially enrich the debate on Indigenous bioeconomies. Cross-regional analyses involving forest-based economies in Africa, Southeast Asia, and other tropical regions could help identify both context-specific dynamics and transferable governance principles for regenerative development pathways.

Taken together, these research directions highlight the importance of interdisciplinary, participatory, and long-term approaches for understanding Indigenous bioeconomies as evolving socio-ecological systems rather than static development models.

REFERENCES

Agarwal, B. (1992). The gender and environment debate: Lessons from India. *Feminist Studies*, *18*(1), 119–158. DOI: 10.2307/3178217

Álava-Núñez, P., Torres, B., Castro, M., & Robles, M. (2025). AGB carbon stock analysis in the Indigenous agroforestry of the Ecuadorian Amazon: Chakra and Aja as Natural Climate Solutions. *Frontiers in Forests and Global Change*, *8*, 1513140. Advance online publication. DOI: 10.3389/ffgc.2025.1513140

Altieri, M. A. (2018). *Agroecology: The science of sustainable agriculture* (2nd ed.). CRC Press. DOI: 10.1201/9780429495465

Andrae, S. (2024). The establishment of a Gross Ecosystem Product: A country comparison. In Boubaker, S., & Le, T. H. (Eds.), *Green finance and sustainable development goals* (pp. 369–407). World Scientific Publishing., DOI: 10.1142/9781800614482_0014

Andrae, S. (2025). Financing mechanisms for the conservation of biodiversity: A multi-country comparison. In Walker, T., Naffa, H., Tharumar, R. K., & Donders, S. (Eds.), *Biodiversity finance: The economic, operational, and societal impacts of biodiversity loss* (pp. 269–299). Springer., DOI: 10.1007/978-3-032-02029-1_11

Antunes, A. P., & Lima Constantino, P. (2025). Healthy forests safeguard traditional wild meat food systems in Amazonia. *Nature*, *648*(8094), 625–633. DOI: 10.1038/s41586-025-09743-z PMID: 41299169

Aragao, S. D., Alfaia, S. S., da Costa Ayres, M. I., & Capelloto Costa, F. R. (2026). Soil fertility and carbon stocks of indigenous agroforestry homegardens in the lower Amazon region. *Agroforestry Systems*, *100*(1), 22. DOI: 10.1007/s10457-025-01398-z

Berkes, F. (2012). *Sacred ecology* (3rd ed.). Routledge. DOI: 10.4324/9780203123843

Berkes, F., Colding, J., & Folke, C. (Eds.). (2003). *Navigating social–ecological systems: Building resilience for complexity and change*. Cambridge University Press.

Bugge, M. M., Hansen, T., & Klitkou, A. (2016). What is the bioeconomy? *Sustainability (Basel)*, *8*(7), 691. DOI: 10.3390/su8070691

Descola, P. (2013). *Beyond nature and culture*. University of Chicago Press. DOI: 10.7208/chicago/9780226145006.001.0001

Downes, D. (2000). *Protecting community rights over traditional knowledge: Implications of customary laws and practices*. Center for International Environmental Law.

Duarte Ritter, C., Muñoz, J., Machado, A. F., Albert, J. S., Ribas, C. C., & Carnaval, A. C.. (2025). Indigenous territories and protected areas are crucial for ecosystem connectivity in the Amazon basin. *Proceedings of the National Academy of Sciences of the United States of America*, *122*(31), e2418189122. DOI: 10.1073/pnas.2418189122 PMID: 40720645

Dürr, J., Carvalho Ribeiro, S. M., Vargas, D., Sellare, J., & Börner, J. (2026). Forest-based Amazonian bioeconomies: New opportunities and old challenges. In Zilberman, D., Zhuang, J., Wesseler, J., & Khanna, M. (Eds.), *Handbook of circular bioeconomy* (pp. 583–603). Springer., DOI: 10.1007/978-3-032-07112-5_27

Escobar, A. (1995). *Encountering development: The making and unmaking of the Third World*. Princeton University Press.

Flores, B. M., Montoya, E., Sakschewski, B., Nascimento, N., Staal, A., Betts, R. A., Levis, C., Lapola, D. M., Esquível-Muelbert, A., Jakovac, C., Nobre, C. A., Oliveira, R. S., Borma, L. S., Nian, D., Boers, N., Hecht, S. B., ter Steege, H., Arieira, J., Lucas, I. L., & Hirota, M. (2024). Critical transitions in the Amazon forest system. *Nature*, *626*(7999), 555–564. DOI: 10.1038/s41586-023-06970-0 PMID: 38356065

Foster, J. B. (2010). *The ecological rift: Capitalism's war on the Earth*. Monthly Review Press.

Garrett, R., Ferreira, J., Abramovay, R., Brandão, J., Brondizio, E., Euler, A., Pinedo, D., Porro, R., Cabrera Rocha, E., Sampaio, O., Schmink, M., Torres, B., & Varese, M. (2024). Transformative changes are needed to support socio-bioeconomies for people and ecosystems in the Amazon. *Nature Ecology & Evolution*, *8*(10), 1815–1825. DOI: 10.1038/s41559-024-02467-9 PMID: 39107371

Gudynas, E. (2011). Buen vivir: Today's tomorrow. *Development*, *54*(4), 441–447. DOI: 10.1057/dev.2011.86

Howard, P. (Ed.). (2003). *Women and plants: Gender relations in biodiversity management and conservation*. Zed Books.

Kohn, E. (2013). *How forests think: Toward an anthropology beyond the human*. University of California Press.

Kopenawa, D., & Albert, B. (2013). *The falling sky: Words of a Yanomami shaman*. Harvard University Press.

Kothari, A., Salleh, A., Escobar, A., Demaria, F., & Acosta, A. (Eds.). (2019). *Pluriverse: A post-development dictionary*. Tulika Books.

Lansbury, M., Mosby, V., Creamer, S. J., Moggridge, B., Evans, G. R., Pecl, G., & Ireland, L. (2025). Indigenous peoples' voices and engagement on climate change: Towards improved health and wellbeing. *Frontiers in Climate*, *7*, 1693544. DOI: 10.3389/fclim.2025.1693544

Lehmann, J., Kern, D. C., Glaser, B., & Woods, W. I. (Eds.). (2003). *Amazonian dark earths: Origin, properties, management*. Springer. DOI: 10.1007/1-4020-2597-1

Levis, C., Flores, B. M., Campos-Silva, J. V., Peroni, N., Staal, A., Padgurschi, M. C. G., Dorshow, W., Moraes, B., Schmidt, M., Kuikuro, T. W., Kuikuro, H., Wauja, K., Kuikuro, K., Kuikuro, A., Fausto, C., Franchetto, B., Watling, J., Lima, H., Heckenberger, M., & Clement, C. R. (2024). Contributions of human cultures to biodiversity and ecosystem conservation. *Nature Ecology & Evolution*, *8*(5), 866–879. DOI: 10.1038/s41559-024-02356-1 PMID: 38503867

Lima, V. P., Ferreira de Lima, R. A., Joner, F., Siddique, I., Raes, N., & ter Steege, H. (2022). Climate change threatens native potential agroforestry plant species in Brazil. *Scientific Reports*, *12*(1), 2267. DOI: 10.1038/s41598-022-06234-3 PMID: 35145191

Machado, R. L. S., Dias, R. R., Deprá, M. C., Schneider, A. T., Dutra, D. A., de Menezes, C. R., Zepka, L. Q., & Jacob-Lopes, E. (2025). Commodities from Amazon biome: A guide to choosing sustainable paths. *Commodities*, *4*(2), 8. DOI: 10.3390/commodities4020008

McMichael, C. N. H., Bush, M. B., ter Steege, H., & Feeley, K. J. (2025). Centuries of compounding human influence on Amazonian forests. *Proceedings of the National Academy of Sciences of the United States of America*, *122*(47), e2514040122. DOI: 10.1073/pnas.2514040122 PMID: 41248280

Morgera, E., Buck, M., & Tsioumani, E. (2012). *Unravelling the Nagoya Protocol: A commentary on the Nagoya Protocol on access and benefit-sharing to the Convention on Biological Diversity*. Brill., DOI: 10.1163/9789004217201

Oarga-Mulec, A., & Skene, K. R. (2026). A dynamic biome-specific governance approach that integrates indigenous knowledge and pluriversal thinking. *Humanities & Social Sciences Communications*, *13*(1), 97. DOI: 10.1057/s41599-025-06402-6

Ostrom, E. (1990). *Governing the commons: The evolution of institutions for collective action*. Cambridge University Press. DOI: 10.1017/CBO9780511807763

Ostrom, E. (2005). *Understanding institutional diversity*. Princeton University Press.

Quijano, A. (2000). Coloniality of power, eurocentrism, and Latin America. *International Sociology*, *1*(3), 533–580. https://muse.jhu.edu/article/23906. DOI: 10.1177/0268580900015002005

Raynolds, L. T. (2004). The globalization of organic agro-food networks. *World Development*, *32*(5), 725–743. DOI: 10.1016/j.worlddev.2003.11.008

Rocheleau, D., Thomas-Slayter, B., & Wangari, E. (Eds.). (1996). *Feminist political ecology: Global issues and local experiences*. Routledge.

Santos, B. de S. (2014). *Epistemologies of the South: Justice against epistemicide*. Routledge.

Schroeder, D., & Pogge, T. (2009). Justice and the convention on biological diversity. *Ethics & International Affairs*, *23*(3), 267–280. DOI: 10.1111/j.1747-7093.2009.00217.x

Schwartzman, S., & Zimmerman, B. (2005). Conservation alliances with Indigenous peoples of the Amazon. *Conservation Biology : The Journal of the Society for Conservation Biology*, *19*(3), 721–727. DOI: 10.1111/j.1523-1739.2005.00695.x

Secretariat of the Convention on Biological Diversity. (2011). *Nagoya Protocol on access and benefit-sharing*. United Nations.

Silva-Junior, C. H. L., Silva, F. B., Maisonnave Arisi, B., Mataveli, G., Pessôa, A. C. M., & Carvalho, N. S.. (2023). Brazilian Amazon indigenous territories under deforestation pressure. *Scientific Reports*, *13*(1), 5851. DOI: 10.1038/s41598-023-32746-7 PMID: 37037850

Stevens, S. (Ed.). (2014). *Indigenous peoples, national parks, and protected areas: A new paradigm linking conservation, culture, and rights*. University of Arizona Press. DOI: 10.2307/j.ctt183pbn5

Viveiros de Castro, E. (2014). *Cannibal metaphysics*. Univocal Publishing.

World Intellectual Property Organization (WIPO). (2017). *Traditional knowledge and intellectual property: Background brief*. WIPO.

Wunder, S. (2015). Revisiting the concept of payments for environmental services. *Ecological Economics*, *117*, 234–243. DOI: 10.1016/j.ecolecon.2014.08.016

KEY TERMS AND DEFINITIONS

Biocentric Bioeconomy: An economic paradigm in which production is embedded within ecological limits and governed through communal institutions, prioritizing regeneration, diversity, and relational value creation over extraction and growth.

Biocultural Rights: Collective rights that link biological resources, traditional knowledge, and territorial governance, emphasizing their inseparability in Indigenous systems.

Conditional Scalability: A form of scaling in which expansion depends on the replication of governance structures, ecological principles, and social relations rather than on standardization or capital intensification.

Financialization (in bioeconomy context): The increasing influence of financial logics and instruments on ecological and productive systems often leads to the commodification of ecosystem functions.

Forest as Economic Infrastructure: A conceptualization of forest ecosystems as integrated systems that simultaneously perform productive, regulatory, and social functions without externalized infrastructure.

Forest Garden / Agroforestry System: Multi-strata, biodiversity-based production systems that integrate food, medicinal, and ecological functions within a single socio-ecological architecture.

Indigenous Knowledge Systems (IKS): Integrated, place-based knowledge systems that combine ecological, social, and cosmological understandings into a unified framework for production, governance, and stewardship.

Relational Governance: Governance systems based on kinship, reciprocity, and moral accountability align individual behavior with collective ecological stewardship.

Socio-Ecological System (SES): A coupled system of human and ecological components characterized by feedback loops, adaptive governance, and resilience dynamics across scales.

Transformation Readiness: The capacity of a socio-ecological system to engage with external economic processes while maintaining ecological integrity, institutional coherence, and cultural continuity.

APPENDIX: ADDITIONAL RESOURCES

To deepen the conceptual and empirical foundations of this chapter, the following works provide key perspectives on Indigenous knowledge systems, socio-ecological governance, bioeconomy debates, and the political ecology of forest-based development. Indigenous Knowledge and Socio-Ecological Systems Berkes, F. (2012). Sacred Ecology (3rd ed.). Routledge. Ostrom, E. (2005). Understanding Institutional Diversity. Princeton University Press.

These works offer foundational insights into Indigenous Knowledge Systems as integrated governance and ecological management frameworks, emphasizing collective action, institutional diversity, and resilience. Amazonian Ecology and Historical Landscapes Levis, C. et al. (2017; updated syntheses 2024). Persistent effects of pre-Columbian plant domestication in Amazonia. McMichael, C. et al. (2025). Long-term human–environment interactions in Amazonian forests. This literature demonstrates that Amazonian ecosystems are historically co-produced landscapes shaped by long-term Indigenous management practices, challenging narratives of pristine wilderness. Bioeconomy and Post-Extractive Development

Bugge, M., Hansen, T., & Klitkou, A. (2016). What is the bioeconomy? Sustainability, 8(7), 691.
Gudynas, E. (2011). Buen vivir: Today's tomorrow. Development, 54(4), 441–447.
Kothari, A., Salleh, A., Escobar, A., Demaria, F., & Acosta, A. (Eds.). (2019). Pluriverse: A Post-Development Dictionary. Tulika Books. These contributions critically examine dominant bioeconomy paradigms and introduce alternative development models rooted in plurality, ecological limits, and social justice. Governance, Law, and Biocultural Rights
Morgera, E., Buck, M., & Tsioumani, E. (2012). The 2010 Nagoya Protocol on Access and Benefit-Sharing. Brill.
Schroeder, D., & Pogge, T. (2009). Ethics, Justice, and the Convention on Biological Diversity. Cambridge University Press.
Bavikatte, K., & Bennett, T. (2015). Community stewardship: The foundation of biocultural rights. Journal of Human Rights and the Environment, 6(1), 109–131.

This body of work explores the legal and ethical dimensions of biodiversity governance, including access and benefit-sharing regimes, intellectual property, and emerging frameworks for biocultural rights.

Scaling, Markets, and Structural Constraints Dürr, J., Ribeiro, S. M. C., Vargas, D., Sellare, J., & Börner, J. (2023/2026). Bioeconomy and NTFP-based development in the Amazon. Garrett, R. D. et al. (2024). Scaling sustainability transitions in tropical landscapes.

These studies highlight structural barriers to scaling forest-based economies and emphasize the conditions under which sustainable bioeconomic pathways can emerge. Political Ecology and Epistemic Justice Escobar, A. (1995). Encountering Development. Princeton University Press.

Quijano, A. (2000). Coloniality of power and Eurocentrism in Latin America. International Sociology, 15(2), 215–232.
Descola, P. (2013). Beyond Nature and Culture. University of Chicago Press.

These works provide critical perspectives on development, power, and knowledge systems, situating Indigenous economies within broader debates on epistemic justice and alternative modernities.

Chapter 16
Ethical Public Finance and Accounting in the AI Era:
Harnessing IKS for Sustainable Financial Future

N. Abhishek
http://orcid.org/0000-0001-9345-8501
Government First Grade College, Halebeedu, India

Sahana Dinesh
http://orcid.org/0000-0002-0933-3245
Government First Grade College, Napoklu, India

Delecta Jenifer Rajendren
http://orcid.org/0000-0002-8179-6383
Saveetha Engineering College, India

Neethu Suraj
Yenepoya (Deemed to be) University, India

P. Shareena
Yenepoya (Deemed to be) University, India

ABSTRACT

This study investigates integrating Indian Knowledge Systems (IKS) principles with Artificial Intelligence (AI) to enhance ethical governance and transparency in public accounting and finance. Through interviews with five experts, key themes were identified regarding the application of IKS concepts, such as Dharma, Karma, and Nyaya, to AI-driven financial systems. The findings highlight consensus on embedding ethical considerations, promoting transparency and explainability, adopting a holistic perspective encompassing economic, social, and environmental factors, and ensuring cultural relevance in AI design. The study explores applications in areas such as financial transparency tools, audit trails, ESG reporting, and public administration. Experts acknowledged potential benefits and identified challenges related to technical complexity, resistance to change, and the need for broader awareness of IKS principles.

DOI: 10.4018/979-8-3373-7976-0.ch016

INTRODUCTION

In the era of rapid technological advancement, particularly with the increasing integration of Artificial Intelligence (AI) into governance, finance, and accounting systems, the need for ethical, transparent, and sustainable practices in public finance has become increasingly paramount (Schweitze, 2024). AI-driven applications are now widely used in areas such as budgeting, auditing, taxation, financial reporting, fraud detection, and policy evaluation, offering significant improvements in efficiency, speed, and analytical capacity. However, the accelerated adoption of AI in public financial systems has also generated complex ethical, accountability, and sustainability challenges that conventional governance frameworks often struggle to address adequately.

Indian Knowledge Systems (IKS), which encompass a vast and diverse body of traditional knowledge and wisdom, provide a distinctive and holistic framework for responding to these emerging challenges. Rooted in ancient texts such as the Vedas and Upanishads and encompassing disciplines including Ayurveda, Yoga, philosophy (notably Sā khya, Nyāya, and Vedānta), mathematics, astronomy, ethics, jurisprudence, and the arts (e.g., Nā yaśāstra), IKS reflect an integrated approach to knowledge, governance, and societal well-being (Mahadevan et al., 2022; K & Kalluraya, 2024). Central to IKS are concepts such as dharma (moral duty and ethical order), karma (the law of cause and effect), balance, harmony, and responsibility, which together offer a value-based foundation for governance and financial management.

Indian Knowledge Systems represent a "treasure trove" of insights that have evolved over thousands of years, emphasising ethical conduct, balance between competing interests, and harmony between human activity, society, and nature (Isser et al., 2024; Mahadevan et al., 2022). These principles are particularly relevant to public finance and accounting, domains that involve stewardship of public resources, long-term societal obligations, and distributive justice. Ethical lapses, opacity, and short-termism in public financial management can have far-reaching social and environmental consequences, making the normative guidance embedded within IKS especially pertinent.

The integration of IKS with modern AI technologies presents an innovative opportunity to enhance ethical governance and ensure sustainable financial futures, particularly for Indigenous peoples and emerging economies (Tharakan, 2015). IKS principles such as dharma and karma emphasise ethical behaviour, moral responsibility, and accountability values that can be meaningfully applied to AI-enabled public finance and accounting systems. Dharma can guide ethical governance by promoting fairness, justice, and rule-based conduct in financial practices, while karma reinforces accountability by emphasising responsibility for outcomes and consequences of financial decisions (Puri, 2007; Mahadevan et al., 2022). When translated into governance mechanisms, these principles align closely with contemporary demands for transparency, auditability, and responsible decision-making in AI systems.

In addition to philosophical ethics, Ayurvedic principles of balance and harmony offer valuable insights for sustainable financial planning. Ayurveda emphasises equilibrium and systemic well-being, suggesting that economic activities should be aligned with environmental sustainability and social welfare rather than pursued in isolation. Applied to public finance and accounting, this holistic perspective encourages decision-making that mitigates risks associated with unsustainable practices, such as environmental degradation, social inequality, and fiscal instability, while supporting long-term societal resilience.

Artificial Intelligence can play a pivotal role in operationalising IKS within modern public finance and accounting systems. AI tools can analyse large volumes of historical texts, policy documents, and financial data to uncover patterns and insights that support ethical and sustainable decision-making.

For example, AI techniques can be used to analyse Vedic and philosophical literature to extract ethical guidelines relevant to governance, or to apply Ayurvedic principles of balance in predictive models for sustainable financial outcomes using machine learning algorithms. Furthermore, AI can enhance efficiency and transparency by automating financial processes, improving real-time monitoring, and strengthening data-driven oversight. These capabilities can help identify areas where IKS principles may be embedded into system design to improve ethical governance and sustainability in public financial management.

Sustainability constitutes a core dimension of IKS, grounded in the recognition of the interconnectedness of human activities, society, and the environment. In the context of public finance and accounting, sustainability involves ensuring that financial decisions support long-term economic development while minimising environmental impacts and promoting social equity. The integration of IKS with AI offers a framework for ethical decision-making that prioritises the well-being of both present and future generations, thereby aligning technological innovation with sustainable development objectives.

Despite the significant potential of integrating IKS with AI in public finance and accounting, several challenges remain. These include the need for cultural sensitivity when applying traditional knowledge systems to modern technological contexts, safeguarding data sovereignty, and preventing biases in AI applications. There is also a risk that AI systems, if inadequately governed, may reinforce existing inequities or obscure accountability. Nevertheless, these challenges also present opportunities for innovation, particularly in designing AI tools that respect, preserve, and enhance IKS while addressing contemporary governance demands.

The rapid expansion of AI in financial systems has already transformed the landscape of public finance and accounting, but it has simultaneously intensified ethical concerns related to bias, transparency, and accountability. AI algorithms can perpetuate discriminatory outcomes when trained on biased datasets, undermining fairness in financial decision-making processes such as credit allocation, subsidy distribution, and procurement. The opacity of many AI models further erodes trust among stakeholders, as decision rationales are often difficult to interpret. Additionally, reduced human oversight in AI-driven systems raises concerns about accountability, while systemic risks such as market disruptions and financial instability pose ethical challenges for public financial governance. Environmental impacts associated with AI's energy-intensive operations further complicate sustainability considerations.

Within this context, IKS principles such as dharma, karma, and balance provide a culturally grounded and ethically robust framework for guiding AI-driven financial decisions toward fairness, accountability, and sustainability (Mahadevan et al., 2022). By integrating these principles, IKS offers a holistic approach to addressing ethical and sustainability challenges posed by AI in public finance. This approach is particularly relevant in emerging economies, where governance deficits and corruption often hinder economic development. Leveraging AI to operationalise IKS principles can contribute to more transparent, accountable, and ethically grounded financial systems, thereby supporting sustainable financial futures.

Despite growing interest in AI ethics and sustainability, existing literature has not sufficiently explored how IKS can be systematically integrated with AI governance in public finance and accounting. Most studies either remain conceptual or focus on Western ethical frameworks, leaving a significant theoretical, empirical, and operational gap. Accordingly, this study seeks to address these gaps by examining how IKS principles can inform ethical governance and sustainable financial planning in the AI era.

Specifically, this study attempts to answer the following research questions:

How can the integration of Indian Knowledge Systems (IKS) improve ethical governance and sustainability in public finance and accounting in the AI era?

What framework can be developed by combining IKS principles to ensure ethical decision-making and sustainability in public finance and accounting in the AI era?

By addressing these questions, the study aims to contribute to the development of a culturally grounded, ethically informed, and practically relevant framework for integrating IKS and AI in public finance and accounting.

LITERATURE REVIEW

Historical context: IKS orientation and implications for public finance

Classical Indian texts especially the Vedas and Upanishads form a long-standing intellectual substrate that combined ritual, cosmology, practical knowledge, and social regulation. Close readings of these texts reveal discussions that extend beyond spiritual guidance to practical matters such as environmental stewardship, systems of knowledge transmission, and proto-technical understandings of natural processes (Mishra & Aithal, 2023). For example, references in the Vedic corpus to water cycles, hydraulic devices, and nature-based solutions point to an applied, problem-solving orientation that historicises IKS as more than metaphysical reflection (Singh et al., 2020). Practices such as yagya traditionally ritualistic have also been interpreted in the literature as techniques for environmental purification with potential contemporary analogues (Mamta et al., 2018). Together these treatments indicate that IKS contains embedded rules, techniques, and value priorities relevant to collective resource management.

Two points are noteworthy for public finance and accounting. First, the Vedic emphasis on "prosperity for all", moderation, and mutual cooperation implies normative constraints on accumulation and distribution that resemble modern notions of public stewardship and social responsibility (Sharma & Talwar, 2005). Second, integrative models such as the Universal Business Excellence Model (UBEM), which explicitly draws on Vedic principles, show how ancient value systems can be translated into organizational metrics and multi-bottom-line measurement frameworks (Sharma & Talwar, 2007). These historical precedents therefore offer conceptual scaffolding for rethinking public finance goals beyond narrow fiscal indices toward sustainability, fairness, and systemic harmony a shift that becomes especially pertinent when AI begins to mediate allocation, auditing, and forecasting decisions.

Ayurveda and Yoga add an explicitly systemic and normative dimension to this historical picture. Both traditions centre balance, prevention, and contextualised interventions principles that can be read as antecedents to risk-aware, precautionary approaches in public policy design (Mamtani & Mamtani, 2005; Morandi et al., 2011). The Bhagavad Gita's exposition of karma-yoga emphasising duty, detachment from reward, and equanimity further extends this repertoire into moral psychology and administrative ethics (Mulla & Krishnan, 2013). For public finance, these orientations suggest governance practices that privilege duty-bound conduct and procedural composure over short-term gains or opportunistic behaviour, an insight that resonates with contemporary concerns about corruption, short-termism, and incentive misalignment in public budgeting and procurement.

That said, the literature also identifies methodological caveats when mapping ancient practices onto modern institutions. Empirical evidence linking religious or philosophical involvement to desirable social outcomes (e.g., improved wellbeing) is mixed and often constrained by measurement and causal identification issues (Chattopadhyay, 2007). Translating IKS into operational public finance instruments therefore requires careful conceptual unpacking and empirical testing rather than uncritical appropriation.

Dharma and Karma: philosophical concepts and direct relevance to accounting and public finance

The twin notions of dharma and karma occupy central positions within Indian ethical and philosophical traditions and have increasingly been examined in organisational ethics, public administration, and governance-related literatures. Dharma, broadly understood as duty, rightful order, and normative obligation, functions as a prescriptive moral principle that guides individual and institutional behaviour toward justice, fairness, and the collective good (Hodge, 2004). In contrast, karma, conceptualised as the ethical logic linking action to consequence, foregrounds responsibility for outcomes and emphasises accountability through the causal relationship between decisions and their social effects (Bhatnagar et al., 2019).

In the context of mental health, the path of selfless action (Karma Yoga) aligns with contemporary psychological concepts such as autonomy, intrinsic motivation, mastery, and a growth mindset (Keshavan et al., 2024).

Beyond individual ethics, dharma has played a significant role in shaping institutional and legal norms in the Indian context. Although often underestimated within Hindu studies, dharma has historically functioned as a foundational category for legal reasoning and normative order, informing legal rhetoric and governance paradigms within the Hindu tradition (Davis, 2007). This institutional interpretation of dharma is particularly relevant for public finance and accounting, where legality, procedural justice, and normative legitimacy underpin fiscal decision-making and financial accountability.

A significant advancement in operationalising these philosophical ideas is found in the Karma-Yoga model, derived from the Bhagavad Gita. Empirical research demonstrates that the three core dimensions of Karma-Yoga duty orientation, indifference to personal rewards, and equanimity are significantly associated with moral sensitivity, moral motivation, and moral character within organisational contexts (Mulla & Krishnan, 2013). This work is especially valuable because it provides a behavioural bridge between abstract philosophical doctrines and observable managerial practice. Within this framework, dharma supplies the normative foundation for role-based responsibility and ethical conduct, while karma introduces an outcomes-oriented accountability logic that aligns closely with mechanisms of audit, evaluation, and ex-post review in accounting and public finance systems.

From an accounting and public finance perspective, the integration of dharma and karma implies a broadened fiduciary conception of professional responsibility. Accountants and public finance officials are not merely technical functionaries implementing rules but are moral stewards of public resources, whose actions should be guided by duty, procedural propriety, and responsibility for long-term societal consequences. This ethical framing contrasts with many Western-derived public sector accounting reform narratives that have primarily emphasised technical harmonisation such as the adoption of accrual accounting—without a corresponding re-articulation of moral or normative foundations (Barton, 2009; Christensen & Parker, 2010; Kober et al., 2010).

Scholars have argued that ethical governance in economic and organisational systems requires the embedding of moral virtues and interdependent ethical principles rather than reliance on compliance-based rules alone (Melé et al., 2016). This perspective aligns closely with dharmic thinking and offers a potentially more holistic and culturally resonant ethical foundation for public finance and accounting practices, particularly in pluralistic and emerging economy contexts.

At the same time, the literature underscores important limitations regarding generalisability and cultural transferability. While the Karma-Yoga model has demonstrated robust empirical relationships within Indian organisational samples, Western-centric models of moral development have often shown limited validity when applied outside their original cultural contexts (Mulla & Krishnan, 2013). Consequently, incorporating dharma–karma frameworks into public finance and accounting requires careful contextualisation, methodological rigor, and comparative validation—especially when such frameworks are extended to AI-enabled governance systems operating across diverse institutional settings.

Bhagavad Gita: ethical precepts and implications for governance practice

Although not an explicit text on public finance or accounting, the Bhagavad Gita offers philosophical resources with direct relevance to governance ethics. Its core teachings on duty, detached action, cognitive discipline, and equanimity map neatly to the ethical desiderata of modern public administration: impartiality, competence, accountability, and service orientation (Bayly, 2010; Dhillon, 2023). For instance, "detached action" can be read as a normative recommendation for public servants to act in the public interest without private gain motives, aligning well with democratic accountability principles (Bovens, 2007).

The Gita's stress on knowledge, emotional regulation, and decisive, duty-centred action has also been linked to managerial competence and ethical resilience (Jeste & Vahia, 2008). Though extant literature rarely conducts direct institutional analyses of the Gita's implications for budgeting, audit, or reporting practice, the conceptual parallels are strong: the Gita's moral psychology supports a model of public finance actors as reflective, duty-bound agents who weigh long-term consequences an orientation that is useful when AI systems automate or assist complex decisions.

Samkhya, Nyaya, Vedanta: philosophical foundations and cognitive/epistemic implications

The classical schools of Sā khya, Nyāya, and Vedānta contribute important epistemic and methodological resources: Samkhya's psychological theorising and Yogic practices illuminate cognition and attention; Nyāya offers a systematic account of inference, evidence, and procedural justice; Vedanta foregrounds ontological and ethical holism (Sedlmeier & Srinivas, 2016; Nicholson, 2010). Collectively, these schools provide intellectual tools for thinking about evidence, justification, and procedural fairness themes highly relevant to design and governance of AI systems in public finance.

Empirical work suggests that Samkhya-Yoga practices have measurable effects on cognition and wellbeing, supporting the plausibility of translating some IKS insights into contemporary behavioural interventions or organisational training (Sedlmeier & Srinivas, 2016). Nyāya's emphasis on reasoned argumentation and valid means of knowledge (pramā a) also resonates with procedural requirements of audit and adjudication in public finance contexts. Yet scholarly engagement with how these schools concretely inform accounting norms and algorithmic governance remains sparse; most treatments are philosophical or psychological rather than institutionally focused. This lacuna highlights the need to connect epistemic principles from IKS to operational governance constructs in AI-mediated financial systems.

Synthesis and critical assessment: what the literature offers and what it omits

Taken together, the review above shows that IKS supplies a rich set of normative concepts (dharma, karma, balance), procedural philosophies (Nyāya), and systemic heuristics (Ayurveda, Yoga) that are potentially valuable for reorienting public finance and accounting toward ethical, sustainable ends. Several studies have begun to operationalise Vedic values into organisational frameworks (e.g., UBEM) and to test moral psychology models (Karma-Yoga) in organisational settings, demonstrating the intellectual and practical promise of this approach (Sharma & Talwar, 2007; Mulla & Krishnan, 2013).

Nevertheless, the literature also reveals persistent limitations. Much of the existing work is descriptive, philosophical, or psychological rather than institutional or technical. Few studies translate IKS concepts into concrete governance mechanisms, measurable indicators, or AI-system design constraints. Empirical work that interrogates stakeholder interpretations of IKS in the specific context of algorithmic public finance remains scarce. Methodologically, there is a reliance on conceptual argumentation and small-sample applications, with limited attempts at piloting, instrument development, or cross-jurisdictional comparison.

Gap in current literature and study aims

From this critical synthesis, three interrelated gaps emerge clearly and motivate the present study:

Theoretical gap: a shortage of frameworks that systematically link IKS philosophical concepts to operational governance constructs for AI in public finance and accounting.

Empirical gap: a lack of evidence on how practitioners and experts interpret IKS principles when asked to apply them to AI-mediated financial systems.

Measurement/implementation gap: an absence of candidate indicators and evaluative criteria to determine when an AI system is meaningfully "IKS-aligned."

Accordingly, the aims of this study are to:

Analyse the relevance of IKS for public finance and accounting in the AI era; and

Develop an exploratory conceptual framework that integrates IKS principles into governance and design features for AI applications in public finance and accounting.

METHODOLOGY

This study adopts an exploratory qualitative design to investigate how Indian Knowledge Systems (IKS) principles might be integrated with Artificial Intelligence (AI) in public accounting and finance. The IMRD structure (Introduction, Methods, Results, Discussion) guides presentation and ensures clarity and rigor (Sollaci & Pereira, 2004). The research addresses a gap in existing literature by empirically exploring the application of IKS principles to AI-driven financial systems and by proposing candidate governance features and indicators that are culturally grounded, ethically informed, and operationally relevant.

Although the research is situated within a broader interdisciplinary and mixed-evidence context (see Creswell & Plano Clark, 2017), the empirical strand reported here is qualitative and exploratory: purposive expert interviews were used to generate rich, theory-building insights that inform an initial conceptual framework. Where the analysis uses frequency counts of codes, these are presented as quasi-quantification for analytic clarity (heuristic counts of emphasis across experts) rather than as inferential statistics.

Exploratory framework

An exploratory framework was developed from a synthesis of IKS literature (e.g., Chakrabarti, 2001; Pradhan, 2008), AI ethics (Floridi & Taddeo, 2016; O'Neil, 2016), and applications of AI in finance (Erel et al., 2015; Jarrow, 2018). The framework foregrounded six IKS-informed domains most relevant to public finance and accounting practice Dharma, Karma, Nyāya, Vedantic holism, Kautilya's Artha-shastra, and the Nā yaśāstra and articulated how each domain could map onto potential AI governance and design features (for example: rule-based constraints, explainability, audit trails, stakeholder deliberation). This framework guided the construction of a semi-structured interview protocol comprising 15 open questions (see Appendix).

Participants and sampling

Purposive sampling selected five expert participants chosen for information richness and domain diversity (Patton, 2015). The intent was to capture multiple vantage points—philosophical, technical, policy, and practice—relevant to the research questions. The expert roles were:

- Sanskrit Professor (SP): specialist in Indian philosophy and ethics.
- Economics Professor (EP): expert in public finance and AI applications.
- Chartered Accountant (CA): practitioner with auditing, compliance, and financial-systems experience.
- Research Scholar (RS): scholar working at the intersection of technology and governance.

Accounting Professor (AP): academic focused on accounting practice and digital transformation.

The decision to interview five experts is justified on the basis of theory-building, expert elicitation protocols, and precedents for focused, information-rich qualitative inquiry (Patton, 2015; Ponterotto, 2013). Given the study's exploratory aims, the sample size is appropriate for generating conceptually dense data that informs subsequent model development and instrument design.

Instrument and interview procedure

A semi-structured interview protocol based on the exploratory framework was used. The 15 open-ended questions explored: (a) potential applications of IKS principles to AI-driven financial systems; (b) ethical considerations and governance mechanisms; and (c) perceived opportunities and constraints for implementing IKS-aligned AI in the public sector. Interviews were conducted remotely (video conferencing), audio-recorded with consent, and transcribed verbatim to ensure accuracy (Yin, 2018). Each interview lasted approximately 60–90 minutes.

Data analysis: thematic analysis and quasi-quantification

Thematic analysis, following Braun and Clarke (2006), was used to identify, organise, and interpret patterns of meaning in the interview transcripts. The analysis proceeded through the following iterative steps:

- Familiarisation: repeated reading of transcripts and listening to recordings to develop immersion in the data.
- Initial coding: open coding generated an initial set of codes derived both inductively from the data and deductively from the exploratory framework and research questions.
- Codebook development: codes were consolidated into a hierarchical codebook. Each code entry in the codebook included a short label, operational definition, inclusion/exclusion criteria, and sample anchor quotes. The codebook is provided in the Appendix.
- Five-point emphasis scale: to capture degree of emphasis across experts, a 5-point ordinal scale was applied during coding (5 = Strong emphasis; 4 = Moderate emphasis; 3 = Neutral/mentioned; 2 = Weak emphasis; 1 = No emphasis). This scale functions as an analytic heuristic to show relative emphasis and was not used for inferential statistics.
- Theme development and refinement: related codes were grouped into higher-order themes and iteratively refined through discussion among researchers.
- Quasi-quantification (NVivo "simulation"): to summarise theme prevalence across the small expert sample, we counted the number of experts who mentioned each theme. This procedure mirrors frequency features available in qualitative software (e.g., NVivo) but does not imply population generalisability; it is presented to indicate emphasis patterns across the interviewed experts. Where useful, simple tables report the counts and are explicitly footnoted with the small-sample caveat.
- Comparative analysis: cross-case comparisons were made to identify convergences and divergences between participants and to surface tensions or trade-offs.

Reliability, validation, and transparency measures

Following established trustworthiness criteria (Lincoln & Guba, 1985), we implemented multiple procedures to enhance credibility, dependability, and confirmability:

Member checking: preliminary thematic summaries were shared with participants for feedback and clarification; their comments were incorporated into the final thematic descriptions.

Peer debriefing / external review: analytic choices, the codebook, and emergent themes were reviewed by an independent peer with expertise in qualitative methods and AI governance; this review contributed to refining code definitions and theme boundaries.

Inter-coder consistency checks: to assess coding consistency, a second coder independently coded a subset of transcripts (two transcripts, representing the heterogeneity of expert perspectives). Discrepancies were resolved in consensus meetings; the codebook was revised accordingly. Cohen's kappa was calculated for the double-coded set and is reported in the Appendix to document inter-coder agreement. (Note: given the exploratory design and small sample, these checks are intended to demonstrate consistency rather than to serve as proof of statistical generalisability.)

Triangulation: analytic triangulation compared interview data against the exploratory framework and relevant secondary sources (textual IKS materials and policy documents referenced in the study) to corroborate interpretations.

Audit trail and appendix materials: a transparent audit trail comprising the codebook, analytic memos, and illustrative quotes is provided in the Appendix to allow readers to follow analytic decisions.

Ethical considerations

All participants gave informed consent prior to interviews. The research protocol, instruments, and analytic plan were presented to the Institutional Review Board (IRB) for scrutiny and ethical clearance; the procedures followed the IRB guidance on participant confidentiality and data protection. Anonymisation was used in reporting (e.g., SP, EP, CA, RS, AP) to protect participant identities while retaining role-specific contextual detail.

Limitations and interpretive boundaries

The study acknowledges several limitations that shape the interpretation of findings. First, the small purposive sample (five experts) is suitable for exploratory theory generation but does not support statistical generalisation. Second, reliance on self-reported expert perspectives introduces the possibility of response biases and aspirational framing; member checking and triangulation mitigate but cannot eliminate this risk. Third, the study's focus on IKS situates findings within the Indian cultural and intellectual context; transferability to other jurisdictions requires careful translation and comparative validation. Fourth, incorporating philosophical and religiously-rooted concepts into secular public policy raises normative and interpretive challenges; we therefore emphasise secular operationalisation (translating dharma/karma into governance constructs such as duty, auditability, and accountability) rather than doctrinal prescriptions. Finally, the quasi-quantification (code counts) is heuristic and presented only as an indicator of emphasis across experts; it should not be interpreted as evidence of prevalence in a broader population.

RESULTS AND PROPOSED FRAMEWORK

This section presents the findings from the thematic analysis of expert interviews exploring the integration of Indian Knowledge Systems (IKS) with Artificial Intelligence (AI) in public finance and accounting. Results are organised around core themes that emerged consistently across expert narratives and that directly address the study's research questions. While limited frequency counts are reported to indicate relative emphasis across experts, the findings are interpreted qualitatively, with emphasis placed on meaning, context, and illustrative quotations rather than numerical comparison or statistical inference.

The presentation prioritises salient themes, areas of convergence, and points of divergence among participants. Where tables summarise themes and emphasis levels, the accompanying narrative highlights only the most analytically significant insights to avoid redundancy. The results should therefore be read as exploratory and interpretive, offering theory-building insights rather than claims of prevalence or generalisability.

The primary purpose of this study was to propose a framework that helps implement ethical governance practices in the public finance and accounting domain. This section of the paper clearly explains the insights derived from the various IKS texts.

Philosophical Foundations from Indian Knowledge Systems (IKS)

The IKS offers a comprehensive governance model rooted in ethics, accountability, and sustainability. The key philosophical principles applicable to public finance and AI-driven decision-making include the following:

Dharma: Ethical Governance and Accountability

Dharma represents moral responsibility, fairness, and duty in governance. In public finance, dharma ensures adherence to ethical accounting standards and transparent fiscal policies. AI algorithms should be aligned with dharma-based governance to mitigate bias, corruption, and unethical financial practices. One of the major problems facing emerging economies today is corruption (Sadik-Zada et al., 2022). Dharma as an IKS principle plays a crucial role in promoting ethical governance and accountability that indirectly supports for the sustainable development of such economies (Isser et al., 2024)

Karma: Cause and Effect in Financial Accountability

Karma, the law of action and consequence, is vital for implementing responsible fiscal policies. AI-driven predictive models can incorporate karma-based frameworks to assess the long-term effects of financial decisions, ensuring sustainability and accountability. Karma, the principle of action and consequence, provides a robust ethical foundation for financial accountability. In AI-driven public finance, applying karma-based accountability can ensure responsible fiscal policies, transparency, and sustainability at both the micro and macro levels (Alonge et al., 2024).

Micro-Level Impact: Organizational and Individual Financial Ethics

At the micro level, Karma fosters financial responsibility among individuals, organisations and regulatory bodies.

AI-Driven Financial Forecasting and Accountability

AI can assess the long-term impact of financial decisions by modelling various scenarios based on past actions (Karmic accumulation). For instance, predictive AI can analyse the consequences of excessive debt accumulation by a government department to ensure fiscal responsibility (Nembe et al., 2024).

Corporate Governance and Ethical Decision-Making

AI-powered financial reporting aligned with Karma ensures that unethical financial practices (e.g. tax evasion, fraudulent reporting, and window dressing) have long-term consequences. Explainable AI (XAI) can make financial decisions transparent and reinforce accountability. This, in turn, reduces the overall cost of financial reporting and compliance with regulatory requirements (Černevičienė & Kabašinskas, 2024).

Incentivizing Ethical Behavior through AI Systems

AI-driven compliance monitoring can reward ethical financial practices by offering higher sustainability scores, better credit ratings and policy incentives. This, in turn, gains the confidence of various stakeholders and promotes an ethical code of conduct (Miralles-Quiros et al., 2017).

Karma-Based Predictive Auditing for Fraud Prevention

AI-based auditing mechanisms can track past fraudulent activities and predict future financial risks, thereby ensuring proactive governance. This indirectly eliminates financial crises and promotes transparent practices (Odeyemi et al., 2024).

Macro-Level Impact: Governance, Sustainability, and Policy Implications

At the macro level, karma provides a framework for ensuring long-term financial sustainability in public governance.

Sustainable Public Finance and AI-Based Fiscal Policies

Karma-based AI models can assess the long-term impacts of government subsidies, tax policies, and public spending, ensuring intergenerational equity. AI-driven fiscal discipline models can prevent reckless financial practices and ensure national economic stability. This aspect can be measured in the long term through intergenerational fairness, subsidy optimisation machine learning model processes, and tax policy simulation neural network model complex interactions.

Long-Term Economic Planning Based on Consequences

AI can simulate future economic conditions based on past financial actions, thereby helping policymakers create sustainable taxation and welfare policies. For example, AI-driven Karma-based simulations can predict the economic impact of environmental taxation policies, guiding sustainable finance decisions.

AI and ESG (Environmental, Social, Governance) Compliance

Karma-driven AI can ensure that businesses remain accountable for their ESG commitments.

Companies with a history of sustainable practices are rewarded with better AI-generated ESG scores, influencing investments and policy incentives.

Policy Implementation for Ethical Governance

AI-powered governance models can integrate Karma by continuously assessing whether public policies have positive or negative consequences over time.

Blockchain-enabled public finance can ensure immutable financial records, thereby making fiscal governance accountable.

Nyaya and Samkhya: Logical Inquiry and Classification for Decision-Making

Nyaya (Logic and Rational Enquiry): Supports the development of AI-driven financial governance models based on rational decision-making and just outcomes.

Samkhya (Enumeration and Classification): Helps classify ethical risks in AI-driven financial decision-making and ensures balanced resource allocation.

Rta and Vedantic Holism: Interconnected Financial Sustainability

Rta, the principle of cosmic order, aligns with sustainable economic policies and ESG (Environmental, Social, and Governance) reporting. AI-driven financial systems should incorporate this principle to balance economic growth and social responsibility.

AI & Digital Technologies in Public Finance and Accounting

Explainable AI (XAI) for Transparency

Transparency in AI-driven financial management can be ensured through Explainable AI (XAI), which aligns with Nyaya's logical enquiry. AI models should be interpretable to prevent bias and ensure ethical governance.

Blockchain for Ethical Financial Reporting

When integrated with IKS principles, blockchain technology ensures truthfulness (Satya) and integrity in financial transactions. It fosters transparency and prevents financial malpractice.

Predictive AI for Sustainable Budgeting

AI-driven fiscal models should incorporate karma-based accountability, ensuring long-term sustainability rather than short-term economic gains.

AI-Driven ESG and Triple Bottom Line (TBL) Reporting

AI can enhance ESG compliance and TBL accounting by integrating holistic governance principles from Vedantic interconnectedness (Sat-Chit-Ananda model: truth-consciousness-bliss).

Table 1. Ethical Governance Framework

IKS Principle	AI Implementation in Public Finance	Expected Outcomes
Dharma (Ethical Responsibility)	AI-driven financial transparency tools and NLP for text analysis to identify ethical guidelines from historical texts.	Corruption-free governance
Karma (Impact & Accountability)	AI audit trails & predictive analysis	Responsible fiscal policies
Nyaya (Logical Inquiry & Justice)	Explainable AI (XAI) for decision-making	Fair and rational financial governance
Vedantic Holism (Interconnected Sustainability)	AI-driven ESG reporting & carbon accounting	Sustainable financial management
Kautilya's management techniques	Adapt Kautilya's management techniques with AI for modern public administration and can use Predictive analytics for forecasting and optimizing resource allocation.	Efficient Management
Natya Shastra	AI based Artistic and Cultural Integration	creativity and innovation in financial management.
Ayurveda and Dharma (Accounting and Compliance)	AI-assisted auditing and financial analysis to ensure transparency and ethical compliance.	Integrated IKS principles into accounting standards for ethical reporting.

Source: Authors compilation

Table 01 outlines the framework outlines how IKS principles can be applied to AI-driven public finance governance(See Table 01). This is the major aspect of the chapter.

QUALITATIVE RESULTS

Table 2. Summary of Qualitative Interview Results

Question	Key Themes	SP	EP	CA	RS	AP
1. AI and Dharma in Financial Transparency	Ethical Algorithms & Data Integrity	5	4	5	4	4
	Transparency & Accountability	5	4	5	4	4
	Corruption Prevention	4	3	4	3	3
2. AI Audit Trails and Karma in Public Finance	Immutable Records	5	4	5	4	4
	Long-Term Impact Assessment	4	3	4	3	3
	Accountability & Responsible Decision-Making	5	4	5	4	4
3. Explainable AI and Nyaya Principles	Logical Justification	5	4	5	4	4
	Unbiased Models	4	3	4	3	3
	Traceability	5	4	5	4	4

continued on following page

Table 2. Continued

Question	Key Themes	SP	EP	CA	RS	AP
4. AI-driven ESG Reporting and Vedantic Holism	Interconnectedness	5	4	4	4	4
	Quantifiable ESG Indicators	4	3	5	4	3
	Holistic Well-being	5	4	4	4	4
5. Kautilya's Techniques in AI for Public Administration	Strategic Planning & Resource Allocation	5	4	4	4	3
	Fraud Detection & Prevention	4	3	5	4	4
	Economic Simulation	5	4	4	4	3
6. Natya Shastra Elements in AI Financial Innovation	Creative Problem-Solving	5	4	3	4	3
	Enhanced Data Visualization	4	3	1	1	1
	Cautious Implementation	4	3	5	4	4
7. AI-assisted Auditing and Ayurvedic Principles	Holistic Assessment	5	4	4	4	4
	Adaptability	4	3	4	3	3
	Real-time Fraud Detection	5	4	5	4	4
8. Ethical Considerations in AI-driven Financial Transparency	Data Privacy	5	4	5	4	4
	Algorithmic Bias Mitigation	4	3	4	3	3
	Accountability	5	4	5	4	4
	Equitable Access	4	3	4	3	3
9. AI Audit Trails: Balancing Accountability and Privacy	Data Encryption	5	4	5	4	4
	Differential Privacy	4	3	4	3	3
	Transparency with Protection	5	4	5	4	4
10. Challenges in Implementing Explainable AI	Technical Complexity	5	4	4	4	4
	Resistance to Change	4	3	5	4	4
	Lack of Understanding	5	4	4	4	4
	Data Security	4	3	5	4	4
11. Culturally Tailored AI-driven ESG Reporting	Integration of Local Values	5	4	4	4	4
	Community Focus	4	3	4	3	3
	Ethical Wealth Creation	5	4	4	4	4
12. Kautilya's Relevance in Modern AI-based Public Finance	Strategic Governance	5	4	4	4	3
	Financial Intelligence	4	3	5	4	4
	Economic Stability	5	4	4	4	3
13. Integrating Natya Shastra in AI Financial Systems	Visual Data Representation	4	3	1	1	1
	Gamification	4	3	4	3	3
	Intuitive Decision Support	5	4	4	4	4
14. Historical Ethical Guidelines	Righteous Conduct	5	4	4	4	4
	Ethical Taxation	4	3	5	4	4
	Non-Exploitation	5	4	4	4	4

continued on following page

Table 2. Continued

Question	Key Themes	SP	EP	CA	RS	AP
15. AI-assisted Auditing: Transparency and Dharmic Principles	Selfless Governance	5	4	4	4	4
	Ethical Algorithms	4	3	5	4	4
	Unbiased Insights	5	4	4	4	4

Source: Interview results

Table 02 summarises the qualitative interview results across the five experts. A qualitative interview was conducted with five experts (SP, EP, CA, RS, and AP) regarding the integration of Indian Knowledge Systems (IKS) principles with Artificial Intelligence (AI) in public finance and accounting.

- **AI and Dharma in Financial Transparency:** All the Experts unanimously emphasised the need to embed Dharma principles (Satya, Ahimsa) into AI-driven financial transparency tools (SP, EP, CA, RS, AP). Prioritising ethical guidelines, data integrity, and accuracy in AI systems can promote fairness, accountability, and responsible technology use, effectively combating corruption (SP, EP, CA, RS, AP).
- **AI Audit Trails and Karma in Public Finance:** Experts emphasised that AI audit trails should create immutable, traceable records aligned with Karmic justice (SP, EP, CA, RS, AP). This definitely helps in tracking the long-term financial and social impacts of decisions and encourages responsible and ethical public finance management (SP, EP, CA, RS, AP). On the other hand, the interviewees opined that predictive analysis should inform decision-makers about potential consequences (EP, RS, AP).
- **Explainable AI (XAI) and Nyaya Principles in Financial Governance:** As per the discussion in the interview, it was found that XAI systems require logical justifications, unbiased models, and traceability, reflecting Nyaya principles (SP, EP, CA, RS, AP). This is because Transparency is crucial. Therefore, AI systems must provide clear and understandable explanations for financial decisions, enabling stakeholders to trace decision-making processes back to the source data and ethical considerations (SP, EP, CA, RS, AP).
- **AI-driven ESG Reporting and Vedantic Holism:** It was found that ESG reporting should integrate interconnected economic, social, and environmental factors, embodying Vedantic holism (SP, EP, RS, AP). Metrics must extend beyond financial profits to incorporate ethical labour practices, community impacts, and environmental degradation risks, fostering long-term well-being (Loka-Samgraha) (SP, EP, CA, RS, and AP). Quantifiable metrics aligned with international standards are important (CA and EP).
- **Kautilya's Techniques in AI for Public Administration:** Interviewees opined that Kautilya's teachings can inform AI systems for public administration and resource allocation (SP, EP, CA, RS, AP). They also pointed out that by Applying Anvikshiki (critical thinking), Danda Niti (law enforcement), and Yukti (strategic reasoning), AI can analyse policy impacts, detect fraud, and optimise resource allocation (SP, EP, CA, RS, AP).
- **Natya Shastra Elements in AI for Financial Innovation:** Interviewees also pointed out that the Natya Shastra can inspire creative problem-solving approaches in AI-driven financial innovation (SP, EP, RS, AP). They recommended that Integrating Navarasa (nine emotions) in data visualisa-

tion and human-AI interaction may improve engagement (SP, RS). However, caution is needed to ensure that accuracy, compliance, and ethical standards are maintained (CA, AP).

- **AI-assisted Auditing and Ayurvedic Principles:** Interestingly, Ayurvedic principles of balance can inform AI-assisted auditing for holistic financial health (SP, EP, RS, AP). Systems should adapt to changing risks, ensure real-time fraud detection, and promote long-term fiscal discipline, aligning with the Tridosha concept of balance (Vata, Pitta, Kapha) (SP, EP, RS, AP).
- **Ethical Considerations in AI-driven Financial Transparency:** The interviewees opined that data privacy, algorithmic bias mitigation, accountability, and equitable access must be prioritised (SP, EP, CA, RS, AP). Ethical considerations aligning with Dharma, Artha, and Nyaya are essential for bias-free decision-making, fair financial disclosures, and robust safeguards in AI design (SP, EP, CA, RS, and AP).
- **AI Audit Trails: Balancing Accountability and Privacy:** According to the experts, AI audit trails should employ advanced encryption, differential privacy, and selective transparency (SP, EP, CA, RS, AP). This helps uphold truth and fairness, while respecting individual privacy requires techniques that balance accountability with data protection principles (SP, EP, CA, RS, AP).
- **Challenges in Implementing Explainable AI (XAI) based on IKS:** Interviewees noted that the challenges include integrating Indian ethical thought, resistance to transparency, and lack of IKS awareness (SP, EP, CA, RS, AP). However, addressing technical complexity, resistance to change, non-expert comprehension, and data security is critical for successful XAI implementation (SP, EP, CA, RS, AP). Education and intuitive interfaces are required (EP, RS, AP).
- **Culturally Tailored AI-driven ESG Reporting:** AI-driven ESG reporting must integrate local cultural and societal values (SP, EP, CA, RS, and AP). Adopting Dharmic financial models, a Vasudhaiva Kutumbakam approach, and community-driven carbon tracking reflects Indian cultural priorities (SP, EP, CA, RS, AP).
- **Kautilya's Relevance in Modern AI-based Public Finance:** Kautilya's management techniques are highly relevant to AI implementation in modern public finance (SP, EP, CA, RS, and AP). It also helps in surveillance for corruption-free governance, dynamic taxation, and predictive AI for economic stability (SP, EP, CA, RS, AP).
- **Integrating Natya Shastra in AI Financial Systems:** It was found that symbolism and visual metaphors can be integrated with gamification (SP, EP, RS, AP). Narrative-based AI models and gamification are useful for intuitive financial decision support and improving financial literacy (SP, EP, RS, and AP). Caution is needed from Accounting Professionals (CA)
- **Ethical Guidelines from Historical Texts:** The experts suggested prioritising righteous conduct, ethical taxation, and non-exploitative practices (SP, EP, CA, RS, AP). They also opined that ancient texts emphasise the balance of Artha (wealth) and Dharma (righteousness), informing ethical AI development (SP, EP, CA, RS, AP).
- **AI-assisted Auditing: Transparency and Dharmic Principles:** Interestingly, they argued that there is a need to promote selfless governance, ethical AI algorithms, and unbiased insights (SP, EP, CA, RS, AP). AI should promote Nishkama Karma (selfless action), Ahimsa (non-harm), and Samyak Drishti (right perspective) in financial governance (SP, EP, CA, RS, AP).

Thus, the integration of IKS principles offers a robust framework for developing ethical, transparent, and culturally aligned AI systems in public finance and accounting. Addressing challenges related to technical complexity, resistance to change, and data security while prioritising ethical considerations is critical for successful implementation (SP, EP, CA, RS, and AP). Further research should explore the development of metrics to quantify the impact of IKS-aligned AI on financial governance and sustainability issues.

Taken together, the findings address RQ1 by demonstrating how IKS principles such as dharma, karma, and balance are interpreted by experts as governance mechanisms for ethical and sustainable AI in public finance. RQ2 is addressed through the identification of specific AI design implications, including explainability, audit trails, and human-in-the-loop controls, which participants consistently associated with IKS-aligned governance.

DISCUSSION AND IMPLICATIONS

This section interprets and synthesises the empirical findings in relation to existing debates on AI governance, ethics, and public finance, highlighting what an Indian Knowledge Systems (IKS) driven framing adds to current understanding. Drawing on expert perspectives, the discussion moves beyond a descriptive summary of themes to critically examine how IKS principles can reshape ethical reasoning, governance mechanisms, and long-term accountability in AI-enabled public accounting and finance.

Interpreting the findings: What does an IKS-driven framing add?

A central insight emerging from the expert interviews is the strong convergence around the view that AI systems used in public finance must be grounded in ethical principles that extend beyond compliance-oriented or purely technical governance models. While much of the existing AI-ethics literature emphasises abstract values such as fairness, transparency, and accountability, the findings of this study suggest that IKS offers a normatively richer and culturally embedded articulation of these values through concepts such as dharma, karma, and nyāya.

Experts consistently framed dharma not as a religious construct but as a secular notion of duty and fiduciary responsibility. In the context of public accounting and finance, this interpretation reframes AI governance as an obligation to act in the public interest, uphold procedural justice, and steward public resources responsibly. This duty-centred perspective contrasts with dominant AI governance approaches that prioritise risk mitigation and regulatory compliance without explicitly articulating moral responsibility. In this sense, IKS provides a deeper ethical grounding that complements existing AI-ethics frameworks rather than replacing them.

Similarly, karma was interpreted as an accountability mechanism that extends ethical evaluation beyond algorithmic inputs to long-term outcomes and societal consequences. This outcome-oriented logic aligns closely with contemporary concerns around auditability, traceability, and ex-post evaluation of AI decisions in public finance. By foregrounding responsibility for consequences, karma offers a lens for assessing not only whether AI systems function correctly, but whether their decisions contribute to sustainable and equitable public financial outcomes over time.

The emphasis on nyāya further reinforces the importance of procedural justice and reason-giving in AI systems. Experts' insistence on explainability and transparency resonates with broader calls for explainable AI, but the IKS framing adds a distinctive dimension: explanations are not merely technical artefacts but moral justifications that enable public scrutiny and legitimacy. This perspective aligns with contemporary discussions on aligning AI systems with human values and societal norms (Ayinla et al.,

2024; Kasirzadeh & Gabriel, 2023), while grounding those discussions in an indigenous philosophical tradition.

Holism, sustainability, and long-term public value

Another significant contribution of the findings lies in the emphasis on holistic and long-term perspectives, particularly through Vedantic holism. Experts highlighted that AI-enabled financial systems should account for the interconnectedness of economic performance, social equity, and environmental sustainability. This aligns with evolving expectations around ESG reporting and sustainable public finance, yet the IKS framing situates these expectations within a worldview that prioritises balance and interdependence rather than trade-offs alone.

By integrating karma with holistic thinking, the findings suggest that AI governance in public finance should explicitly incorporate long-term impact assessment and intergenerational accountability. This challenges short-term efficiency-driven models and supports a shift toward AI systems that evaluate fiscal decisions in terms of their cumulative social and environmental effects.

Cultural relevance and contextual governance

Experts also underscored the importance of cultural relevance in AI design and governance. Rather than advocating for culturally neutral systems, participants argued that AI in public finance should reflect societal values and developmental priorities. This insight is particularly relevant for emerging economies, where imported governance models may not fully align with local institutional realities. The findings suggest that IKS-informed AI governance can enhance legitimacy, public trust, and contextual fit dimensions that are often underemphasised in global AI policy discourse.

At the same time, the interviews revealed important points of divergence. For example, while most experts acknowledged the potential of incorporating creative and communicative elements inspired by the Nā yaśāstra to improve engagement and understanding, concerns were raised—particularly by the Chartered Accountant—about safeguarding financial accuracy, auditability, and regulatory compliance. These tensions highlight the need for careful boundary-setting when translating philosophical or artistic principles into technically constrained financial systems.

Implications for theory

The findings contribute to theory in three key ways. First, they extend AI-ethics discourse by introducing an IKS-based, duty-centred ethical framework that complements dominant rights- and risk-based approaches. Second, they demonstrate how indigenous philosophical concepts can be operationalised as governance mechanisms, linking abstract ethics to concrete AI design features such as explainability, audit trails, and human oversight. Third, the study advances public finance and accounting scholarship by reframing professionals as moral agents and stewards, rather than purely technical implementers of AI systems.

Policy and practical implications

From a policy perspective, the findings suggest that governments and public-sector regulators should consider incorporating IKS-informed principles into AI governance guidelines, procurement frameworks, and capacity-building initiatives. Training programmes for public finance professionals could integrate concepts such as duty orientation, accountability for outcomes, and procedural justice to complement technical AI competencies.

Practically, AI developers and public finance institutions can use the proposed IKS-based framework to design systems that embed ethical constraints at the design stage rather than relying solely on post-hoc regulation. Features such as explainable decision logic, long-term impact monitoring, and structured human oversight align well with the IKS principles identified by experts.

Implementation challenges and future considerations

Despite its promise, the integration of IKS principles into AI-enabled public finance is not without challenges. Experts pointed to limited awareness and understanding of IKS, resistance to institutional change, technical complexity, and data security concerns as significant barriers. These challenges underscore the need for incremental implementation, interdisciplinary collaboration, and pilot testing before large-scale deployment.

Finally, the discussion highlights that an IKS-driven approach does not seek to replace existing AI governance frameworks, but rather to enrich them with culturally grounded ethical reasoning and long-term accountability. By doing so, it offers a pathway toward more legitimate, sustainable, and socially responsive AI systems in public accounting and finance.

CONCLUSION

The primary purpose of this study was to examine the need for integrating Indian Knowledge Systems (IKS) to enhance ethical governance and sustainability in public finance and accounting in the Artificial Intelligence (AI) era, and to develop and validate an exploratory framework to guide such integration. Addressing a clear gap in the existing literature, the study provides an empirically informed, culturally grounded perspective on how traditional ethical principles can inform the governance of AI-enabled public financial systems.

Adopting an exploratory qualitative design supported by expert interviews, the study demonstrates that the integration of IKS with AI has the potential to reshape public finance and accounting in ways that go beyond efficiency and automation. The findings suggest that IKS-informed AI systems can promote ethical conduct, transparency, accountability, and cultural relevance, all of which are essential for public-sector legitimacy and long-term sustainability. Across expert perspectives, there was a strong consensus on the relevance of embedding IKS principles into AI governance for public finance.

Specifically, the study highlights that dharma-based ethics can guide transparency, fiduciary responsibility, and duty-oriented decision-making in AI systems; karma provides a powerful framework for long-term accountability through audit trails and outcome-based evaluation; and nyāya supports explainable, reasoned, and unbiased AI decision processes. Vedantic holism offers a basis for integrating economic, social, and environmental dimensions into ESG reporting and fiscal evaluation, while insights from Kautilya's Arthashastra inform strategic governance, regulatory oversight, and statecraft in AI-enabled public finance. In addition, while primarily symbolic and contextual, elements inspired by the Nā yaśāstra may enhance communication, engagement, and ethical awareness, provided that financial accuracy and regulatory compliance are not compromised. Importantly, experts also emphasised that historical ethical guidelines must be integrated alongside robust data privacy and security safeguards.

Despite these opportunities, the study also underscores several implementation challenges. Technical complexity, limited institutional understanding of IKS, resistance to change, and contextual translation of philosophical concepts into operational AI systems remain significant barriers. These challenges reinforce the need for gradual, well-governed adoption rather than wholesale or symbolic implementation.

Based on the findings, the study identifies several priority initiatives to support the effective integration of IKS into public finance and accounting in the AI era:

- Ethical framework development: Develop comprehensive and operational ethical guidelines for AI in public finance that explicitly draw on IKS principles such as duty, accountability, procedural justice, and balance.
- Interdisciplinary collaboration: Promote sustained collaboration among scholars and practitioners from philosophy, economics, accounting, public administration, and technology to translate IKS concepts into actionable governance mechanisms.
- Educational initiatives: Invest in education and capacity-building programmes to improve awareness and understanding of IKS among policymakers, public finance professionals, and AI developers.
- Case-based and applied research: Encourage further empirical research using case studies, pilot implementations, and comparative analyses to test the feasibility and effectiveness of IKS-informed AI governance frameworks in real-world public finance contexts.

The study contributes to both theory and practice by demonstrating that ethical AI governance in public finance need not rely exclusively on externally derived or purely technical frameworks. Instead, culturally rooted knowledge systems such as IKS can meaningfully inform the design, evaluation, and governance of AI systems when translated into secular, operational principles.

Future research can build on this exploratory work in several directions. First, scholars may develop and validate measurable indicators to assess AI systems' adherence to IKS-aligned ethical principles and to evaluate their effectiveness in preventing financial misconduct. Second, comparative studies may examine how IKS-informed AI governance differs from, or complements, other ethical frameworks across jurisdictions. Third, future work may explore how AI can tailor ESG reporting and sustainability assessment to local values and practices across diverse Indian communities.

Finally, the future of public finance and accounting lies in the thoughtful integration of technological innovation with ethical responsibility. By combining the analytical power of AI with the normative depth of Indian Knowledge Systems, public financial systems can be designed not only to perform efficiently, but also to serve the broader goals of equity, accountability, and societal well-being.

FUTURE RESEARCH DIRECTIONS

One useful future-research direction is to move from a conceptual framework to measurable implementation, develop indicators that quantify how well AI systems follow IKS principles such as dharma, karma, and nyaya; test whether those indicators improve transparency, accountability, and the prevention of financial misconduct; and compare results across public agencies and diverse Indian communities to see how culturally tailored ESG reporting performs in practice. The study also points to the need for case-study and field-based work that addresses the practical barriers it identifies technical complexity, data security, resistance to change, and limited awareness of IKS while examining whether explainable AI, audit trails, and blockchain-enabled controls can strengthen ethical public finance outcomes

REFERENCES

Alonge, E., Alao, O., & Dudu, O. (2024). The impact of digital transformation on financial reporting and accountability in emerging markets. *International Journal of Science and Technology Research Archive, 7*(2), 025–049. .DOI: 10.53771/ijstra.2024.7.2.0061

Ayinla, B., Osasona, F., Abrahams, T., Atadoga, A., Amoo, O., & Farayola, O. (2024). Ethical AI in practice: Balancing technological advancements with human values. *International Journal of Science and Research Archive, 11*(1), 1311–1326. DOI: 10.30574/ijsra.2024.11.1.0218

Barton, A. (2009). The Use and Abuse of Accounting in the Public Sector Financial Management Reform Program in Australia. *Abacus, 45*(2), 221–248. DOI: 10.1111/j.1467-6281.2009.00283.x

Bayly, C. A. (2010). INDIA, THE BHAGAVAD GITA AND THE WORLD. *Modern Intellectual History, 7*(2), 275–295. DOI: 10.1017/S1479244310000077

Bhatnagar, N., Sharma, P., & Ramachandran, K. (2019). Spirituality and Corporate Philanthropy in Indian Family Firms: An Exploratory Study. *Journal of Business Ethics, 163*(4), 715–728. DOI: 10.1007/s10551-019-04394-3

Bovens, M. (2007). *Public Accountability* (pp. 182–208). oxford university. DOI: 10.1093/oxfordhb/9780199226443.003.0009

Braun, V., & Clarke, V. (2006). Using thematic analysis in psychology. *Qualitative Research in Psychology, 3*(2), 77–101. DOI: 10.1191/1478088706qp063oa

Černevičienė, J., & Kabašinskas, A. (2024). Explainable artificial intelligence (XAI) in finance: A systematic literature review. *Artificial Intelligence Review, 57*(8), 216. Advance online publication. DOI: 10.1007/s10462-024-10854-8

Chakrabarti, D. K. (2001). *India: An Archaeological History*. Oxford University Press.

Christensen, M., & Parker, L. (2010). USING IDEAS TO ADVANCE PROFESSIONS: PUBLIC SECTOR ACCRUAL ACCOUNTING. *Financial Accountability & Management, 26*(3), 246–266. DOI: 10.1111/j.1468-0408.2010.00501.x

Creswell, J. W., & Plano Clark, V. L. (2017). *Designing and Conducting Mixed Methods Research*. Sage publications.

Davis, D. R.Jr. (2007). Hinduism as a Legal Tradition. *Journal of the American Academy of Religion, 75*(2), 241–267. DOI: 10.1093/jaarel/lfm004

Dhillon, M. (2023). Weaving Together the Ancient and the Contemporary: Intersections of the Bhagavad Gita with Modern Psychology. *Pastoral Psychology, 72*(4), 525–537. DOI: 10.1007/s11089-023-01070-2 PMID: 37359496

Floridi, L., & Taddeo, M. (2016). What is data ethics? *Philosophical Transactions. Series A, Mathematical, Physical, and Engineering Sciences, 374*(2083), 20160360. DOI: 10.1098/rsta.2016.0360 PMID: 28336805

Hodge, D. R. (2004). Working with Hindu clients in a spiritually sensitive manner. *Social Work*, *49*(1), 27–38. DOI: 10.1093/sw/49.1.27 PMID: 14964516

Isser, S. S., Raj, N., Tomar, M., Marwaha, S. S., & Shastri, S. (2024). Value-based education in NEP 2020: Fostering ethical and moral growth through Dharma. *Asian Education and Development Studies*, *13*(5), 579–597. DOI: 10.1108/AEDS-06-2024-0121

Jarrow, R. A. (2018). *Derivatives & Risk Management*. World Scientific Publishing Company.

Jeste, D. V., & Vahia, I. V. (2008). Comparison of the Conceptualization of Wisdom in Ancient Indian Literature with Modern Views: Focus on the Bhagavad Gita. *Psychiatry*, *71*(3), 197–209. DOI: 10.1521/psyc.2008.71.3.197 PMID: 18834271

K, V., & Kalluraya, D. (2024). HISTORY OF THE INDIAN KNOWLEDGE SYSTEM. *EPRA International Journal of Multidisciplinary Research (IJMR)*, 352–355. .DOI: 10.36713/epra15354

Kasirzadeh, A., & Gabriel, I. (2023). In Conversation with Artificial Intelligence: Aligning language Models with Human Values. *Philosophy & Technology*, *36*(2), 27. Advance online publication. DOI: 10.1007/s13347-023-00606-x

Keshavan, M. S., Hegde, S., & Bhargav, H. (2024). Doing good well (Karma Yoga, the path of selfless action): Psychotherapeutic lessons from the East. *Asian Journal of Psychiatry*, *101*, 104201. DOI: 10.1016/j.ajp.2024.104201 PMID: 39241652

Kober, R., Ng, J., & Lee, J. (2010). MIND YOUR ACCRUALS: PERCEIVED USEFULNESS OF FINANCIAL INFORMATION IN THE AUSTRALIAN PUBLIC SECTOR UNDER DIFFERENT ACCOUNTING SYSTEMS. *Financial Accountability & Management*, *26*(3), 267–298. DOI: 10.1111/j.1468-0408.2010.00502.x

Lincoln, Y. S., & Guba, E. G. (1985). *Naturalistic Inquiry*. Sage publications. DOI: 10.1016/0147-1767(85)90062-8

Mahadevan, B., Bhat, V.R. and NAGENDRA, P.R., 2022. Introduction to Indian knowledge system: concepts and applications.

Mamta, S., Matharu, S., & Kumar, B. (2018). Impact of Yagya on Particulate Matter. *Interdisciplinary Journal of Yagya Research*, *1*(1), 01–08. DOI: 10.36018/ijyr.v1i1.5

Mamtani, R., & Mamtani, R. (2005). Ayurveda and Yoga in Cardiovascular Diseases. *Cardiology*, *13*(3), 155–162. DOI: 10.1097/01.crd.0000128730.31658.36 PMID: 15834238

Melé, D., Fontrodona, J., & Rosanas, J. M. (2016). Ethics in Finance and Accounting: Editorial Introduction. *Journal of Business Ethics*, *140*(4), 609–613. DOI: 10.1007/s10551-016-3328-y

Miralles-Quiros, M. D. M., Arraiano, I. G., & Miralles-Quiros, J. L. (2017). Are Firms that Contribute to Sustainable Development Valued by Investors? *Corporate Social Responsibility and Environmental Management*, *24*(1), 71–84. DOI: 10.1002/csr.1392

Mishra, N., & Aithal, P. S. (2023). Ancient Indian Education: It's Relevance and Importance in the Modern Education System. *International Journal of Case Studies in Business, IT, and Education*, 238–249. .DOI: 10.47992/IJCSBE.2581.6942.0271

Morandi, A., Dalla Libera, D., Roberti Di Sarsina, P., & Tosto, C. (2011). Salutogenesis and Ayurveda: Indications for public health management. *The EPMA Journal*, *2*(4), 459–465. DOI: 10.1007/s13167-011-0132-8 PMID: 23194327

Mulla, Z. R., & Krishnan, V. R. (2013). Karma-Yoga: The Indian Model of Moral Development. *Journal of Business Ethics*, *123*(2), 339–351. DOI: 10.1007/s10551-013-1842-8

Mulla, Z. R., & Krishnan, V. R. (2013). Karma-Yoga: The Indian Model of Moral Development. *Journal of Business Ethics*, *123*(2), 339–351. DOI: 10.1007/s10551-013-1842-8

Mulla, Z. R., & Krishnan, V. R. (2013). Karma-Yoga: The Indian Model of Moral Development. *Journal of Business Ethics*, *123*(2), 339–351. DOI: 10.1007/s10551-013-1842-8

Nembe, J., Daraojimba, A., Olubusola, O., Mhlongo, N., Oguejiofor, B., Falaiye, T., & Atadoga, J. (2024). THE ROLE OF ARTIFICIAL INTELLIGENCE IN ENHANCING TAX COMPLIANCE AND FINANCIAL REGULATION. *Finance & Accounting Research Journal*, *6*(2), 241–251. DOI: 10.51594/farj.v6i2.822

Nicholson, A. J. (2010). *Unifying Hinduism*. columbia university. .DOI: 10.7312/nich14986

O'Neil, C. (2016). *Weapons of Math Destruction: How Big Data Increases Inequality and Threatens Democracy*. Crown.

Odeyemi, O., Nwankwo, E., Soyombo, O., & Mhlongo, N. (2024). Reviewing the role of AI in fraud detection and prevention in financial services. *International Journal of Science and Research Archive*, *11*(1), 2101–2110. DOI: 10.30574/ijsra.2024.11.1.0279

Patton, M. Q. (2015). *Qualitative Research & Evaluation Methods*. Sage publications.

Ponterotto, J. G. (2013). Qualitative research in multicultural psychology: Philosophical underpinnings, popular approaches, and ethical considerations. *Qualitative Psychology, 1*(S), 19–32. .DOI: 10.1037/2326-3598.1.S.19

Puri, S. K. (2007). Integrating scientific with indigenous knowledge: Constructing knowledge alliances for land management in India. *Management Information Systems Quarterly*, *31*(2), 355–379. DOI: 10.2307/25148795

Sadik-Zada, E. R., Gatto, A., & Niftiyev, I. (2022). E-government and petty corruption in public sector service delivery. *Technology Analysis and Strategic Management*, *36*(12), 1–17. DOI: 10.1080/09537325.2022.2067037

Schweitze, B. (2024). Artificial Intelligence (AI) Ethics in Accounting. *Journal of Accounting, Ethics &. Public Policy*, *25*(1). Advance online publication. DOI: 10.60154/jaepp.2024.v25n1p67

Sedlmeier, P., & Srinivas, K. (2016). How Do Theories of Cognition and Consciousness in Ancient Indian Thought Systems Relate to Current Western Theorizing and Research? *Frontiers in Psychology, 7*(613). Advance online publication. DOI: 10.3389/fpsyg.2016.00343 PMID: 27014150

Sharma, A. K., & Talwar, B. (2005). Corporate social responsibility: Modern vis-à-vis Vedic approach. *Measuring Business Excellence, 9*(1), 35–45. DOI: 10.1108/13683040510588828

Sharma, A. K., & Talwar, B. (2007). Evolution of "Universal Business Excellence Model" incorporating Vedic philosophy. *Measuring Business Excellence, 11*(3), 4–20. DOI: 10.1108/13683040710820719

Singh, P. K., Jain, S. K., Mujumdar, P. P., & Dey, P. (2020). Hydrology and water resources management in ancient India. *Hydrology and Earth System Sciences, 24*(10), 4691–4707. DOI: 10.5194/hess-24-4691-2020

Tharakan, J. (2015). Integrating indigenous knowledge into appropriate technology development and implementation. *African Journal of Science, Technology, Innovation and Development, 7*(5), 364–370. DOI: 10.1080/20421338.2015.1085176

Yin, R. K. (2018). *Case Study Research and Applications: Design and Methods*. Sage publications.

ADDITIONAL READING

Mulla, Z. R., & Krishnan, V. R. (2014). Karma-yoga: The Indian model of moral development. *Journal of Business Ethics, 123*(2), 339–351. DOI: 10.1007/s10551-013-1842-8

Sedlmeier, P., & Srinivas, K. (2016). How do theories of cognition and consciousness in ancient Indian thought systems relate to current Western theorizing and research? *Frontiers in Psychology, 7*, 343. DOI: 10.3389/fpsyg.2016.00343 PMID: 27014150

KEY TERMS AND DEFINITIONS

Dharma: ethical duty, fairness, and responsible governance.
Karma: action–consequence logic, accountability, and long-term impact.
Kautilya's Arthashastra: strategic governance, resource allocation, and public administration.
Natya Shastra: creativity, narrative, and culturally informed innovation in AI systems
Nyaya: logical inquiry, justice, and explainable decision-making.
Vedantic Holism: interconnected view of economic, social, and environmental well-being.

Chapter 17
Sustainable Governance Through Ancient Lenses:
Policy Lessons From Traditional Indian Ecological and Resource Management

Ashok Dobhal
http://orcid.org/0009-0006-9077-4190
Uttaranchal University, India

Rajesh Bahuguna
http://orcid.org/0000-0001-8729-5160
Law College Dehradun, India

Nidhi Tyagi
http://orcid.org/0000-0002-3996-7644
Graphic Era University, India

Radhey Shyam Jha
http://orcid.org/0000-0003-0129-4365
Law College Dehradun, India

ABSTRACT

This review examines ancient Indian ecological governance as an integrated system of ethical principles, community institutions, and resource management practices that sustained socio-ecological balance over long time horizons. Drawing on interdisciplinary scholarship, the paper synthesises philosophical foundations such as Dharma, Rta, and Prakriti with traditional governance mechanisms including Panchayat systems, indigenous water and forest management, and culturally embedded conservation practices. Through comparative analysis with contemporary sustainability frameworks, the study demonstrates strong convergence with modern concepts of commons governance, climate resilience, and circular economy. The findings highlight the continuing relevance of ancient Indian governance traditions for informing inclusive, ethical, and community-led sustainability policy in the present.

DOI: 10.4018/979-8-3373-7976-0.ch017

INTRODUCTION

Background

The literature qualifies the modern-day sustainability crisis as one characterized by the increasing rates of climate change, depletion of natural resources, loss of biodiversity, and ecological instability. Empirical research always shows that the leading developmental models, which are based on intensive resource extraction, dependence on fossil fuels, and linear consumption patterns, have exceeded the planetary boundaries, thus threatening the long-term socio-ecological sustainability (Rockström, et al., 2009). Despite the spread of the sustainability models of the world and climate negotiations, the academic research indicates that policy responses are not unified and do not pay enough attention to the frameworks of governance and the ethical aspects of environmental destruction (Chhachhar & Verma, 2024).

The recent studies in relation to sustainability challenges the sufficiency of the strictly technocratic and market-based solutions to deal with the intricate problems of the environment. Rather, there is an increasing awareness of the importance of knowledge systems, which are rooted within the culture and governance practices, which have developed over the course of history and can help to achieve sustainable resource utilization (Berkes, Colding, & Folke, 2000). Having been developed over time as a result of interactions between communities and their ecosystems, traditional ecological knowledge systems focus on ecological balance, intergenerational responsibility, and collective stewardship.

The literatures have often referred to India as a civilization where ecological governance was ingrained in the philosophical, ethical, and institutional framework. The research of ancient Indian scriptures and the ancient Indian philosophy indicate the concepts of Prakriti, Rta, and Dharma. They represent the eco-centric worldview that makes humans the custodians of nature (Ravikanth, 2021). The ecological success of indigenous water-harvesting systems, sacred groves, mixed agricultural systems, and community-managed forests is further studied through empirical research, which helped to maintain biodiversity and adapt to climatic conditions (Pandey, Gupta, & Anderson, 2003). Environmental historians claim that the centralized regime of colonial and post-independent governance changed the protest systems, undermining the stewardship of communities and causing ecological degradation (Agrawal & Gibson, 1999).

Rationale of the Chapter

The motivation of this chapter is based on the new foundations of agreement indicating that sustainable governance should be a combination of ethical values and institutional pluralism. Backed by community involvement, scientific vigour and technological intervention, the current sustainability governance literature is becoming increasingly aware of the fact that traditional and indigenous knowledge systems can be used to supplement contemporary policy tools to contribute to their legitimacy, flexibility, and sustainability (Tengö, Brondizio, Elmqvist, Malmer, & Spierenburg, 2014).

The systems of ecological governance developed in ancient India are highly applicable since they formalized sustainability by means of decentralization of decisions and regulation based on moral and community-based management of resources. Experiences have also shown that conservation activities that are culturally embedded, like sacred groves and traditional water-sharing norms, can work better than externally enforced control systems in safeguarding ecosystems (Ormsby & Bhagwat, 2010). There are several indigenous practices that incorporated the ecological responsibility into the societal norms, which minimized the costs of enforcing practices and promoting voluntary compliance.

In addition, the traditional Indian models of governance are largely compliant with the modern-day sustainability concerns regarding climate adaptation, resilience, and local governance. Climate-resilient development studies have placed the emphasis on the fact that decentralized water systems, agroecology, and community stewardship have the potential to increase adaptive capacity, especially in climate-sensitive areas (Ensor, et al., 2019). The restoration of traditional forms of governance is thus of direct concern to the modern-day climate and environmental policy.

Ancient Indian systems also offer impartial examples of effective commons management from a governance perspective. These systems find a lot of resonance with the contemporary theories of institutions where collective action, social capital, and locally designed rules of sharing and administrating shared resources take precedence (Ostrom, Governing the Commons: The Evolution of Institutions for Collective Action, 1990). Exploring these ancient models in terms of policy, this chapter adds to the current discussions on sustainable governance, environmental ethics, and localization of sustainability policies.

Objectives of the Chapter

The objectives of this chapter are:

1. To examine Indian traditional ecological knowledge as documented in ancient philosophical traditions and indigenous practices.
2. To analyse ancient governance models related to water, forest, agricultural, and urban resource management.
3. To identify ethical and institutional principles that supported long-term ecological sustainability.
4. To evaluate the relevance of ancient Indian governance systems for contemporary sustainability and climate policy.
5. To derive policy-relevant insights for designing community-led and ethically grounded governance frameworks.

LITERATURE REVIEW

Altieri, Nicholls, Henao, & Lana in their work provide evidence that agroecological studies attribute these practices to soil enrichment, low exposure to climate, and sustainable food supply in the long run. They argue that in order to enhance climate resilience, it is essential to transition from vulnerable modern farming culture to traditional farming system. Their evidence-based research embraces the biodiverse farming system by considering them less vulnerable against extreme climatic conditions and offered the risk analysis framework in agroecology. However, this research focuses largely on small farm holders but fails to target large industrial farming sector that covers more than 80% of farmlands (Altieri, Nicholls, Henao, & Lana, 2015). Jules Pretty et al. in their research successfully redefines the sustainable intensification as a course of ecological integration and not a mere method of yield optimization. The research found 163 million farms to be using sustainable practices rather than harmful chemical-based farming approach but the figures mentioned seems to be bit optimistic. The points related to agri-business lobbying, small farm practices and financial subsidies had been largely ignored. While the term redesign threshold is highly subjective (Pretty, et al., 2018).

The empirical research of Ashwini Chhatre and Arun Agrawal 80 forests across 10 developing nations globally, to figure out that in order to sustain mitigation of climate and alleviation of poverty, it is necessary to empower local communities. The authors argued that dependence on social capital, trust, and mutual moral structures can be seen in the old Indian systems, which supported cooperation over long time. The strength of research lies in robust empirical research that challenges the government conservation models and decentralized framework. This paper is a strong research work but fails to prove long term sustainability as there are the chances of high carbon storage and extraction rates of being temporary, that justifies the current momentary situation but not a long-term stability. The research shows the heavy reliance on forest extractions and also ignores the external macro- economic pressures (Chhatre & Agrawal, 2009). Menski in his work, analysed the ancient Indian legal principles and their applicability in modern context. His work is highly pertinent that traced the evolution of Hindu legal principles from Vedic and classical period to post-colonial era. According to him this type of prescription was an informal form of regulation, meaning it influenced behaviour based on morality and associated social sanction instead of through coercion. He depicted inner beliefs to be the reason for following laws. However, there are several other sources and texts that depicts Hindu laws which remained ignored by the author. The author also failed to depict the clarity in its relevance in modern context (Menski, 2012).

The commons theory developed by Elinor Ostrom is commonly utilized as an analytical tool to study the conventional Indian resource systems. In his work, Ostrom has shown that communities can be able to manage efficiently common-pool resources using locally based policies, joint surveillance, and incremental sanctions. Since, the work of author was based on principles derived from localised and geographically confined resources, the framework cannot be considered as an authentic medium to resolve the problems associated to global commons (Ostrom, 2015). Joanna Williams considers circular economy as an essential element for global sustainability. Author argues that the urban waste management was based on decentralized approaches including segregation, reuse, and organic recycling, which were strengthened by civic norms and occupational roles. However, recent scholarship compares these methods to modern-day decentralized waste management and circular economy approaches, highlighting how the ancient city administration is relevant to modern sustainability discussions.

The introduction of RESOLVE technique in circular economy is a key concept of the paper but it seems to be theoretical rather than practical. The paper fails identify the governance gaps which are the key barriers for effective management (Williams, 2019). Sharma and Acharya in this research briefly described the importance of ancient Indian jurisprudence as a good source to attain sustainability. They referred to several ancient texts such as Manusmriti, Arthashastra, Vaha Puran etc. that made people pray environment, which not only maintained the respect towards ecology but also inherently lead them towards sustainability. This paper gives insight towards indigenous sustainable practices and its role in attaining global sustainability. Authors in this paper criticizes the UN SDGs for favouring economic expansions over the ecological balance but fails to create an interlink between ancient spiritually driven indigenous practices and modern statutory frameworks (Sharma & Acharya, 2024).

RESEARCH METHODOLOGY

The research methodology is primarily Qualitative. It includes a critical analysis of ancient wisdom of governance side by side modern aspects of governance too through examining available primary and secondary data both. Ancient models of governance and its adaptability in present era has been explored through various instances and case studies. The research methodology ensures the findings of the study are not only just academically rigorous but are also beneficial and relevant to the participating Indigenous Nations, resulting in operationally viable strategies & actionable policy recommendations.

THEORETICAL FOUNDATIONS OF ANCIENT ECOLOGICAL GOVERNANCE

The governance of ecological system in ancient India was not formulated in terms of official policies of environmental protection as per modern understanding of the world. Rather, it was the outcome of a highly coherent worldview philosophy, ethics, and institutions. This concept of worldview perceived nature, society, and the government as components of a moral-ecological order that were not independent of each other. The recent literature of sustainability scholarship is making the legitimacy of such culturally embedded forms of governance a good source of information on resilience, adaptive management, and long-term sustainability (Tengö, Brondizio, Elmqvist, Malmer, & Spierenburg, 2014). The notions of Prakriti, Rta, and Dharma; the ecological ethics of Vedic and Upanishadic writings; Gandhian ecological philosophy, as a contemporary interpretation of ancient notions; and indigenous knowledge systems as practice-based systems of governance conceptualization are the fundamental philosophical constructs of ancient Indian ecological governance that will be analyzed in this section.

Conceptual basis: The literatures that were based on the interrelated conceptions of Prakriti (nature), Rta (cosmic order), and Dharma (ethical duty) forms the conceptual roots of the Indian science of sustainability in ancient India. Combining all these ideas, human nature relations were viewed as signifying balance, reciprocity, and moral responsibility as opposed to domination and exploitation. Researchers believe that these philosophical constructs were normative forms of governance that determined the use of resources, social behaviour and norms in a sustainable way.

Research explains Prakriti as a self-regulating and dynamic system that is composed of both animate and inanimate constituents. Nature was not viewed as being an object of inert resources but rather a living being of its own value (Shastri, 2018).

This realization is well-congruent with the modern ecological system, which discourages overextraction, supports the ecological restraint, and deeply focuses on complexity, regenerative potential and feedback connections (Folke, Biggs, Norström, Reyers, & Rockström, 2016).

The Rta concept is a derived statement of the moral and natural law of the universe, which dictates the ways of the universe and human conduct. Academic researches indicates that the disturbance of Rta caused by environmental destruction or overconsumption may lead to social disorders and ecological disequilibria. Whereas, Interdisciplinary scholarship finds the comparison between Rta and the modern models of planetary boundaries and Earth system governance, which also stress the need to act within the context of biophysical boundaries to achieve a stable system. As a moral rationale of government, Dharma urged responsibilities rather than rights and group interest instead of individual good. According to scholars, Dharma-based ethics incorporate environmental responsibility into the regular social and religious duty, thus promoting moderation, frugality, and intergenerational responsibility in a way that appeals to modern discourses of sustainability and the circular economy (Geissdoerfer, Savaget, Bocken, & Hultink, 2017).

Ecological Ethics in Vedic and Upanishadic Texts

The ancient Indian literatures unanimously describe Vedic and Upanishadic literature as presenting an ethic system where nature has been perceived to be sacred, alive, and morally pertinent. The rivers, forests, mountains, wind, and soil are described in the hymns of the Rig Veda and Atharva Veda and are regarded as life-giving forces that were revered and needed to be preserved.

The Prithvi Sukta is often mentioned as an important statement of reciprocity between humans and the Earth, focusing on the care, thankfulness, and refraining responsibilities. This orientation is viewed by environmental philosophers as an eco-centric worldview where human beings are integrated in a natural environment. Upanishadic philosophy further elaborated this ethic by the unity of Atman with Brahman, which means that the destruction of nature would eventually amount to the destruction of the self. Such holistic worldviews are also associated with stronger pro-environmental attitudes and behaviours, which are established through empirical research involving environmental psychology.

Another theme that the literature provides is that Vedic and Upanishadic ecological ethics were operationalized on ritual practices, taboos, and customary norms. Moral sanctions were used to preserve sacred trees, rivers, and landscapes, leading to informal systems of governance that proved to be effective. Household research on sacred locations suggests that such culturally imposed safeguards tend to be more successful in enhancing the biodiversity results than externally established regulatory regimes (Ormsby & Bhagwat, 2010).

Indigenous and Tribal Knowledge Systems

Indigenous and tribal knowledge systems in India have been identified in literature as one of the most sustainable and practice-based systems of ecological governance. Such systems are noted to be typified by place-based knowledge, adaptive management, and strong communal institutions. These systems play important roles in conservation of biodiversity, adaptation to climate change, and livelihoods. One of the most often mentioned examples of indigenous ecological governance is sacred groves. These groves are conserved as community controlled and through religious faith and traditional law. Empirical evidence shows that sacred groves tend to have greater abundance of species, genetic diversity, and ecosystem integrity as compared to the adjacent landscapes, and, thus, can demonstrate the functionality of governance norms embodied within cultures (Ormsby & Bhagwat, 2010).

Innovative agricultural methods, including mixed farming, seed banking done by locals, the organic nature of manure, and rotational farming, among others, are also introduced in the literature as being a sign of such advanced ecological knowledge. Equally, literature records water management models developed by the indigenous community, such as rainwater harvesting, tank irrigation, and shifting cultivation with rest and recovery phases, as adaptive mechanisms that enhance groundwater replenishment and community resistance (Pandey, Gupta, & Anderson, 2003).

TRADITIONAL INDIAN MODELS OF RESOURCE GOVERNANCE

The Indian models of governance of resources are traditionally based on local ecological settings, community-based institutions, morality, and collective practices. These governments work without a central authority, making community stewardship, rules, and cultural compulsions a priority. Recent publications in sustainability and commons governance confirm that the models have the key components of resilient social-ecological governance, which include decentralization, adaptive management, and social legitimacy. The section addresses the conventional governance paradigms that are associated with water, forests, agriculture, and urban environments that emphasize their design and institutional framework.

Water Management Systems

The traditional Indian system of water management has been described in the literature as advanced combinations of ecological design, social institutions, and normative governance. Instead of using a centrally managed infrastructure, these systems developed as decentralized and community-controlled systems geomorphologically, climatically, and socially suited. They are increasingly viewed by interdisciplinary research as their initial types of adaptive water governance, which boost resilience, equity, and sustainability in terms of environmental uncertainty (Gowri, Dey, & Mujumdar, 2021).

Research records of stepwells (baolis or vavs) in semi-arid areas show a multitude of functions that include groundwater recharge, buffering the drought, and social utilization. Hydro-ecological studies show that their stepped constructions do minimize the losses of evaporation, equalize a nearby water level, and permit year-long access with administration integrated in community norms, girlish endowments, and guild upkeep. The so-called tank irrigation systems are also a frequent example of peninsular India, which includes vast hydraulic systems capturing monsoon runoff, regulating floods, and sustaining agriculture by maintaining and distributing water on a custom basis (Piplani & Kumar, 2019).

Other examples of locally adapted rainwater management and flood management systems are in western India through the use of the johads and earthen check dams and in eastern India through the Ahar-Pyne system (Koul, Singh, Neelam, & Shukla, 2012). In most instances, principles of decentralization, ecological fit, redundancy, and community stewardship are mentioned as common in the literature, which is consistent with the state of modern commons governance and adaptive water management theory.

Figure 1. Traditional Indian Water Managements, A Socio-Ecological Diagram

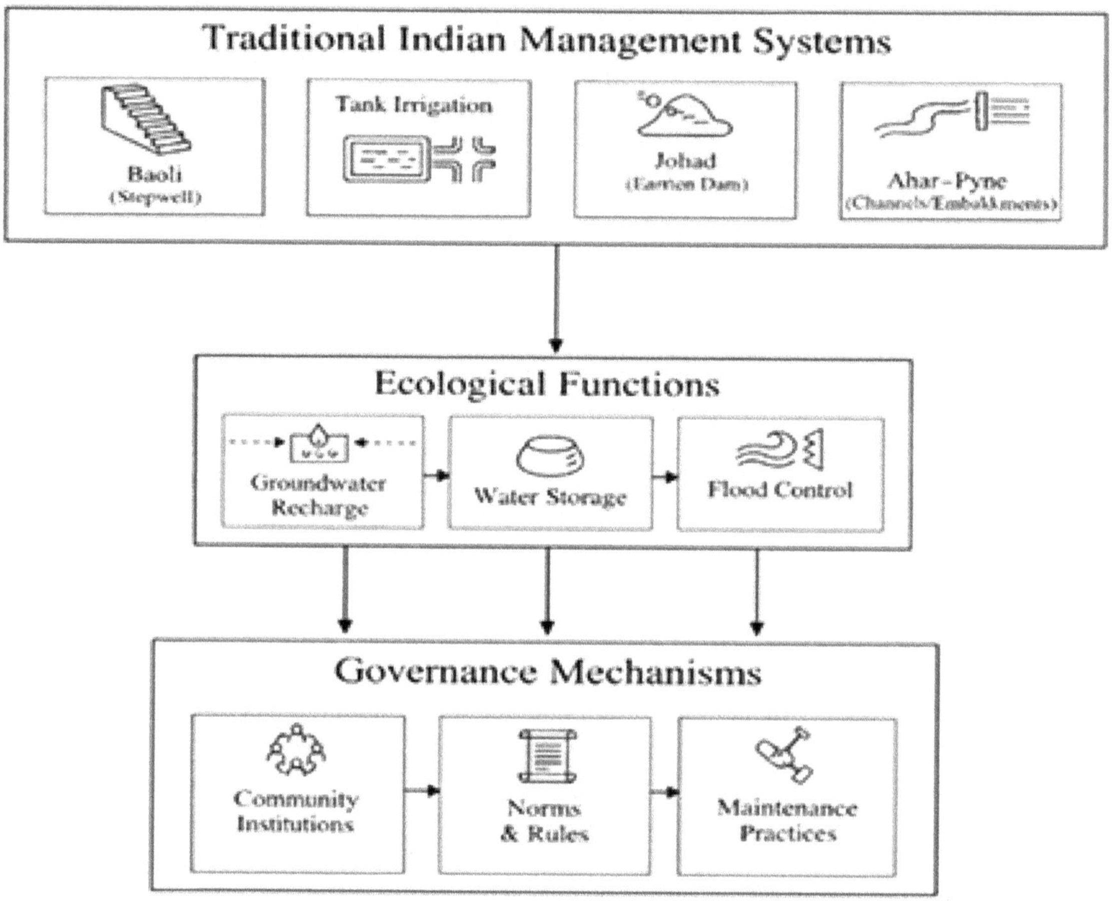

Forest Conservation Traditions

According to the literature, the traditions of forest conservation in India can be said to have developed mostly based on culturally institutionalized practices, ethical guidelines, and communal systems instead of state regulation. Forests were seen as lively socio-ecological structures and therefore fundamental to climatic stability, living and spiritual health, and not just seen as timber or revenue sources. Such traditions are now seen as efficient systems of commons governance that promote ecological stability in the long term by promoting collective action and moral regulation that led to sustainable environmental governance (Chhatre & Agrawal, 2009).

Sacred groves, locally known as Devrai, Devarakadu, or Sarna, are generally known to be one of the earliest expressions of forest conservation by the communities. They were inviolated by religious faith, ritual taboo, and traditional law, hence acting as sacred ecological reserves. Scholarly studies have consistently shown that these groves are more effective in supporting greater species richness and gene pools compared to these managed forests, which is a proven sign of the effectiveness of governance norms embedded within the spirit (Ormsby & Bhagwat, 2010).

The academic history registers Northeast Indian community-forest regimes that are run by village councils and clan institutions. It is found that these forests often perform better than state-run reserves on matters of balancing biodiversity preservation due to defined boundaries, locally formulated policies, and imposed penalties (Ostrom, Governing the Commons: The Evolution of Institutions for Collective Action, 2015). The Chipko Movement has been often seen as a modern reenactment of these moral traditions, bringing the practices of traditional forest management and non-violence to the politics of forestry and commercialization (Guha, 2000).

Figure 2. Chipko movement: Local women and children practicing arboreal hugging to prevent deforestation, 1987. (Photograph courtesy of The Right Livelihood Award).

Urban Environmental Management in Ancient India

According to the literature, sustainability concepts in ancient India were built not only in the rural and agrarian systems but also in the early urban planning and civil governance. It has long been demonstrated in scholarly studies that the antiquarian city embraced comprehensive solutions to water supply, drainage, sanitation, and waste management (Mays, Koutsoyiannis, & Angelakis, 2007).

Examples of cities that appear frequently in archaeological studies of the Indus Valley civilization include Mohenjo-daro and Harappa as pioneering examples of an environmentally friendly city design. It has been proven to have advanced drainage systems, covered sewers, standardized construction materials, and premeditated street patterns that enabled efficient management of stormwater and sanitation at a city level. The standardization and consistency of such infrastructure imply strong civic institutions and well-dispersed norms of environmental order.

The analyses of the history also prove that the administrative systems of Mauryan and Gupta formalized urban environmental rules with the help of the regulatory mechanisms. The interpretation of the Arthashastra and the inscriptions around the same period suggest that authorities have controlled the disposal of waste, safeguarded water, and developed fines due to pollution and failure to perform their sanitation responsibilities, which can be seen as early issues of environmental responsibility.

INSTITUTIONAL & GOVERNANCE MECHANISMS IN ANCIENT INDIA

The ancient Indian institutional and governance mechanisms had close connections with ethics, social organization, and environmental sustainability. Governance was about a set of local self-governing institutions, custom norms, and moral obligations instead of centralized bureaucratic control. Access to land, forests, water, and other commons was controlled by village assemblies, guilds, and councils based on kinship, which ensured shared decision-making and accountability. Local institutions and state arrangements, as expressed in classical literature, supplemented each other in stipulating regulation, control and conservation responsibilities. These mechanisms formed a polycentric form of governance that ensured power equilibrium, civic engagement, and ecology, which allowed the optimal management of resources in various socio-ecological settings.

Panchayat and Local Self-Governance Systems

The academic literature has always found the Panchayat as well as the local self-government as the institutions on which the very nature of environmental management in ancient India was built. Research characterizes village-level groups, or Gram Sabhas, as deliberative assemblies that administered social, economic, and ecological matters via collective decision-making and moral responsibility and not by a bureaucracy. These assemblies allowed communities to turn natural resources in a manner that was organized according to the local environmental circumstances as well as socially embedded standards.

Studies about commons governance emphasize that Gram Sabhas crucially regulated forests, water, and land as the collective resources. The choices in terms of access, restrictions on extraction and seasonal utilization, and maintenance were all negotiated and imposed via customary regulations, which led to the decreased conflict and equalized distribution (Ostrom, Governing the Commons: The Evolution of Institutions for Collective Action, 1990). Village assemblies managed the fuels, grazing, and timber exploitation in the forests through the establishment of remedies that were usually protected and sanctions imposed to facilitate the conservation in the long term.

The management of water is also reported as one of the most important functions of Panchayat, as the local institutions organize the construction, renovation, and distribution of the water resources based on tanks, wells, canals, and harvesting systems. Decentralized management also improved resilience and sharing responsibility and local monitoring (Shah, 2021). The land tenure joined the right of usufruct to individuals and communal management as supported by moral punishment and collective law. These forms are identified in the institutional analyses as anticipatory variants of participatory and polycentric governance that are key to the management of commons (Chhatre & Agrawal, 2009).

Arthashastra: The Kautilyan Model

The Arthashastra is considered by the scholarly literature as one of the original systematic statements of state-based environmental government where the ruler is placed in the role of custodian of natural resources acting on behalf of the societal interest. The text also conceptualizes land, forests, and minerals as strategic resources that need to be regulated, conserved, and ethically managed, instead of treating them as open sources of income, and thus combines economic productivity with the maintenance of ecological processes and social order.

It is noted that according to studies, forests took up a central place in the system of Kautilyan government. The state categorized forests by purpose: production, protection, sacred and controlled extraction, hunting, grazing, and protection of wildlife by elaborate administrative regulations. According to scholars, these prescriptions indicate that at an early stage, the importance of ecological boundaries and regeneration was recognized, and thus they are already related to the principles of sustainable forest management. The same situation was done with the mineral resources, which were tightly controlled by the state to avoid exploitation and harming the environment (Devi, 2025).

The government approach to agriculture focused on productivity and conservation, and the state authorities focused on the maintenance of irrigation, the soil protection, and the assistance to the farmers in the hardship of the ecological conditions. The use of environmental taxation and the imposition of punishments for activities like illegal logging, water contamination, and wildlife destruction are also recorded in the literature, as they served as deterrents and control measures. These steps can be largely viewed as the initial institutional conception of environmental responsibility in administrative law.

Dharmaśāstra-Based Norms

The traditions of Dharma are discarded in the literature as a standard-bearing system where the environmental responsibility was incorporated in the social, legal, and ethical life in ancient India. Instead of colonizing the environment as a distinct sphere and imposing on it a particular regulation, these texts included environmental responsibilities in the basic social behavior. Dharma was viewed as an order of duty over relations between humans, nature, and the order of things. Whereas, environmental protection was conceived as an ethical requirement instead of a legitimate gesture. The social responsibility towards the environment was expressed in prescriptions dictating land use, water conservation, preservation of trees, and animal welfare.

Another hallmark of the Dharma-based government was the emphasis on the communal and collective responsiveness rather than the absolute property rights of individuals. Although individual usufruct rights were accepted, forests, water bodies, grazing lands, and unmanned lands were commons governed by a collective. The framework confined the enclosure of resources and encouraged fair accessibility, which is similar to the modern commons governance theory (Shiva & Bandyopadhyay, 1988).

By making ecological ethics a part of the religious and social identity, ritual punishments and ethical duties solidified compliance. In culturally cohesive societies, such value-based regulation is identified as long-lasting (according to environmental governance scholarship) (Guha, 2000).

Buddhist and Jain Ecological Governance

The academic works name both Buddhist and Jain traditions as full-fledged models of ecological governance, which is based on non-violence (Ahimsa) and compassion as well as restraint towards all forms of life. Along with moral consideration of human society, these traditions applied moral considerations to animals, plants, and ecosystems and thus offered an early normative basis of multispecies ethics. Such views are gaining popularity among environmental philosophers as being critical in resolving the ecological crises based on anthropocentric governance structures. According to Jain philosophy, Ahimsa was a universal moral law of all the activities, and environmental violence was understood as violence, either premeditated or unintended. As in the documents of historical literature, strict Jain guidelines that controlled the use of agriculture, water, and extraction of resources, such as filtered consumption of water, prohibition on farming at nighttime, and prohibitive actions against cutting trees, indicate an increased ethical sensitivity toward ecological interdependence (Laidlaw, 1995).

Monastic laws were institutionalized as the Buddhist ecological governance through the monastic texts in the Vinaya. These regulations controlled the usage of lands and water, waste disposal, and communication with animals and plants. Monasteries were regularly regulated socio-ecological units, with a restricted consumption and with a collective responsibility. According to scholars, these monastic systems represented the ideas of sustainability and sufficiency, and they provided examples of the models of low-impact ways of life. Compassion and ethical self-restraint as a form of governance were central to both traditions, as they internalized effects on the environment in both traditions.

COMPARATIVE ANALYSIS: ANCIENT MODELS VS. MODERN SUSTAINABILITY FRAMEWORKS

The models of sustainability in ancient India were based on restraint, community management, and balance with nature, whereas the current models include more focus on set targets, indicators, and mechanisms regulating it. Although institutional types are different, they both lead to a common core of resilience, responsible usage of resources, and common action, which places the persistence of the culturally grounded governance in dealing with the modern-day

Ancient Indian Approaches vs. SDGs

The ecological literature indicates that the ancient Indian governance of the ecological setting is significantly conceptually aligned with a number of Sustainable Development Goals even though it was produced not via codified policy frameworks but instead through systems of cultural bias and social norms. This alignment is best displayed in water management, climate action, and responsible consumption, whereby sustainability was operationalized in terms of community stewardship, ethical restraint, and ecological regeneration.

Regarding SDG 6 (Clean Water and Sanitation), researchers emphasize that there are considerable similarities between contemporary water security agendas and the traditional Indian water-harvesting models. Tanks, johads, stepwells, and canal networks were built so that there would be recharge of groundwater, uniform distribution, and balancing of the ecosystems. Scholarship of governance is becoming aware that such small-scale community-based governance systems improved water resilience through maintenance and allocation entrenched in local institutions and local role models (Piplani & Kumar, 2019) .

Equally, SDG 13 (Climate Action North) is associated with forest conservation culture, like sacred groves and community stewardship. These culturally safeguarded landscapes that are recorded in the literature acted as long-term ecological buffers, biodiversity preservation, and microclimatic stabilization and helped in carbon sequestration. Their success was based on morality and sacral decree instead of fiat, which in most cases led to more potent conservation delivery than formalized conservation regions (Ormsby & Bhagwat, 2010).

SDG 12 (Responsible Consumption and Production) is echoed by the Indian culture of minimalism, sufficiency, and regeneration of resources, ancient Indian traditions. Farming and household activities focused on the recycling, zero-waste sequence, and restraint in the consumption process, which was surrounded by Dharma-driven ethics where the well-being of society came first. Generally, comparative scholarship suggests that in modern SDGs, governmental indicators and targets are crucial in realizing sustainability goals, whilst in ancient practices, this has been realized by way of ethical internalization, local governance, and cultural legitimacy.

Figure 3. Alignment between Ancient Indian Ecological Practices and Selected Sustainable Development Goals

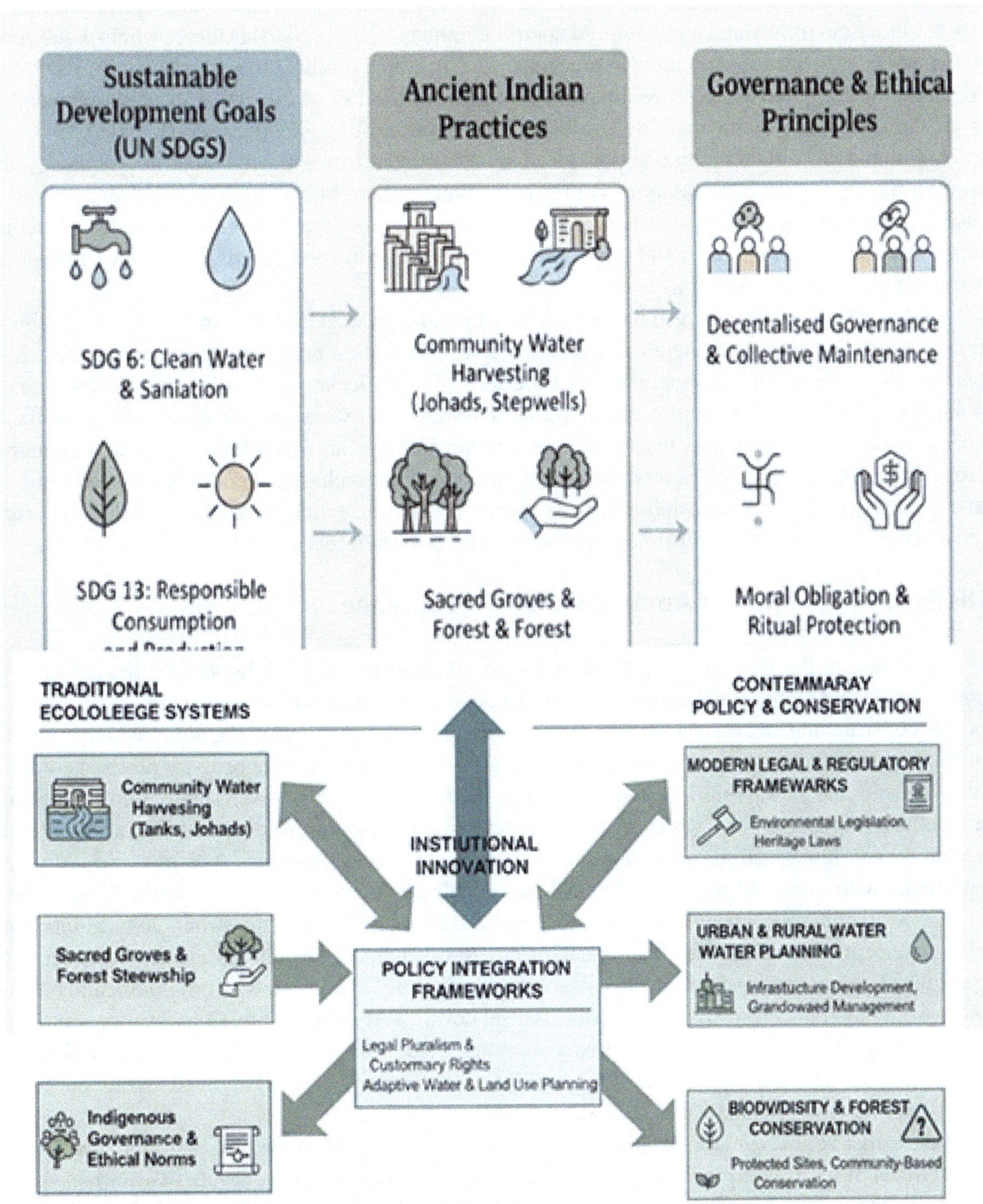

Convergence with Modern Governance Theories

According to the literature, the ancient Indian ecological governance approaches are quite close to a number of modern concepts of governance designed to address the shortcomings of centralized and market-based environmental regulation. Academic commentaries observe that these principles are similar to the practices used in ancient India in managing forests, water, pasture, and irrigation systems whereby regulation was applied based on common agreement and moral standards rather than enforcement (Ostrom, Governing the Commons: The Evolution of Institutions for Collective Action, 2015).

An analogous situation can also be traced in the practices of ancient Indian government, which support the concept of community-based resource management, known as community-based resource management (CBRM) today, that is highly promoted in the modern sustainability policy. Literature reports that village conferences and traditional organizations governed the control of entry, upkeep, and settlement of disputes on natural assets.

Additional convergence is also observed concerning the principles of the circular economy. Modern research highlights waste reduction, their reuse, and regeneration processes as the core issues of sustainable production and consumption (Geissdoerfer, Savaget, Bocken, & Hultink, 2017). The literature points out that the Indian agrarian and household systems that were traditional were the systems that operationalized these principles via practices of zero-waste, recycling of nutrients, and even the integrated crops-livestock systems. Lastly, nature-based solution studies emphasize congruency with the old ways of sacred groves, agroforestry, and water-harvesting landscapes, which interacted with the processes of the ecological environment to increase resilience and ecosystem services.

Distinctive Features of Ancient Indian Governance

According to the literature, the modern system of sustainability has important measurement and coordination and global cooperation tools, but the ancient Indian system of ecological governance had a number of structural and ethical advantages that are usually not so strong in the more modern ones. One of the primary characteristics that are primary in governance studies is the primary place of community ownership and stewardship. Water bodies, forests, and grazing lands were held as commons under the collective management, which discouraged excessive exploitation, especially through individual accumulation. Investigations about commons governance always relate this type of community-based ownership with improved accountability and sustainable ecological results (Agrawal & Gibson, 1999).

The next strength contributing to the scholarship that is identified is the strong cultural integration of the concept of sustainability. The issues of environmental responsibility were not considered as a specialized policy area, and instead, in ancient India, the issues associated with environmental responsibility were integrated into religious practices, social norms, and ordinary life. Conservation values were enhanced by rituals, taboos, and moral education and made less dependency on external enforcement and enhanced self-compliance. This is in contrast to most contemporary systems of regulation that tend to rely on coercion of the law and economic rewards.

The further set of regulations peculiar to ancient forms of government was moral and ethical. The environmental harm was presented as an ethical offense, based on the concepts of Dharma, Ahimsa, and trusteeship. It is becoming more acknowledged in the literature that such ethical internalization facilitates restraint, moderation, and intergenerational responsibility, which are hard to accomplish in a technical regulation setting.

Lastly, researchers remark that the ancient Indian rule had long-range ecological thought, which was based on regenerative processes and intergenerational persistence, as opposed to short-term economic gains. This time richness brings forth lasting teachings to the modern sustainability rules.

POLICY LESSONS FOR 21ST-CENTURY GOVERNANCE

The Indian system of ancient ecological governance provides important policy lessons to modern sustainability problems because the system proves to be effective through decentralized governance, communal management, and morally based systems of governance. A direct approach to this is incorporating the old into the new policy to augment legitimacy, resiliency, and the adaptive ability, especially in those areas that are susceptible to climate changes. Such observations indicate that the notion of sustainable governance in the twenty-first century is the need of scientific novelty and regulation; nevertheless, it must rest on the cultural amalgamation, moral responsibility, and ecological outlook that can be ensured by communal care.

Reintegrating Traditional Knowledge into Policy

The adoption of the reintegration of traditional ecological knowledge into modern policy frameworks is being increasingly accepted by the literature as a decisive route towards the reinforcement of sustainability governance. The understanding of indigenous and traditional systems is quite prolific; they are not fixed cultural artifacts but are empirically based knowledge frameworks, developed over extended periods of time as a result of interaction with the local ecosystems. According to policy and sustainability scholarship, such systems should not be marginalized because they minimize adaptive capacity, especially in areas where the climate varies and resources are stressed (Berkes, Colding, & Folke, 2000).

The development of a formal status of indigenous ecological systems within legal and regulatory processes has been mentioned in the literature as one of the key policy priorities. Research underlines how integration needs to go beyond symbolic acknowledgement to enforceable rights and well-spelled-out duties of communities that use forests, water bodies, and agricultural sceneries. The existing literature on comparative governance shows that legal pluralism (as the co-existence of customary norms with statutory law) can be associated with increased levels of legitimacy, equity, and ecology in case it is properly institutionalized.

An alternative priority area is the revival of traditional water-harvesting structures. Systems like tanks, johads, stepwells, and canal networks are repeatedly articulated to be ecologically pertinent and economical in order to the prevailing climate. It is reported in the literature that the policy encouragement of restoration, community management, and incorporation of these systems into water planning systems enhances groundwater replenishment and water security, which is achieved when technocratic substitution is suppressed and local power is empowered.Recognition of sacred groves as heritage and conservation sites is also a potential policy mechanism that is identified by academic literature. The formal recognition may strengthen the results of biodiversity, and cultural integrity can be maintained when the community custodianship is central to the process of governance (Mays, Koutsoyiannis, & Angelakis, 2007).

Figure 4:.Integration of Traditional Ecological Knowledge and Contemporary Conservation Policy Through Institutional Innovation Frameworks

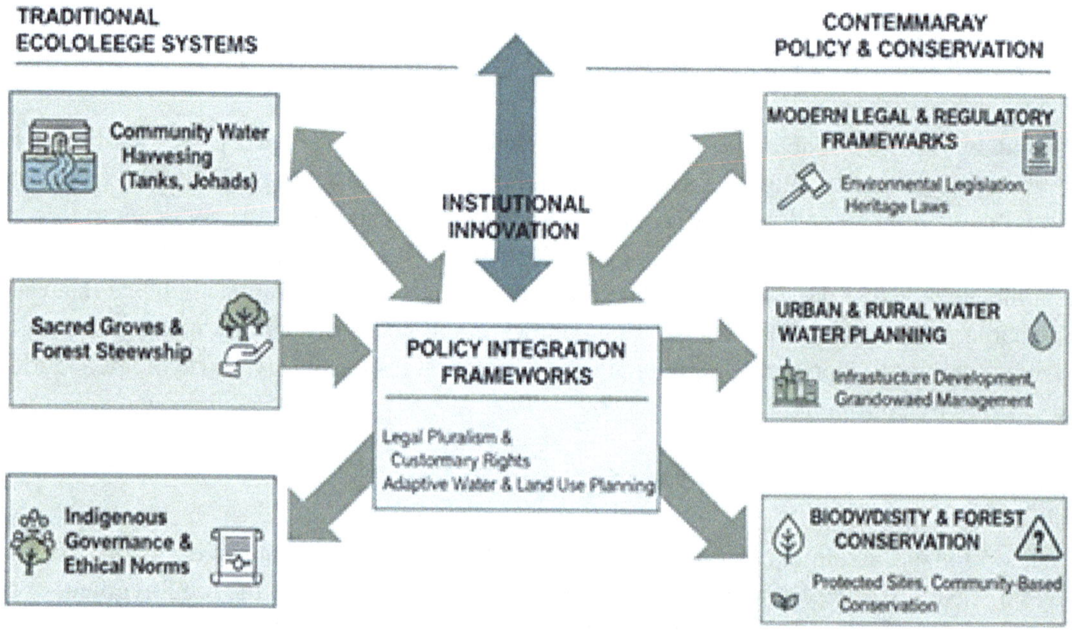

Community-Led Environmental Governance

It is established in literature that community-led environmental governance may be pivotal in establishing sustainability in complex and socio-ecologically diverse environments. Historical research on the governance systems of ancient India shows that having these local institutions like Gram Sabhas and empowerment helped to improve ecological consequences due to the increase in establishing an overlap between the decision-making power and local knowledge and openness to experience. This perspective is also widely endorsed in the contemporary governance scholarship, stressing the explicit benefit of decentralized institutions when dealing with the common-pool resources in the uncertainty and ecological variability conditions (Ostrom, Governing the Commons: The Evolution of Institutions for Collective Action, 2015). Studies have also indicated that successful empowering of Gram Sabhas has to do with giving the bodies substantive powers in governing forest, water, and land and not the consultative roles, which are limiting roles. Empirically, it is found that where local assemblies are given a legally binding authority to make decisions, there is an increased record of fairer resource utilization, higher levels of adherence to the rules and regulations, as well as a reduction in the rate of ecological erosion (Chhatre & Agrawal, 2009). Community ownership of water-harvesting systems, small-scale irrigation networks, and forest commons according to the Indian context is linked with increased capacity to respond to climate stress and improved resource conflict.

Digital tools can be discussed as a modern development of community-led governance through participatory monitoring. Research has shown that data collection, remote sensors, and mapping the community on mobile can enhance openness and ecological surveillance when incorporated as part of effective local institutions instead of being foreign control systems.

Embedding Ethical–Ecological Principles

The principles of ethical-ecological values need to be entrenched in governmental systems to solve the sustainability problems that are not necessarily limited to technical regulations and economic stimuli. The analysis of ancient Indian thought identifies as a central theme the emphasis on a solid ethical basis of the environmental governance that is based on moral responsibility and restraint and intergenerational justice. The modern literature in sustainability strongly suggests the need to supplement policy tools with ethical forms of internalization in order to bring long-lasting ecological change.

One of the relevant ethical tools that is often mentioned as a reference to modern corporate and institutional governance is Gandhian trusteeship. Trusteeship is conceptualized by scholars as the administration of the issue of wealth and natural resources as an asset entrusted to society and future generations instead of being a tool of unrestricted personal acquisition. Policy of governance and ethics in the business economy can associate this principle with the discussion of corporate environmental responsibility issues, stakeholder governance, and the establishment of sustainable value, and this claim implies that the use of the trusteeship-oriented approach and could contribute to the more accountable consideration of the ecological effects.

The element of Dharma-based environmental responsibility helps to supplement this ethical orientation, as the environmental stewardship turns out to be a responsibility rather than an optional activity. Authors focuses on living in balance, moderation, and the common good to incorporate sustainability in moral and social identity. It holds that legal scholarship argues that duty-based orders of thought might be strengthening them on a stronger platform, compared to rights-based systems, in instances where social norms are used to create a powerful level of regulatory control (Menski, 2012).

Another strategy of long-term governance in the literature is sustainability education that is based on Indian philosophical traditions. The establishment of ethical thinking and eco-literacy in the curricula correlates with the enhancement of pro-environmental behavior and enhanced commitment to systemic sustainability transitions.

Policy Recommendations

According to the literature, the interpretation of the ancient environmental governance principles into the modern-day policy requires specific institutional changes that address the equilibrium between the law, community, and the responsibilities of ethics. Governance research focuses on the legalization of the common's custodianship, the strengthening of legal pluralism, and the granting of legal status to the legal institutions governing forests, water bodies, and agrarian landscapes, aligning statutory law and social norms.

The scholarship goes further to highlight the significance of redesigned funding structures that reinstate decentralized and community-led initiatives. Financial instruments for restoring historic water systems, sacred groves, and indigenous agriculture through long-term and flexible funding are linked to more ecological continuity as compared to short-term project-based funding (Geissdoerfer, Savaget, Bocken, & Hultink, 2017).

Partnerships between the public, the private, and the community are also found to be good policy instruments whereby roles, accountability measures, and accountability safeguards against elite capture have been clearly defined. Last but not least, the study of knowledge governance emphasizes the importance of national collections of indigenous ecological practices and states that institutionalization of knowledge increases policy learning and resilience when the communities maintain authority over representation and usage (Berkes, Colding, & Folke, 2000).

CONCEPTUAL FRAMEWORK

Figure 5. Triple layered Ancient-Modern Sustainability Framework

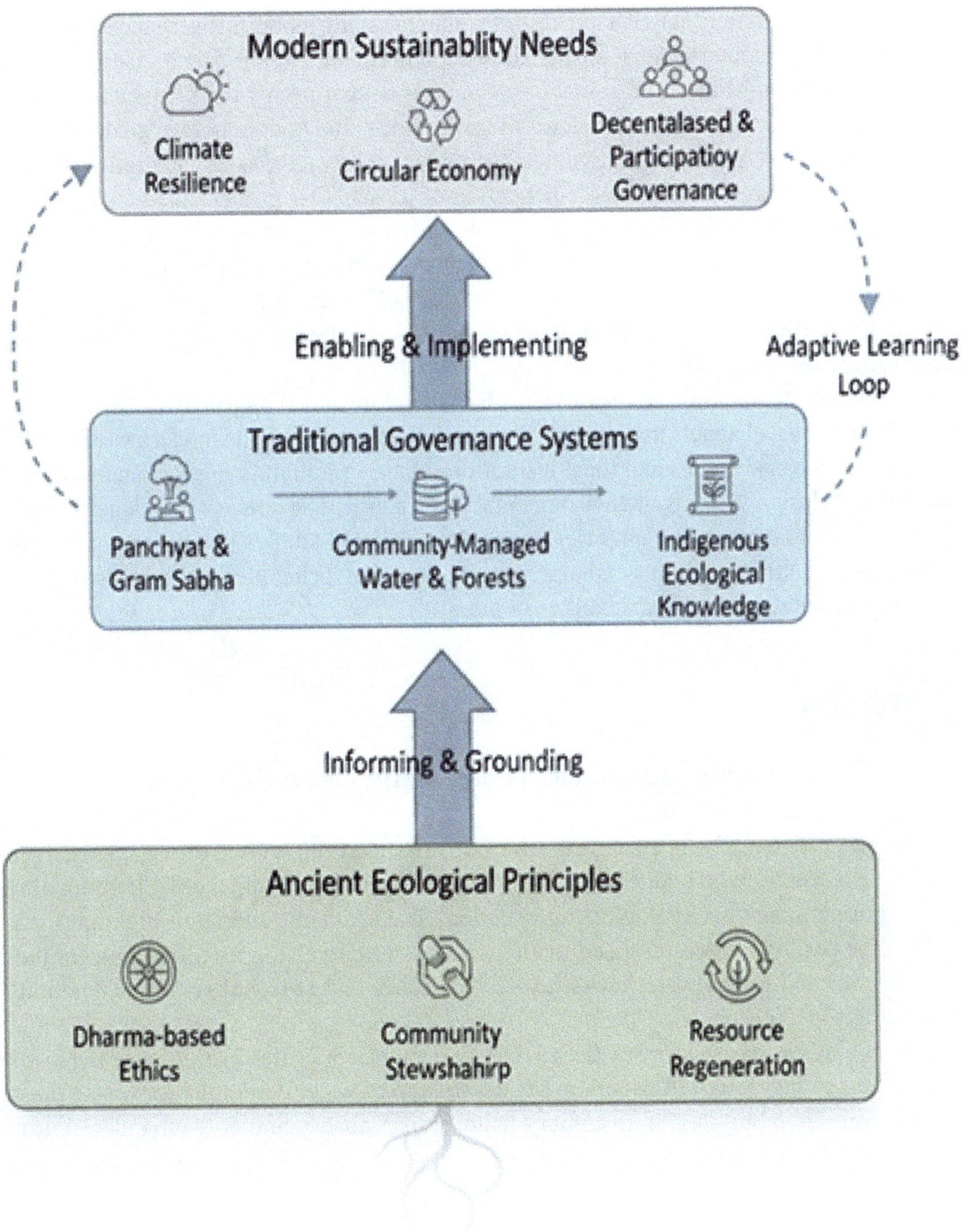

The conceptual framework integrates these strands to clarify how culturally embedded principles and traditional governance systems inform contemporary sustainability needs, providing an analytical bridge between historical practices and modern governance discourse.

These strands are converged into the conceptual framework to explain how principles that are embedded in the culture and traditional systems of governance give meaning to the modern sustainability requirement, thus offering an analytical process of contextualizing modern governance discourse between past practices and the contemporary sustainability requirement. *Figure 5* is a visual depiction of the Triple-Layered Ancient-Modern Sustainability Framework that grants the visual connection between ancient ecological values, traditional systems of governance, and contemporary goals of sustainability. The diagram can be interpreted as the synthesis of reviewed literature as an integrative synthesis of the conceptual continuity and not as a recommendation of a particular governance model. Dharma-based ethics, community stewardship, and regeneration are placed in the framework as its principles. These are informational systems of governance like Panchayats, community-based water and forest networks, water and forest governance, and indigenous knowledge. They complement one another to contribute to the modern sustainability requirements, such as climate resilience, practices of a circular economy, and decentralized governance.

The framework helps to bring the various threads of literature into a congruent framework, emphasizing the ethical-institutional channels that connect the ancient governance to the modern sustainability issues. It shows that norm-based values and local institutions of the community are still analysable in relation to resilient and inclusive systems of environmental governance. The conceptual framework provides an analytical prism of understanding empirical examples. The case studies below employ this framework in various regional settings, and this explains why some ancient principles and governance systems still determine modern practices of sustainability in India.

CASE STUDIES

Rajasthan – Community Johad Revivals (Tarun Bharat Sangh)

Communities active in Tarun Bharat Sangh (TBS) have led the revival of Johad-based water management in Rajasthan, of which much of the literature recounts the resurgence of traditional ecological governance but not a new technological intervention. The academic literature highlights that the use of Johads in the past was a decentralized mode of rainwater harvesting modified to meet the needs of semi-arid terrains and groundwater recharge, soil moisture, and seasonal stream restoration (Pandey, Gupta, & Anderson, 2003).

For review studies, the success of the revival was more about the institutional reconstruction as opposed to engineering design. Restoration of stewardship of water commons was made through community mobilization, rebuilding of customary maintenance norms, and collective decision-making in village councils. There are first positive changes in groundwater levels and reappearance of perennial flows in rivers that were seasonal in previous climates, as shown by the ecological sustainability of the traditional systems in modern climatic conditions (Singh et al., 2020).

The revival of the Rajasthan *Johad* is understood to be an example of adaptive commons governance through the lens of governance scholarship, in which moral legitimacy, a collective duty, and locally important rules minimized any reliance on external force systems. Critically, the literature warns that there was variability of outcomes in each village, which highlights the critical role of social cohesion and institutional continuity. Being a review case, Rajasthan exemplifies the role of ancient principles of water harvesting ingrained in the framework of restored community governance in enhancing climate resilience, water security, and decentralized environmental controls that are not dependent on centralized infrastructure.

Maharashtra – Sacred Groves (Devrai) Conservation

In Maharashtra, sacred groves (Devrai) are widely studied within both the ecological literature and governance literature due to their long-standing use as an illustration of forest conservation being embedded within culture. The purpose of these groves, which are safeguarded by religious faith, ritual taboos, and traditional practices, is to preserve the community-collaborated conservation galleries and not to form formally protected areas. The area reviews of ecological research have always indicated that the Devrai sites have been found to have high species richness, endemic plant diversity, and ecosystem integrity as compared to the adjacent managed forests (Ormsby & Bhagwat, 2010).

It is pointed out in scholarly reviews that it is the moral internalization of sacred groves that leads to conservation effectiveness but not regulatory control. Social norms, spiritual sanctions, and collective control are maintained to sustain extraction restrictions, which lead to low monitoring costs and high compliance. The governance research also adds that sacred groves are usually used as ecological refuges and seed banks and promote landscape-level resiliency and regeneration.

Latest multiple policy literature discusses the attempts to establish sacred groves in the framework of biodiversity and heritage. Although formal recognition is likely to reinforce legal protection, the scholars warn that externally inspired management regimes will jeopardize the maintenance of community custodianship in the event the cultural power is displaced. As such, the Maharashtra case is often referred to as an example of a thin line between legal assimilation and cultural autonomy.

Devrai conservation illustrates the combination of an ethical belief system, community-level governance, and ecology functionality to give lasting conservation effects.

It reinforces arguments that culturally embedded institutions can outperform top-down forest governance models when social legitimacy and local knowledge remain central.

Odisha – Traditional Farming and Seed Sovereignty

Integrated farming is practiced as part of Odisha, which includes farming-related knowledge and practices such as seed sovereignty strategies, knowledge ethics, land governance models, and the creation of practice-based service economies Integrated farming is practiced in Odisha, which entails farming-related experiences and insights like the strategies of seed sovereignty, knowledge moralities, land government plans, and establishment of practice-based service economies.

In Odisha, Odisha has received increasing interest among agroecological and governance scholarly literature as an alternative system to input-intensive, market-based agriculture in terms of agricultural systems and seed sovereignty programs. The continuation and renewal of traditional types of crops, intercropping systems, and organic management of the soil are regularly recorded in literature that is adapted to local climatic variability and socio-economic conditions (Pretty, et al., 2018).

Academic tests highlight that seed-sovereignty programs stipulate community management of the choices of seed and the storage and exchange of seed in specific areas and so restrict some type of dependence on produced inputs and commercial seed systems. These activities work to create agro-biodiversity, increase soil fertility, and improve the ability to endure climatic shocks, especially drought and unpredictable rain. Governance scholarship also defines these systems as decentralized food-governance models based on communal decision-making and customary norms.

It is also noted in the literature that the traditional methods of farming in Odisha exist within ethical frameworks where sufficiency, regeneration, and community interests are of primary priority in comparison with the aim to maximize yield. The institutional mechanisms involved include seed festivals, ritual exchanges, and community seed banks, which help in the passing of knowledge and maintaining social cohesion.

The Odisha case is an exemplar illustrating how indigenous systems of agricultural governance can be reconciled with modern sustainability goals, such as food security, biodiversity conservation, and climate adaptation. It is highlighted in the case that seed sovereignty is not only a technical intervention but also a form of governance that incorporates ecological information, cultural beliefs, and community independence.

Ladakh – Ice Stupas and Traditional Water Principles

The ice stupas of Ladakh are often discussed in the field of sustainability studies as creative variations that refer to the classic Himalayan water-management concepts. The reviews state that despite the common description of ice stupas as technological breakthroughs, they expand the existing traditions of the seasonal storage of water, the flow regulation through gravity, and collaboration within communities in cold desert areas.

Ice stupas trap frozen water in winter stream flows, which are then melted slowly over the course of the spring season, thus sustaining agricultural irrigation expressed in the early spring. Hydrological research findings indicate that this approach enhances the supply of water and also conserves delicate mountain biosystems. The governance examination results indicate that to be successful, ice stupas require maintenance to be community-based, the area to be knowledgeable, and to be consistent with conventional water-sharing principles.

Within the context of the literature, the role of ice stupas is placed in the broader discourses about climate adaptation and nature-based solutions due to its low-energy design and eco-friendliness. Nonetheless, researchers warn such multiplication of interventions requires sensitivity to the local governance structures to prevent the occurrence of technocratic local governance over people.

Ladakh shows the ways of reconceptualizing traditional ecological ideals in modern forms, such as regeneration, seasonal synchronization, and collective stewardship. The case supports the argument that proper climate adaptation can be achieved through a combination of indigenous knowledge, moral accountability, and decentralization as opposed to technological solutions alone.

Disaster Risk Management in Bangladesh

Due to unplanned engineering, Bangladesh was prone to several problems associated to floods, water logging and drainage congestion. The affected indigenous people tried to solve these problems by traditional methods and focused on adaptation rather than development, which promoted disaster risk management in Bangladesh. They adopted the Coastal Embankment Project (CEP) to prevent non-saline tidal flooding and saline water intrusion on the arable lands in the southwest coastal zone. In which they adopted the alternative options suggested by the local people to solve the problem of drainage congestion in the Beel area which resulted in being technically sound, economically viable and ecofriendly (Chowdhooree, 2019).

Cultural Burning in Australia

Australia's Cultural Burning practice, primarily by Aboriginal and Torres Strait Islander peoples, is an effective indigenous practice that promotes the resilience during high intensity forest fires. This is not just a socio-cultural practice but a spiritual approach to teach youngsters the linguistics and their ethical responsibility towards the land. Under the supervision of elderly peoples this reflects the effectiveness of the ancient governance in maintaining ecological sustainability (ANU Communications & Engagement, 2024).

Awajun women pottery art of Peru

The Pottery making art of Awajun women from Peru is a globally recognized effort that shows the strength of social value and tradition. Inspired by the spiritual visions and environmental observations, the indigenous women through this art represents the harmonious and sustainable relationship between the nature and Awajun People. They not only symbolizes pottery as a performative expression of prestige, gender identity and gendered governance but also an exemplar expression of culture, tradition and sustainable governance (Oviedo, Pantoja, Guevara, & López, 2025).

FUTURE RESEARCH DIRECTIONS

This research opens multiple pathways for enquiries related to ancient practices of governance that may suit to the present governance structure and its complexities. Data collection from various stake holders of the politically organized societies through surveys and questionnaires could be used for validation of the ancient principles and their relevancy in the present socio-legal scenario. Another promising research direction involve comparative study of ancient wisdom of different regions of the world for regulating modern societal concerns such as environment, good governance, citizen-centric governmental approaches etc. and their primacy over recent trends and practices. Apart from this it could make a base of future researches on the advantages of ancient model governances rooted on traditional cultural practices and to explore a harmonious scheme of governance mixed up with traditional and modern approaches and aspects of governances.

CONCLUSION AND SUGGESTIONS

This review has summarized the cross-disciplinary research on the ancient Indian ecological governance, which reveals that sustainability in the Indian context has in historical times been anchored in an overall paradigm of ethics, institutions, and community practice. In the philosophical traditions, government systems, and even case studies, it has always been evident in the literature that ecological stewardship was not a specialized sectoral issue but rather had always been incorporated into moral norms, social organization, and everyday life.

The ideas of Dharma, Rta, and Prakriti provided normative leadership, and the Panchayat institutions, customary resource-management structures, and indigenous knowledge applied the ethical ideas operationally to the realistic governance structures. The combination of these factors made a complete system of socio-ecological order, which was able to support resources on the generational level.

One of the key lessons drawn during this review is that, rather than being nostalgic and stagnant, there is a need to revitalize the old governance principles as frameworks that are analytically relevant in dealing with the modern challenges in sustainability. Indian systems based on tradition have shown that decentralized governance, community stewardship, and moral regulation can bring long-lasting ecological effects even in ecologically constrained and climatically unpredictable systems.

The case studies analyzed show that, in case the traditional principles are reinvigorated utilizing community institutions in connection with water getting, forest protection, agriculture, or climate adjustment, one will observe that resilience, equity, and eco-regeneration are strengthened further. These results discredit the notion that sustainability solutions can only be of modern or technocratic origin.

On the policy front, the review highlights the need to move beyond the approach of narrow regulation and market-focused approaches and adopt those models of governance that inject ethical values, institutional pluralism, and community control. The policy agenda to incorporate traditional ecological knowledge needs legalization of community custodianship, institutionalization of a decentralized management system, and protection of community-conservation practices.

It is also crucial that the concept of embedding ethical-ecological premises, i.e., trusteeship, restraint, intergenerational responsibility, and so on, be reflected in the modern system of governance, including governance of corporations, state institutions, and school policies. It is possible that such integration will enhance legitimacy, reduce the enforcement cost, and encourage long-run behaviour change through it.

The avenue of advancement in policymaking lies in selective adaptation instead of the wholesale copying of the ancient systems. The classical governance should be invoked in a critical manner, keeping in view modern socioeconomic realities, needs of the population, and technological environments.

As a result, policies must be focused on facilitating the state of a secure tenure, participatory decision-making, and institutional continuity as well as ethical orientation, which do underlie community-based and culturally-based sustainability practices. The systems can be enhanced by digital tools and scientific knowledge combined with modern infrastructure but not to replace it.

Over and above the Indian context, the review is finding its way into the global sustainability arguments in the sense that it proves that old-fashioned and native regimes of governance have structurally credible alternatives to the current models of environmental governance.

The fact that ancient Indian tradition touches upon the new theories, including common governance and the circular economy, as well as nature-based solutions, hints that sustainability is not an innovation of the future but a rediscovery of long-standing human-nature relations.

In order to attain global recognition, it is essential to promote the knowledge, skills and practices of indigenous communities. In this digital era, merging of the indigenous knowledge and technology can find solutions to various modern concerns and can create a positive social resilience.

REFERENCES

Agrawal, A., & Gibson, C. C. (1999, April). Enchantment and Disenchantment: The Role of Community in Natural Resource Conservation. *World Development*, *27*(4), 629–649. DOI: 10.1016/S0305-750X(98)00161-2

Altieri, M. A., Nicholls, C. I., Henao, A., & Lana, M. A. (2015, May 1). Agroecology and the design of climate change-resilient. *Agronomy for Sustainable Development*, *35*(3), 869–890. DOI: 10.1007/s13593-015-0285-2

ANU Communications & Engagement. (2024, November 1). *Indigenous cultural burning has protected Australia's landscape for millennia*. Retrieved from Australian National University: https://www.anu.edu.au/news/all-news/indigenous-cultural-burning-has-protected-australias-landscape-for-millennia

Berkes, F., Colding, J., & Folke, C. (2000, October). Rediscovery of Traditional Ecological Knowledge as Adaptive Management. *Ecological Applications : A Publication of the Ecological Society of America*, *10*(5), 1251–1262. DOI: 10.1890/1051-0761(2000)010[1251:ROTEKA]2.0.CO;2

Chhachhar, V., & Verma, K. (2024). Climate Change and Negotiating Challenges: A Lesson for Success and Failure. In Rai, P. K., & Rai, S. (Eds.), *Climate, Environment and Agricultural Development* (pp. 1–20). Springer., DOI: 10.1007/978-981-97-8363-2_1

Chhatre, A., & Agrawal, A. (2009, October 20). Trade-offs and synergies between carbon storage and livelihood benefits from forest commons. In E. Ostrom (Ed.), *Proceedings of the National Academy of Sciences of United States of America*. 106, pp. 17667–17670. Bloomington: Sustainability Science. DOI: 10.1073/pnas.0905308106

Chowdhooree, I. (2019). Indigenous knowledge for enhancing community resilience: An experience from the south-western coastal region of Bangladesh. *International Journal of Disaster Risk Reduction*, *40*, 101259. Advance online publication. DOI: 10.1016/j.ijdrr.2019.101259

Devi, S. (2025). Kautilya's Perspective On Ecological Development. [IJCRT]. *International Journal of Creative Research Thoughts*, *13*(2), b827–b839.

Ensor, J. E., Wennström, P., Bhatterai, A., Nightingale, A. J., Eriksen, S., & Sillmann, J. (2019, April). Asking the right questions in adaptation research and practice: Seeing beyond climate impacts in rural Nepal. *Environmental Science & Policy*, *94*, 227–236. DOI: 10.1016/j.envsci.2019.01.013

Folke, C., Biggs, R., Norström, A. V., Reyers, B., & Rockström, J. (2016). Social-ecological resilience and biosphere-based sustainability science. *Ecology and Society*, *21*(3), art41. Advance online publication. DOI: 10.5751/ES-08748-210341

Geissdoerfer, M., Savaget, P., Bocken, N. M., & Hultink, E. J. (2017, February 1). The Circular Economy – A new sustainability paradigm? *Journal of Cleaner Production*, *143*, 757–768. DOI: 10.1016/j.jclepro.2016.12.048

Gowri, R., Dey, P., & Mujumdar, P. P. (2021). A hydro-climatological outlook on the long-term availability of water resources in Cauvery river basin. *Water Security*, *14*, 100102. Advance online publication. DOI: 10.1016/j.wasec.2021.100102

Guha, R. (2000). The unquiet woods: Ecological change and peasant resistance in the Himalaya. Banglore: Oxford University Press.

Koul, D. N., Singh, S., Neelam, G., & Shukla, G. (2012, April). Traditional water management systems - An overview of Ahar-pyne system in South Bihar plains of India and need for its revival. *Indian Journal of Traditional Knowledge*, *11*(2), 266–272. https://www.researchgate.net/publication/263168629_Traditional_water_management_systems_-_An_overview_of_Ahar-pyne_system_in_South_Bihar_plains_of_India_and_need_for_its_revival

Laidlaw, J. (1995). *Riches and renunciation: Religion, economy, and society among the Jains*. Oxford University Press., Retrieved from https://archive.org/details/richesrenunciati0000laid DOI: 10.1093/oso/9780198280316.001.0001

Mays, L. W., Koutsoyiannis, D., & Angelakis, A. N. (2007). A brief history of urban water supply in antiquity. *Water Science and Technology: Water Supply*, *7*(1), 1–12. Advance online publication. DOI: 10.2166/ws.2007.001

Menski, W. F. (2012). *Hindu law: Beyond tradition and modernity*. Oxford University Press., Retrieved from https://global.oup.com/academic/product/hindu-law-beyond-tradition-and-modernity-9780195699210

Ormsby, A., & Bhagwat, S. A. (2010). Sacred Forests of India: A Strong Tradition Of Community-Based Natural Resource Management. *Environmental Conservation*, *37*(3), 320–326. DOI: 10.1017/S0376892910000561

Ostrom, E. (1990). *Governing the Commons: The Evolution of Institutions for Collective Action*. Cambridge University Press. DOI: 10.1017/CBO9780511807763

Ostrom, E. (2015). *Governing the Commons: The Evolution of Institutions for Collective Action*. Cambridge University Press. DOI: 10.1017/CBO9781316423936

Oviedo, A., Pantoja, A. M., Guevara, S. Z., & López, E. A. (2025). Awajún Pottery, Gender, and Transformation: From Social Prestige to Market Logic. *Journal of Intercultural Communication*, *25*(3), 104–115. DOI: 10.36923/jicc.v25i3.1157

Pandey, D. N., Gupta, A. K., & Anderson, D. M. (2003, July 10). Rainwater harvesting as an adaptation to climate change. *Current Science*, *85*(1), 46–59. https://www.currentscience.ac.in/Volumes/85/01/0046.pdf

Piplani, M., & Kumar, T. (2019). Stepwells: Reviving India's Cultural and Traditional Water Storage Systems. *Conference: World Resources Forum* (pp. 135-142). Geneva, Switzerland: World Resources Forum Association.

Pretty, J., Benton, T. G., Bharucha, Z. P., Dicks, L. V., Flora, C. B., Godfray, H. C., Goulson, D., Hartley, S., Lampkin, N., Morris, C., Pierzynski, G., Prasad, P. V. V., Reganold, J., Rockström, J., Smith, P., Thorne, P., & Wratten, S. (2018). Global assessment of agricultural system redesign for sustainable intensification. *Nature Sustainability*, *1*(8), 441–446. DOI: 10.1038/s41893-018-0114-0

Ravikanth, G. (2021, May). Indian Philosophy and Environmental Ethics. *GNOSI: An Interdisciplinary Journal of Human Theory and Praxis*, *4*(1 (Special Issue)), 47–63.

Rockström, J., Steffen, W., Noone, K., Persson, Å., Chapin, F. S.III, Lambin, E. F., Lenton, T. M., Scheffer, M., Folke, C., Schellnhuber, H. J., Nykvist, B., de Wit, C. A., Hughes, T., van der Leeuw, S., Rodhe, H., Sörlin, S., Snyder, P. K., Costanza, R., Svedin, U., & Foley, J. A. (2009). A safe operating space for humanity. *Nature*, *461*(7263), 472–475. DOI: 10.1038/461472a PMID: 19779433

Sharma, R., & Acharya, K. (2024). *ANCIENT WISDOM AND INDIGENOUS KNOWLEDGE: A Holistic Approach to Sustainable Development and Environmental Stewardship*. Retrieved from Himachal Pradesh National Law University: https://www.hpnlu.ac.in/PDF/b9813e07-5d12-4710-ac96-fd487f0a4b56.pdf

Shastri, V. (2018, March 2). *What is Prakriti? Unfolding the Mystery of Life!* Retrieved from Vedanet: https://www.vedanet.com/what-is-prakriti-unfolding-the-mystery-of-life/

Shiva, V., & Bandyopadhyay, J. (1988). The Chipko Movement. In Ives, J., & Pitt, D. C. (Eds.), *Deforestation: Social Dynamics in Watersheds and Mountain Ecosystems* (p. 18). Routledge.

Tengö, M., Brondizio, E. S., Elmqvist, T., Malmer, P., & Spierenburg, M. (2014). Connecting Diverse Knowledge Systems for Enhanced Ecosystem Governance: The Multiple Evidence Base Approach. *Ambio*, *43*(5), 579–591. DOI: 10.1007/s13280-014-0501-3 PMID: 24659474

Williams, J. (2019). Circular cities. *Urban Studies (Edinburgh, Scotland)*, *56*(13), 2746–2762. DOI: 10.1177/0042098018806133

ADDITIONAL READING

Chennells, R. (2013). Traditional Knowledge and Benefit Sharing After the Nagoya Protocol: Three Cases from South Africa. *Law, Environment & Development Journal*, •••, 163–184. https://doaj.org/article/1c0ddc79f0fb405284e1e748847e7700

Convention on Biological Diversity. (2026). *Article 8(j) - Traditional Knowledge, Innovations and Practices*. Retrieved from cbd.int: https://www.cbd.int/traditional/intro.shtml

Davis, D. R.Jr. (2004). Traditional Hindu Law in The Guise of 'Postmodernism': A Review Article. *Michigan Journal of International Law*, *25*(3), 735–749.

Jansen, L., & Mzwakali, S. (2025, April 25). *The Indigenous World 2025: South Africa*. Retrieved from IWGIA: https://iwgia.org/en/south-africa/5636-iw-2025-south-africa.html

UNESCO. (2025, March 26). *Local and Indigenous Knowledge Systems and Climate Change*. Retrieved from unesco.org: https://www.unesco.org/en/climate-change/links

United Nations. (2026). *United Nations Declaration on the Rights of Indigenous Peoples*. Retrieved from Department of Economic and Social Affairs: https://social.desa.un.org/issues/indigenous-peoples/united-nations-declaration-on-the-rights-of-indigenous-peoples

KEY TERMS AND DEFINITIONS

Dharma: A self-regulatory code of conduct that is considered as a divine law and acts as a guiding principle for Hinduism, Buddhism and Jainism. This was one of the major principles on which governance system of ancient India was based.

Dharmashstras: Ancient Indian texts that contains ethical, legal and moral obligations for regulating human conduct in society. They contained the knowledge of governance and statecraft in ancient India.

Governance: A regulatory framework by which a country or organization is regulated, controlled and directed. It includes system of rules, regulations and processes to govern.

Indigenous Knowledge: A community-based knowledge that is passed from generation to generation of that community.

Sustainability: It is a practice of using the resources in such a way that the requirements of present is met without compromising the future supplies.

Traditional knowledge: A knowledge that is acquired by the specific local or cultural community due to their traditions and habitual practices.

Chapter 18

Scalable IoT Solutions for Sustainable Tourism and Urban Heritage Protection

Hitesh Mohapatra

http://orcid.org/0000-0001-8100-4860

School of Computer Engineering, KIIT (Deemed to be) University, Bhubaneswar, India

ABSTRACT

Smart cities leverage IoT platforms to deliver intelligent services, making security, flexibility, and scalability essential design considerations. Robust security mechanisms, including advanced encryption and granular access control, are vital to mitigate rising cyber threats. To manage urban infrastructure effectively, IoT platforms must remain adaptable to diverse network protocols and device types. Scalability is equally critical, as IoT devices generate massive datasets, particularly within applications like environmental monitoring and cultural heritage preservation. However, deploying IoT in heritage conservation presents significant security challenges. Successfully integrating these systems requires balancing technical benefits with a commitment to inclusivity. This study proposes an IoT-based framework for sustainable tourism, specifically prioritizing interoperability and security to protect historical sites. By fostering digital literacy, this approach empowers individuals to actively participate in the interpretation and safeguarding of their cultural legacy.

1. INTRODUCTION

Historic buildings and ancient structures are priceless assets of cultural heritage that are essential to safeguarding for future generations the rich cultural, social, and historical resources. Enhancing conservation efforts, the Internet of Things (IoT) paradigm emerges as a highly effective instrument for monitoring and safeguarding these cultural assets (Addabbo et al., 2019). In a related context, the word "IoT" is changing the travel and hospitality business, impacting several stakeholders in this intricate field. By combining IoT with big data, blockchain, cloud computing, mobile, artificial intelligence, and mobile communication, travel becomes a smart, connected experience. The study examines how the Internet of Things is affecting travel and offers a blueprint for improved visitor experiences and efficient operations. It draws attention to IoT applications in smart tourism projects, emphasising predictive maintenance

DOI: 10.4018/979-8-3373-7976-0.ch018

and individualised services. With its focus on intangible services, the hospitality sector is one where the Internet of Things significantly impacts client satisfaction, cost savings, and higher revenues. The proposed model looks at how the tourism sector is being impacted by the Internet of Things (IoT) and other technologies of the Fourth Industrial Revolution (Alletto et al., 2016).

Every society's cultural history serves as a symbol of its past and current values and customs, and it is essential to the formation of belief systems and the development of a sense of self. Tangible cultural assets are classified by the United Nations Educational, Scientific, and Cultural Organization (UNESCO) into two groups: immovable objects such as monuments and archaeological sites, and submerged features such as shipwrecks. Tangible artifacts include mobile items like paintings and manuscripts. Apart from its intrinsic value, safeguarding these cultural heritage sites stimulates the local economy by attracting tourists. Notwithstanding this potential, UNESCO has voiced worry over policy indifference in the field of preservation, particularly as tourism increases (Al-Omari et al., 2013). The Internet of Things (IoT), which enables objects to efficiently interact with their surroundings, offers itself as a practical solution as the need to monitor these areas grows. IoT applications can remotely monitor and control environmental conditions, identify structural changes, and notify users of unusual presences by utilizing wireless sensor networks.

Numerous Internet of Things (IoT) technologies, such as WLAN, LPWAN, and WPAN, provide various infrastructures for various preservation use cases. With an emphasis on preventive conservation, this paper critically evaluates the body of research on Internet of Things (IoT) applications to cultural assets. Lastly, the study explores issues and starts a conversation about open-ended questions about the use of IoT technology for distant historical building conservation (Brandt et al., 2017). Figure.1 illustrates the graded Importance or Impact levels assigned to important categories, and highlights how initiatives such as Smart Tourism, Augmented Reality for Cultural Sites, and Identification of IoT Applications contribute to the development of technologically enhanced tourism and cultural heritage experiences.

Figure 1. Strategic Significance of IoT in Tourism and Cultural Heritage

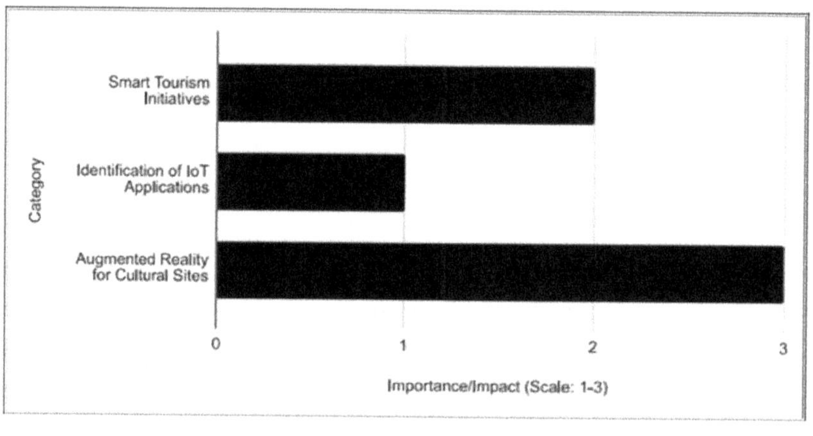

Any community's cultural legacy is its cornerstone; it contains the core of its past as well as its present customs and beliefs. Its importance in the formation of belief systems and identities cannot be overstated. The United Nations Educational, Scientific, and Cultural Organization (UNESCO) places cultural legacy in a broad category that encompasses both material and immaterial components. As per UNESCO, the material component of cultural heritage can be further subdivided into three groups: immovable cultural assets, such as historical or sacred buildings, monuments, and archaeological sites; movable cultural heritage, which includes artwork, numismatics, manuscripts, and shipwrecks; and undersea cultural heritage, which includes phenomena like underwater ruins, cities, and artwork. This complex categorization emphasises the various shapes that cultural legacies can take, stressing the complex interplay between the material components that make up a community's rich history (Costantini et al., 2008).

The tourism business has seen significant changes with the introduction of Information and Communication Technology (ICT), including the incorporation of digital tools, platforms, and mobile apps. These innovations have allowed the sector to provide new services and better experiences to travellers. For example, dynamic forms of information transmission, digital tools for destination promotion, and immersive experience technology have become indispensable in the business. These include augmented reality, knowledge distribution tactics, and learning procedures that help people grasp the cultural and ecological qualities of locations (Diego et al., 2015). Simultaneously, an IoT-driven smart tourism ecosystem has been made possible by the fusion of state-of-the-art technologies such as big data, smart sensors, cloud computing, machine learning, artificial intelligence, wearables, mobile apps, augmented reality, and virtual reality. These technologies have a wide range of applications, from leveraging embedded sensors to streamline data collection to using cloud computing for essential services like analytics and data storage. Upon deeper examination, the Internet of Things (IoT) is contributing to safety, efficiency, and personalized services in smart tourism through applications such as electronic ticketing, smart cities, recommender systems, and payment systems. Strong protections must be included to solve issues with software complexity, interoperability, security, privacy, and scalability (Diwakar et al., 2025).

This chapter is organized as follows. Section 1 introduces the motivation for using IoT in smart tourism and cultural heritage preservation, along with the key challenges related to security, interoperability, scalability, and sustainability. Section 2 reviews the related literature on IoT-enabled tourism, preventive conservation, and smart city infrastructures, and highlights the research gap addressed by this study. Section 3 presents the proposed model, including the IoT architecture for sustainable tourism applications, the role of key technologies for heritage sites, and the security-oriented platform design. Finally, Section 4 concludes the paper by summarizing the main findings and outlining the potential of IoT to enhance visitor experiences while supporting the protection of cultural assets.

2. LITERATURE REVIEW

As a major worldwide industry, tourism contributes significantly to the social and economic fabric of many nations by creating jobs and promoting business expansion. The number of visitors to the tourist industry increased dramatically, from 528 million in 2005 to 1.19 billion in 2015. Estimates indicate that the number will climb to over 1.8 billion by 2030. In 2017, the sector made a direct contribution of 2.57 trillion US dollars and had a significant role in the development of millions of employments worldwide (Fernández-Navajas et al., 2013). Given the significant advantages that the travel sector can offer a nation, there is a growing need to improve the allure of tourist destinations regardless of how well-known or less travelled they are highlighting the significance of consistently enhancing and advertising these places.

The reverse of employing the Internet of Things (IoT) for cultural assets is that it plays a crucial role in use cases pertaining to preventative conservation. The core concept of "monitoring things" aligns well with the need to maintain an eye on objects within cultural heritage sites to optimize environmental conservation conditions and prevent further harm. A key tactic for preventing the loss of cultural heritage is preventive conservation, which places a focus on long-term observation and proactive upkeep of the physical attributes of assets. By being proactive, this method has the potential to greatly minimize damage and lower the expenses related to subsequent restoration activities (Figueredo et al., 2017). The reasons of degradation vary depending on the physical characteristics of each cultural property site. For example, artworks might be subjected to stress from radiation, temperature, humidity, or chemical agents, whereas stone-based archaeological constructions are particularly vulnerable to changes in temperature and humidity (García-Diego et al., 2013). Table.1 presents the major themes found in the literature. It shows that research on smart tourism is not limited to digital tools only. It also includes cultural heritage, sustainability, security, privacy, and visitor experience. These themes are closely connected because smart tourism systems must support services, protect data, and improve the way visitors interact with heritage sites.

Table 1. Main literature themes

Theme	Main focus	Relevance to the study
Smart tourism	Digital services, visitor support, and destination management	Gives the broader context of the study
IoT in tourism	Sensors, real-time monitoring, and service automation	Supports the technical base of the model
Cultural heritage	Conservation, monitoring, and interpretation	Matches the main application area
Security and privacy	Data protection, access control, and trust	Supports the need for secure design
Sustainability	Efficient resource use and responsible tourism	Links the study to long-term planning
User experience	Personalization, accessibility, and engagement	Strengthens the visitor-centered view

The impact of abrupt changes in temperature and relative humidity can be severe, causing damage to diverse objects, including non-isotropic material deformation, detachment in multi-layered materials, and mechanical changes in hygroscopic materials like wood panels commonly used in artworks (González et al., 2020). Frescoes, for instance, are particularly susceptible to deterioration caused by soluble salts and moisture, emphasising the critical importance of early detection of dangerous levels of these physical parameters to prevent damage (Hardilla & Nugroho, 2018). Achieving a fully preventive conservation necessitates not only long-term monitoring of environmental variables but also the ability to predict and anticipate these conditions with sufficient time to react, such as in cases of frosts or heatwaves. Thus, IoT preservation systems should incorporate data analytics capabilities to discern patterns and detect potentially dangerous oscillations, ensuring a comprehensive and timely preventive conservation approach (Konev et al., 2019). Table.2 summarizes the representative studies and their main contributions. It shows that earlier research has explained smart tourism platforms, IoT-based service delivery, and heritage monitoring in different ways. Some studies focus on tourism management, while others focus on conservation or digital interaction. However, many of them remain limited in practical deployment or in detailed discussion of heritage-specific implementation.

Table 2. Representative studies

Study group	Main contribution	Limitation
Smart destination reviews	Show the growth of smart tourism and related themes	Often broad and not heritage-specific
IoT in tourism studies	Explain how sensors and connected systems improve services	Do not always cover practical deployment
Cultural heritage IoT studies	Show how technology supports conservation and monitoring	Limited discussion on large-scale implementation
Smart city IoT studies	Explain platform design, scalability, and connectivity	Less focus on cultural heritage use
Visitor experience studies	Highlight personalization and digital interaction	Often ignore security and infrastructure issues

In the expansive domain of IoT and preventive conservation for cultural heritage, a diverse array of examples showcases the effective application of preservation technologies to surmount the challenges inherent in safeguarding invaluable artefacts and historical sites. The majority of these creative methods rely on Wireless Sensor Network (WSN) designs, which include sensor-based communication networks, multi-modal gateways that are interconnected, and cloud-based platforms for data analytics and central monitoring. Notably, these preservation technologies boast features such as the low power consumption of sensor nodes, contributing to sustainable and energy-efficient operations, and the self-forming and self-healing capabilities of wireless sensor networks, ensuring robust and resilient communication for reliable data acquisition and monitoring. The integration of cloud-based platforms provides a pseudo-infinite storing capacity, allowing for extensive data storage and analysis crucial for comprehensively understanding environmental conditions affecting cultural artefacts. Furthermore, the utilization of open or standard technologies like Wi-Fi, LoRaWAN, and Sigfox, as well as commercial products like Hobo data-loggers, enhances the cost-effectiveness of both network setup and operation.

Figure 2. Literature survey of past works

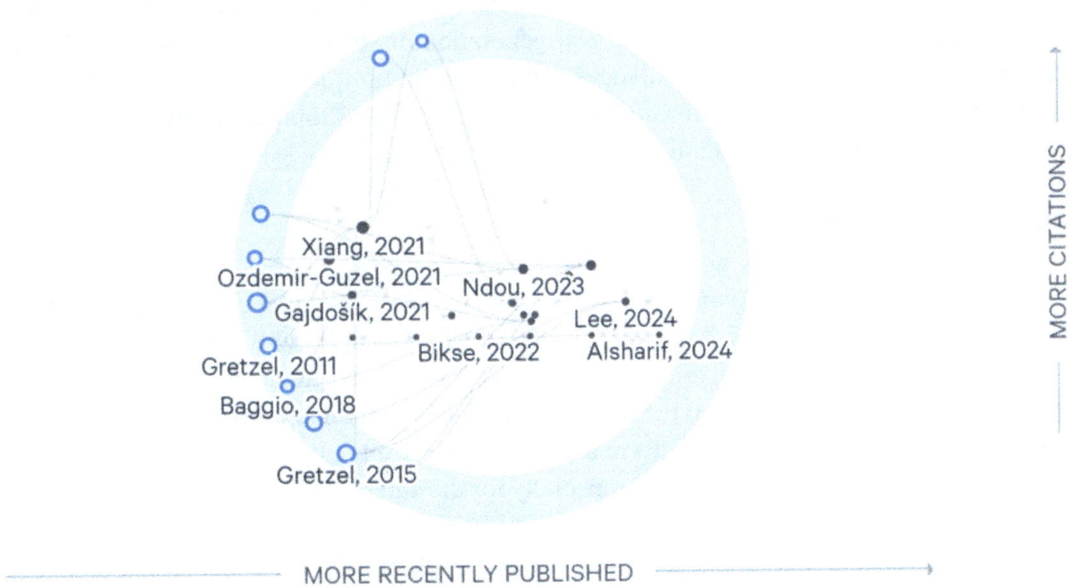

This combination of technologies improves the resilience of preventative conservation measures and streamlines the deployment process, making them more widely available and flexible in cultural heritage contexts (Kumar et al., 2024). Importantly, WSNs, being non-invasive, eliminate the need for cumbersome wired network installations, rendering these solutions scalable, flexible, and particularly suitable for preventive conservation efforts by enabling unobtrusive monitoring of artefacts and historical sites. While wireless approaches offer significant advantages, instances where a hybrid architecture incorporating wired solutions becomes necessary emphasise the adaptability and versatility of IoT technologies in cultural heritage conservation. This dynamic field sees continuous enrichment with new devices and functionalities, underscoring the evolving nature of these innovative conservation approaches (Leccese et al., 2014). Figure.2 illustrates the most recently published and most cited paper graph.

Key Requirements for an IoT Platform in a Smart City Environment

There are myriad interpretations and approaches to implementing the increasingly prevalent concept of "smart cities." However, as elucidated in the introduction, from an infrastructure standpoint, a robust ICT system, epitomised by an IoT platform, serves as the linchpin. This system, bolstered by a comprehensive network of sensors and wired and wireless broadband connectivity, forms the cornerstone upon which smart cities are built, facilitating the seamless integration of data analytics and the creation of intelligent services and applications for the populace at large (Lillie et al., 2008). Yet, amidst this burgeoning landscape, the absence of fully delineated standards for IoT architectures poses a formidable challenge. Indeed, the dynamic and multifaceted nature of smart city environments renders the precise delineation of key requirements for their utilisation a complex endeavour. As such, navigating the intricacies of IoT deployment within the context of smart cities necessitates a nuanced understanding of the evolving technological landscape and a keen awareness of the diverse array of factors influencing their efficacy and functionality. In this regard, the pursuit of interoperability, scalability, security, and sustainability emerges as paramount considerations, underscoring the imperative for stakeholders to collaboratively chart a course toward the harmonisation of standards and the development of robust frameworks capable of underpinning the next generation of smart city infrastructure. Thus, while the path forward may be fraught with challenges and uncertainties, it is also imbued with boundless potential and opportunity, heralding a transformative era in urban development characterised by innovation, efficiency, and inclusivity (Marshall, 2018).

Security Conditions

There are currently more connected devices than people in the globe, with an expected 50 billion interconnected devices by 2020 (Mishra et al., 2021). These gadgets will have the ability to be read and controlled over the Internet as they are a part of IoT systems. IoT environments are similar to other networks in that they are always connected to the Internet, which leaves them open to attacks if their networks are improperly set up or built. An IoT platform must thus have strong integrated security. This implies that it must fulfil the subsequent prerequisites, especially for the smart city setting and the chosen application.

Data encryption and end-to-end security mechanisms

Standard-based encryption is necessary to prevent data theft from hardware devices and the Internet of Things platform. However, data is more vulnerable when it's in transit. To ensure data encryption automatically, developers must allow full authentication without requiring user input. The connections that exist between relevant virtual items and actual objects are the most significant linkages in this architecture. Real-world items exchange keys and IDs with virtual objects during the association phase to prevent unauthorized access to users' data. This prevents data theft and ensures a secure environment for data storage (Mohapatra, 2021).

Flexible, Configurable Access and Authorization Control

Different access profiles for users in Smart City settings should be provided by the proposed IoT platform depending on factors like ownership, security level, visibility, and frequency of data polling. Three different kinds of permission are available for virtual objects: "private," which limits access to the owner key alone, "public," which permits access to anybody without a key, and "friend," which necessitates a friend-key that the virtual object is aware of. The cloud environment that is being utilised to implement the suggested architecture can meet all security criteria. This adaptability guarantees effective data access for a variety of user types (Moropoulou & Bisbikou, 1995).

Adaptability

Even with its vast reach, the market for IoT-enabled gadgets is still in its infancy. Creators of the upcoming wave of interconnected gadgets and goods must rely on a certain level of adaptability from their software rivals who combine their goods. An IoT platform should adhere to the following flexibility guidelines for these reasons. IoT Platforms for Smart cities including hardware developers can continuously design devices without specified settings or compatibility difficulties because of networking and device agnosticism. Device manageability includes operating system patches, profiles, analytics algorithms, and parameter key management, among other tools for remotely managing smart devices and sensors. Use of Open APIs facilitates the creation of novel applications by providing access to common, shareable data from networks of sensors and devices. Cloud-based architectural approaches meet flexibility needs, virtual things are separated from physical objects (Nitti et al., 2017).

When creating an IoT platform, there are a number of important things to keep in mind to make sure it works well and can be adjusted to meet changing needs. First of all, there should be no complicated limitations that come with SQL databases when it comes to creating and configuring new devices on the platform. This adaptability makes it possible to quickly integrate additional devices, which improves the platform's usability. Furthermore, scalability becomes critical as the amount of data created by IoT devices is predicted to grow quickly, especially in scenarios like environmental monitoring in smart cities. This problem can be solved with a cloud-based implementation, which makes it easy to add more storage and processing power as needed. Encouraging component reusability inside the IoT platform also expedites development processes and boosts productivity. Time and effort can be saved by developers by using templates of objects and services that can be quickly modified for various uses (Ordóñez-Martínez et al., 2023).

The Role of Traditional Architecture in Urban Resilience

Comprehending the complex web of regional traditions and culture is not an add-on, but rather a necessary prerequisite for appreciating the significant contribution that traditional architecture makes to urban resilience. This research delves into a thorough investigation, classification, and assessment of the notion of spatial change in the setting of developing communities, shifting socioeconomic landscapes, and dynamic living cultures. Fundamentally, interpretive historical research is a crucial qualitative method that provides direction for the morphological analysis of traditional architectural forms and their inherent relationship to native American culture and domestic settlement environments. This methodology, which delves deeply into historical accounts, involves a methodical process of searching, collecting, evaluating, and piecing together different pieces of information to create cohesive narratives that clarify the essence of local identity and legacy (Pareek et al., 2025).

The understanding that every architectural structure represents a diverse tapestry of customs, beliefs, and values and acts as a silent witness to the cultural ethos and societal dynamics of its era is fundamental to this undertaking. Therefore, it is crucial to comprehend and synthesise these historical artefacts in a comprehensive way in order to clarify the causal explanations and relationships that underlie the development of urban landscapes. Examining structural variation and alteration is an integral part of this architectural investigation and should be further integrated into the analytical framework used to examine traditional dwellings (Rosário & Dias, 2024).

Cultural contents delivery and Cloud services

The key to obtaining the intended cultural resources rests in the complex results produced by an image processing algorithm that has been painstakingly created to distinguish and identify individual artworks in the museum's collection. This algorithm is the central nervous system that facilitates the smooth transfer of data from wearable devices to a central processing unit across the local Wi-Fi network. The core of this processing centre is a specific service that has been painstakingly designed to accept and carefully assess user requests, thereby triggering the start of the relevant operations in response to each individual inquiry (Sharma et al., 2024). When we delve deeper into the finer points of how this process works, we find that the interpretation of artwork identification produces two very different but equally compelling results, each carefully designed to meet the various needs and tastes of museum visitors: first, a wearable device with a novel audio commentary feature seamlessly incorporates users into an immersive cultural experience, made possible by the seamless operation of an audio-streaming server programme; and second, interactive museum walls featuring a wealth of multimedia cultural information engage visitors on a profoundly deep and enriching level, greatly improving their comprehension and appreciation of the wide range of artworks that adorn the museum's halls.

In order to guarantee the fair distribution of cultural knowledge to all users who take part, the processing centre uses a state-of-the-art threshold algorithm that constantly monitors and carefully evaluates the collective viewing experiences of museum visitors in real time. Based on constantly changing visitor engagement measurements, our algorithm dynamically coordinates the manner of multimedia material distribution, smoothly switching between broadcasting multimedia content via interactive museum walls and delivering audio commentary via wearable devices. The programme skilfully switches to audio commentary delivery when visitor numbers fall below a predetermined level, improving customised experiences catered to each user's specific interests. On the other hand, if a sizable group of visitors have a similar "visiting experience," the algorithm will automatically transition to a multimedia broadcast through interactive displays, creating a collaborative and immersive learning environment that goes beyond what is typically found at a museum (Valero et al., 2014).

On the other hand, a significant computational and storage challenge is presented by the seamless transmission of data and audiovisual elements customised to each user's cultural trip. Because these requirements are resource-intensive, using cloud infrastructure's unmatched capabilities is a practical answer. The cloud's innate computing and storage capabilities enable previously unheard-of processing speeds for enormous amounts of data and multimedia content, making it the ideal medium for efficient data processing and distribution inside the system under consideration. Innovation and efficiency come together in this symbiotic relationship between local processing and cloud-based computing and storage to provide users with unmatched immersive cultural experiences, which epitomises the state-of-the-art developments and transformative potential inherent in the field of smart city technologies (Verma et al., 2021). Table.3 provides an overview of the primary sensors applicable to these advanced preventive conservation systems.

Table 3. Application and types of sensors in the context of tourism and the IoT

Type of Sensor	Objectives	Applications
Temperature	Calculates the thermal energy.	Control of the cooling system. Applications: creation of agricultural systems, manufacturing procedures
Proximity	Determine whether any surrounding items are present or not. They are also able to recognize the attributes of the object.	Command of presence. Applications include traffic flow, visitor presence, security, and alarm systems. information about deals available to travellers online. Parking spots that are available
Pressure	Identify pressures and changes.	Upkeep of heating and water systems
Water quality	Determine the water quality and keep an eye on the ions (pH, chlorine, organic carbon, etc.).	Regulates water distribution networks Swimming-quality water
Chemical/Gas	Determine whether the chemical composition of liquids and the air have changed.	Contamination control (CO2, CO, O3, O2, NO2, NO3, particulates, etc.) and environmental monitoring. Systems for security and safety
Smoke	Smoke detectors	Fire safety and security systems
IR	An infrared sensor that recognizes features of the surroundings	Medical care (blood pressure, arterial flow, etc.)
Level sensors	Ascertain the volume of liquid flowing through an open or closed system.	Water deposit, fuel levels
Image sensors	Take pictures and videos	Safety. Car, medical images
Movement sensors	Identify any movement in a specific area	Security. Toll booths, air conditioning, parking, fans, lights, air conditioning, and intrusions
Accelerometer sensors	Determine an object's physical acceleration. They pick up acceleration, tilt, and vibrations.	Systems that prevent stealing. tracking sports
Gyroscope sensors	Calculate the angular velocity. The speed about an axis	Navigation of cars and drones
Humidity sensors	Very similar to temperature sensors	Air conditioning control. Hospitals and pharmaceutical industries

continued on following page

Table 3. Continued

Type of Sensor	Objectives	Applications
Optic sensors	Quantity of electromagnetic energy and light rays	Regulation of electric energy. Control of the environment and energy
Positioning sensors	Provide geographical coordinates to position objects	

Table.4 highlights the research gaps and explains why the present study is needed. It shows that the literature still lacks enough real-world cases, strong security discussion, and clear solutions for interoperability and scalability. These issues are important because cultural heritage systems handle sensitive information and often involve many connected devices. The table also shows how the current study addresses these gaps through a more integrated and practical IoT-based framework.

Table 4. Research gap matrix

Gap in literature	Why it matters	How this study helps
Limited real-world heritage cases	Theory is stronger than practice	This study connects IoT with a practical heritage setting
Weak security discussion	Heritage systems need strong protection	This study places security at the center
Interoperability issues	Many devices must work together	This study supports connected architecture
Scalability challenges	Data grows quickly in smart tourism	This study emphasizes scalable design
Fragmented review structure	Studies are spread across fields	This study brings them into one frame

3. PROPOSED MODEL

3.1 Research Gap

This study explores the crucial relationship between preventive conservation for cultural heritage and the Internet of Things (IoT), concentrating on the difficulties, uses, and consequences in the context of smart tourism. In contrast to previous research, it carries out a thorough analysis of the particular challenges faced when utilising IoT technology for the conservation of cultural assets in practical settings. While noting the ongoing problems in this field, the study highlights the integration of various sensor technologies, data analytics, and long-term monitoring. It makes a significant contribution by addressing the lack of thorough real-world case studies that show how to successfully use IoT in cultural contexts. The research aims to provide a thorough understanding of the useful applications and effectiveness of IoT in preserving cultural assets through an examination of particular projects, technologies used, difficulties encountered, and results obtained. This sets it apart from previous literature, which frequently lacks in-depth insights into practical implementations (Zonta et al., 2010).

This study explores the critical relationship between preventive conservation for cultural heritage and the Internet of Things. Unlike prior research, we conduct a thorough analysis of challenges encountered when applying IoT to the conservation of assets in practical, real-world settings. While current work touches on monitoring, it often lacks in-depth exploration of security frameworks and interdisciplinary cooperation. Our research fills these gaps by proposing a robust architecture that integrates diverse sensor technologies and data analytics to support long-term heritage protection. Additionally, the study adopts a proactive approach towards privacy and security concerns, which are touched upon in passing in the body of current work. It calls for a thorough examination and suggests strong frameworks and solutions to deal with any possible flaws in the IoT-based procedure for preserving cultural assets. The report stands out due to its emphasis on privacy and security issues, which acknowledges how crucial it is to have robust safeguards in place when using IoT technologies to protect cultural assets. The report also emphasises how crucial it is for specialists in data analytics, technology, and heritage conservation to collaborate across disciplinary boundaries in order to use IoT effectively. Technologists, environmentalists, and policymakers can all benefit from the research's examination of the dynamics of this kind of cooperation and identification of optimal approaches. The paper stands out due to its comprehensive viewpoint on interdisciplinary teamwork, which recognises the necessity of a cooperative approach to effectively utilise the potential of IoT in cultural asset protection. Essentially, this study not only points out gaps in the literature but also provides a path for filling them, expanding our knowledge of IoT applications for the protection of cultural heritage, and providing useful information for a range of stakeholders (Zubiaga et al., 2019).

As previously noted, the proliferation of information and communication technologies (ICTs) represents one of the most significant changes in 21st-century civilization. Recent technological advancements have had a major impact on tourists in particular as well as on people, companies, and society in general when it comes to metropolitan settings. The phrase "smart tourism" refers to the use of technology (such as mobile devices, augmented reality, and the internet) to collect copious amounts of data and provide real-time support to all destination stakeholders. By giving tourists one-of-a-kind experiences, information and communication technology (ICT) has the potential to improve tourism. It might also help related businesses automate processes and increase productivity. It said the following about smart tourism:

- connectivity through web-based applications with location capabilities;
- tourists as co-producers of destination content;
- enhancing experiences through new technologies (augmented reality);
- connecting with local communities and other tourists in the destination; and
- improving social and environmental sustainability.

3.2 IoT Architecture for a Sustainable Tourism Application

In the context of Cagliari's smart city development, the integration of Platform as a Service (PaaS) capability into the Internet of Things architecture for sustainable tourism applications represents a noteworthy breakthrough. Through the establishment of defined communication protocols, this integration enhances interoperability by overcoming the challenges presented by a variety of IoT devices and services. It creates a unified and networked environment in which information moves freely, promoting efficiency and collaboration throughout the city's infrastructure. Adoption of PaaS also encourages innovation since it optimises resource allocation and service delivery through enhanced mechanisms for service discovery and search. The architecture enables more flexibility and scalability by separating service requests from physical IoT items. This allows resources to be dynamically allocated to meet the changing demands of a bustling metropolis. Additionally, PaaS makes it possible to create personalised services that are tailored to each user's wants and tastes by utilising machine learning and data analytics. This strengthens a person's sense of community connection and belonging in addition to increasing satisfaction and engagement.

Future smart city development is built on the resilient and scalable infrastructure that PaaS platforms offer, which can handle exponential growth and changing requirements. This infrastructure accelerates innovation and moves Cagliari closer to a better, more sustainable future by streamlining the development, deployment, and management processes. The architecture is divided into multiple strata, each of which has a distinct purpose. Application Layer interfaces enable users to interact with the system by processing queries, displaying results, and initiating actions. The service layer acts as a link between the application and virtualization layers, assessing user requests and providing context to ensure that the appropriate services are called upon. The virtualization layer allows for the smooth administration and interaction of physical items in an Internet of Things environment by abstracting them into virtual objects (VOs). Real-World Object Layer describes the capabilities and operations of physical entities by converting them into ICT objects in the virtual world. The Engine for Trust and Security layer handles issues with data integrity and access control by putting security mechanisms in place to guarantee dependable communication throughout the architecture. The Trust and Security Engine ensure data protection and access policy enforcement by tackling security concerns at every level and improving the overall integrity and trustworthiness of the IoT architecture. The IoT architecture for sustainable tourism applications in Cagliari, with its integration of PaaS features and strong security measures, has the potential to transform visitor experiences and encourage sustainable practices. The IoT architecture's layers are shown in Figure.3.

Figure 3. Basic Cloud-Based IoT Architectural Solution

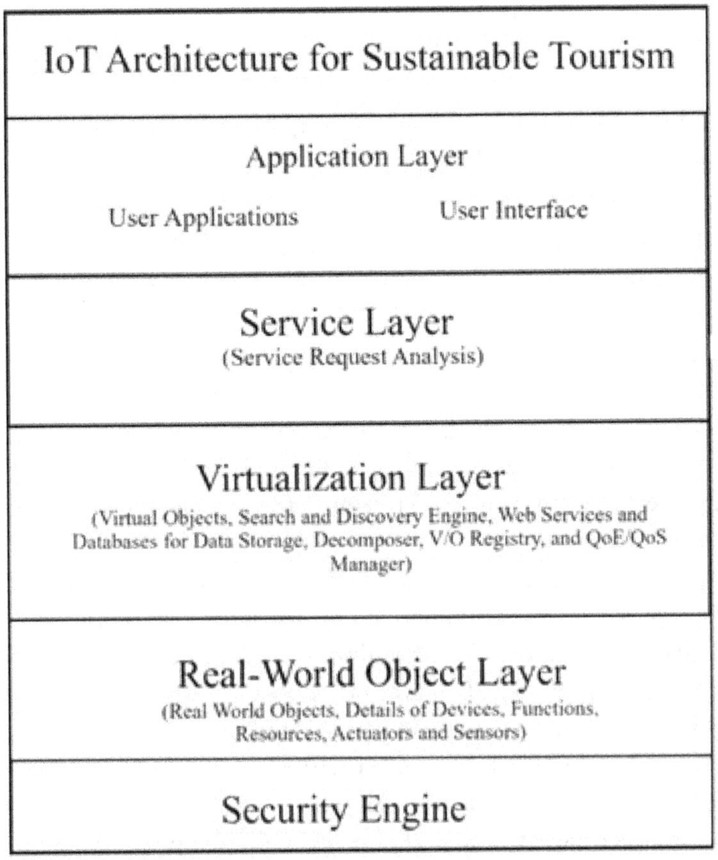

The proposed architecture utilizes a cloud-based implementation to ensure scalability, interoperability, and flexibility. The design is structured into five functional layers that work together to facilitate seamless data flow. The Real-World Object Layer serves as the foundation by converting physical assets, such as monuments and artifacts, into digital entities. The Virtualization Layer then abstracts these physical items into virtual objects to allow for easier management and standardized interaction. Above this, the Trust and Security Engine acts as a critical backbone by managing data integrity and enforcing access control across the entire system. The Service Layer links the virtualized assets with external requests, assessing user needs to ensure that the correct services are triggered. Finally, the Application Layer provides the interface where users can interact with the system, process specific queries, and access real-time results.

Through shedding light on the complex interactions between architectural morphology and social change, this method improves our understanding of the built environment and provides priceless information that may guide modern urban planning and design practices. As a result, the combination of interpretive historical study and careful structural evolution analysis is a powerful tool for revealing the many layers of culture, tradition, and resiliency woven throughout urban environments. Examining the complex web of regional traditions and culture reveals that traditional architecture plays a crucial role in determining the adaptability of metropolitan areas. Knowing how architectural forms interact with indigenous cultural practices and socioeconomic contexts is crucial for comprehending spatial transformation in the context of changing urban settings. The qualitative methodology and narrative building of interpretive historical research provide a sophisticated lens through which to examine the connection between traditional architecture and cultural heritage. Through examining historical records and assembling disjointed proof, scholars can reveal the underlying stories that give architectural structures their value and meaning.

Recognising that architectural legacy is complex and includes both tangible and intangible elements, such as customs, ceremonies, and communal activities, is fundamental to our investigation. Every piece of architecture, be it a house of worship, a typical home, or a place for people to congregate, is a physical manifestation of the social and cultural ideals of the people who designed it. Studying traditional architecture involves more than just documenting it; it also involves examining how structures change and adapt over time. Researchers can learn more about how societies adapt to social, economic, and environmental challenges by tracking the morphological changes and spatial dynamics of built environments. This comprehensive method of researching traditional architecture enhances our knowledge of the past and offers insightful guidance for modern urban planning and design. The symbiotic relationship between architecture and culture, as well as the inherent durability of traditional building practices, allow planners and designers to draw inspiration from long-standing customs to create more inclusive, resilient, and sustainable urban settings.

A thorough foundation for removing the many layers of tradition, culture, and resiliency ingrained in urban environments is provided by the combination of structural evolution analysis and interpretive historical study. Through the integration of historical study and architectural analysis, scholars can reveal latent themes and fundamental dynamics that have influenced the constructed surroundings throughout history. The durability of cultural heritage and the tenacity of urban populations are both demonstrated by traditional architecture. Through the use of interpretive historical research and an investigation into the structural development of architectural forms, scholars can get essential knowledge about the adaptive tactics used by societies across time. In addition to deepening our appreciation of the past, this newfound knowledge offers a guide for building more resilient and sustainable urban settings in the future.

3.3. Implementation and Security Design

To address security and interoperability, our model incorporates a dedicated Trust and Security Engine. This layer manages data integrity and access control, ensuring secure communication across all system strata. Additionally, the platform is designed to be device-agnostic, allowing for the integration of various sensors including temperature, humidity, and image sensors, without requiring specific settings or compatibility configurations. Table.5 summarizes the primary sensor types integrated into the proposed IoT architecture along with their specific roles in heritage preservation. Each sensor type serves a distinct purpose, ranging from environmental and chemical monitoring to structural stress detection and visitor flow management. By leveraging these diverse data inputs, the system ensures comprehensive protection of cultural assets while simultaneously enhancing the efficiency of preventive conservation efforts. This technical configuration allows for real-time adjustments and proactive maintenance, ultimately creating a robust and responsive infrastructure for smart tourism applications.

Table 5. Technical Specification Summar

Sensor Type	Purpose	Application
Temperature	Thermal energy calculation	Environmental control for delicate artifacts
Humidity	Moisture level tracking	Preventing material deformation and decay
Pressure	Structural stress detection	Monitoring stability of ancient constructions
Movement	Activity tracking	Detecting unauthorized presence and intrusions
Image	Visual documentation	Real-time monitoring and safety assessment
Proximity	Presence detection	Managing visitor flow and facility usage
Gas	Chemical monitoring	Detecting pollution and harmful contaminants
Water Quality	Contamination tracking	Protecting submerged heritage and water systems

3.4. IoT Technologies for Heritage Sites

Ever since the 1990s, museums and historical sites have led the way in utilising digital technology to improve visitor engagement. This has sparked a continuous improvement trajectory that aims to provide more immersive and educational experiences by overcoming the limitations of traditional text labels. Many digital technologies and platforms have been carefully incorporated into museum operations throughout time. Examples include smartphone museum guides, social media-driven visit recording mechanisms, and indoor navigation systems. These forward-thinking advancements have not only transformed the way cultural narratives are experienced and presented, but they have also greatly increased the understanding and appreciation of the displayed displays among visitors. The focus on offering visitors tangible encounters with museum displays has undergone a significant change in recent years, as research indicates that these kinds of interactions lead to more meaningful and long-lasting visitor experiences. Museums have set out to design new interactive solutions that let users connect with physical-digital interfaces using the principles of embodied interaction, palpable manipulation, physical data representation, and integration into tangible space. Through the use of these interfaces, visitors are able to access digital content that enhances their comprehension of the exhibitions while also maintaining a tangible connection to the genuine historical relics.

Even with the advancements in digital interaction, a lot of these experiences still mostly stand alone and have little link to other interactive components inside an exhibition. Nonetheless, the emergence of the Internet of Things (IoT) presents a promising opportunity to transform visitor experiences through the facilitation of interconnected displays. Museums might provide more complex and customised experiences for visitors by implementing IoT technologies. Although networked artefacts and the Internet of Things are generally seen by museum professionals as future developments rather than present realities, preliminary studies have already started to examine the potential of IoT in cultural heritage settings. This paper synthesises prior research and suggests new IoT-based technologies to augment digital capabilities in an effort to investigate the enormous potential of IoT for cultural heritage. It also explores the difficulties and potential paths ahead in utilising IoT for cultural assets, especially with regard to providing interactive experiences in historical locations.

Museums have been pioneers in experimenting with and incorporating digital technology into their operations and exhibits for a number of decades. Researchers studying human-computer interaction and related subjects have been investigating the creation and usage of interactive technology in museum settings concurrently. This line of research has moved away from the traditional model of discrete devices strewn all over an exhibition and towards integrated visitor experiences that smoothly blend interactive technology into the exhibition area. Numerous technologies, including NFC, Bluetooth LE, Wi-Fi, Arduino, Raspberry Pi, and telephones, have been utilised in these attempts. But this strategy frequently leads to a disjointed environment, which forces developers to combine several devices and connect them to create interesting interactions. Furthermore, every exhibition usually requires a new development effort; technology is used for each exhibition from scratch, with no existing systems extended or reused.

The Internet of Things (IoT) holds promise for enabling a more seamless incorporation of technology into museum operations, both in the context of individual exhibitions and in a wider range of environments. With the use of IoT, networked interactive displays that can dynamically adjust their responses based on visitor demographics or foot traffic, for example, can be created. Moreover, these displays have the ability to connect visitors from different places, remember them over time, and provide pre- and post-visit conversations. This increased degree of IoT connectivity has the potential to completely change how museums engage with their visitors and improve the entire experience.

4. CONCLUSION AND FUTURE SCOPE

Internet of Things (IoT) technology has great potential to revolutionise tourist experiences while promoting sustainable practices in the fields of smart tourism and cultural property preservation. Stakeholders are better equipped to take preventative action and make educated decisions by having real-time monitoring of structural integrity and environmental conditions thanks to IoT. In this perspective, IoT's ability to provide travellers with immersive and customised experiences is one of its greatest advantages. Through the utilisation of Internet of Things technologies, such as sensors, data analytics, and augmented reality, stakeholders can create captivating stories that connect with tourists more deeply, improving their comprehension and admiration of cultural heritage places. IoT adoption in smart tourism and cultural preservation is not without its own set of difficulties, though. The most important of them are issues with privacy, security, interoperability, and complexity of software. One major obstacle is the intricacy of integrating various IoT systems and guaranteeing smooth communication between them.

In addition, the additional degree of complexity that comes with implementing IoT in these settings is the requirement to protect sensitive data and guarantee visitor privacy. Collaboration between researchers, policymakers, and industry stakeholders is essential to effectively addressing these difficulties. Together, these parties may create strong frameworks and strict protocols to reduce the risks involved in the adoption of IoT. This cooperative strategy also promotes creativity and the creation of original ideas to remove obstacles that already exist. In the end, it is impossible to overestimate the revolutionary potential of IoT in protecting cultural treasures and raising tourist engagement. We can use IoT to protect our collective legacy for future generations by collaborating across disciplines and remaining innovative. We can guarantee that cultural heritage remains a fundamental component of our collective identity and cultural history by welcoming technology innovations and cultivating a collaborative culture.

Table 6. Future Scope

Future Research Area	Focus and Objective	Potential Impact
Advanced AI Integration	Incorporating deep learning for predictive maintenance	Higher accuracy in forecasting damage risks
Cross-Platform Interoperability	Developing universal standards for diverse IoT networks	Smoother data exchange between different heritage sites
Edge-Cloud Synergy	Shifting critical processing to the network edge	Faster response times and reduced latency
Energy Harvesting	Using sustainable power for remote sensor nodes	Increased system lifespan in isolated locations
Blockchain Security	Enhancing data integrity and ownership transparency	Stronger protection against unauthorized access

Table.6 above outlines critical areas for future development to enhance the IoT-based conservation framework. Advanced AI integration serves to move beyond simple monitoring by enabling predictive analytics, which allows for early intervention before structural damage occurs. Cross-platform interoperability addresses the need for universal standards, ensuring that data can be shared seamlessly across diverse global heritage networks. Edge-cloud synergy is essential for improving system response times by processing data closer to the sensor nodes, while energy harvesting solutions are necessary to maintain continuous operation in remote or sensitive historical sites. Finally, blockchain security provides an immutable layer for data integrity, which is vital for maintaining trust and transparency in the management of priceless cultural assets.

REFERENCES

Addabbo, T., Fort, A., Mugnaini, M., Panzardi, E., Pozzebon, A., & Vignoli, V. (2019). A city-scale IoT architecture for monumental structures monitoring. *Measurement : Journal of the International Measurement Confederation*, *131*, 349–357. DOI: 10.1016/j.measurement.2018.08.058

Al-Omari, A., Brunetaud, X., Beck, K., & Al-Mukhtar, M. (2013). Effect of thermal stress, condensation and freezing–thawing action on the degradation of stones on the castle of Chambord, France. *Environmental Earth Sciences*, *71*(9), 3977–3989. DOI: 10.1007/s12665-013-2782-4

Alletto, S., Cucchiara, R., Del Fiore, G., Mainetti, L., Mighali, V., Patrono, L., & Serra, G. (2016). An indoor location-aware system for an IoT-based smart Museum. *IEEE Internet of Things Journal*, *3*(2), 244–253. DOI: 10.1109/JIOT.2015.2506258

Brandt, T., Bendler, J., & Neumann, D. (2017). Social media analytics and value creation in urban smart tourism ecosystems. *Information & Management*, *54*(6), 703–713. DOI: 10.1016/j.im.2017.01.004

Costantini, S., Mostarda, L., Tocchio, A., & Tsintza, P. (2008). DALICA: Agent-based ambient intelligence for cultural-heritage scenarios. *IEEE Intelligent Systems*, *23*(2), 34–41. DOI: 10.1109/MIS.2008.24

Diego, F., Esteban, B., & Merello, P. (2015). Design of a hybrid (Wired/Wireless) acquisition data system for monitoring of cultural heritage physical parameters in smart cities. *Sensors (Basel)*, *15*(4), 7246–7266. DOI: 10.3390/s150407246

Diwakar, R., Sharma, R., Nayak, D., & Mohapatra, H. (2025). Optimizing load distribution in big data ecosystems. *AI and the Revival of Big Data*, 177-200. DOI: 10.4018/979-8-3693-8472-5.ch008

Fernández-Navajas, Á., Merello, P., Beltrán, P., & García-Diego, F. (2013). Software for storage and management of Microclimatic data for preventive conservation of cultural heritage. *Sensors (Basel)*, *13*(3), 2700–2718. DOI: 10.3390/s130302700

Figueredo, M., Cacho, N., Thome, A., Cacho, A., Lopes, F., & Araujo, M. (2017). Using social media photos to identify tourism preferences in smart tourism destination. *2017 IEEE International Conference on Big Data (Big Data)*. DOI: 10.1109/BigData.2017.8258423

García-Diego, F., Fernández-Navajas, Á., Beltrán, P., & Merello, P. (2013). Study of the effect of the strategy of heating on the mudejar church of Santa Maria in Ateca (Spain) for preventive conservation of the altarpiece surroundings. *Sensors (Basel)*, *13*(9), 11407–11423. DOI: 10.3390/s130911407

González, E. M., Municio, E., Alemán, M. N., & Marquez-Barja, J. M. (2020). Cultural heritage and Internet of things. *Proceedings of the 6th EAI International Conference on Smart Objects and Technologies for Social Good*, 248-251. DOI: 10.1145/3411170.3411267

Hardilla, D., & Nugroho, A. C. (2018). The role of Internet of things to support cultural heritage inventory in urban resiliency approach: Tradisional house in bandar Lampung case. *2018 International Conference on Information Technology Systems and Innovation (ICITSI)*, 193-198. DOI: 10.1109/ICITSI.2018.8695935

Konev, A., Khaydarova, R., Lapaev, M., Feng, L., Hu, L., Chen, M., & Bondarenko, I. (2019). CHPC: A complex semantic-based secured approach to heritage preservation and secure IoT-based Museum processes. *Computer Communications*, *148*, 240–249. DOI: 10.1016/j.comcom.2019.08.001

Kumar, A., Bag, A., Anand, A., Saha, S., Mohapatra, H., & Kolhar, M. (2024). Examining healthcare services utilizing cloud technology in intelligent urban environments. *Advances in Healthcare Information Systems and Administration*, •••, 77–98. DOI: 10.4018/979-8-3693-7225-8.ch004

Leccese, F., Cagnetti, M., Calogero, A., Trinca, D., Pasquale, S., Giarnetti, S., & Cozzella, L. (2014). A new acquisition and imaging system for environmental measurements: An experience on the Italian cultural heritage. *Sensors (Basel)*, *14*(5), 9290–9312. DOI: 10.3390/s140509290

Lillie, M., Smith, R., Reed, J., & Inglis, R. (2008). Southwest Scottish crannogs: Using in situ studies to assess preservation in wetland archaeological contexts. *Journal of Archaeological Science*, *35*(7), 1886–1900. DOI: 10.1016/j.jas.2007.11.029

Marshall, M. T. (2018). Interacting with heritage: On the use and potential of IoT within the cultural heritage sector. *2018 Fifth International Conference on Internet of Things: Systems, Management and Security*, 15-22. DOI: 10.1109/IoTSMS.2018.8554899

Mishra, P., Thakur, P., & Singh, G. (2021). Enabling technologies for IoT based smart city. *2021 Sixth International Conference on Image Information Processing (ICIIP)*, 99-104. DOI: 10.1109/ICIIP53038.2021.9702673

Mohapatra, H. (2021). Socio-technical challenges in the implementation of smart city. *2021 International Conference on Innovation and Intelligence for Informatics, Computing, and Technologies (3ICT)*, 57-62. DOI: 10.1109/3ICT53449.2021.9581905

Moropoulou, A. I., & Bisbikou, K. G. (1995). Environmental monitoring and damage assessment at the ancient sanctuary of Demeter in Eleusis, Greece. *MRS Proceedings, 352*. DOI: 10.1557/PROC-352-745

Nitti, M., Pilloni, V., Giusto, D., & Popescu, V. (2017). IoT architecture for a sustainable tourism application in a smart city environment. *Mobile Information Systems*, *2017*, 1–9. DOI: 10.1155/2017/9201640

Ordóñez-Martínez, D., Seguí-Pons, J. M., & Ruiz-Pérez, M. (2023). Conceptual framework and prospective analysis of EU tourism data spaces. *Sustainability (Basel)*, *16*(1), 371. DOI: 10.3390/su16010371

Pareek, P., Hajra, R., & Mohapatra, H. (2025). Empowering public officials by bridging the digital divide with smart technology for public services. *Advances in Public Policy and Administration*, •••, 419–452. DOI: 10.4018/979-8-3693-6547-2.ch016

Rosário, A. T., & Dias, J. C. (2024). Exploring the landscape of smart tourism: A systematic bibliometric review of the literature of the Internet of things. *Administrative Sciences*, *14*(2), 22. DOI: 10.3390/admsci14020022

Sharma, A., Anand, N., Mukherjee, S., Kumar, S., Kumar, S., & Mohapatra, H. (2024). Analyzing territorial variations in the implementation of IoT-based smart homes. *Advances in Electronic Government, Digital Divide, and Regional Development*, •••, 305–334. DOI: 10.4018/979-8-3693-6854-1.ch010

Valero, M., Merello, P., Navajas, Á., & García-Diego, F. (2014). Statistical tools applied in the characterisation and evaluation of a thermo-hygrometric corrective action carried out at the Noheda archaeological site (Noheda, Spain). *Sensors (Basel)*, *14*(1), 1665–1679. DOI: 10.3390/s140101665

Verma, A., Shukla, V. K., & Sharma, R. (2021). Convergence of IOT in tourism industry: A pragmatic analysis. *Journal of Physics: Conference Series*, *1714*(1), 012037. DOI: 10.1088/1742-6596/1714/1/012037

Zonta, D., Wu, H., Pozzi, M., Zanon, P., Ceriotti, M., Mottola, L., Picco, G. P., Murphy, A. L., Guna, S., & Corra, M. (2010). Wireless sensor networks for permanent health monitoring of historic buildings. *Smart Structures and Systems*, *6*(5_6), 595-618. DOI: 10.12989/sss.2010.6.5_6.595

Zubiaga, M., Izkara, J. L., Gandini, A., Alonso, I., & Saralegui, U. (2019). Towards smarter management of Overtourism in historic centres through visitor-flow monitoring. *Sustainability (Basel)*, *11*(24), 7254. DOI: 10.3390/su11247254

KEY TERMS AND DEFINITIONS

Artificial Intelligence: Systems that continuously learn by analyzing data and improving decision-making for tourism and heritage applications.

Augmented Reality: The blending of digital information with real-world heritage sites to enrich visitor understanding and engagement.

Cloud Computing: Remote computing resources used to store, process, and manage IoT data for scalable tourism services.

Cultural Heritage: Historical buildings, monuments, and artefacts that preserve the identity, memory, and values of a community.

Data Analytics: The examination of sensor and visitor data to detect patterns, predict risks, and support conservation decisions.

Digital Twin: A digital representation of a physical heritage asset used for monitoring, analysis, and management.

Environmental Monitoring: Continuous observation of temperature, humidity, gases, and other conditions affecting heritage preservation.

Flexible Access Control: A security mechanism that regulates user permissions according to roles, ownership, or trust level.

Heritage Site Monitoring: The use of sensors and connected devices to observe the condition of monuments and cultural assets.

Interoperability: The ability of different devices, platforms, and services to work together in a connected system.

IoT Architecture: The layered structure that links physical objects, virtual objects, services, and user applications.

Preventive Conservation: A proactive method of protecting cultural assets by detecting risks before serious damage occurs.

Real-Time Monitoring: Immediate observation of system or environmental conditions as they change.

Scalability: The capacity of a system to expand and handle increased devices, data, and users efficiently.

Security and Privacy: Protective measures that safeguard data, users, and connected objects from unauthorized access.

Smart Tourism: Technology-enabled tourism that improves services, personalization, sustainability, and visitor experience.

Sustainability: Efficient and responsible use of resources to support long-term tourism and heritage protection.

Virtualization Layer: The layer that converts physical objects into virtual objects for easier management in IoT systems.

Visitor Experience: The overall quality of interaction, personalization, and engagement a tourist receives at a destination.

Wireless Sensor Networks: Networks of distributed sensors used to collect environmental and operational data without wired installation.

Chapter 19
Sustainable Eco-Tourism for Cultural Preservation and Community Development

Lethukuthula Mtshali
http://orcid.org/0009-0009-5362-4988
University of KwaZulu-Natal, South Africa

Sthokozile Mamba
http://orcid.org/0009-0005-3789-4030
University of KwaZulu-Natal, South Africa

Andrisha Beharry Ramraj
http://orcid.org/0000-0001-5674-885X
University of KwaZulu-Natal, South Africa

ABSTRACT

Eco-tourism in KwaZulu-Natal has been widely framed as a strategy to link conservation with rural development, particularly in landscapes marked by historical dispossession. However, a persistent gap remained between policy rhetoric on inclusion and the structural realities limiting indigenous authority over land, revenue, and representation. The objective of the chapter was to explore how indigenous rural communities in KwaZulu-Natal can reclaim, preserve, and revive their cultural heritage through active involvement in eco-tourism. A qualitative secondary data design was adopted, synthesising place-based case studies and governance analyses within a thematic framework. The findings revealed that cultural revitalisation depended less on tourism activity itself and more on secure land rights, shared governance, and epistemic recognition. The study therefore recommended restructuring co-management systems, embedding Indigenous Knowledge Systems (IKS) in decision-making, and strengthening community authority to enable sustainable heritage renewal.

DOI: 10.4018/979-8-3373-7976-0.ch019

INTRODUCTION

Ecotourism has emerged as a prominent strategy for linking environmental conservation with socio-economic development in South Africa, and Phil (2022) states that it seeks to balance economic growth with socio-cultural integrity and environmental stewardship. Nyathi (2025) argues that modern ecotourism in KwaZulu-Natal is built on land with a history of dispossession, permanently linking the industry to unresolved land disputes. In areas such as the Dukuduku Forest, land functions as a living archive of song, ritual, and environmental knowledge, positioning IKS as foundational to tourism meaning making rather than decorative additions.

On paper, post-apartheid community tourism policies promote inclusion, whereas Giampiccoli and Glassom (2021) argue that the reality is quite different. In practice, institutional structures often prevent local communities from exercising meaningful authority over protected areas. This imbalance is especially evident in public-private partnerships, where government and commercial interests typically maintain a tight grip on decision-making, side lining indigenous voices in land use and cultural decisions (Litheko, 2022). For cultures to remain truly sustainable, symbolic participation is no longer enough. Communities need a real stake in both the profits and the way their stories are told (Wang & Hao, 2024).

Persistent poverty in KwaZulu-Natal continues to exclude local communities from meaningful ownership in the ecotourism industry (Mthembu & Mutambara, 2018). This financial disadvantage is further deepened by limited access to critical knowledge, particularly around land rights, leaving communities with little leverage when negotiating their stake in tourism development (Fairer-Wessels, 2017). In light of these barriers, this study critically examines the systemic and governance conditions that shape whether tourism development genuinely empowers local communities or further entrenches their marginalisation.

BACKGROUND OF THE STUDY

Ecotourism in KwaZulu-Natal must be understood within the broader transformation of South Africa's conservation and development agenda. National and provincial tourism strategies increasingly positioned ecotourism as a mechanism for rural development, cultural preservation, and environmental protection (Phil, 2022). Within this policy shift, community-based tourism models gained prominence as instruments intended to redress historical exclusion from conservation economies (Giampiccoli & Glassom, 2021). These models sought to reposition rural communities from passive beneficiaries to active stakeholders within protected landscapes.

Conservation areas in KwaZulu-Natal are not simply ecological spaces; they carry contested histories that still shape how tourism is managed today. Nyathi (2025) points to places like the Dukuduku Forest, where disputes over land ownership and governance have never been fully resolved, leaving a residue of tension in community-government relations that tourism operators cannot easily sidestep. Impey (2002) adds another dimension to this: these landscapes hold ritual significance, oral histories, and indigenous environmental knowledge accumulated over generations. Tourism, then, does not unfold against a neutral backdrop; it moves through spaces that are already deeply inhabited, culturally and historically.

As tourism grew, public–private partnerships became the dominant model for managing conservation economies across South Africa (Litheko, 2022). These arrangements brought market logic into conservation governance, streamlining decision-making, and quietly reshaping who holds authority within protected areas. Jones (2006) cautions that formal participation mechanisms rarely operate on communities' own terms; they tend to function within administrative systems designed elsewhere, which effectively limits how far community voices can reach. Against this backdrop, indigenous knowledge systems began attracting serious attention, not merely as cultural curiosities, but as heritage assets with real currency in tourism marketing and conservation policy (Masuku, 2025).

In rural KwaZulu-Natal, economic conditions do more than influence how communities engage in ecotourism; they dictate its entire structure. With high unemployment and limited access to capital, local involvement naturally stays small and informal. This does not stem from a lack of ambition, but rather from the current economic setup that blocks larger opportunities (Mthembu & Mutambara, 2018).

There are also gaps in information that complicate things further. Issues around land rights and conservation regulations are not always clearly communicated, leaving communities uncertain about where they stand and how to get involved (Fairer-Wessels, 2017). This lack of clarity can discourage engagement or lead to uneven participation, leaving only a few individuals with better access to information to benefit. Comparative studies from Indonesia and Cameroon show that when indigenous authority is structurally embedded within governance frameworks, ecotourism aligns more closely with cultural continuity and local development objectives (Alfitri et al., 2022; Tata et al., 2024). This contextual landscape situates ecotourism in KwaZulu-Natal at the intersection of conservation reform, rural development policy, indigenous heritage recognition, and evolving governance models.

PROBLEM STATEMENT

Ecotourism in KwaZulu-Natal is built on a promise that remains only partially realised. While it is frequently promoted as a means of empowering local communities, in practice, control over land, financial resources, and long-term planning remains largely beyond local community authority (Giampiccoli & Glassom, 2021; Litheko, 2022).

This power imbalance raises serious concerns. The ecotourism market profits directly from indigenous land, cultural memory, and environmental knowledge, yet indigenous systems are seldom placed at the centre of how these projects are designed and managed (Impey, 2002; Masuku, 2025). As a result, complex cultural practices are frequently reduced to marketable commodities (Wang & Hao, 2024).

Communities face significant barriers in challenging these limitations due to persistent economic constraints. High levels of poverty, unemployment, and limited access to clear information about land rights, weaken their negotiating position and increase reliance on external operators (Mthembu & Mutambara, 2018; Fairer-Wessels, 2017; Ramaano, 2025). Moreover, formal administrative frameworks intended to support participation often function according to externally defined priorities, which can restrict meaningful local autonomy (Jones, 2006; Prasetyo et al., 2019; Tata et al., 2024).

At its core, the issue reflects a clear contradiction: the ecotourism industry derives value from indigenous heritage while simultaneously limiting local communities' participation in decision-making. Without addressing this structural imbalance, ecotourism is likely to continue extracting value rather than supporting the preservation and self-determination of the communities on which it depends.

PURPOSE OF THE STUDY

The purpose of this study is to examine whether ecotourism in KwaZulu-Natal can genuinely empower indigenous rural communities to exercise authority over their land, knowledge systems, and cultural representation. To achieve this, the research interrogates the structural, governance, and economic conditions that dictate local participation in areas deeply affected by historical dispossession and unequal development. By analysing how Indigenous Knowledge Systems are currently positioned within tourism planning and marketing, this study aims to clarify the specific conditions required for ecotourism to strengthen cultural continuity rather than reproduce historical marginalisation.

Ultimately, this research seeks to move ecotourism governance in KwaZulu-Natal beyond symbolic inclusion and toward structural transformation by offering actionable insights to key stakeholders. For indigenous rural communities, the study provides an evidence base to support stronger claims for decision-making authority, equitable profit-sharing, and control over their own cultural narratives. At the institutional level, it offers provincial tourism and conservation authorities a pathway to translate inclusive policy promises into substantive reform within protected areas. Furthermore, private operators can draw on this analysis to design more equitable collaboration models that ensure both legitimacy and long-term sustainability. Finally, the research supplies policymakers and development planners with a practical framework to align rural development strategies with governance approaches that genuinely recognise indigenous communities as true rights-holders and knowledge authorities.

OBJECTIVE OF THE STUDY

To explore how indigenous rural communities in KwaZulu-Natal can reclaim, preserve, and revive their cultural heritage through active involvement in ecotourism.

LITERATURE REVIEW

The roots of conservation and tourism in KwaZulu-Natal lie in colonial and apartheid-era land dispossession. These historic events reshaped how indigenous communities related to land, replacing systems of coexistence with exclusionary policies (Nyathi, 2025). In places like the Dukuduku Forest, people were removed through legal systems that ignored how they lived with and understood the land. This approach is commonly known as fortress conservation, treating land as empty wilderness, and removing people in the name of protection (Impey, 2002). Conservation was also about control over land, people, and belonging (Nyathi, 2025).

These removals went beyond physical displacement. They disrupted cultural practices tied to land such as song, ritual, and oral traditions which carry both environmental knowledge and social memory (Impey, 2002). When communities were forced off their land, these connections weakened. The result was not only economic loss but also a breakdown in identity and everyday cultural life (Nyathi, 2025). This history also created lasting mistrust toward conservation authorities, which still affects ecotourism today.

After apartheid, policies began to promote inclusion through community-based tourism. While this suggested progress, many core power structures remained the same. In practice, outside actors often still control decisions and financial benefits (Giampiccoli & Glassom, 2021; Jones, 2006). As a result, community participation is often limited, and inequality continues under the language of inclusion (Litheko, 2022).

The legacy of dispossession is still visible today. High unemployment in rural KwaZulu-Natal makes it difficult for communities to start and sustain tourism businesses (Mthembu & Mutambara, 2018). At the same time, areas rich in cultural and natural resources often lack the funding and institutional support needed to benefit from tourism (Ramaano, 2025). These challenges are closely tied to past land removals that weakened local economies (Nyathi, 2025).

Research from other regions shows a similar pattern. ecotourism tends to work better where indigenous land rights are recognised and local knowledge is included in planning (Prasetyo et al., 2019). Where this is not the case, tourism can end up reinforcing the same exclusion it claims to address (Tata et al., 2024). This remains a concern in KwaZulu-Natal, where decision-making power is still often held outside local communities (Impey, 2002; Litheko, 2022).

There are some promising examples. Ndlovu et al. (2018) highlight the Zulu-Mpophomeni Tourism Experience, where communities take the lead in shaping how their heritage is presented. However, these efforts are still limited by wider governance systems that restrict local control.

In the end, ecotourism cannot support cultural preservation without addressing the legacy of dispossession. Access to land, decision-making power, and representation are still uneven (Nyathi, 2025; Litheko, 2022). Without real change in these areas, ecotourism risks showcasing culture without truly supporting the communities behind it (Wang & Hao, 2024).

IKS as Foundations for Culturally Grounded ecotourism

In KwaZulu-Natal, Indigenous Knowledge Systems are not static traditions put on display; they are living ways of knowing, rooted in land, memory, and everyday practice. This is evident in places like the Dukuduku Forest, where song, ritual, and storytelling carry environmental knowledge and ancestral history, linking identity directly to the landscape (Impey, 2002). As Nyathi (2025) explains, these knowledge systems are not decorative additions to tourism; they guide how communities live, care for the environment, and relate to one another. For tourism to be truly inclusive, it must recognise indigenous knowledge as a valid and central way of understanding the world (Masuku, 2025).

IKS include land-use practices, oral traditions, crafts, and values such as Ubuntu, all of which shape community life and responsibility (Ndlovu et al., 2018). The Zulu-Mpophomeni Tourism Experience shows how tourism can reflect these local realities by focusing on everyday heritage and community stories (Ndlovu et al., 2018). When communities control how their culture is shared and benefit from it, ecotourism can support both cultural and environmental sustainability (Wang & Hao, 2024). In this sense, tourism becomes an expression of living heritage rather than a staged performance (Giampiccoli & Glassom, 2021).

However, this is not always the case. Many ecotourism projects reduce IKS to performances shaped by market expectations (Jones, 2006). When projects are controlled from the outside, commercial goals often take priority over community meaning, weakening local authority over knowledge (Jones, 2006). Public–private partnerships can also limit community influence, reinforcing existing power imbalances (Litheko, 2022). In these cases, culture is presented, but not truly owned by the people it comes from (Wang & Hao, 2024).

As Wang and Hao (2024) note, tourism can separate cultural practices from their lived context, turning them into products. However, studies from other regions show a different path. In Indonesia, for example, embedding indigenous knowledge into governance strengthens both conservation and cultural continuity (Alfitri et al., 2022). Similarly, integrating local knowledge into formal planning improves environmental management and community legitimacy (Prasetyo et al., 2019). These examples show that recognising IKS in decision-making, not just in presentation, is essential.

Evidence from Cameroon supports this further. ecotourism works best when traditional leaders are directly involved in governance (Tata et al., 2024). This reinforces a key point also relevant in KwaZulu-Natal: communities must have real authority over land and cultural narratives (Impey, 2002). As Ramaano (2025) argues, tourism development should build on local knowledge rather than isolate it as something to be performed. ecotourism only supports cultural revival when IKS shapes how tourism is managed, not just what visitors see (Giampiccoli & Glassom, 2021; Tata et al., 2024).

Another important aspect is how knowledge is passed on. Impey (2002) shows that involving young people in documenting oral histories helps keep knowledge alive while strengthening community ownership. This highlights that culture survives through practice, not performance (Masuku, 2025). In the same way, ecotourism businesses rooted in local culture can reflect community values rather than external branding (Rahmawati et al., 2023). When young people are involved, and knowledge is lived, tourism becomes a space for cultural renewal rather than loss (Alfitri et al., 2022).

Still, there are real challenges. Economic constraints limit community ownership of tourism ventures (Mthembu & Mutambara, 2018), while a lack of access to information makes meaningful participation difficult (Fairer-Wessels, 2017). Weak institutional support also allows externally controlled businesses to dominate how culture is presented (Mtshali & Beharry-Ramraj, 2025). As a result, communities are often excluded not only from decision-making but also from shaping their own narratives (Wang & Hao, 2024).

Despite this, research consistently shows that ecotourism can support both conservation and cultural continuity when it is grounded in indigenous knowledge. When local knowledge guides practice, tourism can balance environmental care with social and cultural development (Phil, 2022; Indwar & Muthukumar, 2023). This requires embedding culture within governance, rather than treating it as an attraction (Sihombing et al., 2024).

Overall, the literature makes one thing clear: IKS must shape how tourism is designed, managed, and represented if ecotourism is to support cultural reclamation in KwaZulu-Natal (Giampiccoli & Glassom, 2021). When treated as a core knowledge system, IKS strengthens identity, continuity, and community control (Impey, 2002; Alfitri et al., 2022). However, when reduced to performance within externally controlled systems, its meaning is easily lost (Jones, 2006; Litheko, 2022). IKS is not an add-on to ecotourism; it is the foundation that determines how tourism works and what it represents (Masuku, 2025; Wang & Hao, 2024).

Governance, Power, and Community Authority in ecotourism Development

ecotourism governance in KwaZulu-Natal is shaped by post-apartheid reforms that aimed to be more inclusive, but in many cases, control remains centralised (Giampiccoli & Glassom, 2021). Community-based tourism was introduced to address past exclusion by involving local people in planning and management. However, in practice, many projects are still controlled by external actors, with communities playing mostly advisory roles (Jones, 2006). This gap between inclusion in theory and control in practice raises concerns about whether these models truly empower communities or keep them dependent (Litheko, 2022).

Public–private partnerships are now common in managing protected areas, but they often prioritise investment and conservation targets over community authority (Litheko, 2022). In many cases, communities are treated as beneficiaries rather than rights-holders, limiting their say over land use, income, and cultural representation. When decision-making stays in external hands, local knowledge and social networks are underused, and existing power imbalances continue (Giampiccoli & Glassom, 2021; Litheko, 2022; Jones, 2006).

Stakeholder theory helps explain this issue. For ecotourism to be sustainable, communities must be recognised as key stakeholders with real authority over decisions and benefits (Wang & Hao, 2024). But when participation is limited to consultation, it becomes symbolic rather than meaningful. Limited access to information about land rights and conservation processes further weakens communities' ability to take part in decisions (Fairer-Wessels, 2017). As a result, governance systems can continue to exclude communities, even while claiming to include them (Litheko, 2022).

This is also reflected in how economic benefits are shared. While ecotourism has created jobs in rural KwaZulu-Natal, ownership and control are still largely out of reach for local communities (Mthembu & Mutambara, 2018). In many areas near protected parks, communities have rich natural and cultural resources but lack the capital and authority to manage tourism themselves (Ramaano, 2025). This means people may benefit financially in small ways but remain dependent rather than empowered.

Governance also affects how culture is preserved and represented. Cultural practices like song and ritual are closely tied to land and identity (Impey, 2002). When control over land and storytelling sits with external actors, communities lose the ability to define their own heritage (Wang & Hao, 2024). This often leads to simplified or commercialised versions of culture that appeal to tourists but weaken its original meaning (Jones, 2006; Wang & Hao, 2024).

Studies from other countries show similar patterns. In Cameroon, excluding traditional leaders from governance limits the use of indigenous knowledge in ecotourism (Tata et al., 2024). In contrast, research from Indonesia shows that when indigenous authority is recognised and included in decision-making, both conservation and tourism outcomes improve (Alfitri et al., 2022; Prasetyo et al., 2019). These examples highlight that real inclusion must go beyond jobs and income; it must include decision-making power.

In South Africa, similar findings emerge. Stronger conservation outcomes are linked to meaningful local involvement in both economic and governance processes (Ijatuyi et al., 2025). However, weak institutional support, especially in areas like marketing, often allows externally managed businesses to dominate how tourism is presented (Mtshali & Beharry-Ramraj, 2025). This shows that control over narratives is also a form of power, shaping whose culture is visible and how it is understood.

Overall, governance plays a central role in determining whether ecotourism supports cultural revival or simply extracts value. Without real control over land, income, and representation, communities remain on the margins of systems built around their own heritage (Litheko, 2022). But when communities have ownership, secure rights, and recognition of their knowledge, ecotourism can support both conservation and cultural continuity (Prasetyo et al., 2019; Alfitri et al., 2022). For this to happen in KwaZulu-Natal, governance structures need to shift from symbolic inclusion to genuine community authority.

Structural Constraints, Cultural Commodification, and the Conditions for Heritage Revitalisation ecotourism in KwaZulu-Natal operates under conditions that limit communities' ability to participate as owners rather than low-paid workers (Mthembu & Mutambara, 2018). High unemployment, low incomes, and limited access to capital make it difficult for residents to start or manage tourism businesses. Even in areas rich in culture and biodiversity, weak infrastructure and limited skills keep communities in roles like guiding or craft selling, while ownership remains elsewhere (Ramaano, 2025).

Lack of access to information makes this worse. Many communities are unclear about land rights, tourism policies, and decision-making processes (Fairer-Wessels, 2017). When people are focused on daily survival, they are less able to engage in long-term planning. As a result, participation often reflects necessity rather than real empowerment, reinforcing dependency (Litheko, 2022).

These challenges are tied to governance systems that still centralise control. Public–private partnerships often treat communities as beneficiaries, not decision-makers, limiting their influence over income and representation (Litheko, 2022). When tourism is externally controlled, cultural practices are often reshaped to fit market demands, reducing community authority over meaning (Jones, 2006; Wang & Hao, 2024).

This is where cultural commodification becomes important. Tourism can simplify and package culture for visitors, separating it from everyday life (Wang & Hao, 2024). In places like the Dukuduku Forest, cultural practices such as song and ritual carry deep environmental knowledge and identity (Impey, 2002). When turned into performances, their meaning can be weakened. This is more likely when communities lack control over how their culture is presented (Jones, 2006).

Marketing power also plays a role. Limited support for community-led promotion means externally managed businesses often dominate tourism narratives (Mtshali & Beharry-Ramraj, 2025). This shapes how culture is seen and valued, reinforcing inequality in who gets to tell the story.

Despite these challenges, ecotourism can support cultural revival when conditions change. Locally owned initiatives like the Zulu-Mpophomeni Tourism Experience show that when communities control both ownership and storytelling, tourism can strengthen identity and continuity (Ndlovu et al., 2018). This suggests that outcomes depend less on tourism itself and more on the systems around it (Giampiccoli & Glassom, 2021).

Passing knowledge to younger generations is also key. Involving youth in documenting and practising cultural traditions helps keep knowledge alive and strengthens community ownership (Impey, 2002). At the same time, building skills in business, governance, and access to information allows communities to move beyond service roles (Ramaano, 2025; Fairer-Wessels, 2017).

Overall, ecotourism can either strengthen or weaken cultural heritage, depending on the conditions in which it operates. Where poverty, limited information, and external control persist, tourism often reinforces dependency and commodification. Nevertheless, where communities have ownership, secure rights, and recognition of their knowledge, ecotourism can support cultural revival and long-term sustainability (Alfitri et al., 2022; Giampiccoli & Glassom, 2021).

Insights and Practical Pathways for Community-Led ecotourism

Ecotourism is often assumed to deliver both conservation and development, but the outcomes are uneven. Evidence from KwaZulu-Natal shows that it can create opportunities, especially for women, by improving access to training, business information, and small enterprise development (Xaba & Adanlawo, 2024).

However, these benefits are not evenly shared. Some areas perform better than others, depending on available support systems and local conditions. While many people can access basic information, fewer have access to funding or consistent training (Makhanyela et al., 2024). This shows that ecotourism does not automatically lead to inclusive development.

A key issue is how projects are planned. Many are still designed from the top down, with limited space for communities to shape decisions. As a result, participation is often shallow, and inequalities remain. ecotourism is effective only when strong local systems, meaningful participation, and access to resources support it.

METHODOLOGY

This study uses a qualitative research design based on secondary data analysis to examine how ecotourism governance, Indigenous Knowledge Systems, and structural inequality intersect in KwaZulu-Natal. Secondary analysis allows existing studies to be reinterpreted to generate new insights (Heaton, 2008). Rather than treating past research as fixed evidence, this study connects findings across different case studies to build a broader understanding of cultural reclamation in ecotourism. This approach is suitable given the diverse and historically complex nature of sites such as the Dukuduku Forest and Bergville (Impey, 2002; Nyathi, 2025; Mthembu & Mutambara, 2018).

The study follows a qualitative approach that focuses on interpretation and context rather than numerical data (Walliman, 2021; Khoa et al., 2023). ecotourism, governance, and IKS are socially embedded, making them better suited to qualitative analysis (Mazhar et al., 2021). Data was drawn from peer-reviewed articles, case studies, and policy-focused research to understand how ecotourism operates within specific historical and social conditions.

Sources were selected based on their relevance to key themes such as land, governance, cultural representation, and participation. Priority was given to studies with strong empirical evidence, including work on Dukuduku (Impey, 2002), land dispossession (Nyathi, 2025), community-based tourism (Giampiccoli & Glassom, 2021), and rural tourism participation (Mthembu & Mutambara, 2018). Additional studies on governance, marketing, and cultural sustainability were included to deepen the analysis and provide broader context.

The analysis used a thematic approach. Each study was reviewed to identify patterns related to: (1) land dispossession, (2) governance and decision-making, (3) socio-economic constraints, (4) IKS and cultural representation, and (5) conditions for cultural revitalisation. This allowed the study to move beyond individual cases and identify broader structural patterns shaping ecotourism outcomes (Heaton, 2008). Concepts from stakeholder theory and participatory development were used to assess whether communities are treated as decision-makers or simply consulted.

To strengthen the analysis, insights from other countries were included. Studies from Indonesia and Cameroon were compared with South African cases to see whether similar patterns exist in different contexts (Prasetyo et al., 2019; Tata et al., 2024; Alfitri et al., 2022). This helped place the findings within a wider global perspective.

To ensure rigour, the study remained grounded in the original data and avoided selective interpretation. All sources were used carefully and within their original context (Baldwin et al., 2022). As this research relies on existing studies, it does not involve direct interaction with participants, but it follows ethical guidelines by accurately representing prior work (Tripathy, 2013).

Overall, this approach provides a structured, context-sensitive way to understand how ecotourism operates in KwaZulu-Natal and whether current systems support or limit cultural reclamation.

FINDINGS AND DISCUSSION

The findings show that ecotourism in KwaZulu-Natal is not a neutral development tool. It operates within a history of land dispossession, unequal governance, and socio-economic inequality that still shapes who benefits, who makes decisions, and whose knowledge is valued. Rather than a clear success or failure, ecotourism sits in a space of tension: it can either reinforce existing inequalities or support cultural renewal, depending on who controls land, knowledge, income, and representation (Nyathi, 2025; Litheko, 2022).

History plays a key role. In areas like the Dukuduku Forest, forced removals under conservation policies disrupted not just access to land, but also cultural systems tied to it (Nyathi, 2025; Impey, 2002). Land holds memory, identity, and knowledge, so dispossession weakened these connections. Although post-apartheid policies promote inclusion, land restitution remains slow and uncertain, and trust in conservation systems is still fragile (Fairer-Wessels, 2017). As a result, community participation in ecotourism is often cautious rather than fully engaged.

At the same time, Indigenous Knowledge Systems are widely recognised as essential to meaningful ecotourism but are often sidelined in practice. Cultural knowledge is frequently used to attract tourists, while communities have limited control over how it is presented (Jones, 2006). This creates a gap between recognising culture and giving communities authority over it. Without decision-making power, cultural reclamation remains limited (Wang & Hao, 2024).

Governance is a major factor behind this. While community-based tourism aims to be inclusive, many projects still place control in the hands of government or private actors (Giampiccoli & Glassom, 2021; Litheko, 2022). Communities are often consulted but rarely have real authority over land use, investment, or revenue. This weakens their ability to shape tourism and reinforces dependency, especially in already vulnerable communities (Mthembu & Mutambara, 2018).

Economic constraints add to the problem. High unemployment, low incomes, and limited access to capital make it difficult for communities to own or manage tourism enterprises (Mthembu & Mutambara, 2018; Ramaano, 2025). As a result, many people participate as workers rather than decision-makers. Limited access to information about land and governance further reduces their ability to negotiate fair outcomes (Fairer-Wessels, 2017). In these conditions, participation is often symbolic rather than empowering.

The findings also highlight the risk of cultural commodification. Tourism markets tend to favour simplified and predictable versions of culture, which can pressure communities to reshape their practices (Wang & Hao, 2024). When cultural narratives are controlled externally, heritage can lose its deeper meaning (Impey, 2002). However, when communities control how their culture is shared, tourism can support pride and continuity (Ndlovu et al., 2018). The key issue is not whether culture is shared, but who controls its representation.

There are also positive examples. Research from countries like Indonesia and Cameroon shows that when indigenous knowledge is included in governance, ecotourism can strengthen both conservation and cultural sustainability (Alfitri et al., 2022; Prasetyo et al., 2019; Tata et al., 2024). These findings support the idea that secure land rights and recognised community authority are essential for meaningful change.

Another important factor is intergenerational knowledge transfer. Involving younger generations in cultural practices and documentation helps keep knowledge alive and strengthens community identity (Impey, 2002). When combined with local enterprise development, this can turn ecotourism into a space for cultural renewal rather than loss (Rahmawati et al., 2023).

Overall, ecotourism in KwaZulu-Natal stands at a crossroads. It is shaped by historical and structural challenges, but it also holds potential for positive change. Its impact depends on whether power relations shift. Where land rights are insecure, information is limited, and control remains external, ecotourism risks repeating patterns of exclusion. But where communities have ownership, decision-making power, and control over their knowledge, it can support cultural reclamation and long-term sustainability.

To meet the study's objective, it is clear that community involvement must go beyond jobs or consultation. It requires real changes in governance, secure land rights, fair revenue sharing, and recognition of indigenous knowledge. Only then can ecotourism move from extracting cultural value to supporting cultural revival.

RECOMMENDATIONS

The findings of this study demonstrate that ecotourism in KwaZulu-Natal cannot meaningfully contribute to cultural reclamation unless structural conditions within the province are deliberately reshaped. The recommendations below are therefore grounded in the specific historical, institutional, and socio-economic realities of KwaZulu-Natal, particularly in communities living within or adjacent to protected areas such as Dukuduku, iSimangaliso, and other conservation landscapes shaped by fortress conservation legacies.

First, governance arrangements in KwaZulu-Natal's protected areas must be restructured to reflect the province's specific history of dispossession and contested land restitution. In areas such as Dukuduku, where forced removals disrupted ancestral land relations and created enduring mistrust, shared authority over land-use planning and tourism development is not merely a developmental enhancement but a restorative necessity. Public–private partnership agreements currently operating in provincial reserves and parks should be reviewed to incorporate formal co-management clauses that grant community representatives binding decision-making powers over cultural representation, benefit-sharing models, and tourism expansion plans. This is particularly relevant in KwaZulu-Natal, where tourism branding frequently draws on Zulu cultural identity while governance authority remains concentrated in state agencies and private operators. Institutional recognition of community councils or traditional authorities within tourism boards would help align conservation practice with the constitutional recognition of customary systems, thereby addressing the province's specific historical imbalance between conservation authority and indigenous land rights.

Second, IKS in KwaZulu-Natal should be formally embedded within provincial tourism and conservation frameworks as authoritative knowledge systems rather than supplementary cultural features. This requires that KwaZulu-Natal Tourism and Film Authority and Ezemvelo KZN Wildlife integrate indigenous knowledge holders into tourism planning committees at the earliest design stages of ecotourism initiatives. In contexts such as Dukuduku and rural Bergville, where song, ritual, and place-based environmental practices remain living systems, interpretation strategies should be co-developed with local knowledge custodians to ensure that cultural narratives reflect community-defined meanings. This approach would move beyond the current tendency to extract selected cultural symbols for visitor appeal and instead position indigenous epistemologies as guiding principles for environmental stewardship and visitor engagement. Such embedding is particularly crucial in KwaZulu-Natal, where cultural identity and landscape are inseparable, and where tourism marketing frequently relies on images of tradition that risk simplification without community governance.

Third, capacity development strategies must be contextually tailored to the socio-economic realities of rural KwaZulu-Natal. In municipalities characterised by high unemployment and limited household income, such as Bergville and parks-adjacent settlements, capacity building should extend beyond hospitality training to include governance literacy, land rights education, and financial management skills specific to tourism enterprise ownership. Given documented information asymmetries in communities surrounding iSimangaliso and similar protected areas, provincial departments should prioritise accessible dissemination of information on conservation regulations, tourism licensing, and benefit-sharing frameworks. This could include isiZulu-language workshops, community-based information hubs, and simplified guides to land and tourism policy. Enhancing knowledge of governance processes would reduce reliance on intermediaries and strengthen community bargaining power within tourism partnerships.

Fourth, marketing and branding strategies in KwaZulu-Natal require recalibration to prioritise community-led narratives. Currently, externally managed enterprises dominate tourism visibility in the province, often shaping representations of Zulu culture through commercially driven lenses. Provincial marketing campaigns should therefore allocate dedicated platforms for community-owned ecotourism ventures such as the Zulu-Mpophomeni Tourism Experience, ensuring that locally defined heritage expressions are visible in mainstream tourism circuits. Support mechanisms should include sustainable marketing training, digital storytelling initiatives led by youth within communities, and funding streams that enable small-scale enterprises to compete without compromising cultural integrity. Importantly, promotional frameworks should avoid pressuring communities to standardise or simplify cultural practices for market appeal, recognising instead that authenticity in KwaZulu-Natal is rooted in living practice rather than performance.

Fifth, secure tenure and restitution processes must be accelerated and integrated into ecotourism development planning. In KwaZulu-Natal, where historical dispossession continues to shape participation, unresolved land claims undermine trust and limit long-term investment by communities. Aligning land restitution processes with ecotourism co-management agreements would provide clarity and confidence for community-led enterprise development. Secure land rights would also enable communities to negotiate partnerships from positions of authority rather than from a position of dependency, thereby transforming ecotourism from a site of cultural extraction into a mechanism for land-based cultural renewal.

Finally, intergenerational knowledge transmission should be intentionally supported within ecotourism programming. Given the centrality of oral history, ritual, and environmental knowledge in areas such as Dukuduku, participatory documentation projects involving elders and youth should be incorporated into tourism strategies. Provincial tourism authorities could collaborate with local schools and community organisations to integrate heritage education into tourism activities, ensuring that ecotourism reinforces rather than displaces everyday cultural practice. In KwaZulu-Natal, where cultural continuity is closely tied to land and lineage, strengthening youth engagement within community-led ecotourism enterprises would help secure the long-term vitality of IKS.

Collectively, these recommendations are not generic development prescriptions but context-specific interventions grounded in KwaZulu-Natal's historical and socio-economic landscape. By restructuring governance authority, embedding indigenous epistemologies in policy, strengthening locally relevant capacity, recalibrating marketing practices, accelerating tenure security, and supporting intergenerational knowledge transmission, ecotourism in KwaZulu-Natal can move beyond symbolic inclusion towards substantive cultural reclamation. Under such conditions, indigenous rural communities would not merely participate in ecotourism but actively shape it as a vehicle for preserving and reviving their heritage within landscapes that have long defined their identity.

FUTURE RESEARCH DIRECTIONS

While this study has provided a critical synthesis of secondary data on ecotourism and IKS in KwaZulu-Natal, future research should prioritise deeply contextual, place-based inquiry within specific communities such as Dukuduku, Bergville, and settlements surrounding iSimangaliso. Ethnographic and participatory research approaches conducted in isiZulu and grounded in everyday lived experience would be particularly valuable in capturing how indigenous knowledge holders themselves interpret ecotourism, restitution processes, and governance reforms. Such work could illuminate how mistrust linked to historical dispossession continues to shape decision-making at household and community levels, thereby moving beyond policy narratives to examine how ecotourism is experienced on the ground within KwaZulu-Natal's conservation landscapes.

Further research should also interrogate intra-community dynamics within rural KwaZulu-Natal, particularly along the lines of gender, age, and traditional authority. In many Zulu communities, women are custodians of craft traditions and oral knowledge, while youth increasingly engage with digital platforms and tourism entrepreneurship. Empirical studies focusing on how women and young people participate in, benefit from, or are excluded from ecotourism initiatives would deepen understanding of how IKS are transmitted, adapted, or reshaped in tourism contexts. Such research would prevent homogenising communities and would reveal how cultural revitalisation unfolds unevenly within the specific social structures of KwaZulu-Natal.

CONCLUSION

This study has demonstrated that ecotourism in KwaZulu-Natal operates within a deeply layered context shaped by historical land dispossession, institutional inequality, and persistent socio-economic marginalisation. Although IKS are widely acknowledged as central to the province's cultural identity and tourism appeal, their influence within governance, revenue distribution, and narrative control remains structurally constrained. The evidence makes clear that poverty, information asymmetries, and centralised decision-making are not isolated barriers but interconnected conditions that continue to limit indigenous communities' authority over land and heritage. Under such circumstances, ecotourism risks reproducing forms of exclusion in which culture is used as an economic resource while communities remain peripheral to strategic power.

At the same time, the study has shown that ecotourism holds transformative potential when structural conditions shift towards secure land rights, shared governance, institutional recognition of indigenous epistemologies, and equitable benefit-sharing. Where communities exercise meaningful authority over land, knowledge, and representation, tourism can reinforce intergenerational transmission, strengthen identity, and support sustainable livelihoods. The central argument advanced throughout this study is therefore unequivocal: ecotourism in KwaZulu-Natal will only contribute to cultural reclamation and revival when indigenous rural communities are recognised not as symbolic participants, but as decision-makers and knowledge authorities within tourism systems. Without such structural transformation, claims of cultural sustainability remain rhetorical; with it, ecotourism can become a viable pathway for reclaiming, preserving, and reviving indigenous heritage within the very landscapes from which communities were historically excluded.

REFERENCES

Alfitri, M., Hambali, H., & Yuliani, S. (2022). Indigenous knowledge and social capital in community-based ecotourism: Evidence from Indonesia. *Journal of Sustainable Tourism*, *30*(8), 1920–1938.

Baldwin, J. R., Pingault, J. B., Schoeler, T., Sallis, H. M., & Munafò, M. R. (2022). Protecting against researcher bias in secondary data analysis: Challenges and potential solutions. *European Journal of Epidemiology*, *37*(1), 1–10. DOI: 10.1007/s10654-021-00839-0 PMID: 35025022

Fairer-Wessels, F. A. (2017). Determining the role of information in sustainable tourism development: A case study of communities surrounding the iSimangaliso Wetland Park. *African Journal of Hospitality, Tourism and Leisure*, *6*(4), 1–15.

Giampiccoli, A., & Glassom, D. (2021). Community-based tourism in protected areas: Elaborating a model from a South African perspective. *African Journal of Hospitality, Tourism and Leisure*, *10*(3), 873–889. DOI: 10.46222/ajhtl.19770720-143

Heaton, J. (2008). Secondary analysis of qualitative data: An overview. *Historical Social Research (Köln)*, *33*(3), 33–45.

Ijatuyi, E. J., Ayanlade, A., & Olanrewaju, O. (2025). Sustainable tourism and green space: Exploring how green spaces and natural attractions contribute to local tourism economies and revenue generation in Gauteng Province, South Africa. *Sustainability*, *17*(6), 2878. DOI: 10.3390/su17062878

Impey, A. (2002). Culture, conservation and community reconstruction: Explorations in advocacy ethnomusicology and ecotourism in the Dukuduku Forest, South Africa. *Yearbook for Traditional Music*, *34*, 9–25. DOI: 10.2307/3649187

Indwar, T., & Muthukumar, P. K. (2023). Role of ecotourism in sustainable development: An overview. *Journal of Pharmaceutical Negative Results*, *14*(2).

Jones, S. (2006). Community-based ecotourism: The significance of social capital. *Annals of Tourism Research*, *33*(2), 303–324. DOI: 10.1016/j.annals.2004.06.007

Khoa, B. T., Hung, B. P., & Hejsalem-Brahmi, M. (2023). Qualitative research in social sciences: Data collection, data analysis and report writing. *International Journal of Public Sector Performance Management*, *12*(1–2), 187–209. DOI: 10.1504/IJPSPM.2023.132247

Litheko, A. (2022). Management of ecotourism through public-private partnerships in South Africa. *African Journal of Hospitality, Tourism and Leisure*, *11*(6), 2103–2127.

Makhanyela, D. S., Nkosi, G. S., Tshabalala, S. P., & Gumede, T. K. (2024). Comprehensive participation: A theoretically hailed yet practically disregarded approach to sustainable ecotourism. *Geo Journal of Tourism and Geosites*, *52*(1), 276–285. DOI: 10.30892/gtg.52127-1204

Masuku, X. R. (2025). Indigenizing sustainable tourism: Empowering indigenous voices in tourism education. In *Indigenous Teaching Disciplines and Perspectives for Higher Education* (pp. 221–240). IGI Global Scientific Publishing. DOI: 10.4018/979-8-3693-9296-6.ch011

Mazhar, S. A., Anjum, R., Anwar, A. I., & Khan, A. A. (2021). Methods of data collection: A fundamental tool of research. *Journal of Integrated Community Health, 10*(1), 6–10. DOI: 10.24321/2319.9113.202101

Mthembu, B. A., & Mutambara, E. (2018). Rural tourism as a mechanism for poverty alleviation in KwaZulu-Natal Province of South Africa: A case of Bergville. *African Journal of Hospitality, Tourism and Leisure*, *7*(4), 1–17.

Mtshali, L., & Beharry-Ramraj, A. (2025). Barriers to implementing green marketing strategies for sustainable ecotourism in KwaZulu-Natal. e-BANGI Journal, 22(2), 1–15. DOI: 10.17576/ebangi.2025.2202.54

Ndlovu, T. P., Sibiya, N. P., & Giampiccoli, A. (2018). Assessing local community participation in community-based tourism: The case of the Zulu-Mpophomeni Tourism Experience. *African Journal of Hospitality, Tourism and Leisure*, *7*(2).

Nyathi, P. A. (2025). Exploring forced removals and dispossession in the Dukuduku Forest, KwaZulu-Natal, before and after 1994. *History (London)*, *110*(391), 393–412. DOI: 10.1111/1468-229X.13438

Phil, T. M. (2022). ecotourism: A sustainable development connect to nature and a strategy for balancing economic growth, socio-cultural development and conservation. [Publication details incomplete; journal/source information to be verified by authors.]

Prasetyo, L. B., Kartodihardjo, H., Adiwibowo, S., & Okarda, B. (2019). The complexity of integrating indigenous knowledge for ecotourism planning: A case of customary forests in Indonesia. *Forest Policy and Economics*, *103*, 84–92. DOI: 10.1016/j.forpol.2019.03.002

Rahmawati, R., Handayani, S. R., Suprapti, A. R., Airawaty, D., & Latifah, L. (2023). Green entrepreneurship based on local characteristics and culture to support sustainable ecotourism: A case study. *Journal of Intercultural Communication*, *23*(1), 66–75. DOI: 10.36923/jicc.v23i1.71

Ramaano, A. I. (2025). Community-based tourism advancement and sustainable tourism enterprise development in marginalised rural municipalities: Context for ecotourism development in parks-adjacent communities. *African Journal of Hospitality, Tourism and Leisure*, *14*(1), 1–18.

Sihombing, I. H. H., Suastini, N. M., & Puja, I. B. P. (2024). Sustainable cultural tourism in the era of sustainable development. *International Journal of Sustainable Competitiveness on Tourism*, *3*(02), 100–115. DOI: 10.34013/ijscot.v3i02.1601

Tata, E. S., Ndzi, D. L., & Fonjong, L. N. (2024). Challenges in the application of indigenous knowledge in the management of ecotourism around Mount Cameroon National Park. *Journal of Sustainable Tourism*, *32*(4), 623–640. DOI: 10.1080/09669582.2023.2178946

Tripathy, J. P. (2013). Secondary data analysis: Ethical issues and challenges. *Iranian Journal of Public Health*, *42*(12), 1478–1479. PMID: 26060652

Walliman, N. (2021). *Research methods: The basics*. Routledge. DOI: 10.4324/9781003141693

Wang, J., & Hao, J. (2024). Community-based tourism and cultural sustainability in indigenous contexts. *Journal of Sustainable Tourism*, *32*(2), 215–231. DOI: 10.1080/09669582.2023.2201147

Xaba, F., & Adanlawo, E. F. (2024). The potential role of ecotourism in sustainable development: A systematic review. *Journal of Ecohumanism, 3*(4), 3356–3367. DOI: 10.62754/joe.v3i4.3852

Xaba, F., & Adanlawo, E. F. (2024). Are women empowered through ecotourism projects? A quantitative approach: Ecotourism and women empowerment. *Pakistan Journal of Life and Social Sciences, 22*(1), 3085–3092. DOI: 10.57239/PJLSS-2024-22.1.00224

ADDITIONAL READING

Berkes, F. (2009). Evolution of co-management: Role of knowledge generation, bridging organisations and social learning. *Journal of Environmental Management, 90*(5), 1692–1702. DOI: 10.1016/j.jenvman.2008.12.001 PMID: 19110363

Brockington, D., Duffy, R., & Igoe, J. (2008). *Nature unbound: Conservation, capitalism and the future of protected areas*. Earthscan.

Comaroff, J. L., & Comaroff, J. (2009). *Ethnicity, Inc.* University of Chicago Press. DOI: 10.7208/chicago/9780226114736.001.0001

Kiss, A. (2004). Is community-based ecotourism a good use of biodiversity conservation funds? *Trends in Ecology & Evolution, 19*(5), 232–237. DOI: 10.1016/j.tree.2004.03.010 PMID: 16701261

Rogerson, C. M. (2006). Pro-poor local economic development in South Africa: The role of pro-poor tourism. *Local Environment, 11*(1), 37–60. DOI: 10.1080/13549830500396149

Scheyvens, R. (2002). *Tourism for development: Empowering communities*. Prentice Hall.

Spenceley, A. (Ed.). (2008). *Responsible tourism: Critical issues for conservation and development*. Earthscan.

Wearing, S., & Neil, J. (2009). *Ecotourism: Impacts, potentials and possibilities* (2nd ed.). Butterworth-Heinemann. DOI: 10.4324/9780080940182

West, P., Igoe, J., & Brockington, D. (2006). Parks and peoples: The social impact of protected areas. *Annual Review of Anthropology, 35*(1), 251–277. DOI: 10.1146/annurev.anthro.35.081705.123308

Zeppel, H. (2006). *Indigenous ecotourism: Sustainable development and management*. CABI. DOI: 10.1079/9781845931247.0000

KEY TERMS AND DEFINITIONS

Co-Management: A governance arrangement in which authority over natural resources, land use, and tourism planning is formally shared between the state, private operators, and local communities, with binding decision-making rights extended to all parties rather than concentrated in external actors.

Community-Based Tourism: A tourism model designed to position local communities as active participants and primary beneficiaries in tourism development, rather than passive recipients of externally directed investment or state-managed planning processes.

Cultural Commodification: The process by which cultural practices, symbols, and traditions are extracted from their lived social context and transformed into marketable products for tourist consumption, frequently at the cost of their original meaning and community ownership.

Cultural Sovereignty: The right of a community to define, control, and transmit its own cultural identity, narratives, and practices without external interference, commercial appropriation, or institutional simplification.

Ecotourism: A form of nature-based tourism that seeks to balance environmental conservation with socio-economic development while promoting cultural integrity and meaningful local community participation in tourism planning and governance.

Epistemic Recognition: The formal acknowledgement, within governance frameworks, policy, and institutional structures, that indigenous knowledge systems constitute valid and authoritative ways of understanding the world, on equal standing with externally derived knowledge frameworks.

Fortress Conservation: A conservation model, historically dominant in Southern Africa, that treats natural landscapes as uninhabited wilderness requiring the removal of indigenous communities from their ancestral lands in the name of environmental protection, disregarding existing cultural and ecological relationships.

Indigenous Knowledge Systems (IKS): Living, place-based ways of knowing that encompass land-use practices, oral traditions, ritual, craft, environmental stewardship, and social values, rooted in the historical and cultural experience of a specific community and transmitted across generations through practice rather than formal documentation.

Land Restitution: The legal and administrative process through which communities dispossessed of their land during colonial or apartheid-era policies may reclaim ownership, use rights, or equitable compensation, and through which secure tenure can be established as a foundation for genuine community authority in development planning.

Compilation of References

Abosede, A. J., Ebie, S., Somoye, R. O. C., & Ashidi, J. S. (2025). Decolonising Entrepreneurship by Embracing the Global Dichotomy. In *The Palgrave Handbook of Decolonising Entrepreneurship* (pp. 587–603). Springer Nature Switzerland., DOI: 10.1007/978-3-031-92310-4_22

Abrahams, S. (2024). Ethical content creation in the digital age. *Media Update*. Available at: https://www.mediaupdate.co.za/media/155772/ethical-content-creation-in-the-digital-age [Accessed: 20 January 2026]

Absolon, K. (2022). *Kaandossiwin: How we come to know*. Fernwood Publishing.

Achieng, M. S. (2022). Digitalisation of indigenous knowledge of agricultural practices: A strategy for food security in Africa. *Journal of Public Administration*, *57*(4), 802–818. https://hdl.handle.net/10520/ejc-jpad_v57_n4_a5

Addabbo, T., Fort, A., Mugnaini, M., Panzardi, E., Pozzebon, A., & Vignoli, V. (2019). A city-scale IoT architecture for monumental structures monitoring. *Measurement : Journal of the International Measurement Confederation*, *131*, 349–357. DOI: 10.1016/j.measurement.2018.08.058

Adekunle, O. (2022). Hierarchy and respect in African indigenous communication systems. *Journal of African Cultural Studies*, *34*(2), 145–158.

Afolabi, O. S. (2020). Globalisation, decoloniality and the question of knowledge production in Africa. *Journal of Higher Education in Africa/Revue de l'enseignement supérieur en Afrique, 18*(1), 93-110. https://www.jstor.org/stable/48618319

Agarwal, B. (1992). The gender and environment debate: Lessons from India. *Feminist Studies*, *18*(1), 119–158. DOI: 10.2307/3178217

Agarwal, B. (2010). *Gender and green governance: The political economy of women's presence within and beyond community forestry*. Oxford University Press. DOI: 10.1093/acprof:oso/9780199569687.001.0001

Agrawal, A. (1995). Dismantling the divide between Indigenous and scientific knowledge. *Development and Change*, *26*(3), 413–439. DOI: 10.1111/j.1467-7660.1995.tb00560.x

Agrawal, A., & Gibson, C. C. (1999, April). Enchantment and Disenchantment: The Role of Community in Natural Resource Conservation. *World Development*, *27*(4), 629–649. DOI: 10.1016/S0305-750X(98)00161-2

Agrawal, A., & Ostrom, E. (2001). Collective action, property rights, and decentralization in resource use in India and Nepal. *Politics & Society*, *29*(4), 485–514. DOI: 10.1177/0032329201029004002

Aguessy, A. N. J. (2023). Integration of traditional healing practices in modern medicine in South America: Insights for African development. *International Journal on Studies in English Language and Literature, 11*(12), 36–47. DOI: 10.20431/2347-3134.1112005

Ahluwalia, A. (2025, January 29). *Ima Keithel: Manipur's Mothers Are Leading A 500-Year Revolution Through Commerce & Unity*. Retrieved from Homegrown: https://homegrown.co.in/homegrown-voices/ima-keithel-manipurs-mothers-are-leading-a-500-year-revolution-through-commerce-unity

Aikin, J. (2007). Preparing for a National disaster: The committee on conservation of cultural resources, 1939-1944. *The Library Quarterly, 77*(3), 257–285. DOI: 10.1086/519416

Akano, O. A., Hanson, E., Nwakile, C., & Esiri, A. E. (2024). Improving worker safety in confined space entry and hot work operations: Best practices for high-risk industries". Global. *Journal of Advanced Research and Reviews, 2*(2), 31–39. DOI: 10.58175/gjarr.2024.2.2.0056(GJARR-2024-0056)

Akinwale, A. A. (2013). Digitisation of indigenous knowledge for natural resources management in Africa. *A Paper Presented at the 20th Anniversary Summit of the African Educational Research Network at North Carolina State University, Raleigh, USA on 19th May 2012.*

Álava-Núñez, P., Torres, B., Castro, M., & Robles, M. (2025). AGB carbon stock analysis in the Indigenous agroforestry of the Ecuadorian Amazon: Chakra and Aja as Natural Climate Solutions. *Frontiers in Forests and Global Change, 8*, 1513140. Advance online publication. DOI: 10.3389/ffgc.2025.1513140

Alberts, R., Khalala, G., & Molefe, O. (2011). Collecting structured indigenous knowledge for enhanced exploration: The National Recordal System (NRS), South Africa. *Knowledge Management for Development Journal, 7*(3), 291–307. DOI: 10.1080/19474199.2011.649009

Alfitri, M., Hambali, H., & Yuliani, S. (2022). Indigenous knowledge and social capital in community-based ecotourism: Evidence from Indonesia. *Journal of Sustainable Tourism, 30*(8), 1920–1938.

Alletto, S., Cucchiara, R., Del Fiore, G., Mainetti, L., Mighali, V., Patrono, L., & Serra, G. (2016). An indoor location-aware system for an IoT-based smart Museum. *IEEE Internet of Things Journal, 3*(2), 244–253. DOI: 10.1109/JIOT.2015.2506258

Al-Omari, A., Brunetaud, X., Beck, K., & Al-Mukhtar, M. (2013). Effect of thermal stress, condensation and freezing–thawing action on the degradation of stones on the castle of Chambord, France. *Environmental Earth Sciences, 71*(9), 3977–3989. DOI: 10.1007/s12665-013-2782-4

Alonge, E., Alao, O., & Dudu, O. (2024). The impact of digital transformation on financial reporting and accountability in emerging markets. *International Journal of Science and Technology Research Archive, 7*(2), 025–049. .DOI: 10.53771/ijstra.2024.7.2.0061

Altieri, M. A. (2018). *Agroecology: The science of sustainable agriculture* (2nd ed.). CRC Press. DOI: 10.1201/9780429495465

Altieri, M. A., & Koohafkan, P. (2008). *Enduring farms: Climate change, smallholders and traditional farming communities*. FAO.

Altieri, M. A., & Nicholls, C. I. (2017). The adaptation and mitigation potential of traditional agriculture in a changing climate. *Climatic Change, 140*(1), 33–45. DOI: 10.1007/s10584-013-0909-y

Altieri, M. A., Nicholls, C. I., Henao, A., & Lana, M. A. (2015, May 1). Agroecology and the design of climate change-resilient. *Agronomy for Sustainable Development, 35*(3), 869–890. DOI: 10.1007/s13593-015-0285-2

Altieri, M. A., & Toledo, V. M. (2011). The agroecological revolution in Latin America. *The Journal of Peasant Studies, 38*(3), 587–612. DOI: 10.1080/03066150.2011.582947

Amankwah-Amoah, J., & Medase, S. K. (2024). Extracting innovation value from intellectual property: Evidence from sub-Saharan Africa. *Journal of the Knowledge Economy, 15*(2), 8933–8967. DOI: 10.1007/s13132-023-01225-9

Amusan, M. A. (2016). Cultivating Effective Pedagogical Skills in In-Service Teachers: The Role of Some Teacher Variables. *Journal of the International Society for Teacher Education, 20*(1), 83–89.

Anane-Donkor, L., Dei, D. J., & Ocloo, P. E. D. (2026). Indigenous knowledge for innovation and sustainable livelihood in Ghana's informal economy. *Discover Global Society, 4*(1), 17. Advance online publication. DOI: 10.1007/s44282-026-00352-3

Anderson, T. (2008). Towards a theory of online learning. In Flexible learning. DOI: 10.15215/aupress/9781897425084.004

Anderson, J. (2005). Protocols for libraries, archives and information services. *The Australian Library Journal, 54*(4), 370–382.

Andrae, S. (2024). The establishment of a Gross Ecosystem Product: A country comparison. In Boubaker, S., & Le, T. H. (Eds.), *Green finance and sustainable development goals* (pp. 369–407). World Scientific Publishing., DOI: 10.1142/9781800614482_0014

Andrae, S. (2025). Financing mechanisms for the conservation of biodiversity: A multi-country comparison. In Walker, T., Naffa, H., Tharumar, R. K., & Donders, S. (Eds.), *Biodiversity finance: The economic, operational, and societal impacts of biodiversity loss* (pp. 269–299). Springer., DOI: 10.1007/978-3-032-02029-1_11

Anquet, R., & Girard, F. (2022). *Biocultural community protocol and the Ethic of stewardship: the sovereign steward of biodiversity. Biocultural rights, Indigenous Peoples and Local Communities*. Routledge. DOI: 10.4324/9781003172642-14

Antunes, A. P., & Lima Constantino, P. (2025). Healthy forests safeguard traditional wild meat food systems in Amazonia. *Nature, 648*(8094), 625–633. DOI: 10.1038/s41586-025-09743-z PMID: 41299169

ANU Communications & Engagement. (2024, November 1). *Indigenous cultural burning has protected Australia's landscape for millennia.* Retrieved from Australian National University: https://www.anu.edu.au/news/all-news/indigenous-cultural-burning-has-protected-australias-landscape-for-millennia

Aragao, S. D., Alfaia, S. S., da Costa Ayres, M. I., & Capelloto Costa, F. R. (2026). Soil fertility and carbon stocks of indigenous agroforestry homegardens in the lower Amazon region. *Agroforestry Systems, 100*(1), 22. DOI: 10.1007/s10457-025-01398-z

Archibald, J. (2008). *Indigenous storywork: Educating the heart, mind, body, and spirit.* UBC Press. DOI: .59962/9780774855204

Archibald, J. A. 2008. Indigenous storywork: Educating the heart, mind, body, and spirit. UBC press. Acorn (n.d) Available at: https://acorn.works/capability/digital-content-creation [Accessed: 26 December 2025

Archibald, J. A. (2008). *Indigenous storywork: Educating the heart, mind, body, and spirit*. UBC press. DOI: 10.59962/9780774855440

Asante, M. K. (2003). *Afrocentricity: A theory of social change* (revised and expanded ed.). Chicago, IL: African American Images. *Urban Education, 41*(1), 101–111. DOI: 10.1177/0042085905284322

Asante, M. K. (2007). *An Afrocentric manifesto: Toward an African renaissance*. Polity Press.

Asante, M. K. (2012). Afrocentric idea and communication theory. *International Journal of Intercultural Relations, 36*(6), 760–769.

Ayandibu, E. O. (2025). Integrating indigenous knowledge systems into entrepreneurship education: A South African perspective. *International Journal of Research in Business & Social Science, 14*(6), 380–392. Advance online publication. DOI: 10.20525/ijrbs.v14i6.4263

Ayinla, B., Osasona, F., Abrahams, T., Atadoga, A., Amoo, O., & Farayola, O. (2024). Ethical AI in practice: Balancing technological advancements with human values. *International Journal of Science and Research Archive, 11*(1), 1311–1326. DOI: 10.30574/ijsra.2024.11.1.0218

Bagwasi, M. M. (2012). The effect of gender and age in Setswana greetings. *Southern African Linguistics and Applied Language Studies, 30*(1), 93–100. DOI: 10.2989/16073614.2012.693717

Baldassar, L., Nedelcu, M., Merla, L., & Wilding, R. (2016). ICT-based co-presence in transnational families and communities: Challenging the premise of face-to-face proximity in sustaining relationships. *Global Networks, 16*(2), 133–144. DOI: 10.1111/glob.12108

Baldwin, J. R., Pingault, J. B., Schoeler, T., Sallis, H. M., & Munafò, M. R. (2022). Protecting against researcher bias in secondary data analysis: Challenges and potential solutions. *European Journal of Epidemiology, 37*(1), 1–10. DOI: 10.1007/s10654-021-00839-0 PMID: 35025022

Balogun, T. (2023). Digitization of Indigenous knowledge systems in Africa: The case of South Africa's National Recorded System (NRS). *Records Management Journal, 33*(1), 88–102. DOI: 10.1108/RMJ-10-2022-0029

Balogun, T. (2023b). Data management of digitized indigenous knowledge systems in repositories. *Information Development, 39*(3), 425–438. DOI: 10.1177/02666669231186777

Balogun, T., & Kalusopa, T. (2020). A framework for digital preservation of indigenous knowledge system in repositories in South Africa. *Records Management Journal, 31*(2), 176–196. https://www.emerald.com/rmj/article-pdf/31/2/176/2298008/rmj-12-2020-0042. DOI: 10.1108/RMJ-12-2020-0042

Balogun, T., & Kalusopa, T. (2022). Web archiving of indigenous knowledge systems in South Africa. *Information Development, 38*(4), 658–671. DOI: 10.1177/02666669211005522

Barasa, D. W. (2007). Indigenous knowledge systems and sustainable development in Africa: Case study on Kenya. *Tribes Tribals, 2007*, 141–156.

Barber, K. (2007). *The anthropology of texts, persons and publics*. Cambridge University Press. DOI: 10.1017/CBO9780511619656

Bartlett, C., Marshall, M., & Marshall, A. (2012). Two-eyed seeing and other lessons learned within a co-learning journey of bringing together indigenous and mainstream knowledges and ways of knowing. *Journal of Environmental Studies and Sciences*, *2*(4), 331–340. DOI: 10.1007/s13412-012-0086-8

Barton, A. (2009). The Use and Abuse of Accounting in the Public Sector Financial Management Reform Program in Australia. *Abacus*, *45*(2), 221–248. DOI: 10.1111/j.1467-6281.2009.00283.x

Bat, M., Kilgariff, C., & Doe, T. (2014). Indigenous tertiary education–we are all learning: Both-ways pedagogy in the Northern Territory of Australia. *Higher Education Research & Development*, *33*(5), 871–886. DOI: 10.1080/07294360.2014.890575

Battiste, M. (2002). *Indigenous knowledge and pedagogy in First Nations education: A literature review with recommendations*. Indian and Northern Affairs Canada.

Battiste, M. (2005). Indigenous knowledge: Foundations for First Nations. *WINHEC: International Journal of Indigenous Education Scholarship*, *1*(1), 1–17.

Battiste, M. (2013). *Decolonizing education: Nourishing the learning spirit*. Purich Publishing.

Battiste, M. (2014). Decolonizing education: Nourishing the learning spirit. *The Alberta Journal of Educational Research*, *60*(3), 615–618.

Battiste, M., & Henderson, J. Y. (2000). *Protecting Indigenous knowledge and Heritage: A Global Challenge*. Purich Publishing. DOI: 10.59962/9781895830439

Bawack, R., Roderick, S., Badhrus, A., Dennehy, D., & Corbett, J. (2025). Indigenous knowledge and information technology for sustainable development. *Information Technology for Development*, *31*(2), 233–250. DOI: 10.1080/02681102.2025.2472495

Bayly, C. A. (2010). INDIA, THE BHAGAVAD GITA AND THE WORLD. *Modern Intellectual History*, *7*(2), 275–295. DOI: 10.1017/S1479244310000077

Bebbington, C., Wilson, E., Smith, E., & Van Alstine, J. (2017). *Community Liaison officers: exploring the frontline of corporate practice in the oil and gas sector*. Audire and ECW Energy LTD.

Bender, E. M., Gebru, T., McMillan-Major, A., & Shmitchell, S. (2021). On the dangers of stochastic parrots: Can language models be too big? In *Proceedings of the 2021 ACM Conference on Fairness, Accountability, and Transparency* (pp. 610-623). ACM. DOI: 10.1145/3442188.3445922

Berkes, F. (2012). *Sacred Ecology* (3rd ed.). Routledge. DOI: 10.4324/9780203123843

Berkes, F. (2017). *Sacred ecology* (4th ed.). Routledge. DOI: 10.4324/9781315114644

Berkes, F., Colding, J., & Folke, C. (2000, October). Rediscovery of Traditional Ecological Knowledge as Adaptive Management. *Ecological Applications : A Publication of the Ecological Society of America*, *10*(5), 1251–1262. DOI: 10.1890/1051-0761(2000)010[1251:ROTEKA]2.0.CO;2

Berkes, F., Colding, J., & Folke, C. (Eds.). (2003). *Navigating social–ecological systems: Building resilience for complexity and change*. Cambridge University Press.

BESNet. (2024). *Languages, Cultures, Knowledge: UNESCO's*. Retrieved from www.besnet.world: https://www.besnet.world/library/languages-cultures-knowledge-unescos-action-for-indigenous-peoples/#:~:text=UNESCO%20is%20committed%20to%20the%20preservation%20of%20Indigenous,management%20of%20natural%20and%20cultural%20World%20Heritage%20sites.

Bhanye, J., & Bhanye, A. (2023). A blessing or a curse? The role of social media during the COVID-19 pandemic in Africa. In *The Palgrave handbook of global social problems* (pp. 1–23). Springer International Publishing.

Bhatia, R., (2025). Bridging Generations: The Role of Grandparents in Transmitting Sustainable Development Values to Grandchildren.

Bhatnagar, N., Sharma, P., & Ramachandran, K. (2019). Spirituality and Corporate Philanthropy in Indian Family Firms: An Exploratory Study. *Journal of Business Ethics*, *163*(4), 715–728. DOI: 10.1007/s10551-019-04394-3

Bhusari, K. (24, February 2024). *Ut res magis valeat quam pereat*. Retrieved from LAW ARTICLE: https://lawarticle.in/ut-res-magis-valeat-quam-pereat/

Bihari, S. (2023). Cultural heritage and Indigenous knowledge: Reviving traditions for future generations. *Sustainable Development Goals in SAARC Countries: Key Issues. Opportunities and Challenges*, *1*, 24–32.

Bird, S. (2020). Decolonising speech and language technology. In *Proceedings of the 28th International Conference on Computational Linguistics* (pp. 3504-3519). DOI: 10.18653/v1/2020.coling-main.313

Blackie, M. A. (2024). Indigenous knowledge systems and science education. *South African Journal of Science*, *120*(5-6), 1–4.

Boamah, E., & Liew, C. L. (2017). Conceptualising the digitisation and preservation of Indigenous knowledge: The importance of attitudes. In, Choemprayong, S., Crestani, F., & Cunningham, S. (eds), *International Conference on Asian Digital Libraries* (pp. 65-80). Cham: Springer International Publishing. DOI: 10.1007/978-3-319-70232-2_6

Boatbil, C. S., Baada, F. N.-A., Akonga, S., & Ayoung, D. A. (2025). Preserving indigenous knowledge for development in northern Ghana: Exploring the views of artisans in the art and craft industry in the Upper East Region. *African Quarterly Social ScienceReview*, *2*(3), 251–262. DOI: 10.51867/AQSSR.2.3.22

Bogopa, D. L. (2012). The importance of Indigenous games: The selected cases of Indigenous games in South Africa. *Indilinga*, *11*(2), 245–256.

Bolake, R. N. (2025). INDIAN KNOWLEDGE SYSTEM AND RURAL DEVELOPMENT: A PATHWAY TO SUSTAINABLE DEVELOPMENT. *IIP: International Multidisciplinary Research Journal*, *2*(Issue-IV (October-December)), 9-9.

Bongaarts, J. (2019). IPBES, 2019. Summary for policymakers of the global assessment report on biodiversity and ecosystem services of the Intergovernmental Science-Policy Platform on Biodiversity and Ecosystem Services. *Wiley*, 680-686. DOI: 10.1111/padr.12283

Borrini-Feyerabend, G.. (2013). *Governance of protected areas: From understanding to action*. IUCN.

Bovens, M. (2007). *Public Accountability* (pp. 182–208). oxford university. DOI: 10.1093/oxfordhb/9780199226443.003.0009

Bowrey, K., & Anderson, J. (2009). The politics of global intellectual property regulation and access to knowledge. *Law in Context (Bundoora, Vic.)*, *26*(1), 1–25.

Brandt, T., Bendler, J., & Neumann, D. (2017). Social media analytics and value creation in urban smart tourism ecosystems. *Information & Management*, *54*(6), 703–713. DOI: 10.1016/j.im.2017.01.004

Braun, V., & Clarke, V. (2006). Using thematic analysis in psychology. *Qualitative Research in Psychology*, *3*(2), 77–101. DOI: 10.1191/1478088706qp063oa

Bray, D. B., Merino-Pérez, L., & Barry, D. (2006). *The community forests of Mexico*. University of Texas Press.

Breidlid, A. (2009). Culture, indigenous knowledge systems and sustainable development: A critical view of education in an African context. *International Journal of Educational Development*, *29*(2), 140–148. DOI: 10.1016/j.ijedudev.2008.09.009

Brugnach, M., Craps, M., & Dewulf, A. (2017). Including Indigenous people in climate mitigation: Addressing issues of scale, knowledge and power. *Climatic Change*, *140*(1), 19–32. DOI: 10.1007/s10584-014-1280-3

Bryan, C. J. O., Garnett, S. T., Fa, J. E., Leiper, I., Rebein, J. A., Fernandez-Llamazares, A., Jackson, M. V., Jonas, H. D., Brondizio, E. S., Burgess, N. D., Rosinson, C. J., Zander, K. K., Molner, Z., Venter, O., & Watson, J. E. M. (2019). The importance of indigenous people's land for the conservation of terrestrial mammals. *Conservation Biology : The Journal of the Society for Conservation Biology*, *35*(3), 1002–1008. DOI: 10.1111/cobi.13620

Bugge, M. M., Hansen, T., & Klitkou, A. (2016). What is the bioeconomy? *Sustainability (Basel)*, *8*(7), 691. DOI: 10.3390/su8070691

Byscov. M.F. (2021). "What makes Epistemic Injustice an 'injustice'?" Journal of Social Philosophy, 52(1): 114-131. Date of access: 10 Jan. 2026.DOI: 10.1111/josp.12348

Cajete, G. (2000). *Native science: Natural laws of interdependence*. Clear Light Publishers.

Carroll, S. R., Garba, I., Figueroa-Rodríguez, O. L., Holbrook, J., Lovett, R., Materechera, S., Parsons, M., Raseroka, K., Rodriguez-Lonebear, D., Rowe, R., Sara, R., Walker, J. D., Anderson, J., & Hudson, M. (2020). The CARE Principles for Indigenous Data Governance. *Data Science Journal, 19*(1), 43. DOI: .5334/dsj-2020-043

Carroll, S. R., Garba, I., Figueroa-Rodríguez, O. L., Holbrook, J., Lovett, R., Materechera, S., Parsons, M., Raseroka, K., Rodriguez-Lonebear, D., Rowe, R., Sara, R., Walker, J. D., Anderson, J., & Hudson, M. (2020). The CARE principles for Indigenous data governance. *Data Science Journal*, *19*(1), 43. Advance online publication. DOI: 10.5334/dsj-2020-043

Carroll, S. R., Garba, I., Plevel, R., Small-Rodriguez, D., Hiratsuka, V. Y., Hudson, M., & Garrison, N. A. (2022). Using indigenous standards to implement the CARE principles: Setting expectations through tribal research codes. *Frontiers in Genetics, 13*, 823309. DOI: 10.3389/fgene.2022.823309 PMID: 35386282

Caswell, C., & Jang, N. (2024). Mainstreaming gender equality and social inclusion in nature-based solutions for climate change adaptation. http://icsfarchives.net/id/eprint/21582

Černevičienė, J., & Kabašinskas, A. (2024). Explainable artificial intelligence (XAI) in finance: A systematic literature review. *Artificial Intelligence Review, 57*(8), 216. Advance online publication. DOI: 10.1007/s10462-024-10854-8

Chakrabarti, D. K. (2001). *India: An Archaeological History*. Oxford University Press.

Chambers, R. (1997). *Whose reality counts? Putting the first last*. Intermediate Technology Publications. DOI: 10.3362/9781780440453

Chanda, A., & Alfred, K. (2025). Ethical Integration of Indigenous and Scientific Knowledge Systems for a Sustainable Mitigation of Environmental and Climate Change Challenges. *International Journal of Global Community, VIII*(1), 1. DOI: 10.33473/ijgc-ri.v8i1.222

Chawane, M. (2016). The development of Afrocentricity: A historical survey. *Yesterday and Today*, (16), 78–99. DOI: 10.17159/2223-0386/2016/n16a5

Chennels, R. (2013). "Traditional knowledge and benefit-sharing after the Nakoya Protocol: Three cases from South Africa''. *Law, Environment, and Development Journal:* 163. https://www.lead-journal.org/content/13163 Date of access: 03 Feb.2026.

Chennels, R., & Schoeder, D. 2018. The San Code of Research Ethics: Its Origins and History. TRUST Equitable Research Partnerships. South Africa: The South African San Institute.

Chhachhar, V., & Verma, K. (2024). Climate Change and Negotiating Challenges: A Lesson for Success and Failure. In Rai, P. K., & Rai, S. (Eds.), *Climate, Environment and Agricultural Development* (pp. 1–20). Springer., DOI: 10.1007/978-981-97-8363-2_1

Chhatre, A., & Agrawal, A. (2009, October 20). Trade-offs and synergies between carbon storage and livelihood benefits from forest commons. In E. Ostrom (Ed.), *Proceedings of the National Academy of Sciences of United States of America.* 106, pp. 17667–17670. Bloomington: Sustainability Science. DOI: 10.1073/pnas.0905308106

Chigevenga, R. (2022). Decolonising research methodologies in the Global South: Experiences of an African social scientist. *African Journal of Social Work, 12*(4), 199–206.

Chilisa, B. (2019). *Indigenous research methodologies* (2nd ed.). SAGE Publications., DOI: 10.4135/9781071802686

Chirima, R. (2025). Harnessing Digital Technologies to Preserve and Promote Indigenous Knowledge Systems in Africa. *International Journal of Innovative Science and Research Technology, 10*(5), 4411–4416. DOI: 10.38124/ijisrt/25may1906

Chisa, K., & Hoskins, R. (2016). African customary law and the protection of indigenous cultural heritage: Challenges and issues in the digitization of indigenous knowledge in South Africa. *Indilinga, 15*(1), 1–15. https://hdl.handle.net/10520/EJC192833

Chiwawa, N., & Wissink, H. (2023). Sustainable tourism for local and regional development in South Africa: Unlocking economic potential through responsible tourism strategies. *African Journal of Hospitality, Tourism and Leisure, 12*, 1163. DOI: 10.46222/ajhtl.19770720.423

Chowdhooree, I. (2019). Indigenous knowledge for enhancing community resilience: An experience from the south-western coastal region of Bangladesh. *International Journal of Disaster Risk Reduction, 40*, 101259. Advance online publication. DOI: 10.1016/j.ijdrr.2019.101259

Chowdhury, S. J., Wahab, H. A., Saad, M. M., Roy, P. K., Hamidi, M., & Ahmad, M. M. (2021). Ubuntu Philosophy: 'I Am Because We Are' – A Road to 'Individualism' to Global Solidarity. In I. Global, *Handbook of Research on the Impact of COVID-19 on Marginalized Populations and Support for the Future* (pp. 361-381). IGI Global Scientific Publishing.

Christen, K. (2011). Opening archives: Respectful repatriation. *The American Archivist, 74*(1), 185–210. DOI: 10.17723/aarc.74.1.4233nv6nv6428521

Christen, K. (2012). Does information really want to be free? Indigenous knowledge systems and the question of openness. *International Journal of Communication, 6*, 2870–2893.

Christen, K., & Anderson, J. (2019). Toward slow archives. *Archival Science, 19*(2), 87–116. DOI: 10.1007/s10502-019-09307-x

Christen, K., Merrill, A., & Wynne, E. (2017). A community of relations: Mukurtu hubs and spokes. *D-Lib Magazine : The Magazine of the Digital Library Forum, 23*(5/6). Advance online publication. DOI: 10.1045/may2017-christen

Christensen, M., & Parker, L. (2010). USING IDEAS TO ADVANCE PROFESSIONS: PUBLIC SECTOR ACCRUAL ACCOUNTING. *Financial Accountability & Management, 26*(3), 246–266. DOI: 10.1111/j.1468-0408.2010.00501.x

Cockburn, J. J., Khumalo-Seegelken, B., & Villet, M. H. (2014). IziNambuzani: IsiZulu names for Insects. *South African Journal of Science*, 2014, 110 (9/10), 2013-0292, 1-13. https://dx.doi.org/DOI: 10.1590/sajs.2014/20130292

Comaroff, J., & Comaroff, J. (2009). *Ethnicity,* Inc. University of Chicago Press.

Convention on Biological Diversity. (2011). *Nagoya Protocol on access to genetic resources and the fair and equitable sharing of benefits arising from their utilization.* CBD Secretariat.

Costantini, S., Mostarda, L., Tocchio, A., & Tsintza, P. (2008). DALICA: Agent-based ambient intelligence for cultural-heritage scenarios. *IEEE Intelligent Systems, 23*(2), 34–41. DOI: 10.1109/MIS.2008.24

Couldry, N., & Mejias, U. A. (2019). Data colonialism: Rethinking big data's relation to the contemporary subject. *Television & New Media, 20*(4), 336–349. DOI: 10.1177/1527476418796632

Couldry, N., & Mejias, U. A. (2019). *The costs of connection: How data is colonizing human life and appropriating it for capitalism*. Stanford University Press.

Cox, G. R., FireMoon, P., Anastario, M. P., Ricker, A., Escarcega-Growing Thunder, R., Baldwin, J. A., & Rink, E. (2021). Indigenous standpoint theory as a theoretical framework for decolonizing social science health research with American Indian communities. *Alternative*, *17*(4), 460–468. DOI: 10.1177/11771801211042019 PMID: 38680293

Creswell, J. W., & Clark, V. P. (2007). *Mixed methods research*.

Creswell, J. W., & Plano Clark, V. L. (2017). *Designing and Conducting Mixed Methods Research*. Sage publications.

CSIR. (2025). *Traditional Knowledge Digital Library Unit (TKDL)*. Retrieved from Council of Scientific and Industrial Research, Ministry of Science & Technology, Govt. of India: https://www.csir.res.in/en/documents/tkdl

Cummings, S., Munthali, N., & Sittoni, T. (2024). Epistemic justice as a "New Normal?" Integrating the contributions of Communities of Practice to decolonization of knowledge. *Sustainable Development (Bradford)*, *33*(3), 3228–3245. DOI: 10.1002/sd.3285

Dahlberg, A. C., & Trygger, S. B. (2009). Indigenous medicine and primary health care: The importance of lay knowledge and use of medicinal plants in rural South Africa. *Human Ecology: an Interdisciplinary Journal*, *37*(1), 79–94. DOI: 10.1007/s10745-009-9217-6

David-Chavez, D. M., & Gavin, M. C. (2018). A global assessment of Indigenous community engagement in climate research. *Environmental Research Letters*, *13*(12), 1–17. DOI: 10.1088/1748-9326/aaf300

David, J. O. (2025). Beyond the Extractive Paradigm: Appraising Indigenous Knowledge in South African Environmental Governance. *Environment and Ecology Research*, *13*(4), 459–471. DOI: 10.13189/eer.2025.130401

Davies E, (1987). A contrastive approach to the analysis of politeness formulas. *Applied Linguistics.8, 76:88.*

Davis, D. R.Jr. (2007). Hinduism as a Legal Tradition. *Journal of the American Academy of Religion*, *75*(2), 241–267. DOI: 10.1093/jaarel/lfm004

Davis, F. D. (1989). Perceived usefulness, perceived ease of use, and user acceptance of information technology. *Management Information Systems Quarterly*, *13*(3), 319–340. DOI: 10.2307/249008

Davis, J. L., & Wagner, J. R. (2003). Who knows? On the importance of identifying "experts" when researching local ecological knowledge. *Human Ecology: an Interdisciplinary Journal*, *31*(3), 463–489. DOI: 10.1023/A:1025075923297

Dawson, P. C. (2019). Repatriation, virtual repatriation and the Inuvialuit living history project. In Fforde, C., McKeown, C. J., & Keeler, H. (Eds.), *The Routledge companion to indigenous repatriation* (pp. 515–534). Routledge.

de Sousa Santos, B. (2015). *Epistemologies of the South: Justice against epistemicide.* Routledge. DOI: 10.4324/9781315634876

Dei, G. J. S. (2000). Rethinking the role of Indigenous knowledges in education. *International Journal of Inclusive Education, 4*(2), 111–132. DOI: 10.1080/136031100284849

Dei, G. J. S., Karanja, W., & Erger, G. (2022). The role of elders and their cultural knowledges in schools. In *Elders' cultural knowledges and the question of Black/African indigeneity in education* (pp. 127–151). Springer International Publishing., DOI: 10.1007/978-3-030-84201-7_6

Denny-Smith, G., Loosemore, M., Sunindijo, R. Y., & Williams, M. (2025). Policy-ledd innovations: the impact of indigenous procurement policies on contactors 'behaviour towards indigenous people's participation in construction projects. In Dulaini, M. F. (Ed.), *Research Companion to Innovation in Construction* (pp. 178–202)., DOI: 10.4337/9781035326594.00017

Denzin, N. K., & Lincoln, Y. S. (2018). *The SAGE handbook of qualitative research.* Sage.

Department of Arts and Culture, (2010). National Policy on the Digitisation of Heritage Resources: Final draft for Public Review.

Department of Arts and Culture, Government of the Republic of South Africa. (2010). "National policy on the digitization of heritage resources. Final draft for public review (V8)". www.mcnulty.co.za/wp-content/uploads/2010/12/NATIONAL _POLICY_ON-DIGITIZATION_V8.pdf

Department of Arts and Culture. (2013). National Archives and Records Service: Digitisation Strategy. Available at: https://www.nationalarchives.gov.za/sites/default/files/u59/National%20Archives%20 Digitisation%20Strategy%20-%20approved%20November%202013.pdf [Accessed: 14 February 2026].

Department of Basic Education. (2021). Curriculum assessment policy statement (CAPS), viewed 17 December 2025 from https://www.education.gov.za/curriculum/CurriculumAssessmentPolicyStatements(CAPS).aspx#:~:text=A%20National%20Curriculum%20and%20Assessment%20Policy%20 Statement%20is20a%20single,Curriculum%20Statement%20Grades%20R%20%2D%2012

Department of Science and Innovation. (2022). New system to promote access to indigenous knowledge across South Africa, viewed 14 December 2025 from www.dst.gov.za/index.php/media-room/latestnews/ 3598-new-system-to-promote-access-to-indigenous-knowledge-access-across-South-Africa

Descola, P. (2013). *Beyond nature and culture.* University of Chicago Press. DOI: 10.7208/chicago/9780226145006.001.0001

Devi, S. (2025). Kautilya's Perspective On Ecological Development. [IJCRT]. *International Journal of Creative Research Thoughts, 13*(2), b827–b839.

Dhillon, M. (2023). Weaving Together the Ancient and the Contemporary: Intersections of the Bhagavad Gita with Modern Psychology. *Pastoral Psychology, 72*(4), 525–537. DOI: 10.1007/s11089-023-01070-2 PMID: 37359496

Dhir, A., Yossatorn, Y., Kaur, P., & Chen, S. (2018). Online social media fatigue and psychological wellbeing—A study of compulsive use, fear of missing out, fatigue, anxiety and depression. *International Journal of Information Management, 40*, 141–152. DOI: 10.1016/j.ijinfomgt.2018.01.012

Diego, F., Esteban, B., & Merello, P. (2015). Design of a hybrid (Wired/Wireless) acquisition data system for monitoring of cultural heritage physical parameters in smart cities. *Sensors (Basel), 15*(4), 7246–7266. DOI: 10.3390/s150407246

DiPrete Brown, L., Atapattu, S., Stull, V. J., Calderón, C. I., Huambachano, M., Houénou, M. J. P., Snider, A., & Monzón, A. (2020). From a three-legged stool to a three-dimensional world: Integrating rights, gender and indigenous knowledge into sustainability practice and law. *Sustainability (Basel), 12*(22), 9521. DOI: 10.3390/su12229521

Diwakar, R., Sharma, R., Nayak, D., & Mohapatra, H. (2025). Optimizing load distribution in big data ecosystems. *AI and the Revival of Big Data*, 177-200. DOI: 10.4018/979-8-3693-8472-5.ch008

Dixon, D. H., Dixon, T., & Jordan, E. (2022). Second language (L2) gains through digital game-based language learning (DGBLL): A meta-analysis.'. *Language Learning & Technology, 26*(1), 1–25. DOI: 10.64152/10125/73464

Dladla, N. (2024). Copyright Law and the Ruse of Culture:'Traditional Cultural Expressions and Expressions of Folklore'as a Conception of Racial Difference. *Phronimon, 25*(1), 1–26. DOI: 10.25159/2413-3086/14898

Dorji, T., Rinchen, K., Morrison-Saunders, A., Blake, D., Banham, V., & Pelden, S. (2024). Understanding how indigenous knowledge contributes to climate change adaptation and resilience: A systematic literature review. *Environmental Management, 74*(6), 1101–1123. DOI: 10.1007/s00267-024-02032-x PMID: 39215837

Downes, D. (2000). *Protecting community rights over traditional knowledge: Implications of customary laws and practices.* Center for International Environmental Law.

DRONAH. (2021). *Challenges faced in Heritage Management in India and Policy Imperatives.* Gurugram: NITI Aayog.

Drouin-Gagné, M. E. (2021). Beyond the "Indigenizing the academy" trend: Learning from Indigenous higher education land-based and intercultural pedagogies to build trans-systemic education. *Engaged Scholar Journal, 7*(1), 45–65. DOI: 10.15402/esj.v7i1.69978

Drucker, P. F. (1987). *Innovation and entrepreneurship.* Harper & Row.

Drucker, P. F. (1987). *Social innovation – management's new dimension.* Long.

Drugge, A. L. (Ed.). (2016). *Ethics in Indigenous research: Past experiences, future challenges.* Centre for Sami Research, Umeå University.

Du Plessis, P. (2021). Decolonisation of education in South Africa: Challenges to decolonize the university curriculum. *South African Journal of Higher Education, 35*(1), 54–69. DOI: 10.20853/35-1-4426

Duarte Ritter, C., Muñoz, J., Machado, A. F., Albert, J. S., Ribas, C. C., & Carnaval, A. C.. (2025). Indigenous territories and protected areas are crucial for ecosystem connectivity in the Amazon basin. *Proceedings of the National Academy of Sciences of the United States of America, 122*(31), e2418189122. DOI: 10.1073/pnas.2418189122 PMID: 40720645

Duchlova, H., & Leventon, J. (2023). *Epistemic justice: How can a diverse forms of knowledge improve policy making?* SSH Centre.

Dudgeon, R. C., & Berkes, F. (2003). Local understandings of the land: Traditional ecological knowledge and indigenous knowledge. In *Nature across cultures: Views of nature and the environment in non-Western cultures* (pp. 75–96). Springer Netherlands. DOI: 10.1007/978-94-017-0149-5_4

Dürr, J., Carvalho Ribeiro, S. M., Vargas, D., Sellare, J., & Börner, J. (2026). Forest-based Amazonian bioeconomies: New opportunities and old challenges. In Zilberman, D., Zhuang, J., Wesseler, J., & Khanna, M. (Eds.), *Handbook of circular bioeconomy* (pp. 583–603). Springer., DOI: 10.1007/978-3-032-07112-5_27

Dyantyi, N. (2025, November). University curriculum reform and employability: addressing the needs of the post-colonial workforce. In *Frontiers in Education* (Vol. 10, p. 1591654). Frontiers Media SA. https://doi.org/DOI: 10.3389/feduc.2025.1591654

Dzvimbo, M. A., Tirivangasi, H. M., Mawonde, A., & Ncube, C. T. (2023). The role of local people and Indigenous Knowledge in participatory development approach in rural Zimbabwe. https://www .researchgate.net Date of access: 30 Dec. 2025.

Ebbe, K. (1993, September). Traditional Knowledge and Sustainable Development. In *Conference held at The World Bank Washington.*

Editorial, L. (n.d.). *Understanding the Ancient Egypt Legal System: Structure and Function.* Retrieved July 12, 2025, from Laws Pulse: https://lawspulse.com/ancient-egypt-legal-system/

Eke, C., & Adeyemi, M. O. (2024). Utilising digital media tools to foster indigenous communication systems in Nigeria. *African Journal of Social and Behavioural Sciences, 14*(6).

Ekene, A. K., Lee, K. S., Decide, M. T., & Thobejane, T. D. (2023). Placing Indigenous Knowledge at the Core of Production and Marketing of Medicinal Plant (Mpesu) in Zimbabwe and South Africa. https://doi.org/DOI: 10.1108/978-1-80262-251-520231017

Ekobi, G. A., Mboh, L., & Tanga, P. (2023). Contribution of indigenous knowledge to agricultural growth in South Africa: A case of Disaneng community in the Ratlou Local Municipality. *African Journal of Inter/Multidisciplinary Studies, 5*(1), 1-12. https://hdl.handle.net/10520/ejc-ajims_v5_n1_a27

Ensor, J. E., Wennström, P., Bhatterai, A., Nightingale, A. J., Eriksen, S., & Sillmann, J. (2019, April). Asking the right questions in adaptation research and practice: Seeing beyond climate impacts in rural Nepal. *Environmental Science & Policy, 94*, 227–236. DOI: 10.1016/j.envsci.2019.01.013

Ermine, W., Sinclair, R., & Jeffery, B. 2004. The ethics of research involving Indigenous peoples. Saskatoon, SK: Indigenous Peoples' Health Research Centre. *Report of the Indigenous Peoples Health Research Centre to the Interagency Advisory Panel on Research Ethics,* pp. 1-272.

Escobar, A. (2018). 3. Transition Discourses and the Politics of Relationality: Toward Designs for the Pluriverse. In *Constructing the pluriverse* (pp. 63-89). Duke University Press. https://doi.org/DOI: 10.1515/9781478002017-005

Escobar, A. (2012). *Encountering development: The making and unmaking of the Third World* (2nd ed.). Princeton University Press. DOI: 10.1515/9781400839926

Eze, U. T. (2013). Integrating African indigenous knowledge In Nigeria's formal education system: It's potential for sustainable development. *Journal of Education and Practice*, *4*(6).

Fahmy, H., & URHC . (2013). *Egyptian legislations in relation to the rehabilitation of Historic Cairo*. Cairo: UNESCO, World Heritage Centre.

Fairer-Wessels, F. A. (2017). Determining the role of information in sustainable tourism development: A case study of communities surrounding the iSimangaliso Wetland Park. *African Journal of Hospitality, Tourism and Leisure*, *6*(4), 1–15.

Fernández-Navajas, Á., Merello, P., Beltrán, P., & García-Diego, F. (2013). Software for storage and management of Microclimatic data for preventive conservation of cultural heritage. *Sensors (Basel)*, *13*(3), 2700–2718. DOI: 10.3390/s130302700

Figueredo, M., Cacho, N., Thome, A., Cacho, A., Lopes, F., & Araujo, M. (2017). Using social media photos to identify tourism preferences in smart tourism destination. *2017 IEEE International Conference on Big Data (Big Data)*. DOI: 10.1109/BigData.2017.8258423

Finkelman, P., & Garrison, T. A. (2008). Archaeological Resources Protection Act (ARPA; 1979). In Finkelman, P., & Garrison, T. A. (Eds.), *Encyclopedia of United States Indian Policy and Law*. SAGE., DOI: 10.4135/9781604265767.n40

Finnegan, R. (2012). *Oral literature in Africa*. Open Book Publishers. DOI: 10.11647/OBP.0025

Flores, B. M., Montoya, E., Sakschewski, B., Nascimento, N., Staal, A., Betts, R. A., Levis, C., Lapola, D. M., Esquível-Muelbert, A., Jakovac, C., Nobre, C. A., Oliveira, R. S., Borma, L. S., Nian, D., Boers, N., Hecht, S. B., ter Steege, H., Arieira, J., Lucas, I. L., & Hirota, M. (2024). Critical transitions in the Amazon forest system. *Nature*, *626*(7999), 555–564. DOI: 10.1038/s41586-023-06970-0 PMID: 38356065

Floridi, L., Cowls, J., Beltrametti, M., Chatila, R., Chazerand, P., Dignum, V., Luetge, C., Madelin, R., Pagallo, U., Rossi, F., Schafer, B., Valcke, P., & Vayena, E. (2018). AI4People—An ethical framework for a good AI society: Opportunities, risks, principles, and recommendations. *Minds and Machines*, *28*(4), 689–707. DOI: 10.1007/s11023-018-9482-5 PMID: 30930541

Floridi, L., & Taddeo, M. (2016). What is data ethics? *Philosophical Transactions. Series A, Mathematical, Physical, and Engineering Sciences*, *374*(2083), 20160360. DOI: 10.1098/rsta.2016.0360 PMID: 28336805

Folke, C., Biggs, R., Norström, A. V., Reyers, B., & Rockström, J. (2016). Social-ecological resilience and biosphere-based sustainability science. *Ecology and Society*, *21*(3), art41. Advance online publication. DOI: 10.5751/ES-08748-210341

Food and Agriculture Organization (FAO)., Alliance of Biodiversity., & CIAT. (2022). Labelling and certification schemes for Indigenous peoples' foods- generating income while protecting and promoting Indigenous peoples' values. Rome. https://doi.org/DOI: 10.4060/cc0155en

Food and Agriculture Organization. (2019). *Indigenous peoples' food systems: Insights on sustainability and resilience*. FAO.

Foster, J. B. (2010). *The ecological rift: Capitalism's war on the Earth.* Monthly Review Press.

Foster, L. A. (2016). Decolonizing patent law: Postcolonial technoscience and indigenous knowledge in South Africa. *Feminist Formations, 28*(3), 148–173. DOI: 10.1353/ff.2016.0047

Foyer, J., Mauri, M. M., Filoche, G., & Castillo, G. (2024). The institutionalization of biculturalism through community protocols: The case of Guna yala. *International Indigenous Policy Journal, 15*(1). Advance online publication. DOI: 10.18584/iipj.2024.15.1.15872

Fricker, M. (2007). *Epistemic injustice: Power and the ethics of knowing*. Oxford university press. DOI: 10.1093/acprof:oso/9780198237907.001.0001

Galla, C. (2016). Indigenous language revitalization, promotion, and education: Function of digital technology. *Computer Assisted Language Learning, 29*(7), 1137–1151. DOI: 10.1080/09588221.2016.1166137

García-Diego, F., Fernández-Navajas, Á., Beltrán, P., & Merello, P. (2013). Study of the effect of the strategy of heating on the mudejar church of Santa Maria in Ateca (Spain) for preventive conservation of the altarpiece surroundings. *Sensors (Basel), 13*(9), 11407–11423. DOI: 10.3390/s130911407

Garrett, R., Ferreira, J., Abramovay, R., Brandão, J., Brondizio, E., Euler, A., Pinedo, D., Porro, R., Cabrera Rocha, E., Sampaio, O., Schmink, M., Torres, B., & Varese, M. (2024). Transformative changes are needed to support socio-bioeconomies for people and ecosystems in the Amazon. *Nature Ecology & Evolution, 8*(10), 1815–1825. DOI: 10.1038/s41559-024-02467-9 PMID: 39107371

Garutsa, T. C., & Nekhwevha, F. H. (2018). Indigenous knowledge, food production and food security in rural Khambashe in the Eastern Cape, South Africa. *South African Review of Sociology, 49*(3-4), 16–31. DOI: 10.1080/21528586.2018.1564692

Gasca-Alvares, H. J., & Costa-Neto, E. M. (2022). Insects as a food source for indigenous communities in Columbia: A review and research perspective. *Journal of Insects as Food and Feed, 8*(6), 593–603. DOI: 10.3920/JIFF2021.0148

Gefen, D., Karahanna, E., & Straub, D. W. (2003). Trust and tam in online shopping: An integrated model1. *Management Information Systems Quarterly, 27*(1), 51–90. DOI: 10.2307/30036519

Geissdoerfer, M., Savaget, P., Bocken, N. M., & Hultink, E. J. (2017, February 1). The Circular Economy – A new sustainability paradigm? *Journal of Cleaner Production, 143*, 757–768. DOI: 10.1016/j.jclepro.2016.12.048

Ghimire, P. (2021). Digitalization of indigenous knowledge in Nepal–Review article. *Acta Informatica Malaysia, 5*(2), 42–47. DOI: 10.26480/aim.02.2021.42.47

Giampiccoli, A., & Glassom, D. (2021). Community-based tourism in protected areas: Elaborating a model from a South African perspective. *African Journal of Hospitality, Tourism and Leisure, 10*(3), 873–889. DOI: 10.46222/ajhtl.19770720-143

Gibson, J., & Martin, D. (2019). Re-situating Indigenous knowledge in the digital age. *The Journal of Documentation, 75*(4), 902–918. DOI: 10.1108/JD-07-2018-0108

Giunta, J. (2024, November 14). *The Concept of Ma'at: Balance, Truth, and Its Role in Egyptian Society*. Retrieved July 12, 2025, from ancient pedia: https://ancientpedia.com/the-concept-of-maat-balance-truth-and-its-role-in-egyptian-society/

Godfrey, B., & Siraje, K. (2019). Community development in the rural areas through traditional indigenous knowledge. International Journal of Research in Sociology and Anthropology, 5(2): 8-17. http://dx.org./10.20431/2454/-8677.0502002 Date of access: 01 Feb. 2026.

González, E. M., Municio, E., Alemán, M. N., & Marquez-Barja, J. M. (2020). Cultural heritage and Internet of things. *Proceedings of the 6th EAI International Conference on Smart Objects and Technologies for Social Good*, 248-251. DOI: 10.1145/3411170.3411267

Government of India. (2025). *Ministry of Minority Affairs*. Retrieved from Ministry of Minority Affairs: https://minorityaffairs.gov.in/show_content.php?lang=1&level=2&ls_id=104&lid=48

Gowri, R., Dey, P., & Mujumdar, P. P. (2021). A hydro-climatological outlook on the long-term availability of water resources in Cauvery river basin. *Water Security, 14*, 100102. Advance online publication. DOI: 10.1016/j.wasec.2021.100102

Green, L. J. (2012). Beyond South Africa's' indigenous knowledge-science'wars. *South African Journal of Science, 108*(7), 1–10. DOI: 10.4102/sajs.v108i7/8.631

Grey, S., & Kuokkanen, R. (2020). Indigenous governance of cultural heritage: Searching for alternatives to co-management. *International Journal of Heritage Studies, 26*(10), 919–941. DOI: 10.1080/13527258.2019.1703202

Grisel, F. (2023). *Arbitration as a dispute resolution process: historical developments*. Cambridge Compendium on International Commercial and Investment Arbitration.

Guba, E. G., & Lincoln, Y. S. (1985). Naturalistic inquiry. *Sage (Atlanta, Ga.)*.

Gudeman, S. (2001). *The anthropology of economy: Community, market, and culture*. Blackwell.

Gudynas, E. (2011). Buen vivir: Today's tomorrow. *Development, 54*(4), 441–447. DOI: 10.1057/dev.2011.86

Guha, R. (2000). The unquiet woods: Ecological change and peasant resistance in the Himalaya. Banglore: Oxford University Press.

Gupta, A. D. (2013). Does indigenous knowledge have anything to deal with sustainable development. *Cultural Anthropology, 57*, 64.

Haines, J. (2022). Researching the knowledge journey practices of Indigenous elders relevant to the younger generation: A community-based participatory study. *Proceedings of the Association for Information Science and Technology, 59*(1), 426–430. DOI: 10.1002/pra2.646

Hardilla, D., & Nugroho, A. C. (2018). The role of Internet of things to support cultural heritage inventory in urban resiliency approach: Tradisional house in bandar Lampung case. *2018 International Conference on Information Technology Systems and Innovation (ICITSI)*, 193-198. DOI: 10.1109/ICITSI.2018.8695935

Harrison, R., DeBruijn, L., Morely, I., Beltran, M., & Smith, C. (2020). Heritage and data justice. *Heritage & Society*, *13*(1-2), 78–102. DOI: 10.1080/2159032X.2020.1847890

Hay, D. (2008). African craft production and markets. *Development Southern Africa*, *25*(3), 343–359.

Heaton, J. (2008). Secondary analysis of qualitative data: An overview. *Historical Social Research (Köln)*, *33*(3), 33–45.

Heleta, S. (2018). Decolonizing knowledge in South Africa: Dismantling the 'pedagogy of big lies'. *Ufahamu. Journal of African Studies*, *40*(2). Advance online publication. DOI: 10.5070/F7402040942

Heleta, S., & Chasi, S. (2024). Curriculum decolonization and internationalization: A critical perspective from South Africa. *Journal of International Students*, *14*(2), 75–90. DOI: 10.32674/jis.v14i2.6383

Herrera, M. I. H. (2022). The Use of Technologies in Language Revitalisation Projects. *Journal of Global Indigeneity*, *6*(1), 1–17.

Hettiarachchi, H., & Mhaddolkar, N. (2022). Mysuru: The story of a clean city in India. In Hettiarachchi, H., & Mhaddolkar, N. (Eds.), *SDG11, Sustainable Cities and Communities* (pp. 37–57). Routledge India. DOI: 10.4324/9781003205975-3

Heuheu, T. t., Kawharu, M., & Tuheiava, R. A. (2012). World heritage and indigeneity. *World heritage review*, *62*, pp. 6-16. Retrieved from https://unesdoc.unesco.org/ark:/48223/pf0000215774

Hill, R., Adem, C., Alungui, W. V., Molnar, Z., Aumeeruddy-Thomas, Y., & Bridgewater, P.. (2020). 'Working with indigenous, local and scientific knowledge is assessments of nature and nature's linkages with people'. *Current Opinion in Environmental Sustainability*, *43*, 8–20. DOI: 10.1016/j.cosust.2019.12.006

Hodge, D. R. (2004). Working with Hindu clients in a spiritually sensitive manner. *Social Work*, *49*(1), 27–38. DOI: 10.1093/sw/49.1.27 PMID: 14964516

Holotescu, M. C., & Cre u, V. I. (2015). *Emerging technologies in education; Conceiving and building a microblogging platform for formal and informal learning*. Editura Politehnica.

Holton, G., Leonard, W. Y., & Pulsifer, P. L. (2012). *The Arctic Borderlands Ecological Knowledge Co-op*. NSERC.

Horna-Saldaña, C. J., Perez Perez, J. E., & Toro Galeano, M. L. (2025). Artificial intelligence in the preservation of native languages and bridging the information access gap for Indigenous peoples. *Journal of Enabling Technologies*, *19*(1), 63–75. DOI: 10.1108/JET-09-2024-0063

Howard, P. (Ed.). (2003). *Women and plants: Gender relations in biodiversity management and conservation*. Zed Books.

Huambachano, M., Arulingam, I., Bowness, E., Korzenszky, A., Mungai, C., Termine, P., & Wittman, H. (2022). *Knowledge networks to support youth engagement in sustainable food systems. Frontiers in Sustainable Food Systems, 6*, 867344. DOI: 10.3389/fsufs.2022.867344

Hughes, M., & Dallwitz, J. (2009). Ara Irititja: Towards Culturally Appropriate IT Best Practice in Remote Indigenous Australia. In Erickson, J. (Ed.), *Database Technologies: Concepts, Methodologies, Tools, and Applications* (pp. 1462–1471). IGI Global Scientific Publishing., DOI: 10.4018/978-1-60566-058-5.ch088

Hunter, J. (2005). The role of information technologies in indigenous knowledge management. *Australian Academic and Research Libraries, 36*(2), 109–124. DOI: 10.1080/00048623.2005.10721252

Ige, A. B., Kupa, E., & Ilori, O. (2024). Aligning sustainable development goals with cybersecurity strategies: Ensuring a secure and sustainable future. *GSC Advanced Research Reviews, 19*(3), 344–360. DOI: 10.30574/gscarr.2024.19.3.0236

Ijatuyi, E. J., Ayanlade, A., & Olanrewaju, O. (2025). Sustainable tourism and green space: Exploring how green spaces and natural attractions contribute to local tourism economies and revenue generation in Gauteng Province, South Africa. *Sustainability, 17*(6), 2878. DOI: 10.3390/su17062878

Ijatuyi, E. J., Lamm, A., Yessoufou, K., Suinyuy, T., & Patrick, H. O. (2025). Integration of indigenous knowledge with scientific knowledge: A systematic review. *Environmental Science & Policy, 170*, 104119. DOI: 10.1016/j.envsci.2025.104119

Ikuenobe, P. (2018). Oral tradition, epistemic dependence, and knowledge in African cultures. *Synthesis Philosophica, 65*(1), 23–40. DOI: 10.21464/sp33102

Impey, A. (2002). Culture, conservation and community reconstruction: Explorations in advocacy ethnomusicology and ecotourism in the Dukuduku Forest, South Africa. *Yearbook for Traditional Music, 34*, 9–25. DOI: 10.2307/3649187

Indwar, T., & Muthukumar, P. K. (2023). Role of ecotourism in sustainable development: An overview. *Journal of Pharmaceutical Negative Results, 14*(2).

International Cooperative Alliance. (2016). *Cooperative identity, values and principles*. ICA.

International Labour Organization. (1989). *Indigenous and Tribal Peoples Convention, 1989 (No. 169)*. ILO.

International Labour Organization. (2013). Handbook of the ILO tripartite constituents: Understanding the indigenous and tribal peoples convention, 1998 (Number 169), viewed 15 December 2025 viewed from https://www.ilo.org/wcmsp5/groups/public/---ed_norm/norms/document/publications/wcms/_205225.pdf

IPBES. (2019). *Global assessment report on biodiversity and ecosystem services*. Intergovernmental Science-Policy Platform on Biodiversity and Ecosystem Services.

IPBES. W. (2019). Intergovernmental science-policy platform on biodiversity and ecosystem services. *Summary for policy makers of the global assessment report on biodiversity and ecosystem services of the intergovernmental science-policy platform on biodiversity and ecosystem services. IPBES Secretariat, Bonn, Germany.*

Isser, S. S., Raj, N., Tomar, M., Marwaha, S. S., & Shastri, S. (2024). Value-based education in NEP 2020: Fostering ethical and moral growth through Dharma. *Asian Education and Development Studies, 13*(5), 579–597. DOI: 10.1108/AEDS-06-2024-0121

IWGIA. (2025, April 25). *Indigenous peoples in India.* Retrieved from IWGIA: https://iwgia.org/en/india.html

Jarrow, R. A. (2018). *Derivatives & Risk Management.* World Scientific Publishing Company.

Jefferson, D. J. (2025). The World Intellectual Property Organization Treaty on genetic resources and traditional knowledge: Implications for plant science. *Plants, People, Planet, 7*(4), 901–905. DOI: 10.1002/ppp3.10615

Jeste, D. V., & Vahia, I. V. (2008). Comparison of the Conceptualization of Wisdom in Ancient Indian Literature with Modern Views: Focus on the Bhagavad Gita. *Psychiatry, 71*(3), 197–209. DOI: 10.1521/psyc.2008.71.3.197 PMID: 18834271

Jimenez, A., Delgado, D., Merino, R., & Argumedo, A. (2022). A decolonial approach to innovation? Building paths towards Buen Vivir. *The Journal of Development Studies, 58*(9), 1633–1650. DOI: 10.1080/00220388.2022.2043281

Jobin, A., Ienca, M., & Vayena, E. (2019). The global landscape of AI ethics guidelines. *Nature Machine Intelligence, 1*(9), 389–399. DOI: 10.1038/s42256-019-0088-2

Jogesh, A., Pal, U., Bugler, W., Wade, G., & Messling, L. (2019, October). *Case study: For the community, by the community: Building resilience through infrastructure in rural India.* Retrieved from ResearchGate: DOI:DOI: 10.13140/RG.2.2.33606.80965

Johnson, D. W., & Johnson, R. T. (2014). *Joining together: Group theory and group skills.* Pearson.

Johnson, J. T., Louis, R. P., & Pramono, A. H. (2017). Indigenous environmental knowledge and participatory GIS. *Geographical Review, 107*(3), 382–397. DOI: 10.1111/gere.12215

Jones, S. (2006). Community-based ecotourism: The significance of social capital. *Annals of Tourism Research, 33*(2), 303–324. DOI: 10.1016/j.annals.2004.06.007

Juma, F. A. (2022). Recapturing the oral tradition of storytelling in spiritual conversations with older adults: An Afro-Indigenous approach. *Religions, 13*(6), 563. DOI: 10.3390/rel13060563

K, V., & Kalluraya, D. (2024). HISTORY OF THE INDIAN KNOWLEDGE SYSTEM. *EPRA International Journal of Multidisciplinary Research (IJMR)*, 352–355. .DOI: 10.36713/epra15354

Kabeer, N. (1999). Resources, agency, achievements: Reflections on the measurement of women's empowerment. *Development and Change, 30*(3), 435–464. DOI: 10.1111/1467-7660.00125

Kalafatic, C. (2007). Indigenous peoples' sustainable livelihoods. *Food and Agriculture Organization of the United Nations Thematic brief.* ftp://ftp. fao. org/docrep/fao/010/aj033e/aj033e02. pdf

Kamassah, S. S. (2018). *The Impact of Post-Secondary Educational Institution Policies and Practices on Indigenous Staff Recruitment and Retention.* University of Toronto.

Kane, P. V. (1968). *History of dharmaśāstra (ancient and mediæval religious and civil law in India).* Bhandarkar Oriental Research Institute.

Kapuire, G. K. (2013). An Explorative Action Research Study toward the Design of a Digital Knowledge Organisation as Part of an Indigenous Knowledge Management System with a Herero Community. MA thesis, Dept. of Computer Science, University of Cape Town.

Kasih, D. P. D., Dharmawan, N. K. S., Putra, I. B. W., Sudiarawan, K. A., & Rakhima, A. S. (2021). The Exploitation of Indigenous Communities by Commercial Actors. *Journal of Ethnic and Cultural Studies*, *8*(4), 91–108. DOI: 10.29333/ejecs/898

Kasirzadeh, A., & Gabriel, I. (2023). In Conversation with Artificial Intelligence: Aligning language Models with Human Values. *Philosophy & Technology*, *36*(2), 27. Advance online publication. DOI: 10.1007/s13347-023-00606-x

Kater, I. (2022). Natural and Indigenous sciences: Reflections on an attempt to collaborate. *Regional Environmental Change*, *22*(4), 1–15. DOI: 10.1007/s10113-022-01967-3

Katuu, S. (2016). Managing digital heritage in times of technological change and innovation. In *Managing Digital Cultural Objects: Analysis, Discovery and Retrieval* (pp. 27-48). Facet Publishing. https://doi.org/DOI: 10.29085/9781783301270.003

Keane, M. (2006). "Understanding science curriculum and research in research in rural Kwa-Zulu Natal." Accessed https://wiredspace.wits.ac.za/bitstream/handle/10539/1508/10Part%202%20Ch10%20Indigenous %20Knowledge.pdf?sequesnce=16

Keovorabouth, S. T. (2025). 'Coming In'as Indigenous, Queer, and Urban–Utilising Indigenous Queer Standpoint Theory as Methodology. *JGI, 9*(3).

Keshavan, M. S., Hegde, S., & Bhargav, H. (2024). Doing good well (Karma Yoga, the path of selfless action): Psychotherapeutic lessons from the East. *Asian Journal of Psychiatry*, *101*, 104201. DOI: 10.1016/j.ajp.2024.104201 PMID: 39241652

Khalala, G. (2021). Protecting and benefiting from indigenous knowledge: The national indigenous knowledge management system. *CSIR Science Scope*, *17*(1), 56–57.

Khan, F. A. (2011). The immortal life of Henrietta Lacks. *The Journal of IMA / Islamic Medical Association of North America*, *43*(2), 93. DOI: 10.5915/43-2-8609

Khoa, B. T., Hung, B. P., & Hejsalem-Brahmi, M. (2023). Qualitative research in social sciences: Data collection, data analysis and report writing. *International Journal of Public Sector Performance Management*, *12*(1–2), 187–209. DOI: 10.1504/IJPSPM.2023.132247

Kiage, O. E. (2019). The Ogiek peoples' Indigenous Knowledge: A pathway towards sustainable natural resource management in the Mau Forest, Kenya. *African Journal of Hospitality, Tourism, and Leisure*,8(1):1-19. https://www.ajhtl.com Date of access: 10 Jan. 2026.

Kimmerer, R. W. (2013). *Braiding Sweetgrass: Indigenous wisdom, scientific knowledge and the teachings of plants*. Milkweed Editions.

Kisambira, A., Khadijah, B., & Ahmed, M. M. (2024). The role of university community engagement programs in influencing higher education outcomes and community development: An insight from Uganda. *Extensive Reviews*, *4*(1), 15–28. DOI: 10.21467/exr.4.1.8028

Kitchenham, A. (2013). The preservation of Canadian Indigenous language and culture through educational technology. *Alternative*, *9*(4), 351–364. DOI: 10.1177/117718011300900406

Kober, R., Ng, J., & Lee, J. (2010). MIND YOUR ACCRUALS: PERCEIVED USEFULNESS OF FINANCIAL INFORMATION IN THE AUSTRALIAN PUBLIC SECTOR UNDER DIFFERENT ACCOUNTING SYSTEMS. *Financial Accountability & Management*, *26*(3), 267–298. DOI: 10.1111/j.1468-0408.2010.00502.x

Kohn, E. (2013). *How forests think: Toward an anthropology beyond the human*. University of California Press.

Kom, Z., Nicolau, M. D., & Nenwiini, S. C. (2024). The use of indigenous knowledge systems practices to enhance food security in Vhembe District, South Africa. *Agricultural Research*, *13*(3), 599–612. DOI: 10.1007/s40003-024-00716-8

Konev, A., Khaydarova, R., Lapaev, M., Feng, L., Hu, L., Chen, M., & Bondarenko, I. (2019). CHPC: A complex semantic-based secured approach to heritage preservation and secure IoT-based Museum processes. *Computer Communications*, *148*, 240–249. DOI: 10.1016/j.comcom.2019.08.001

Kopenawa, D., & Albert, B. (2013). *The falling sky: Words of a Yanomami shaman*. Harvard University Press.

Kothari, A., Salleh, A., Escobar, A., Demaria, F., & Acosta, A. (Eds.). (2019). *Pluriverse: A post-development dictionary*. Tulika Books.

Koul, D. N., Singh, S., Neelam, G., & Shukla, G. (2012, April). Traditional water management systems - An overview of Ahar-pyne system in South Bihar plains of India and need for its revival. *Indian Journal of Traditional Knowledge*, *11*(2), 266–272. https://www.researchgate.net/publication/263168629_Traditional_water_management_systems_-_An_overview_of_Ahar-pyne_system_in_South_Bihar_plains_of_India_and_need_for_its_revival

Kovach, M. (2021). *Indigenous methodologies: Characteristics, conversations, and contexts*. University of Toronto press.

Krueger, R. A., & Casey, M. A. (2015). Focus groups: A practical guide for applied research. *Sage (Atlanta, Ga.)*.

Kudumo, P., & Ngcoza, K. M. (2023). Prospects of using indigenous knowledge as culturally responsive pedagogy to teach science concepts regarding iron smelting. *Indilinga African Journal of Indigenous Knowledge System*, *22*(2), 204–221.

Kugara, S. L., & Mokgoatšana, S. (2022). Challenges presented by digitisation of VhaVenda oral tradition: An African Indigenous knowledge systems perspective. *HTS Teologiese Studies*, *78*(3), 7428. DOI: 10.4102/hts.v78i3.7428

Kukutai, T. (2023). Indigenous data sovereignty: A new take on an old theme. *Science, 382*(6674), eadl4664. Advance online publication. DOI: 10.1126/science.adl4664 PMID: 38033063

Kukutai, T., & Taylor, J. (Eds.). (2016). *Indigenous data sovereignty: Toward an agenda* (Vol. 38). ANU press. DOI: 10.22459/CAEPR38.11.2016

Kulundu, I., McGarry, D. K., & Lotz-Sisitka, H. (2020). Think Piece: Learning, Living and Leading into Transgression–A reflection on decolonial praxis in a neoliberal world. *Southern African Journal of Environmental Education, 36*. Advance online publication. DOI: 10.4314/sajee.v36i1.14

Kumar, A., Bag, A., Anand, A., Saha, S., Mohapatra, H., & Kolhar, M. (2024). Examining healthcare services utilizing cloud technology in intelligent urban environments. *Advances in Healthcare Information Systems and Administration*, •••, 77–98. DOI: 10.4018/979-8-3693-7225-8.ch004

Kuo, T. T., Jiang, X., Tang, H., & Ohno-Machado, L. (2023). Enabling precision medicine via standard communication of HTS provenance, analysis, and results. *PLoS Biology, 21*(3), e3002023. DOI: 10.1371/journal.pbio.3002023 PMID: 36917574

Kupangwa, W., Venter, E., & Farrington, S. M. (2026). Manifestations of ethical values in indigenous African family businesses: An application of the Ubuntu philosophy. [South African Journal of Business Management.]. *South African Journal of Business Management, 57*(1), 1–11. DOI: 10.4102/SAJBM.v57i1.5592

Lachapelle, P. R., & Austin, E. K. (2014). Community Participation. In Michalos, A. C. (Ed.), *Encyclopedia of Quality of Life and Well-Being Research* (pp. 1073–1078). Springer., DOI: 10.1007/978-94-007-0753-5_471

Lahiri, S. (2024). Integrating Biocultural Protocol in tourism and biodiversity policies: An advocacy strategy for the Maasai indigenous communities in East Africa. https://globalforsetcoalition.org Date of access: 02 Feb. 2026.

Laidlaw, J. (1995). *Riches and renunciation: Religion, economy, and society among the Jains*. Oxford University Press., Retrieved from https://archive.org/details/richesrenunciati0000laid DOI: 10.1093/oso/9780198280316.001.0001

Lansbury, M., Mosby, V., Creamer, S. J., Moggridge, B., Evans, G. R., Pecl, G., & Ireland, L. (2025). Indigenous peoples' voices and engagement on climate change: Towards improved health and wellbeing. *Frontiers in Climate, 7*, 1693544. DOI: 10.3389/fclim.2025.1693544

Lasky, S. (2005). A sociocultural approach to understanding teacher identity, agency and professionalism vulnerability in a context of secondary school reform.

Lassen, B., Jansen, L., Rasolojaona, J., Githaiga, C., Fey, L., & Bossou, B. (2018). Community protocols in Africa. https://naturaljustice.org Date of access: 25 June 2026

Lawson, C., Rourke, M., & Humphries, F. (2023). *Access and benefit sharing of genetic resources, information and traditional knowledge*. Routledge.

Lebaka, M. E. K. (2022). Informal learning of Indigenous music and dance through observation and imitation: The case of Bapedi children's musical arts. *European Journal of Language and Literature Studies, 8*(1), 55–67. DOI: 10.26417/658qju31

Leccese, F., Cagnetti, M., Calogero, A., Trinca, D., Pasquale, S., Giarnetti, S., & Cozzella, L. (2014). A new acquisition and imaging system for environmental measurements: An experience on the Italian cultural heritage. *Sensors (Basel), 14*(5), 9290–9312. DOI: 10.3390/s140509290

Legal response International (2019). Legal status and mandate of local communities and indigenous peoples' platform. Viewed from: legalresponce.org/resource/legal-status-and-mandate-of-the-local-communities-and indigenous-peoples-platform

Lehmann, J., Kern, D. C., Glaser, B., & Woods, W. I. (Eds.). (2003). *Amazonian dark earths: Origin, properties, management*. Springer. DOI: 10.1007/1-4020-2597-1

Levis, C., Flores, B. M., Campos-Silva, J. V., Peroni, N., Staal, A., Padgurschi, M. C. G., Dorshow, W., Moraes, B., Schmidt, M., Kuikuro, T. W., Kuikuro, H., Wauja, K., Kuikuro, K., Kuikuro, A., Fausto, C., Franchetto, B., Watling, J., Lima, H., Heckenberger, M., & Clement, C. R. (2024). Contributions of human cultures to biodiversity and ecosystem conservation. *Nature Ecology & Evolution, 8*(5), 866–879. DOI: 10.1038/s41559-024-02356-1 PMID: 38503867

Lewis, J. E., Arista, N., Pechawis, A., & Kite, S. (2023). *Making kin with the machines. Feminist AI*. Oxford University Press.

Li, J., Brar, A., & Roihan, N. (2021). The use of digital technology to enhance language and literacy skills for Indigenous people: A systematic literature review. *Computers and Education Open, 2*, 100035. DOI: 10.1016/j.caeo.2021.100035

Lillie, M., Smith, R., Reed, J., & Inglis, R. (2008). Southwest Scottish crannogs: Using in situ studies to assess preservation in wetland archaeological contexts. *Journal of Archaeological Science, 35*(7), 1886–1900. DOI: 10.1016/j.jas.2007.11.029

Lima, V. P., Ferreira de Lima, R. A., Joner, F., Siddique, I., Raes, N., & ter Steege, H. (2022). Climate change threatens native potential agroforestry plant species in Brazil. *Scientific Reports, 12*(1), 2267. DOI: 10.1038/s41598-022-06234-3 PMID: 35145191

Lincoln, Y. S., & Guba, E. G. (1985). *Naturalistic Inquiry*. Sage publications. DOI: 10.1016/0147-1767(85)90062-8

Lindstrom, J., & Christen, K. (2019). Decolonial archives and Indigenous data sovereignty. *Archival Science, 19*(2), 101–120. DOI: 10.1007/s10502-019-09313-1

Lith, R. 2023. WhatsApp for business opens doors to 21 million South Africans. Independent Online. Available at: https://www.iol.co.za/business-report/opinion/whatsapp-forbusiness-opens-doors-to-21-million-south-africans-19a5339a-ee20-43af-9315- 92125906ce8f. [Accessed on 10 April 2026].

Litheko, A. (2022). Management of ecotourism through public-private partnerships in South Africa. *African Journal of Hospitality, Tourism and Leisure, 11*(6), 2103–2127.

Li, Y., & Li, Y. (2019). Cultural differences in everyday conversation between the East and the West. In *5th International Conference on Education Technology, Management and Humanities Science (ETMHS 2019)* (pp. 1070-1073).

Lohdi, A. Y. (1993). The language situation in Africa today. *Nordic Journal of African Studies, 2*(1), 79–86.

Lwoga, E. T., Ngulube, P., & Stilwell, C. (2010). Managing indigenous knowledge for sustainable agricultural development in developing countries: Knowledge management approaches in the social context. *The International Information & Library Review, 42*(3), 174–185. DOI: 10.1080/10572317.2010.10762862

Lyndon, N., Mohd, N. L., Jali, F. M., & Rose, A. R. C. (2020). Indigenous knowledge and socio- economic development of a Malaysian Indigenous Community. *Journal of Southeast Asian Studies, 25*(2), 109–129. DOI: 10.22452/jati.vol25no2.6

Machado, R. L. S., Dias, R. R., Deprá, M. C., Schneider, A. T., Dutra, D. A., de Menezes, C. R., Zepka, L. Q., & Jacob-Lopes, E. (2025). Commodities from Amazon biome: A guide to choosing sustainable paths. *Commodities, 4*(2), 8. DOI: 10.3390/commodities4020008

MacLeod, L. (2021). More than personal communication: Templates for citing Indigenous elders and knowledge keepers. *Kula, 5*(1), 1-5. DOI: DOI: 10.18357/kula.135

Madaan, M., & Mohanty, A. (2025). Constitutional and Legal Safeguards for Indigenous Communities in India: A Critical Analysis of Rights, Realities and Reforms. *INTERNATIONAL JOURNAL OF HUMAN RIGHTS LAW REVIEW, 4*(5), 156-185. Retrieved from https://humanrightlawreview.in/wp-content/uploads/2025/10/Constitutional-and-Legal-Safeguards-for-Indigenous-Communities-in-India-A-Critical-Analysis-of-Rights-Realities-and-Reforms.pdf

Madlela, B. (2023). Prospects and challenges of integrating Indigenous Knowledge System into the Natural Science curriculum in schools. *EUREKA: Social and Humanities, 3*(3), 3–19. DOI: 10.21303/2504-5571.2023.002867

Madzivhandila, T. S. (2024). Integrating indigenous knowledge system to climate change adaptation in South Africa. *South African Journal of Higher Education, 38*(5), 159–168. DOI: 10.20853/38-5-6417

Mafongoya, P. L., & Ajayi, O. C. (2017). Indigenous knowledge systems: Their history, development over time and role in sustainable development and climate change management. *Indigenous knowledge systems and climate change management in Africa*, 29.

Mafongoya, P., & Ajayi, O. C. (2017). Indigenous knowledge systems and climate change adaptation. *Climate and Development, 9*(5), 1–12.

Magaldi, D., & Berler, M. (2020). Semi-structured Interviews. In V. Zeigler-Hill & T. K. Shackelford (Eds.), Encyclopedia of Personality and Individual Differences. Springer., Retrieved April 15, 2026, from, DOI: 10.1007/978-3-319-24612-3_857

Magare, D. A., & Patil, M. B. (2025). Ethnobotany in the digital age: Opportunities and challenges of traditional knowledge digitization. *World Journal of Biology Pharmacy and Health Sciences, 21*(1), 235–242. DOI: 10.30574/wjbphs.2025.21.1.0043

Mager, M., Gutierrez-Vasques, X., Sierra, G., & Meza-Ruiz, I. (2021). Findings of the AmericasNLP 2021 shared task on open machine translation for indigenous languages of the Americas. In *Proceedings of the First Workshop on Natural Language Processing for Indigenous Languages of the Americas* (pp. 118-138). ACL. DOI: 10.18653/v1/2021.americasnlp-1.23

Magni, G. (2017). Indigenous knowledge and implications for the sustainable development agenda. *European Journal of Education*, *52*(4), 437–447. DOI: 10.1111/ejed.12238

Mahadevan, B., Bhat, V.R. and NAGENDRA, P.R., 2022. Introduction to Indian knowledge system: concepts and applications.

Maimela, G., & Nedombeloni, R. (2026). Decolonising the mainstream education and preserving Indigenous knowledge systems: learning through African Indigenous games. *AlterNative: An International Journal of Indigenous Peoples*, 1-9. DOI: 10.1177/11771801251403201

Maimela, G., Madima, T., & Makhanikhe, J. (2024). Overcoming digital barriers to African indigenous games in local communities. *Bussecon Review of Social Sciences (2687-2285), 6*(1), 29-39.

Maimela, G., Madima, T., & Makhanikhe, J. (2024). Overcoming digital barriers to African indigenous games in local communities. *Bussecon Review of Social Sciences*, *6*(1), 29–39. DOI: 10.36096/brss.v6i1.552

Makgamatha, M. G. (2025). Bridging Heritage and Enterprise to Address Socio-Economic Issues: A Policy Analysis of Indigenous Knowledge System Businesses in South Africa. In *Revaluation and Preservation of Indigenous Knowledge Systems in Modern Society* (pp. 367-384). IGI Global Scientific Publishing. https://doi.org/DOI: 10.4018/979-8-3693-7964-6.ch018

Makhanyela, D. S., Nkosi, G. S., Tshabalala, S. P., & Gumede, T. K. (2024). Comprehensive participation: A theoretically hailed yet practically disregarded approach to sustainable ecotourism. *Geo Journal of Tourism and Geosites*, *52*(1), 276–285. DOI: 10.30892/gtg.52127-1204

Makhloufi, L. (2024). *Tangible and Intangible Heritage in the Age of Globalisation*. OpenBook Publishers. DOI: 10.11647/OBP.0388

Makokotlela, M. V., & Gumbo, M. T. (2025). Decolonisation of the curriculum through the integration of indigenous knowledge in higher education. *South African Journal of Higher Education*, *39*(1), 237–254. https://hdl.handle.net/10520/ejc-high_v39_n1_a12. DOI: 10.20853/39-1-6004

Mak, S., & Thomas, A. (2022). An introduction to scoping reviews. *Journal of Graduate Medical Education, 561-564.* Advance online publication. 10.4300JGME PMID: 36274758

Makwara, E. C. (2013). Indigenous Knowledge System weather and modern forecasting: Exploring the linkages. *Journal of Agriculture and Sustainability*, *2*(1), 98–141.

Malapane, O. L., Chanza, N., & Musakwa, W. (2024). Transmission of indigenous knowledge systems under changing landscapes within the vhavenda community, South Africa. *Environmental Science & Policy*, *161*, 103861. DOI: 10.1016/j.envsci.2024.103861

Malatji, E., & Lesame, C. (2019). The use of South African languages by youth on social media: The case of Limpopo Province. *Communicare*, *38*(1), 76–95.

Malhotra, A., & Dobriyal, R. (2023). ADAPTIVE REUSE OF HERITAGE ARCHITECTURE AS A SUSTAINABLE DEVELOPMENT TOOL: OPPORTUNITIES AND CHALLENGES. *ShodhKosh: Journal of Visual and Performing Arts*, 21-27.

Maluleka, R. J. (2020). The status of traditional healing in the Limpopo province of South Africa. *HTS Teologiese Studies*, *76*(4), 1–12. DOI: 10.4102/hts.v76i4.6103

Mamta, S., Matharu, S., & Kumar, B. (2018). Impact of Yagya on Particulate Matter. *Interdisciplinary Journal of Yagya Research, 1*(1), 01–08. DOI: 10.36018/ijyr.v1i1.5

Mamtani, R., & Mamtani, R. (2005). Ayurveda and Yoga in Cardiovascular Diseases. *Cardiology*, *13*(3), 155–162. DOI: 10.1097/01.crd.0000128730.31658.36 PMID: 15834238

Manathunga, C., Qi, J., Raciti, M., Stanton, S., Tuxworth, J., Whop, J., & Gilbey, K. (2026). Entangled Relationality in Doctoral Education. In Lyons, D., Singh, K., & Southcott, J. (Eds.), *Shaping Doctoral Experiences: Nurturing the Relational* (pp. 61–82). Bloomsbury Academic. DOI: 10.5040/9798881899103.0008

Mangla, M., Akhare, R., Deokar, S., & Mehta, V. (2020). Employing machine learning for multi-perspective emotional health analysis. In *Emotion and information processing: a practical approach* (pp. 199–211). Springer International Publishing. DOI: 10.1007/978-3-030-48849-9_13

Manning, J. (2014). Definition and Classes of Social Media.Available at www.researchgate.net

Mann-Lang, J. B., Branch, G. M., Mann, B. Q., Sink, K. J., Kirkman, S. P., & Adams, R. (2021). Social and economic effects of marine protected areas in South Africa, with recommendations for future assessments. *African Journal of Marine Science*, *43*(3), 367–387. DOI: 10.2989/1814232X.2021.1961166

Manyaga, M. B., Goldman, G. A., & Thomas, P. (2024). Sustaining SMEs through Indigenous Knowledge Systems: Exploring opportunities and challenges. *The Southern African Journal of Entrepreneurship and Small Business Management*, *16*(1), a882. DOI: 10.4102/SAJESBM.v16i1.882

Mapara, J. (2021). 11 The future of indigenous knowledge system (IKS) in the face of technological disruption. Culture and rural–urban revitalisation in South Africa: Indigenous Knowledge. *Policies, and Planning*, *189*, 189–208. Advance online publication. DOI: 10.4324/9781003145912-11-14

Marshall, M. T. (2018). Interacting with heritage: On the use and potential of IoT within the cultural heritage sector. *2018 Fifth International Conference on Internet of Things: Systems, Management and Security*, 15-22. DOI: 10.1109/IoTSMS.2018.8554899

Masenya, T. M. (2023). Revitalization and digital preservation of Indigenous knowledge systems for sustainable development of Indigenous communities in South Africa. *The Serials Librarian*, *84*(5-8), 86–102. DOI: 10.1080/0361526X.2023.2277962

Masenya, T. M., & Ngulube, P. (2019). Digital preservation practices in academic libraries in South Africa in the wake of digital revolution. *South African Journal of Information Management*, *21*(1), 1–9. DOI: 10.4102/sajim.v21i1.1011

Mashiyane, D. M. (2023). Looting of Indigenous Knowledge Systems through Patents and Intellectual Property Policies. *The Afrocentricity Trajectories of Looting in South Africa*, *17*. Advance online publication. DOI: 10.5040/9781978728943.ch-2

Masuku, X. R. (2025). Indigenizing sustainable tourism: Empowering indigenous voices in tourism education. In *Indigenous Teaching Disciplines and Perspectives for Higher Education* (pp. 221–240). IGI Global Scientific Publishing. DOI: 10.4018/979-8-3693-9296-6.ch011

Matindike, F., & Ramdhany, V. (2024). Incorporating Indigenous Knowledge perspectives in integrated STEM education: A systematic review. *Research in Science & Technological Education*. Advance online publication. DOI: 10.1080/02635143.2024.2413675

Maunganidze, L. (2016). A moral compass that slipped: Indigenous knowledge systems and rural development in Zimbabwe. *Cogent Social Sciences*, *2*(1), 1266749. Advance online publication. DOI: 10.1080/23311886.2016.1266749

Mavhunga, C. C. (2014). *Translating knowledge between indigenous and modern systems*. Palgrave Macmillan.

Maxwell, J. A. (2012). Qualitative research design: An interactive approach. *Sage (Atlanta, Ga.).*

Mays, L. W., Koutsoyiannis, D., & Angelakis, A. N. (2007). A brief history of urban water supply in antiquity. *Water Science and Technology: Water Supply*, *7*(1), 1–12. Advance online publication. DOI: 10.2166/ws.2007.001

Mazhar, S. A., Anjum, R., Anwar, A. I., & Khan, A. A. (2021). Methods of data collection: A fundamental tool of research. *Journal of Integrated Community Health*, *10*(1), 6–10. DOI: 10.24321/2319.9113.202101

Mazzocchi, F. (2020). A deeper meaning of sustainability: Insights from Indigenous knowledge. *The Anthropocene Review*, *7*(1), 77–93. DOI: 10.1177/2053019619898888

Mbah, M., Ajaps, S., & Molthan-Hill, P. (2021). A systematic review of the deployment of Indigenous knowledge systems towards climate change adaptation in developing world contexts: Implications for climate change education. *Sustainability (Basel)*, *13*(9), 4811. DOI: 10.3390/su13094811

Mbaye, J. (2015). Culture and sustainable development in Africa. *African Development Review*, *27*(4), 334–346.

Mc Mahon, B., & Portelli, J. (2012). *Student engagements in urban schools: Beyond neoliberal discussions*. Information Age., https://www.tcrecord.org DOI: 10.1108/978-1-61735-733-6

McDonald, M. (2023). *Indigenous land-based education in theory and practice*. Yellowhead Institute.

McGregor, D. (2004). Coming full circle: Indigenous knowledge, environment, and our future. *American Indian Quarterly*, *28*(3/4), 385–410. DOI: 10.1353/aiq.2004.0101

McKnight, M. (2015). Teacher's experiences of indigenous knowledge found in life sciences curriculum: Access study of life science teachers at high school in Pinetown District, Msc dissertation, University of KwaZulu-Natal.

McKnight, D. H., Choudhury, V., & Kacmar, C. (2002). Developing and validating trust measures for e-commerce: An integrative typology. *Information Systems Research*, *13*(3), 334–359. DOI: 10.1287/isre.13.3.334.81

McMichael, C. N. H., Bush, M. B., ter Steege, H., & Feeley, K. J. (2025). Centuries of compounding human influence on Amazonian forests. *Proceedings of the National Academy of Sciences of the United States of America, 122*(47), e2514040122. DOI: 10.1073/pnas.2514040122 PMID: 41248280

Mdhluli, T. D., Mokgoatšana, S., Kugara, S. L., & Vuma, L. (2021). Knowledge management: Preserving, managing and sharing Indigenous knowledge through digital library. *HTS Teologiese Studies, 77*(2). Advance online publication. DOI: 10.4102/hts.v77i2.6795

Meighan, P. J. (2021). Decolonizing the digital landscape: The role of technology in Indigenous language revitalization. *Alternative, 17*(3), 397–405. DOI: 10.1177/11771801211037672 PMID: 41924568

Melé, D., Fontrodona, J., & Rosanas, J. M. (2016). Ethics in Finance and Accounting: Editorial Introduction. *Journal of Business Ethics, 140*(4), 609–613. DOI: 10.1007/s10551-016-3328-y

Menski, W. F. (2012). *Hindu law: Beyond tradition and modernity*. Oxford University Press., Retrieved from https://global.oup.com/academic/product/hindu-law-beyond-tradition-and-modernity-9780195699210

Mheta, G., Lungu, B. N., & Govender, T. (2018). Decolonisation of the curriculum: A case study of the Durban University of Technology. *South African Journal of Education, 38*(4), 1635–1641. DOI: 10.15700/saje.v38n4a1635

Mhlope, M. M. (2026). Navigating the Digital Divide: Preserving Indigenous Knowledge in the Information Age. *Indigenous Knowledge Management in the Open Science Framework,* 49-76.

Mignolo, W. D. (2007). Delinking: The rhetoric of modernity, the logic of coloniality and the grammar of de-coloniality. *Cultural Studies, 21*(2-3), 449–514. DOI: 10.1080/09502380601162647

Miralles-Quiros, M. D. M., Arraiano, I. G., & Miralles-Quiros, J. L. (2017). Are Firms that Contribute to Sustainable Development Valued by Investors? *Corporate Social Responsibility and Environmental Management, 24*(1), 71–84. DOI: 10.1002/csr.1392

Mishra, N., & Aithal, P. S. (2023). Ancient Indian Education: It's Relevance and Importance in the Modern Education System. *International Journal of Case Studies in Business, IT, and Education*, 238–249. .DOI: 10.47992/IJCSBE.2581.6942.0271

Mishra, R. (2026). Bureau Indian Standards: Purpose, standardization, & application. https://www.testronixinstrumets.com Date of access: 29 April 2026.

Mishra, P., Thakur, P., & Singh, G. (2021). Enabling technologies for IoT based smart city. *2021 Sixth International Conference on Image Information Processing (ICIIP)*, 99-104. DOI: 10.1109/ICIIP53038.2021.9702673

Mkenda, B. K., & Aikaeli, J. (2019). I*ndigenous Knowledge and Prospects for Income and Employment Generation: The Case of Handicraft Production among Rural Women in Tanzania*. IK: Other Ways of Knowing, Vol. 5 https://journals.psu.edu/ik/issue/view/3036

Mkhwebane, L. N. (2024). Life Science teacher's integration of indigenous knowledge: A revision for making life science classes culturally responsive. EURACIA *Journal of Mathematics. Science and Technology Education, 20*(8), em2483. Advance online publication. DOI: 10.29333/ejmste/14859

Mncwango, E. M. (2009). Language and the current challenges in the South African schools system. *Inkanyiso*, *1*(1), 51–54. DOI: 10.4314/ijhss.v1i1.62110

Modiba, M. A., & Emsley, M. R. (2024b). E-pedagogies and teacher efficacy: A case of rapid adjustment. In Akinlolu, M., Makua, M., Ngubane, N. (eds) online teaching and learning in HE. Springer, Cham. https://doi.org/DOI: 10.1007/978-3-031-56953-16

Modiba, M.A. & Emsley, M.R. (2024a). Decolonising education in institution of higher learning: A review of literature. *African perspectives of research in teaching and learning Journal* 3(8), Special Issue

Mogashoa, T. (2014). Understanding critical discourse analysis in qualitative research. *International Journal of Humanities Social Sciences and Education*, *1*(7), 104–113.

Mogolane, M. J., Neille, J., & Sher, J. (2025). Pragmatics: Exploring language use by younger generations in Pedi families. *The South African Journal of Communication Disorders = Die Suid-Afrikaanse Tydskrif vir Kommunikasieafwykings*, *72*(1), a1097. DOI: 10.4102/sajcd.v72i1.1097 PMID: 40922622

Mohamed, S., Png, M. T., & Isaac, W. (2020). Decolonial AI: Decolonial theory as sociotechnical foresight in artificial intelligence. *Philosophy & Technology*, *33*(4), 659–684. DOI: 10.1007/s13347-020-00405-8

Mohapatra, H. (2021). Socio-technical challenges in the implementation of smart city. *2021 International Conference on Innovation and Intelligence for Informatics, Computing, and Technologies (3ICT)*, 57-62. DOI: 10.1109/3ICT53449.2021.9581905

Moloko-Phiri, S. S., Sepeng, N. V., & Sebaeng, J. M. (2022). Communication discordance among health care professionals and indigenous patients. In Mulaudzi, F. M., & Lebese, R. T. (Eds.), *Working with indigenous knowledge: Strategies for health professionals* (pp. 5–20). AOSIS Books. DOI: 10.4102/aosis.2022.BK296.01

Mongolo, N.I., Mashele, S.S. & Makhafola, T.J. (2020). Ziziphus mucronate wild. (Rhamnaceae): its botany, toxicity, phytochemistry and pharmacological activities. Heliyon, 6(4), 2020, e03708, 1-20. DOI: 10.1016/j.heliyon.2020.e03708

Morandi, A., Dalla Libera, D., Roberti Di Sarsina, P., & Tosto, C. (2011). Salutogenesis and Ayurveda: Indications for public health management. *The EPMA Journal*, *2*(4), 459–465. DOI: 10.1007/s13167-011-0132-8 PMID: 23194327

Morgera, E., Buck, M., & Tsioumani, E. (2012). *Unravelling the Nagoya Protocol: A commentary on the Nagoya Protocol on access and benefit-sharing to the Convention on Biological Diversity*. Brill., DOI: 10.1163/9789004217201

Moropoulou, A. I., & Bisbikou, K. G. (1995). Environmental monitoring and damage assessment at the ancient sanctuary of Demeter in Eleusis, Greece. *MRS Proceedings, 352*. DOI: 10.1557/PROC-352-745

Mtapuri, O., & Giampiccoli, A. (2019). Tourism, community-based tourism and ecotourism: a definitional problematic. *South African Geographical Journal= Suid-Afrikaanse Geografiese Tydskrif, 101*(1), 22-35. https://hdl.handle.net/10520/EJC-13ae94668e

Mthembu, B. A., & Mutambara, E. (2018). Rural tourism as a mechanism for poverty alleviation in KwaZulu-Natal Province of South Africa: A case of Bergville. *African Journal of Hospitality, Tourism and Leisure*, *7*(4), 1–17.

Mtshali, L., & Beharry-Ramraj, A. (2025). Barriers to implementing green marketing strategies for sustainable ecotourism in KwaZulu-Natal. e-BANGI Journal, 22(2), 1–15. DOI: 10.17576/ebangi.2025.2202.54

Mudau, O. (2023). The Digital Dilemma: Ethics in content creation. *PDBY*. Available at: https://pdby.co.za/the-digital-dilemma-ethics-in-content-creation/ [Accessed: 20 January 2026]

Mudau, P. P., Mandende, I. P., & Mushaathoni, M. (2024). Through our indigenous greeting terms, we are identified: A study of Tshiven a from Tshimbupfhe, Vhembe District Municipality, Limpopo, South Africa. *African Identities*, •••, 1–15. DOI: 10.1080/14725843.2024.2381484

Mudau, T. J., & Sikhwari, M. G. (2025). The Influence of Indigenous Knowledge Systems on Youth Behaviourin Mutale Municipality, South Africa. *E-Journal of Humanities, Arts and Social Sciences*, *6*(10), 2486–2502. DOI: 10.38159/ehass.202561018

Mulla, Z. R., & Krishnan, V. R. (2013). Karma-Yoga: The Indian Model of Moral Development. *Journal of Business Ethics*, *123*(2), 339–351. DOI: 10.1007/s10551-013-1842-8

Munn, Z., Peters, M. D., Stern, C., Tufanaru, C., McArthur, A., & Aromataris, E. (2018). Systematic review or scoping review? Guidance for authors when choosing between a systematic or scoping review approach. *BMC Medical Research Methodology*, *18*(1), 143. DOI: 10.1186/s12874-018-0611-x PMID: 30453902

Mushaathoni, M. (2023). Exploring Intercultural Communication as a Means to Promote Inclusivity in Diverse Organisations: A Study in a South African University. *IAFOR Journal of Education*, *11*(3), 161–184. DOI: 10.22492/ije.11.3.08

Muyambo, F., Bahta, Y. T., & Jordaan, A. J. (2017). The role of indigenous knowledge in drought risk reduction: A case of communal farmers in South Africa. *Jàambá*, *9*(1). Advance online publication. DOI: 10.4102/jamba.v9i1.420 PMID: 29955345

Naidoo, P. D., & Vithai, R. (2014). Teacher approaches in introducing indigenous knowledge system in school science classrooms. *African Journal of Research Science Technology Education*, *18*(3), 253–263. DOI: 10.1080/10288457.2014.956407

Nakata, M., & Langton, M. (2005). Australian indigenous knowledge and libraries [Internet]. Available: https://www.alia.org.au/publishing/aarl/AARL_vol36_no2_2005(reprint).pdf Accessed 14 April 2026.

Nakata, M. (2007). *Disciplining the savages: Savaging the disciplines*. Aboriginal Studies Press.

Nakata, M., Nakata, V., Gardiner, G., McKeough, J., Byrne, A., & Gibson, J. (2008). *Australian Indigenous digital collections: First generation issues*. Australian Academic and Research Network.

National Research Foundation. (2010). "Managing digital collections: a collaborative initiative on the South African framework". https://wiki.lib.sun.ac.za/images/5/51/Managing_Digital_Collections.pdf

Nayak, C., Chakraborty, D. K., & Chakrabarti, D. (2025). Preserving Indigenous Knowledge Systems by Employing Digital Twin Technologies in the Disruptive Era. In Chakraborty, A. K., Singh, A. P., & Banerjee, S. (Eds.), *Indian Knowledge System (IKS) and Viksit Bharat by 2047: fusion in LIS and Allied Discipline*. Cambridge Scholars Publishing.

Ncube, I., Chigwada, J., Ngulube, P., & Maluleka, J. R. (2025). Government support for indigenous knowledge for sustainability in Southern Africa. *South African Journal of Information Management, 27*(1), 1901. DOI: 10.4102/sajim.v27i1.1901

Ndalilo, L., Swiderska, K., & Wekesa, C. (2023). Establishing a biocultural heritage territory to protect Kenya's Kaya Forests; mid-term project workshop. IIED, London. https://www.iid.org/21556iied Date of access: 28 April 2026.

Ndhlovu, P. T., Omotayo, A. O., Olagunju, K. O., Otang-Mbeng, W., & Aremu, A. O. (2025). Assessing the impacts of commercializing medicinal plants on livelihood outcomes: Evidence from indigenous knowledge holders in South Africa. *Environment, Development and Sustainability, 27*(2), 4489–4511. DOI: 10.1007/s10668-023-04087-y

Ndlovu, L., & Svodziwa, M. (2026). *The Levels Of Participation in Non Governmental Organizations: Ndlovu, Loveness, Svodziwa, Mpathew: Amazon.ie: Books*. Amazon.ie. https://www.amazon.ie/Levels-Participation-Non-Governmental-Organizations/dp/3659917672

Ndlovu-Gatsheni, S. J. (2021). The cognitive empire, politics of knowledge and African intellectual productions: Reflections on struggles for epistemic freedom and resurgence of decolonisation in the twenty-first century. *Third World Quarterly, 42*(5), 882–901. DOI: 10.1080/01436597.2020.1775487

Ndlovu, T. P., Sibiya, N. P., & Giampiccoli, A. (2018). Assessing local community participation in community-based tourism: The case of the Zulu-Mpophomeni Tourism Experience. *African Journal of Hospitality, Tourism and Leisure, 7*(2).

Ndwandwe, S., & Mudhara, M. (2014). Contribution of indigenous knowledge practices to household food security: A case study of rural households in KwaZulu-Natal. *Indilinga, 13*(2), 271–282. https://hdl.handle.net/10520/EJC166449

Nembe, J., Daraojimba, A., Olubusola, O., Mhlongo, N., Oguejiofor, B., Falaiye, T., & Atadoga, J. (2024). THE ROLE OF ARTIFICIAL INTELLIGENCE IN ENHANCING TAX COMPLIANCE AND FINANCIAL REGULATION. *Finance & Accounting Research Journal, 6*(2), 241–251. DOI: 10.51594/farj.v6i2.822

Nenungwi, T. G. (2015). Non-verbal communication in Tshiven a: a sociolinguistic and discourse analysis. Doctoral thesis. Retrieved from http://hdl.handle.net/ 10386/1542

Nevondo, P. U., Madima, T., & Makhanikhe, J. (2024). The role of ICT in preserving proverbs among Vhavenda communities. *Bussecon Review of Social Sciences, 6*(3), 41–49. DOI: 10.36096/brss.v6i3.614

Ngubane, N. I., & Makua, M. (2021). Ubuntu pedagogy-transforming educational practices in South Africa through an African philosophy: From theory to practice. *Inkanyiso, 13*(1), 1–12. https://hdl.handle.net/10520/ejc-uz_inka-v13-n1-a2. DOI: 10.4102/ink.v13i1.9

Ngulube, P. (2018). Preserving Indigenous Knowledge in the knowledge economy. *Information Development, 34*(3), 1–12.

Nhlapo, J. V. (2023). Participatory Action Research as Value-Laden Practice: Values that Matter to the Participants. *Noyam-Journal of Education and Technology, 4*, 330–344. DOI: 10.38159/jelt.2025614

Nichols, J., & McKay, P. (2018). Indigenous stewardship and digital knowledge infrastructures. *Journal of Cultural Heritage, 29*, 56–64. DOI: 10.1016/j.culher.2017.07.002

Nicholson, A. J. (2010). *Unifying Hinduism*. columbia university. .DOI: 10.7312/nich14986

Nitti, M., Pilloni, V., Giusto, D., & Popescu, V. (2017). IoT architecture for a sustainable tourism application in a smart city environment. *Mobile Information Systems, 2017*, 1–9. DOI: 10.1155/2017/9201640

Nurfitri, D., Bustani, S., & Yunari, S. B. (2025). Protecting genetic resources and traditional knowledge in indigenous communities and governance through inclusive moral rights. *Scholars International Journal of Law. Crime and Justice*. Advance online publication. DOI: 10.36348/sijlcj

Nwagbara (2025), *Social Media Pressures on Cultural Authenticity in Traditional Igbo Wedding*.

Nwokocha, G., & Legg-Jack, D. (2024). Reimagining STEM education in South Africa: Leveraging indigenous knowledge system through M-know model for curriculum enhancement. *International Journal of Social Science Research and Review, 7*(2), 173–189. DOI: 10.47814/ijssrr.v7i2.1951

Nyahunda, L. (2024). Integration of indigenous knowledge systems (IKS) into climate change mitigation and adaptation endeavours: Milestones and gaps in South Africa and Zimbabwe's climate policy frameworks. *Climatic Change, 177*(11), 162. DOI: 10.1007/s10584-024-03822-2

Nyathi, I. (2025). *Knowledge System and the Mopane Worm Economy Amongst the Kalanga of Mangwe in Southwestern Zimbabwe*: Harnessing Indigenous Epistemologies for Sustainable Progress. IGI Global Scientific Publishing. *Page, 16*, 273–288. Advance online publication. DOI: 10.4018/979-8-3693-8447-3.ch013

Nyathi, P. A. (2025). Exploring forced removals and dispossession in the Dukuduku Forest, KwaZulu-Natal, before and after 1994. *History (London), 110*(391), 393–412. DOI: 10.1111/1468-229X.13438

Nyoni, T. E. (2023). *The co-creation of Afrocentric ethical guidelines for digital content creators producing indigenous knowledge-based content* (Doctoral dissertation).

Nyoni, T., & Maimela, G. (2025). Digital content creation as a tool for safeguarding and promoting indigenous knowledge systems in South Africa. *International Journal of Research in Business and Social Science (2147-4478), 14*(8), 395-402.

Nyoni, T., & Maimela, G. (2025). Digital content creation as a tool for safeguarding and promoting Indigenous knowledge systems in South Africa. *International Journal of Research in Business and Social Science (2147-4478), 14*(8), 395-402. DOI: 10.20525/ijrbs.v14i8.4408

O'Neil, C. (2016). *Weapons of Math Destruction: How Big Data Increases Inequality and Threatens Democracy*. Crown.

Oarga-Mulec, A., & Skene, K. R. (2026). A dynamic biome-specific governance approach that integrates indigenous knowledge and pluriversal thinking. *Humanities & Social Sciences Communications*, *13*(1), 97. DOI: 10.1057/s41599-025-06402-6

Ocholla, D. N. (2025). The complex marriage: Interfacing open science and open indigenous knowledge. Emerald Publishing Limited. *Library Management*, *46*(3/4), 252–357. DOI: 10.1108/LM-11-2024-0125

Odayar, S. (2024). Biocultural community protocols, a powerful tool to affirm and protect Indigenous Peoples 'rights. www://internationalrivers.org Date of access: 20 February 2026.

Odeyemi, O., Nwankwo, E., Soyombo, O., & Mhlongo, N. (2024). Reviewing the role of AI in fraud detection and prevention in financial services. *International Journal of Science and Research Archive*, *11*(1), 2101–2110. DOI: 10.30574/ijsra.2024.11.1.0279

Odi, A. C., Nwogbaga, N. E., & Chukwuka, N. O. (2015). The proposed roles of VLAN and inter-VLAN routing in effective distribution of network services in Ebonyi State University. *International Journal of Science and Research*, *4*, 2608–2615.

Odora Hoppers, C. A. (2002). *Indigenous knowledge and the integration of knowledge systems*. New Africa Books.

Odora Hoppers, C. A. (2017). *Decolonisation of knowledge, epistemicide and justice*. Routledge.

Office of the Comptroller General Ministry of Finance. (2021). Indigenous Procurement and Contract Management Guidelines: Ministry guidelines for the practical applications of the Province of the Province of British Columbia's procurement policies, in the delivery of culturally- appropriate programs and services to Indigenous peoples and communications. www2.gov.bc/assets/gov/british-columbian

Oguamanam, C. (2020). Indigenous peoples, data sovereignty, and self-determination: Current realities and imperatives. *The African Journal of Information and Communication*, *26*(26), 1–20. DOI: 10.23962/10539/30360

Oguamanam, C. (2022). Transition to the Fourth Industrial Revolution: Africa's Science, Technology and Innovation Framework and Indigenous Knowledge Systems. *African Journal of Legal Studies*, *15*(1), 1–37. DOI: 10.1163/17087384-bja10058

Ogunyiola, A., Gardezi, M., & Vij, S. (2022). Smallholder farmers' engagement with climate smart agriculture in Africa: Role of local knowledge and upscaling. *Climate Policy*, *22*(4), 411–426. DOI: 10.1080/14693062.2021.2023451

Okedele, P. O., Aziza, O. R., Oduro, P., & Ishola, A. O. (2024). Integrating indigenous knowledge systems into global climate adaptation policies. *International Journal of Engineering Research and Development*, *20*(12), 223–231.

Olaitan, Z. M. (2020). The relevance of African indigenous knowledge systems for development in Africa: Policy brief. https://africaindigenousresearch.com Date of access: 28 April 2026.

Olaopa, O. R., & Ayodele, O. A. (2022). Building on the strengths of African Indigenous knowledge and innovation (AIK&I) for sustainable development in Africa. *African Journal of Science, Technology, Innovation and Development*, *14*(5), 1313–1326. DOI: 10.1080/20421338.2021.1950111

Olawuyi, S. O., Ijila, O. J., Adegbite, A., Olawuyi, T. D., & Farayola, C. O. (2024). Smallholder farmers' use of indigenous knowledge practices in agri-food systems: Contribution of food security attainment drive. *Research on World Agricultural Economy*, *5*(2), 45–67. DOI: 10.36956/rwae.v5i2.1056

Oli, K. P., Rana, P., Shah, R., Rawal, R. S., Chaudhary, R., Yongping, Y., Yao, F., Kotru, R., & Bhatta, L. D. (2015). Tools for developing a Biocultural Community Protocol. International Centre for Integrated Mountain Development. Nepal. https://lib.icidmod.org/record/30771/filesWP15 Date of access: 16 Jan. 2016. (intext)

Oloko, M. Reed, M. G., & Robson, J. P. (2022). *Engaging youth in food preservation: Examining knowledge and practice on Canada's West Coast*. Canadian Food Studies La Revue Canadienne Des études Sur l'alimentation, 9(3), 61–86. .DOI: 10.15353/cfs-rcea.v9i3.523

Omipidan, I. A., & Sanusi, B. O. (2024). Rise of social media in the digital age: Whatsapp a threat to effective communication. *IMSU Journal of Communication Studies*, *8*(1), 142–153.

Onfra. (2024, November 9). *Indigenous People: Cultural Heritage, Resilience, and Impact*. Retrieved from Onfra: https://onfra.io/indigenous-people-cultural-heritage-resilience-and-impact/

Ordóñez-Martínez, D., Seguí-Pons, J. M., & Ruiz-Pérez, M. (2023). Conceptual framework and prospective analysis of EU tourism data spaces. *Sustainability (Basel)*, *16*(1), 371. DOI: 10.3390/su16010371

Orlikowski, W. J. (2000). Using technology and constituting structures: A practice lens for studying technology in organizations. *Organization Science*, *11*(4), 404–428. DOI: 10.1287/orsc.11.4.404.14600

Ormsby, A., & Bhagwat, S. A. (2010). Sacred Forests of India: A Strong Tradition Of Community-Based Natural Resource Management. *Environmental Conservation*, *37*(3), 320–326. DOI: 10.1017/S0376892910000561

Osborne, R. (2009). *Greece in the Making 1200-479 BC*. Routledge. DOI: 10.4324/9780203880173

Ostrom, E. (1990). *Governing the commons: The evolution of institutions for collective action*. Cambridge University Press. DOI: 10.1017/CBO9780511807763

Ostrom, E. (2005). *Understanding institutional diversity*. Princeton University Press.

Ostrom, E. (2010). Beyond markets and states: Polycentric governance of complex economic systems. *The American Economic Review*, *100*(3), 641–672. DOI: 10.1257/aer.100.3.641

Ostrom, E. (2015). *Governing the Commons: The Evolution of Institutions for Collective Action*. Cambridge University Press. DOI: 10.1017/CBO9781316423936

Outakoski, H., Cocq, C., & Steggo, P. (2018). Strengthening Indigenous languages in the digital age: Social media–supported learning in Sápmi. *Media International Australia, Incorporating Culture & Policy*, *169*(1), 21–31. DOI: 10.1177/1329878X18803700

Oviedo, A., Pantoja, A. M., Guevara, S. Z., & López, E. A. (2025). Awajún Pottery, Gender, and Transformation: From Social Prestige to Market Logic. *Journal of Intercultural Communication*, *25*(3), 104–115. DOI: 10.36923/jicc.v25i3.1157

Owiny, S., & Mehta, K & Maretzki. (2014). *The Use of Social Media Technologies to Create, Preserve, and Disseminate Indigenous Knowledge and Skills to Communities in East Africa*. International Journal of Communication. 8. 14. *Range Planning*, *20*(6), 29–34.

Owusu-Ansah, F. E., & Mji, G. (2013). African indigenous knowledge and research. *African Journal of Disability*, *2*(1), 1–5. DOI: 10.4102/ajod.v2i1.30 PMID: 28729984

Oyinlade, E. O. (2024). Technological culture and the challenge of erosion of Yorùbá moral values. *African Journal of Culture, History. Religion and Traditions*, *7*(3), 1–9.

Page, S., Hall, M., & Naepi, S. (2025). Indigenous leadership in higher education: Baya, Eiseni, Rangatiratanga (voice, agency, sovereignty). *Journal of Higher Education Policy and Management*, *47*(5), 563–566. DOI: 10.1080/1360080X.2025.2546752

Pandey, A. (2025). The Transformative Potential of Artificial Intelligence in Building Sustainable Agricultural Systems: Innovations, Challenges, and Future Directions. In *AI and Ecological Change for Sustainable Development* (pp. 161-182). IGI Global Scientific Publishing.

Pandey, A. (2026). The Intersection of Industrial Policy and Environmental Regulation: Balancing Growth and Green Goals. In *Industrial Policy for Resilience, Equity, and Sustainability* (pp. 339-362). IGI Global Scientific Publishing.

Pandey, A., Joseph, J., & Mishra, A. (2026). Advancing Sustainable Agriculture Through Artificial Intelligence: Harvesting Greener Future. In *Artificial Intelligence Enabled Real Time Environmental Monitoring* (pp. 19–41). Springer Nature Singapore. DOI: 10.1007/978-981-95-3767-9_2

Pandey, D. N., Gupta, A. K., & Anderson, D. M. (2003, July 10). Rainwater harvesting as an adaptation to climate change. *Current Science*, *85*(1), 46–59. https://www.currentscience.ac.in/Volumes/85/01/0046.pdf

Pandey, K. (2014). Indigenous Knowledge System: Concepts and Discourses. *Anthropological Bulletin*, *2*(1), 17–28.

Pandit, S. D. (2025). Reviving Indian Knowledge Systems (IKS): Bridging Tradition with Modernity. *International Studies*, *62*(1), 11–29. DOI: 10.1177/00208817251382323

Panezi, M. (2021, November/December). *A Description of the Structure of the Hellenic Republic, the Greek Legal System, and Legal Research*. Retrieved July 11, 2025, from Globalex: https://www.nyulawglobal.org/globalex/greece1.html

Papaioannou, K. (2017). THE INTERNATIONAL LAW ON THE PROTECTION OF CULTURAL HERITAGE. *IJASOS- International E-journal of Advances in Social Sciences*, 257-262.

Parahakaran, S. (2022). Digital Ethnography Within Indigenous Contexts: An Exploratory Perspective. In Chowdhury, J. S., Wahab, H. A., Saad, R. M., Roy, P. K., & Wronka, J. (Eds.), *Practices, Challenges, and Prospects of Digital Ethnography as a Multidisciplinary Method* (pp. 138–147). IGI Global Scientific Publishing. DOI: 10.4018/978-1-6684-4190-9.ch010

Pareek, P., Hajra, R., & Mohapatra, H. (2025). Empowering public officials by bridging the digital divide with smart technology for public services. *Advances in Public Policy and Administration*, •••, 419–452. DOI: 10.4018/979-8-3693-6547-2.ch016

Patalauskaitė, E. (2024). Ethical Aspects of Content Creation. *Filosofija. Sociologija*, *35*(3), 369–377.

Patel, M., Mohammed, T. A. S., & Koen, R. (2024). Ubuntu in post-apartheid South Africa: Educational, cultural and philosophical considerations. [MDPI Philosophies Journal]. *Philosophies*, *9*(1), 1–15. DOI: 10.3390/philosophies9010021

Pathak, T., & Das, S. (2024, June). CULTURAL HERITAGE PRESERVATION IN INDIA: SAFEGUARDING A LIVING LEGACY. *ShodhKosh: Journal of Visual and Performing Arts*, 1286-1291. DOI: 10.29121/shodhkosh.v5.i1.2024.4184

Patil, P. V., & Deshpande, K. R. (2025). Digital Preservation of Indigenous Knowledge Systems: Challenges and Opportunities in India. *International Journal of Creative and Open Research in Engineering and Management, 01*(01), 1-9. https://doi.org/https://doi.org/10.55041/ijco pe.v1i2.005

Patowary, K. (2022, July 7). *Code of Ur-Nammu: The Oldest Law in History*. Retrieved July 13, 2025, from Amusing Planet: https://www.amusingplanet.com/2022/07/code-of-ur-nammu-oldest-law-in-history .html#:~:text=Some%20of%20the%20earliest%20legal%20codes%20concerning%20crimes,Kassites %2C%20and%20Hittites%2C%20each%20had%20their%20own%20laws

Patton, M. Q. (2015). *Qualitative Research & Evaluation Methods*. Sage publications.

Peters, E. J., & Andersen, C. (Eds.). (2013). *Indigenous in the city: Contemporary identities and cultural innovation*. UBC Press. DOI: 10.59962/9780774824668

Phanvisetsak, S., & Chung, H.-L. (2025). Bridging Heritage and Technology: Rethinking Cultural Asset Classification through Digitalisation. *Journal of Cultural Analysis and Social Change*, *10*(4), 285–298. DOI: 10.64753/jcasc.v10i4.2803

Phil, T. M. (2022). ecotourism: A sustainable development connect to nature and a strategy for balancing economic growth, socio-cultural development and conservation. [Publication details incomplete; journal/ source information to be verified by authors.]

Phinney, J. S. (1992). The multigroup ethnic identity measure: A new scale for use with diverse groups. *Journal of Adolescent Research*, *7*(2), 156–176. DOI: 10.1177/074355489272003

Phirke, P. (2024). Reviving ancient knowledge: Integrating indigenous knowledge system in biological education. *International Journal of Research Culture Society*, *9*(4), 16–20. www.ijrcs.org

Photo, P., & McKnight, M. (2024). Investigating indigenous knowledge awareness among South African science teachers for developing a guideline. *Curriculum Perspectives*, *44*(1), 61–71. DOI: 10.1007/ s41297-023-00224-9

Pidgeon, M., & Riley, T. (2021). Understanding the Application and Use of Indigenous Research Methodologies in the Social Sciences by Indigenous and Non-Indigenous Scholars. *International Journal of Education Policy and Leadership*, *17*(8), n8. DOI: 10.22230/ijepl.2021v17n8a1065

Pihama, L., Tiakiwai, S. J., & Southey, K. (Eds.). (2015). *Kaupapa rangahau: A reader* (2nd ed.). Te Kotahi Research Institute.

Piplani, M., & Kumar, T. (2019). Stepwells: Reviving India's Cultural and Traditional Water Storage Systems. *Conference: World Resources Forum* (pp. 135-142). Geneva, Switzerland: World Resources Forum Association.

Pittz, T. G., Claw, C. M., & Adler, T. R. (2026). Sovereignty as a site of innovation: Institutional entrepreneurship in Native American tribal nations. *Research Policy, 55*(4), 105431. Advance online publication. DOI: 10.1016/j.respol.2026.105431

Pittz, T., Adler, T., & Claw, C. (2024). The institution of tribal sovereignty: From historical challenges to modern expressions. *Journal of Management History, 32*(3), 507–521. DOI: 10.1108/JMH-06-2024-0096

Ponterotto, J. G. (2013). Qualitative research in multicultural psychology: Philosophical underpinnings, popular approaches, and ethical considerations. *Qualitative Psychology, 1*(S), 19–32. .DOI: 10.1037/2326-3598.1.S.19

Porri, F., McConnachie, B., van der Walt, K. A., Wynberg, R., & Pattrick, P. (2023). Eco-creative nature-based solutions to transform urban coastlines, local coastal communities and enhance biodiversity through the lens of scientific and Indigenous knowledge. *Cambridge Prisms: Coastal Futures, 1*, e17. DOI: 10.1017/cft.2022.10

Posey, D. A., & Dutfield, G. (1996). *Beyond intellectual property: Toward traditional resource rights for indigenous peoples and local communities*. IDRC.

Prasetyo, L. B., Kartodihardjo, H., Adiwibowo, S., & Okarda, B. (2019). The complexity of integrating indigenous knowledge for ecotourism planning: A case of customary forests in Indonesia. *Forest Policy and Economics, 103*, 84–92. DOI: 10.1016/j.forpol.2019.03.002

Pretty, J., Benton, T. G., Bharucha, Z. P., Dicks, L. V., Flora, C. B., Godfray, H. C., Goulson, D., Hartley, S., Lampkin, N., Morris, C., Pierzynski, G., Prasad, P. V. V., Reganold, J., Rockström, J., Smith, P., Thorne, P., & Wratten, S. (2018). Global assessment of agricultural system redesign for sustainable intensification. *Nature Sustainability, 1*(8), 441–446. DOI: 10.1038/s41893-018-0114-0

Puri, S. K. (2007). Integrating scientific with indigenous knowledge: Constructing knowledge alliances for land management in India. *Management Information Systems Quarterly, 31*(2), 355–379. DOI: 10.2307/25148795

Putnam, R. D. (1993). *Making democracy work: Civic traditions in modern Italy*. Princeton University Press.

Quijano, A. (2000). Coloniality of power and Eurocentrism in Latin America. *International Sociology, 15*(2), 215–232. DOI: 10.1177/0268580900015002005

Quijano, A. (2024). Aníbal Quijano: Foundational essays on the coloniality of power. In *Aníbal Quijano*. Duke University Press., DOI: 10.1515/9781478059356

Rahmawati, R., Handayani, S. R., Suprapti, A. R., Airawaty, D., & Latifah, L. (2023). Green entrepreneurship based on local characteristics and culture to support sustainable ecotourism: A case study. *Journal of Intercultural Communication, 23*(1), 66–75. DOI: 10.36923/jicc.v23i1.71

Rainie, S. C., Kukutai, T., Walter, M., Figueroa-Rodríguez, O. L., Walker, J., & Cariou, E. (2017). Issues in Open Data: Indigenous Data Sovereignty. In Davies, T., Walker, S., Rubinstein, M., & Perini, F. (Eds.), *The State of Open Data: Histories and Horizons* (pp. 300–319). African Minds and International Development Research Centre.

Rainie, S. C., Lee Schultz, J., Briggs, E., Riggs, P., & Palmanteer-Holder, N. L. (2017). Data as a Strategic Resource. *International Indigenous Policy Journal, 8*(2), 1–29. DOI: 10.18584/iipj.2017.8.2.1

Rajani, M. B. (2025). Regulation Boundaries for Preservation of Cultural Heritage Sites. In Nayak, S. (Ed.), *Policy Recommendations for Sustainable Development* (Vol. 2, pp. 17–24). Springer., DOI: 10.1007/978-981-96-0718-1_3

Rajasekaran, B., & Martin, R. A. (1992). Incorporating Indigenous Knowledge Systems into Agricultural and Extension Education Programs: A Study of the Perceptions of Extension Professionals.

Raji, T., & Dube, B. (2024). Social media and African crises: A comparative study of Nigeria and South Africa. [RSIS International]. *International Journal of Research and Innovation in Social Science, 8*(2), 120–132.

Rakotondrabe, M., & Girard, F. (2021). Protecting traditional knowledge through biocultural community protocols in Madagascar: Do not forget the "B" in BCP. *Sustainability (Basel), 13*(18), 10255. DOI: 10.3390/su131810255

Ramaano, A. I. (2025). Community-based tourism advancement and sustainable tourism enterprise development in marginalised rural municipalities: Context for ecotourism development in parks-adjacent communities. *African Journal of Hospitality, Tourism and Leisure, 14*(1), 1–18.

Rankoana, S. A. (2017). The use of indigenous knowledge in subsistence farming: Implications for sustainable agricultural production in Dikgale community in Limpopo Province, South Africa. *Toward a sustainable agriculture: Farming practices and water use, 63*.

Rathor, V. (2024). Preserving Indian Culture Amidst Modernization: Challenges and Solutions. *International Journal of Research and Analytical Reviews*, •••, 778–781.

Raven, M. M., & Robinson, D. (2022). Biocultural rights and protocols in the Pacific. In *Biocultural rights, Indigenous Peoples and local communities* (pp. 203–220). Routledge. DOI: 10.4324/9781003172642-10

Ravikanth, G. (2021, May). Indian Philosophy and Environmental Ethics. *GNOSI: An Interdisciplinary Journal of Human Theory and Praxis, 4*(1 (Special Issue)), 47–63.

Rawat, G., & Mishra, S. (2021). SPIRITUALITY AND ENVIRONMENT: SIGNIFICANCE OF INDIGENOUS KNOWLEDGE SYSTEMS IN AFRICA AND CHALLENGES. *Literary Endeavour, 12*(1).

Raynolds, L. T. (2004). The globalization of organic agro-food networks. *World Development, 32*(5), 725–743. DOI: 10.1016/j.worlddev.2003.11.008

Reid-Merritt, P. (2015). Afrocentric approaches to social work. In *International Encyclopedia of the Social & Behavioral Sciences* (2nd ed., pp. 132–136). Elsevier. DOI: 10.1016/B978-0-08-097086-8.28003-3

Republic of South Africa. (2004). Indigenous Knowledge System viewed 15 December 2025 from https://www.dst.gov.za/images/pdfs/IKS_Policy%20PDF.pdf

Republic of South Africa. (2004). *Indigenous knowledge systems*, viewed 15 January 2026, from https://www.dst.gov.za/images/pdfs/IKS_Policy%20PDF.pdf

Republic of South Africa. (2012). National Development Plan 2030: Our future-make it work viewed 17 December 2025, from https://www.gov.za/sites/default/files/gcis/_document/201409/ndp-2030-our-future-make-it-workr.pdf

Republic of South Africa. (2016). 2004 IKS Policy, Role of Department of Science and Technology in Policy Implementation, and key achievements of the department's IKS Programme. A presentation to the Portfolio Committee on Science and Technology, 16 August 2016. https://pmg.org.za>daily schedule

Republic of South Africa. (2019). Protection, Promotion, Development, and Management of Indigenous Knowledge Act, 2019, viewed 15 December 2025, from https://www.gov.za/sites/default/gcis_document/201908/4264719-8actof6of2019protectpromodevelopmentindigenousknowledgeact.pdf

Republic of South Africa. 2019, *Protection, Promotion, Development and Management of Indigenous Knowledge* Act. viewed 15 January 2026, from https://www.gov.za/sites/default/files/gcis_document/201908/4264719-8act6of2019protectpromodevelopmanagementindigenousknowledgeact.pdf

Research Data Alliance International Indigenous Data Sovereignty Interest Group. (September 2019). CARE Principles for Indigenous Data Governance. *The Global Indigenous Data Alliance.* DOI: http://GIDA-global.org

Ridge, M. (2013). From tagging to theorizing: Deepening engagement with cultural heritage through crowdsourcing. *Curator (New York, N.Y.)*, *56*(4), 435–450. DOI: 10.1111/cura.12046

Rieke, N., Hancox, J., Li, W., Milletari, F., Roth, H. R., Albarqouni, S., Bakas, S., Galtier, M. N., Landman, B. A., Maier-Hein, K., Ourselin, S., Sheller, M., Summers, R. M., Trask, A., Xu, D., Baust, M., & Cardoso, M. J. (2020). The future of digital health with federated learning. *NPJ Digital Medicine*, *3*(1), 119. DOI: 10.1038/s41746-020-00323-1 PMID: 33015372

Rights and Resources Initiative. (2015). *Who owns the world's land?* RRI.

Rocheleau, D., Thomas-Slayter, B., & Wangari, E. (Eds.). (1996). *Feminist political ecology: Global issues and local experiences*. Routledge.

Rockström, J., Steffen, W., Noone, K., Persson, Å., Chapin, F. S.III, Lambin, E. F., Lenton, T. M., Scheffer, M., Folke, C., Schellnhuber, H. J., Nykvist, B., de Wit, C. A., Hughes, T., van der Leeuw, S., Rodhe, H., Sörlin, S., Snyder, P. K., Costanza, R., Svedin, U., & Foley, J. A. (2009). A safe operating space for humanity. *Nature*, *461*(7263), 472–475. DOI: 10.1038/461472a PMID: 19779433

Rodger, D., Admani, A., & Thomas, M. (2024). What is a scoping review? BM Journal of Evidence Based Nurse, 27(3). https://ebn.bmj.com Date of access: 27 Jan. 2026.

Roger, E., Kellie, D., Slatyer, C., Brenton, P., Torresan, O., Wallis, E., & Zerger, A. (2023). Open access research infrastructures are critical for improving the accessibility and utility of citizen science: A case study of Australia's national biodiversity infrastructure, the Atlas of Living Australia (ALA). *Citizen Science: Theory and Practice*, *8*(1), 56. Advance online publication. DOI: 10.5334/cstp.564

Rogerson, C. M. (2007). Cultural industries and local economic development in South Africa. *Urban Forum*, *18*(2), 49–71. DOI: 10.1007/s12132-007-9006-5

Rogerson, C. M. (2012). Creative industries and youth employment. *African Journal of Business Management*, *6*(39), 10435–10448.

Rosário, A. T., & Dias, J. C. (2024). Exploring the landscape of smart tourism: A systematic bibliometric review of the literature of the Internet of things. *Administrative Sciences*, *14*(2), 22. DOI: 10.3390/admsci14020022

Roya, W., & Ngcobo, S. (2023). The role of information communication technologies on African indigenous knowledge systems: Folktales. *International Journal of Research in Business and Social Science*, *12*(6), 46–55. DOI: 10.20525/ijrbs.v12i6.2662

Rutaremwa, C. B., & Shirindi, M. L. (2025). Fathers' perspectives on the preparation of sons for family responsibilities. *Social Work/Maatskaplike Werk, 61*(2), 286-317.

Saatchi, S. G., Wahed, M. K. Y. A., Alqaraleh, M. K. S., Al-Shorman, H. M., Alanazi, T., Alka'awneh, S. M. N., & Alzyoud, M. (2025). The influence of compatibility on the acceptance of artificial intelligence in Kuwaiti universities. In *Intelligence-Driven Circular Economy: Regeneration Towards Sustainability and Social Responsibility–Volume 1* (pp. 103–117). Springer Nature Switzerland. DOI: 10.1007/978-3-031-73899-9_9

Sadik-Zada, E. R., Gatto, A., & Niftiyev, I. (2022). E-government and petty corruption in public sector service delivery. *Technology Analysis and Strategic Management*, *36*(12), 1–17. DOI: 10.1080/09537325.2022.2067037

Sanchez, M. (2025). Sustaining the future through addressing the past: the transitional justice functions of biocultural community protocols. *Law Environmental and Development Journal*, 21(1). doi: https://doi.org. Date of access: 27 Jan. 2026.DOI: 10.2550/.SOAS.00506365

Schelenz, L., & Schopp, K. (2018). Digitalization in Africa: Interdisciplinary perspective on technology, development, and justice. [IJDS]. *International Journal of Digital Society*, *9*(4), 1412–1420. DOI: 10.20533/ijds.2040.2570.2018.0175

Schindler, D. a. (1988). The Laws of armed conflicts: a collection of conventions, resolutions and other documents. Nijhoff, Kluwer Academic Publishers.

Schroeder, D., & Pogge, T. (2009). Justice and the convention on biological diversity. *Ethics & International Affairs*, *23*(3), 267–280. DOI: 10.1111/j.1747-7093.2009.00217.x

Schwartzman, S., & Zimmerman, B. (2005). Conservation alliances with Indigenous peoples of the Amazon. *Conservation Biology : The Journal of the Society for Conservation Biology*, *19*(3), 721–727. DOI: 10.1111/j.1523-1739.2005.00695.x

Schweitze, B. (2024). Artificial Intelligence (AI) Ethics in Accounting. *Journal of Accounting, Ethics &. Public Policy, 25*(1). Advance online publication. DOI: 10.60154/jaepp.2024.v25n1p67

Secretariat of the Convention on Biological Diversity. (2011). *Nagoya Protocol on access and benefit-sharing*. United Nations.

Sedlmeier, P., & Srinivas, K. (2016). How Do Theories of Cognition and Consciousness in Ancient Indian Thought Systems Relate to Current Western Theorizing and Research? *Frontiers in Psychology, 7*(613). Advance online publication. DOI: 10.3389/fpsyg.2016.00343 PMID: 27014150

Seehawer, M. K. (2018). Decolonising research in a Sub-Saharan African context: Exploring Ubuntu as a foundation for research methodology, ethics and agenda. *International Journal of Social Research Methodology, 21*(4), 453–466. DOI: 10.1080/13645579.2018.1432404

Sefa Dei, G. J., & Cacciavillani, I. (2024). *Decolonizing the university: Toward pluriversal knowledge*. Peter Lang.

Sefoka, T. S., & Chuene, K. J. (2025). Life Science learner views on the integration of indigenous knowledge into indigenous knowledge related topics using cooperative learning approach: A case of South African grade 10 classroom. *EURACIA Journal of Mathematics. Science and Technology, 21*(5), em2633. Advance online publication. DOI: 10.29333./ejemste/16311

Selepe, K., Ramsaroop, S., & Carolin, A. (2022). Incorporating indigenous knowledge in the teaching of weather and climate in the primary schools in Kwa-ZuluNatal, South Africa. *Journal of Geography Education in Africa, 5*, 1–25. DOI: 10.46622/jogea.v5i1.3953

Sen, A. (1999). *Development as freedom*. Oxford University Press.

Senanayake, S. G. J. N. (2006). Indigenous knowledge as a key to sustainable development.

Senanayake, S. G. J. N. (2006). Indigenous knowledge as a key to sustainable development. *Journal of Agricultural Sciences, 2*(1), •••.

Senegal, Q. & Lenz. R. (2020). South African Higher Education Curriculum: An investigation into the challenges. *International Journal of Social Sciences and Humanity Studies*, 12(1), 146-160. Received from degipark. org.tr/en/pub/ijsshs/52543/676852

Sewdass, N. (2014). Identifying knowledge management processes and practices used for decision-making and knowledge sharing in the modern San community. *TD. The Journal of Transdisciplinary Research in Southern Africa, 10*(4), 1–18. DOI: 10.4102/td.v10i4.85

Sharma, R., & Acharya, K. (2024). *ANCIENT WISDOM AND INDIGENOUS KNOWLEDGE: A Holistic Approach to Sustainable Development and Environmental Stewardship*. Retrieved from Himachal Pradesh National Law University: https://www.hpnlu.ac.in/PDF/b9813e07-5d12-4710-ac96-fd487f0a4b56.pdf

Sharma, A. K., & Talwar, B. (2005). Corporate social responsibility: Modern vis-à-vis Vedic approach. *Measuring Business Excellence, 9*(1), 35–45. DOI: 10.1108/13683040510588828

Sharma, A. K., & Talwar, B. (2007). Evolution of "Universal Business Excellence Model" incorporating Vedic philosophy. *Measuring Business Excellence, 11*(3), 4–20. DOI: 10.1108/13683040710820719

Sharma, A., Anand, N., Mukherjee, S., Kumar, S., Kumar, S., & Mohapatra, H. (2024). Analyzing territorial variations in the implementation of IoT-based smart homes. *Advances in Electronic Government, Digital Divide, and Regional Development*, •••, 305–334. DOI: 10.4018/979-8-3693-6854-1.ch010

Sharma, N. K., & Niharika, N. (2025). Roots and Resonance: The African Paradigm of Afrocentricism. *International Journal of English Literature and Social Sciences, 10*(1), 593321. Li, Sithole, J. (2007). The challenges faced by African libraries and information centres in documenting and preserving indigenous knowledge. *IFLA Journal*, *33*(2), 117–123.

Shastri, V. (2018, March 2). *What is Prakriti? Unfolding the Mystery of Life!* Retrieved from Vedanet: https://www.vedanet.com/what-is-prakriti-unfolding-the-mystery-of-life/

Shay, M., Sarra, G., & Woods, A. (2022). Grounded ontologies: Indigenous methodologies in qualitative cross-cultural research. In Liamputtong, P. (Ed.), *Handbook of qualitative cross-cultural research methods* (pp. 26–39). Edward Elgar Publishing. DOI: 10.4337/9781800376625.00011

Shiva, V. (2007). *Biopiracy: The plunder of nature and knowledge*. South End Press.

Shiva, V., & Bandyopadhyay, J. (1988). The Chipko Movement. In Ives, J., & Pitt, D. C. (Eds.), *Deforestation: Social Dynamics in Watersheds and Mountain Ecosystems* (p. 18). Routledge.

Shopify, (2024). 8 Digital Content Types: How To Create Digital Content. Available at: https://www.shopify.com/za/blog/digital-content [Accessed: 26 December 2025]

Sihombing, I. H. H., Suastini, N. M., & Puja, I. B. P. (2024). Sustainable cultural tourism in the era of sustainable development. *International Journal of Sustainable Competitiveness on Tourism*, *3*(02), 100–115. DOI: 10.34013/ijscot.v3i02.1601

Silva-Junior, C. H. L., Silva, F. B., Maisonnave Arisi, B., Mataveli, G., Pessôa, A. C. M., & Carvalho, N. S.. (2023). Brazilian Amazon indigenous territories under deforestation pressure. *Scientific Reports*, *13*(1), 5851. DOI: 10.1038/s41598-023-32746-7 PMID: 37037850

Simpson, A. (2014). *Mohawk interruptus: Political life across the borders of settler states*. Duke University Press.

Simpson, L. (2004). Anticolonial strategies for the recovery and maintenance of indigenous knowledge. *American Indian Quarterly*, *28*(3), 373–384. DOI: 10.1353/aiq.2004.0107

Simpson, L. B. (2014). Land as pedagogy: Nishnaabeg intelligence and rebellious transformation. *Decolonization*, *3*(3), 1–25.

Sinclair, O. (2025). The Ethics of Content Creation: Responsibility and Accountability. *Planksip*. Available at: https://www.planksip.org/the-ethics-of-content-creation-responsibility-and-accountability/ [Accessed: 20 January 2026]

Singh, P. K., Jain, S. K., Mujumdar, P. P., & Dey, P. (2020). Hydrology and water resources management in ancient India. *Hydrology and Earth System Sciences*, *24*(10), 4691–4707. DOI: 10.5194/hess-24-4691-2020

Slikkerveer, L. J. (2019). The Indigenous knowledge systems perspectives on sustainable development. In L. Slikkerveer, G. Baourakis, & K. Saefullah (eds.). Integrated Community-Managed Development, Corporative management, 33-66, Springer, Cham. South Africa.

Smith, L. T. (2021). Decolonizing methodologies: Research and indigenous peoples.

Smith, L. T. (1999). *Decolonizing methodologies: Research and Indigenous peoples* (1st ed.). Zed Books.

Smith, L. T. (1999). *Decolonizing Methodologies: Research and Indigenous Peoples.* Zed.

Smith, L. T. (2012). *Decolonizing methodologies: Research and indigenous peoples* (2nd ed.). Zed Books.

Snapp, S. S., Kerr, R. B., Bybee-Finley, A., Chikowo, R., Dakishoni, L., Grabowski, P., Lupafya, E., Mhango, W., Morrone, V. L., Shumba, L., & Kanyama-Phiri, G. (2023). Participatory action research generates knowledge for sustainable development goals. *Frontiers in Ecology and the Environment, 21*(7), 341–349. DOI: 10.1002/fee.2591

Solo-Anaeto, M., & Jacobs, B. (2015). Exploring social media as channels for sustaining African culture. *International Journal of Humanities and Social Science, 5*(4), 37–42.

Sraku-Lartey, M., Acquah, S., Brefo, S. & Djagbletey, G., (2017). Digitisation of indigenous traditional knowledge on forest foods and medicine in Ghana, 341 - 349.

Srinivas, S. (2013, September 10). *Dharma Shastras- Legal Literature of Ancient India.* Retrieved July 8, 2025, from ithihas: https://ithihas.wordpress.com/2013/09/10/dharma-shastras-legal-literature-of -ancient-india/

Statista (2024). *Global Social Media Usage Statistics.* https://www.statista.com

Statistics South Africa. (2025). *Quarterly Labour Force Survey (QLFS)* (Vol. Q1). Statistics South Africa.

Stein, S. (2022). *Unsettling the university: Confronting the colonial foundations of US higher education.* Johns Hopkins University Press. DOI: 10.56021/9781421445052

Stevens, S. (Ed.). (2014). *Indigenous peoples, national parks, and protected areas: A new paradigm linking conservation, culture, and rights.* University of Arizona Press. DOI: 10.2307/j.ctt183pbn5

Sultana, R., Muhammad, N., & Zakaria, A. K. M. (2018). Role of indigenous knowledge in sustainable development. *International Journal of Development Research,* 8(2). https://www.journalijrd.com Date of access: 20 April 2026.

Sultana, R., Muhammad, N., & Zakaria, A. K. M. (2018). Role of indigenous knowledge in sustainable development. *International Journal of Developmental Research, 8*(2), 18902–18906.

Swiderska, K. (2006). *Protecting Traditional Knowledge: A framework based on Customary Laws and Bio-Cultural Heritage. International Conference on Endogenous Development and Bio-Cultural Diversity, 3-5 October 2006.* IIED., Retrieved from https://www.iied.org/sites/default/files/pdfs/migrate/G01069.pdf

Talašová, Z., & Souralová, A. (2025). Meanings and Roles of Great-Grandparenthood: A Scoping Review. *The Gerontologist, 65*(4), 1–14. PMID: 39679945

TallBear. K. (2013). *Native American DNA: Tribal belonging and the false promise of genetic science.* University of Minnesota Press.

Tarugarira, G. (2014). Interrogating generational connections and conflicts on the utility of indigenous knowledge systems as a cultural dimension of development in Zimbabwe.

Tata, E. S., Ndzi, D. L., & Fonjong, L. N. (2024). Challenges in the application of indigenous knowledge in the management of ecotourism around Mount Cameroon National Park. *Journal of Sustainable Tourism, 32*(4), 623–640. DOI: 10.1080/09669582.2023.2178946

Taylor, J., & Pagliari, C. (2018). Comprehensive scoping review of health research using social media data. *BMJ Open, 8*(12), e022931. Advance online publication. DOI: 10.1136/bmjopen-2018-022931 PMID: 30552260

Te Hiku Media. (2020). *Indigenous data sovereignty and Māori language revitalisation.* https://tehiku.nz

Te Mana Raraunga. (2018). *Principles of Māori data sovereignty.* Te Mana Raraunga.

Teele, T., Nkoane, M. M., & Mahlomaholo, M. G. (2020). Corollaries of pedagogic praxis on technological skills transfer for emerging farmers. *Gender & Behaviour, 18*(1), 14749–14760. DOI: 10.17159/2413-3221/2020/v48n2a541

Tella, A., Jatto, E., & Yusuf, A. (2025). Preserving indigenous knowledge: Leveraging digital technology and artificial intelligence. *IFLA Journal, 51*(3), 1–19. DOI: 10.1177/03400352251342505

Tengö, M., Brondizio, E. S., Elmqvist, T., Malmer, P., & Spierenburg, M. (2014). Connecting Diverse Knowledge Systems for Enhanced Ecosystem Governance: The Multiple Evidence Base Approach. *Ambio, 43*(5), 579–591. DOI: 10.1007/s13280-014-0501-3 PMID: 24659474

Tharakan, J. (2015). Integrating indigenous knowledge into appropriate technology development and implementation. *African Journal of Science, Technology, Innovation and Development, 7*(5), 364–370. DOI: 10.1080/20421338.2015.1085176

Thompson, K. L., Lantz, T., & Ban, N. C. (2020). A review of indigenous knowledge and participation environmental monitoring. *Ecology and Society, 52*(2), art10. Advance online publication. DOI: 10.5751/ES-11503-250210

Thompson, M., Vervoort, J., & Frantzeskaki, N. (2020). Relationality and ethics in transformative sustainability research. *Sustainability Science, 15*(3), 849–861. DOI: 10.1007/s11625-020-00786-5

Tibuhwa, D. D. (2012). Folk taxonomy and use of mushroom in communities around Ngorongoro and Serengeti National Park, Tanzania. *Journal of Ethnobiology and Ethnomedicine, 2012*(8), 36. DOI: 10.1186/1746-4269-8-36 PMID: 22999253

Tippeconnic, J. W., & Faircloth, S. C. (2010). The Education of Indigenous Students. In P. Peterson, E. Baker, & B. McGaw, *International Encyclopedia of Education (Third Edition)* (pp. 661-666). Elsevier. DOI: 10.1016/B978-0-08-044894-7.00123-8

Tirivangasi, H. M., & Tayengwa, D. (2017). Indigenous knowledge systems (IKS) and food security in South Africa: Is land reform a prerequisite. *Journal of Human Ecology (Delhi, India), 57*(3), 118–124.

Todd, Z. (2016). An indigenous feminist's take on the ontological turn: 'Ontology' is just another word for colonialism. *Journal of Historical Sociology*, *29*(1), 4–22. DOI: 10.1111/johs.12124

Tondi, P. (2019). The Significance of Indigenous Knowledge Systems (IKS) for Africa's Socio- Cultural and Economic Development in the dawn of the Fourth Industrial Revolution (4IR). *Journal of Gender, Information and Development in Africa*, (pp 237-243). DOI: 10.31920/2050-4284/2019/S1n1a14

Tong, L. A. (2019). South Africa adopts sui generis indigenous knowledge protection legislation. *Journal of Intellectual Property Law & Practice*, *14*(12), 935–937. DOI: 10.1093/jiplp/jpz124

Toumaras, N. (2025). Networked Hyperlocal Activists: Digital Democracy and Engagement in Sub-Saharan Africa. *Social Media + Society*, *11*(2), 20563051251345945. DOI: 10.1177/20563051251345945

Traditional Knowledge Digital Library. (2014). *TKDL framework and practices*. Government of India.

Tripathi, S. (2021, November 17). *Monumental!* Retrieved from The Hindu: https://www.thehindu.com/features/friday-review/history-and-culture/Monumental/article16298184.ece

Tripathy, J. P. (2013). Secondary data analysis: Ethical issues and challenges. *Iranian Journal of Public Health*, *42*(12), 1478–1479. PMID: 26060652

Truby, J. (2018). Decarbonizing Bitcoin: Law and policy choices for reducing the energy consumption of Blockchain technologies and digital currencies. *Energy Research & Social Science*, *44*, 399–410. DOI: 10.1016/j.erss.2018.06.009

Truth and Reconciliation Commission of Canada. (2015). *Honouring the truth, reconciling for the future: Summary of the final report of the Truth and Reconciliation Commission of Canada*. Truth and Reconciliation Commission of Canada.

Tuck, E., & Yang, K. W. (2012). Decolonization is not a metaphor. *Decolonization*, *1*(1), 1–40.

Tume, S. J. P., Kimengsi, J. N., & Fogwe, Z. N. (2019). Indigenous knowledge and farmer perceptions of climate and ecological changes in the Bamenda Highlands of Cameroon: Insights from the Bui Plateau. *Climate (Basel)*, *7*(12), 138. DOI: 10.3390/cli7120138

Tussay, Á. (2024, September 26). Natural law and unwritten law in Classical Greek thought. *Hungarian Journal of Legal Studies*, *65*(1), 114–134. DOI: 10.1556/2052.2024.00534

Udeh, C.A., Daraojimba, R.E., Odulaja, B.A., Afolabi, J.O.A., Ogedembe, D.E. & James, O.O. (2023). "Youth empowerment in Africa: Lessons for US youth development programs."

Uma, N., Velayutham, A., Magudeeswaran, T., & Mcpe, S. (2024). Empowering Indigenous Heritage: Preserving, Promoting, and Connecting Through Technology. *National Conference*. Coimbatore: ResearchGate.

UN DESA. (2021). *Indigenous peoples and the digital divide*. United Nations Department of Economic and Social Affairs.

UNDP. (2021). *Indigenous peoples and COVID-19: From education to crisis recovery*. United Nations Development Programme.

UNESCO. (1970). *Convention on the Means of Prohibiting and Preventing the Illicit Import, Export and Transfer of Ownership of Cultural Property*. Retrieved from UNESCO: https://www.unesco.org/en/legal-affairs/convention-means-prohibiting-and-preventing-illicit-import-export-and-transfer-ownership-cultural

UNESCO. (2002). *Local and Indigenous Knowledge Systems (LINKS) programme*. https://en.unesco.org/links

UNESCO. (2009). *Convention on the Protection of the Underwater Cultural Heritage*. Retrieved from UNESCO: https://leblancf.com/documents/charters%20&%20conventions/UNESCO%20conventions/underwater_e.pdf

UNESCO. (2010). *Atlas of the world's languages in danger*. UNESCO Publishing.

UNESCO. (2011). *Heritage of Religious Interest*. Retrieved from UNESCO: https://whc.unesco.org/en/religious-sacred-heritage/

UNESCO. (2011). *Preparing World Heritage Nominations*. Retrieved from UNESCO: https://whc.unesco.org/en/preparing-world-heritage-nominations/

UNESCO. (2016, August 27). *The World Heritage Convention*. Retrieved from UNESCO World Heritage: https://web.archive.org/web/20160827065310/https://whc.unesco.org/en/convention/

UNESCO. (2017). *Culture for sustainable development*. UNESCO Publishing.

UNESCO. (2017). *Local and Indigenous Knowledge System*. UNESCO.

UNESCO. (2017). *Local and Indigenous Knowledge Systems* (LINKS) Programme. Paris: UNESCO.

UNESCO. (2018). *Pimachiowin Aki*. Retrieved from UNESCO: https://whc.unesco.org/en/list/1415/

UNESCO. (2018). *Safeguarding intangible cultural heritage*. UNESCO Publishing.

UNESCO. (2022). *Cultural diversity and the impact of digital media*. UNESCO.

UNESCO. (2025). *UNESCO Database of National Cultural Heritage Laws*. UNESCO. Retrieved from https://www.unesco.org/en/cultnatlaws

UNESCO. (2025, August 8). UNESCO's tracker training helps safeguard Indigenous knowledge. https://www.unesco.org/en/articles/unescos-tracker-training-helps-safeguard-indigenous-knowledge

UNFCCC. (2025). *The Paris Agreement*. Retrieved from unfccc: https://unfccc.int/process-and-meetings/the-paris-agreement

United Nations Development Programme. (2023). *Indigenous peoples and sustainable development*. UNDP.

United Nations Environmental Programme and Environmental Defenders Office New South Wales. (2013). *Community protocols for environmental sustainability: A guideline for policymakers*. UNEP, Nairobi and EDO NSW.

United Nations. (2007). *United Nations Declaration on the Rights of Indigenous Peoples (UNDRIP)*. https://www.un.org/development/desa/indigenouspeoples

United Nations. (2007). *United Nations Declaration on the Rights of Indigenous Peoples*. UN General Assembly.

University of Saskatchewan. (2019). *Guidelines for the ethical conduct of research involving Indigenous peoples*. University of Saskatchewan Research Ethics Office.

Valero, M., Merello, P., Navajas, Á., & García-Diego, F. (2014). Statistical tools applied in the characterisation and evaluation of a thermo-hygrometric corrective action carried out at the Noheda archaeological site (Noheda, Spain). *Sensors (Basel)*, *14*(1), 1665–1679. DOI: 10.3390/s140101665

Van Pixteren, P. (2022). The elephant in the room: why and how medium of instruction and decolonization of education are linked. *Journal of Contemporary issues in Education*, 17(2), 79-91. https://doi.orgDOI: 10.20355/jcie29496

Van Rooyen, B. (2011). Safeguarding the future of indigenous knowledge through ICT: ICT and society. *CSIR Science Scope*, *5*(2), 24–26.

Venkatesh, V., Thong, J. Y., & Xu, X. (2012). Consumer acceptance and use of information technology: Extending the Unified Theory of Acceptance and Use of Technology1. *Management Information Systems Quarterly*, *36*(1), 157–178. DOI: 10.2307/41410412

Verma, A., Shukla, V. K., & Sharma, R. (2021). Convergence of IOT in tourism industry: A pragmatic analysis. *Journal of Physics: Conference Series*, *1714*(1), 012037. DOI: 10.1088/1742-6596/1714/1/012037

Viveiros de Castro, E. (2014). *Cannibal metaphysics*. Univocal Publishing.

Vorster, C., Kritzinger, A., Lekganyane, M., Taljard, E., & Van Der Linde, J. (2022). Cultural adaptation and Northern Sotho translation of the Modified Checklist for Autism in Toddlers. *South African Journal of Childhood Education*, *12*(1), a968. DOI: 10.4102/sajce.v12i1.968

Vygotsky, L. S. (1986). *Thought and language*. MIT Press.

Walderman, A., & Branova, S. 2025. Indigenous Knowledge: Connecting Data, Land and Story. Available at: https://asm.org/articles/2025/november/indigenous-knowledge-connecting-data,-land-and-sto [Accessed: 20 April 2026].

Walker-Jones, A. (2021, November 27). *Trial, Error, And Success: The Indigenous Origins Of The Scientific Method*. Retrieved July 12, 2025, from Medium: https://medium.com/sofi-stories/trial-error-and-success-indigenous-origins-of-the-scientific-method-d020829afda1

Walliman, N. (2021). *Research methods: The basics*. Routledge. DOI: 10.4324/9781003141693

Walter, M., Kukutai, T., Carroll, S. R., & Rodriguez-Lonebear, D. (Eds.). (2021). *Indigenous data sovereignty and policy*. Routledge., DOI: 10.4324/9780429273957

Walter, M., & Suina, M. (2019). Indigenous data, Indigenous methodologies and global data governance. *International Journal of Social Research Methodology*, *8*(1), 1–12. DOI: 10.1080/13645579.2018.1531228

Wang, J., & Hao, J. (2024). Community-based tourism and cultural sustainability in indigenous contexts. *Journal of Sustainable Tourism*, *32*(2), 215–231. DOI: 10.1080/09669582.2023.2201147

Warren, D. M. (1991). Using Indigenous knowledge in agricultural development. *World Bank Discussion Papers*, No. 127.

Waterton, E., & Smith, L. (2010). The recognition and misrecognition of community heritage. *International Journal of Heritage Studies, 16*(1-2), 4–15. DOI: 10.1080/13527250903441671

Weir, J. K. (2009). *The Gunditjmara Land Justice Story*. National Library of Australia.

Whyte, K. (2017). Indigenous climate change studies: Indigenizing futures, decolonizing the Anthropocene. *English Language Notes, 55*(1), 153–162. DOI: 10.1215/00138282-55.1-2.153

Wierzbicka, A. (2003). *Cross-cultural pragmatics*. Mouton de Gruyter. DOI: 10.1515/9783110220964

Williams, J. (2019). Circular cities. *Urban Studies (Edinburgh, Scotland), 56*(13), 2746–2762. DOI: 10.1177/0042098018806133

Wilson, A. (2007). N'tacimowin inna nah': Our coming in stories. *Canadian Women's Studies, 26*(3-4), 193–199.

Wilson, S. (2008). *Research is ceremony. Indigenous Research Methods*. Fernwood.

Wilson, S. (2020). *Research is ceremony: Indigenous research methods*. Fernwood Publishing.

WIPO. (2003). *Convention for the Safeguarding of the Intangible Cultural Heritage*. Retrieved from WIPO: https://www.wipo.int/wipolex/en/text/594740

Woldegiorgis, E. T. (2021). Decolonising a higher education system which has never been colonized. *Educational Philosophy and Theory, 53*(9), 894–906. DOI: 10.1080/00131857.2020.1835643

Woods, S. (2012). *Eurogames: The Design, Culture and Play of Modern European Board Games*. McFarland & Company, Inc, Publishers.

World Intellectual Property Organization (WIPO). (2017). *Traditional knowledge and intellectual property: Background brief*. WIPO.

World Intellectual Property Organization. (2017). *Intellectual property and genetic resources, traditional knowledge and traditional cultural expressions*. WIPO.

World Intellectual Property Organization. (2020). *Protection of traditional knowledge: Draft articles*. WIPO Intergovernmental Committee on Intellectual Property and Genetic Resources.

Wunder, S. (2015). Revisiting the concept of payments for environmental services. *Ecological Economics, 117*, 234–243. DOI: 10.1016/j.ecolecon.2014.08.016

Wynberg, R. (2017). Making sense of access and benefit sharing in the rooibos industry: Towards a holistic, just and sustainable framing. *South African Journal of Botany, 110*, 39–51. DOI: 10.1016/j.sajb.2016.09.015

Xaba, F., & Adanlawo, E. F. (2024). Are women empowered through ecotourism projects? A quantitative approach: Ecotourism and women empowerment. *Pakistan Journal of Life and Social Sciences, 22*(1), 3085–3092. DOI: 10.57239/PJLSS-2024-22.1.00224

Xaba, F., & Adanlawo, E. F. (2024). The potential role of ecotourism in sustainable development: A systematic review. *Journal of Ecohumanism, 3*(4), 3356–3367. DOI: 10.62754/joe.v3i4.3852

Yang, X. (2010). Address Forms of English: Rules and Variations. *Journal of Language Teaching & Research, 1*(5). Advance online publication. DOI: 10.4304/jltr.1.5.743-745

Yazidi, R. E., & Rijal, K. (2024). Science Learning in the Context of 'Indigenous Knowledge' for Sustainable Development. *International Journal of Ethnoscience and Technology in Education, 1*(1), 28–41. DOI: 10.33394/ijete.v1i1.10880

Yin, R. K. (2018). *Case Study Research and Applications: Design and Methods*. Sage publications.

Zeppel, H. (2006). *Indigenous ecotourism: Sustainable development and management*. CABI. DOI: 10.1079/9781845931247.0000

Zonta, D., Wu, H., Pozzi, M., Zanon, P., Ceriotti, M., Mottola, L., Picco, G. P., Murphy, A. L., Guna, S., & Corra, M. (2010). Wireless sensor networks for permanent health monitoring of historic buildings. *Smart Structures and Systems, 6*(5_6), 595-618. DOI: 10.12989/sss.2010.6.5_6.595

Zubiaga, M., Izkara, J. L., Gandini, A., Alonso, I., & Saralegui, U. (2019). Towards smarter management of Overtourism in historic centres through visitor-flow monitoring. *Sustainability (Basel), 11*(24), 7254. DOI: 10.3390/su11247254

About the Contributors

Monicca Thulisile Bhuda is an indigenous scholar, editor, children's author, multi award winning academic, research fellow awardee and Indigenous Knowledge Systems expert. At the University of Mpumalanga, Dr. Bhuda coordinates a course on African identities and lectures on culture and heritage studies. She graduated from North-West University with a Bachelor's degree (hons) in Indigenous Knowledge Systems (with distinction), a Master's degree also in Indigenous Knowledge Systems (with distinction) and a PhD in Social Sciences with Indigenous Knowledge Systems (IKS). She has been featured on numerous media platforms such as TV, Radio, Magazines and Newspapers that were based on her interests as an indigenous scholar and expertise in the Indigenous Knowledge Systems discipline. She has also published over 40 academic works in accredited journals, including books indexed with SCOPUS. Her research is on various subjects related to Indigenous Knowledge Systems, which is multidisciplinary in nature.

Tebogo Maria Mothiba is a distinguished academic and research leader who, as of late 2025, serves as the Acting Deputy Vice-Chancellor for Research, Innovation, and Partnerships, and as Executive Dean of the Faculty of Health Sciences at the University of Limpopo. Her work centers on health service management and the management of non-communicable diseases, such as diabetes and hypertension, with extensive experience both as an academic and a registered nurse. Professor Mothiba holds a PhD in Health Service Management from the University of Johannesburg and an MCur from the University of Limpopo. She began her university career in 2006, rising to professorship in 2014, and was promoted through multiple leadership roles including Faculty Research Professor and Executive Dean. She is recognized as a South African National Research Foundation (NRF) C3-rated researcher and has published over 120 peer-reviewed articles and nine book chapters. Her research encompasses healthcare management, indigenous health practices, maternal and child health, and digital health innovations.

***

Rajesh Bahuguna is a distinguished legal academician currently serving as the Pro Vice Chancellor of Uttaranchal University and the Dean of Law College Dehradun. With over 45 years of professional experience, his career is marked by a unique blend of military and academic service, including 15 years in the Indian Air Force. He holds an LL.M. from Jai Narain Vyas University and a Ph.D. from Kurukshetra University, specializing in Constitutional Law and Alternative Dispute Resolution (ADR). As a trained mediator and the Editor-in-Chief of the Dehradun Law Review, Dr. Bahuguna has contributed extensively to legal scholarship, publishing nearly 50 research papers and supervising numerous doctoral candidates. Beyond his administrative and teaching roles, he is a recognized legal expert who frequently appears as a panelist on television and radio to discuss contemporary legal issues.

Parag Chatterjee received the B.E. degree in Computer Science & Engineering from the Karnatak University, Dharwad, India, in 1998 and received M.Tech. degree from IIEST Shibpur, Kolkata, India, in 2007. He has an academic experience of 20 years. Currently he is working as an Assistant Professor in the Department of CSE at Global Institute of Management and Technology, Krishnanagar, West Bengal, India.His current research interests include Data ware Housing, Data mining and Wireless sensor. He has 15 publication in national and international journals and 3 Patents.

Ashok Dobhal is a PhD Scholar and an Assistant Professor of Law at Law College Dehradun, Uttaranchal University. Prior to joining academia, he has established a background in litigation as a practicing advocate before the Hon'ble Supreme Court of India. Currently, Mr. Dobhal actively contributes to legal education and research in India. Known for his dedication and domain expertise, he plays a pivotal role in shaping future legal professionals and advancing the field of legal studies in India.

Soumen Ghosh,currently persuing a Bachelor of Technology (B.Tech) in Computer Science and Engineering from the Global Institute of Management & Technology which is located in Krishnanagar, India. With a forvest passion for technology and problem-solving. Soumen is engaged in coding competitions, hackathons, and workshops.

Johannes Bastira Ginting, M.K.M. is an academic in Public Health with a primary expertise in Epidemiology. He holds a Master's degree in Public Health and a doctoral degree, with research interests focused on disease epidemiology, health risk factors, and population-based disease analysis. His academic work includes teaching, research, and community service, particularly in communicable and non-communicable disease epidemiology, health surveillance, and epidemiological study designs such as cross-sectional and case-control studies, supporting evidence-based public health decision-making.

Piali Haldar earned her MBA from the Institute of Rural Management Anand and completed her PhD at Birla Institute of Technology, Mesra. Her core areas of specialization include rural marketing, rural development, branding strategy, and consumer behaviour. With over 20 years of experience in academia and research, she has published extensively in renowned journals listed in ABDC and Scopus. She has authored one textbook on Marketing and edited three academic books. Dr. Haldar has presented her research at numerous national and international conferences and has contributed more than 50 papers to reputed journals, edited volumes, and conference proceedings.

Rupam Hazra, currently pursuing a B.Tech in Computer Science and Engineering (C.S.E) at the Global Institute of Management & Technology, Krishnanagar, India. Driven by a deep passion for technology and a strong curiosity for innovation, Rupam is committed to exploring and mastering the ever-evolving world of software development and computer systems. With a focus on continuous learning and problem-solving, Rupam strives to turn complex challenges into meaningful solutions through code.

Kristiawan Indriyanto holds Ph.D. in American Studies from Universitas Gadjah Mada on the topic of postcolonial ecocriticism in Hawai'ian-American literature. His main research interests are analyzing literature with a postcolonial ecocriticism perspective, primarily focusing on Native Hawaiian literature and the decolonizing discourse of Hawaiian indigene through Aloha 'āina. His other research interests include environmental ethics, econarratology, space in literature and world literature. He is the author of over 85 publications on various topics, a number of which are indexed in Scopus or Web of Science. He is also active as Editorial Board members, Section Editors and Reviewers of several International (SCOPUS-Indexed) and Indonesian nationally accredited journals.

Radhey Shyam Jha is a distinguished legal academic currently serving as Professor and Head at Law College Dehradun, Uttaranchal University. With a career spanning over 25 years, he blends a decade of trial court advocacy with 15 years of higher education leadership. His academic credentials include a Ph.D. in Law, LL.M., and an M.B.A. in International Business, which informs his interdisciplinary approach to modern legal challenges. Dr. Jha is a prolific researcher with over 70 publications, focusing on the intersection of law and technology, including Artificial Intelligence, Blockchain, and Digital Rights. He is a key administrative figure at the college, serving on the Editorial Board of the Dehradun Law Review and chairing student bodies like the ADR Centre. Recognized with the "Best Teacher Award," he is equally committed to social justice, having organized over 40 Legal Awareness Camps focusing on POCSO and Cyber Laws to bridge the gap between legal theory and community welfare.

Gedong Maulana Kabir is a lecturer and serves as a Subject Content Specialist at the Character Building and Development Center (CBDC) at Bina Nusantara University. He is an alumnus of the Center for Religious and Cross-cultural Studies (CRCS) at Gadjah Mada University. His academic interests center on the relationship between religion, the state, and society. Specifically, his research explores issues related to indigenous religion, customary traditions (adat), and minority religions, examining their intersection with environmental, gender, cultural, and state policy aspects.

Saad Ullah Khan brings over a decade of distinguished academic and research experience in higher education, with an interdisciplinary expertise that bridges communication studies and its broader societal dimensions. A Ph.D. graduate from the prestigious Aligarh Muslim University, Dr. Khan has contributed extensively to scholarly discourse, with multiple publications indexed in SCOPUS and the Web of Science, alongside works featured in nationally and internationally reputed journals. His research excellence has been recognized through highly competitive fellowships, including the Rajat Jayanti Vigyan Sancharak Fellowship (2009), awarded by the Department of Science and Technology, Government of India, for his pioneering work on communication and its social implications, and the Maulana Azad National Fellowship (2011) by the University Grants Commission in support of his doctoral studies. Dr. Khan has presented more than 90 research papers at national and international forums, reflecting both scholarly depth and global engagement. His academic achievements have been further acknowledged through multiple accolades, including the Best Researcher Award (2020 & 2023) and the Best Journalism Teacher Award (2021), underscoring his dual contributions as both a researcher and educator.

Sadaf Khan has an extensive teaching experience of more than ten years. A core academician and a prolific researcher, she has done incredible work in the domain of English Literature and Afghanistan Studies. Having an international teaching experience, she has been felicitated multiple times as 'Best Teacher' by various academic institutions in which she worked. Besides, she has also served as the Joint Editor of Women's College Annual magazine of Aligarh Muslim University for the session 2007-2008.

Motheo Koitsiwe short bio Dr Motheo Koitsiwe is currently the Director of the DSTI - NRF Center of Excellence in Indigenous Knowledge Systems Centre (CEIKS), Faculty of Natural and

Agricultural Sciences (FNAS), North - West University (NWU) in South Africa. He is responsible for the Teaching, Learning, Research, Innovation, and Community Engagement in Indigenous Knowledge Systems at the North West University. Dr Koitsiwe is mainly driving the decolonial and transformation agenda at the North West University and using IKS as a strategy to make impact not only among the academia, students but also to the communities, government and industry. He is also science diplomat, supervisor for Masters and PhD candidates in IKS at NWU. Furthermore, in partnership with the DSTI and the Innovation Hub, Dr Koitsiwe established the IKS Coachlab to ensure that the undergraduates students are equipped with entrepreneurial skills and prepare them for the future. Dr Motheo Koitsiwe is the first African indigenous scholar to have completed successfully the IKS PhD thesis in African indigenous astronomy (bolepa - dinaledi) of Batswana in Botswana and South Africa. He has done and continues to conduct indigenous research with indigenous astronomy experts, host conferences, seminars, communitybased engagement (Maitiso) sessions in indigenous astronomy (Bolepa Dinaledi). He is currently working on the book titled "Bolepa Dinaledi" of Batswana in South Africa and Botswana. Dr Koitsiwe was the first coordinator of the Bachelor of Indigenous Knowledge Systems (BIKS) which is a four (04) years professional degree registered with the South African Qualification Authority (SAQA) and approved by the Council for Higher Education (CHE). He is currently working on the Programme Qualification Mix (PQM) Review of the Bachelor of Indigenous Knowledge Systems to revise the curriculum, make it relevant, responsive to society, economy, industry and to be futuristic. He also established reputation with engaging and working with African indigenous health practitioners, traditional leaders in the Recognition of Prior Learning which is a project sponsored by the Department of Science and Innovation. Dr Koitsiwe also runs the project for the Indigenous Knowledge Systems Documentation Center in the North West province, including the Indigenous Knowledge Regulations, Bio Cultural Community Protocols funded by the Department of Science, Technology and Innovation. Dr Koitsiwe has also successfully completed the Resistance Liberation Heritage Route Project which focused on the Feasibility Study of Dr Ruth Segomotsi Mompati (Vryburg), Onkgopotse Tiro (Dinokana), and Moses Kotane (Matlhako) funded by the National Heritage Council.

Ushanandhini M. is an academician in the Department of Artificial Intelligence and Data Science at Vivekanandha Arts and Science College for Women, Veerachipalayam, Sankari, India. She holds postgraduate qualifications in Computer Science and specializes in artificial intelligence, data science, digital ethics, and emerging technologies. Her research interests include ethical AI, indigenous knowledge preservation, data governance, digital heritage systems, and technology-enabled social sustainability. She has published scholarly works in peer-reviewed journals and edited volumes and actively engages in interdisciplinary research that bridges technology, ethics, and cultural studies.

Mothusiotsile Edwin Maditsi is a PhD candidate at the North-West University (Mahikeng campus). He holds a BA (Hons) in Indigenous Knowledge Systems (IKS) degree from the North-West University (NWU), a Master of IKS degree also from the NWU and currently a PhD candidate at the NWU. During his Master's degree journey he served as a full member of PRIMCO in which he was appointed as a researcher for the project leader. He has participated and facilitated some community engagement activities for and on behalf of the IKS Centre at National and local level.Ssome of the community engagement activities entails; IK Act of 2019 and BCP public awareness workshops across the country; IK Act regulations public consultations across the country and a host of other IKS related activities. He is now appointed as a lecturer under the nGAP at the Indigenous Knowledge Systems Centre (NWU

Mahikeng). In his PhD he is focusing on decolonizing community engagement using the NWU as a case study and he is supervised/promoted by Prof SA Materechera. Otsile is a member of South African Higher Education Community Engagement Forum (SAHECEF). Since assuming his position as a lecturer, he has shown dedication and commitment in decolonizing the teaching and learning activities within the higher education sector. This has seen him attend and present papers at the NWU's annual teaching and learning conference. In 2022, Otsile has since been appointed by the IKS Centre Director to coordinate the third leg of the centre which is community engagement and stakeholder relations, a position that he is still currently serving in.

Glen Maimela is a Research Assistant in the Research and Knowledge Production Unit at the National Heritage Council. He earned his Bachelor's Degree in Indigenous Knowledge Systems in 2020 and his Master of Arts in African Studies (Cum Laude) in 2024 from the University of Venda. He is currently a PhD candidate at the University of the Free State (UFS) under the Centre for Gender and African Studies (CGAS). He is an Indigenous knowledge practitioner and multidisciplinary scholar with research interests in participatory action research, research decoloniality, Indigenous knowledge, community engagement, digitisation, environmental justice, just energy transition, heritage preservation and management. He is a recently 2025 trained leadership coach trained through the Global Enterprise Experience (GEE), which is facilitated by the Wind-eaters and the Victoria University of Wellington.

Dev Kumar Mandal currently serves as an Assistant Professor in the Department of Management at Brainware University, Kolkata. Before this, he worked at ICRA Analytics Ltd and Shriram Finance Ltd. He holds an MBA in Finance and a Bachelor of Commerce degree from Vidyasagar University. Additionally, he has successfully cleared the UGC NET and WB SET examinations in Management.

Nkosinamandla Mbopheni is a highly motivated and accomplished scholar in Indigenous Knowledge Systems (IKS). He earned his Bachelor of Indigenous Knowledge Systems degree from North-West University, graduated on 16 July 2025, and is now in his second year of a master's program at the same institution, where his research centers on indigenous agricultural practices aimed at promoting sustainability and cultural preservation.

Mokgadi Asnath Modiba is a lecturer in the Department of Biodiversity, School of Molecular and Life Sciences, University of Limpopo, South Africa. She holds an MSc in Zoology, Postgraduate Diploma in Higher Education (Durban University of Technology) and a Postgraduate Certificate in Education (University of Limpopo). Awarded as the best-established teacher at the school level in 2023.

Hitesh Mohapatra received the B.E. degree in Information Technology from Biju Patnaik University of Technology (BPUT), Odisha in 2006, and the MTech. Degree in CSE from Odisha University of Technology and Research (OUTR) University, Odisha in 2009. He received his Ph.D. in Computer Science & Engineering in 2021 from Veer Surendra Sai University of Technology (VSSUT), Burla, India. He has contributed 50+ SCI and Scopus indexed research articles. He has authored two books on Software Engineering and C Programming respectively. He has 15 years of teaching experience both in industry and academia. He has served the research community with various capacities like session chair, technical chair, keynote speaker etc. His research interests include wireless sensor network, smart city, smart grid and smart water. Currently he is working as Associate Professor at School of Computer Engineering, KIIT (Deemed to be) University, Bhubaneswar, Odisha, India.

Diane Thelma Molokwa is currently serving as a Junior Lecturer at the Indigenous Knowledge Systems Centre, Mahikeng Campus. She possesses a Master's degree in Indigenous Knowledge Systems from North-West University. Her research interests focus on the realms of indigenous food security, climate change, and preservation of wild genetic resources. Notably, Molokwa has received recognition from esteemed International GoldenKey Organization as a high-achieving student. She has actively participated in scholarly gatherings, including a symposium themed "Taking Indigenous Knowledge Seriously" hosted by the University of Johannesburg and the Indigenous Knowledge Conference hosted by Nelson Mandela University in South Africa, thereby demonstrating her commitment to advancing the field of indigenous knowledge systems.

Lindiwe Carol Mthethwa is a Senior Lecturer at Educational Foundations, Faculty of Education, University of Zululand, South Africa. She is a leader of the digital championship in the Faculty of Education and has written articles based on infusing technology into teaching and learning. Her passion is on handling case studies in philosophy, sociology, educational leadership and management, decolonisation strategies, indigenous knowledge and climate change education together with innovative strategies for teaching and learning.

Srishti Muralidharan is a UGC NET qualified postgraduate in Clinical Psychology and an Assistant Professor of Psychology in PES University, RR Nagar Bangalore. Her areas of interest include Clinical Psychology, Indian Psychology, Philosophy, Gender and Sexuality, Social Psychology and Neuropsychology. She has published a paper titled 'Mental Health Literacy for Depression and Schizophrenia among Undergraduate Students' in the International Journal of Indian Psychology. She has also co-authored several book chapters. She is a practicing psychologist and believes in making mental health care accessible to all. She has also obtained Level I Certifications in Pranic Healing and Reiki and strongly believes in integrating alternative healing practices into modern day psychotherapy, customising it according to clients and delivering quality treatment.

Divhani Muvhango is a junior lecturer at the University of Venda, specializing in Indigenous Knowledge Systems and heritage studies. She is pursuing a master's degree exploring traditional Vhavenda paintings focusing on their cultural meaning and potential to create sustainable income opportunities for youth. Her research explores the intersections of Indigenous knowledge, heritage preservation, and community wellbeing, including traditional food systems and culturally grounded approaches to health and development. Divhani is committed to promoting African cultural knowledge, empowering youth, and integrating Indigenous practices into contemporary education and development initiatives.

Rudzani Nedombeloni holds a Master's in Rural Development, Honours in Rural Development, and a Bachelor of Arts in Development Studies, all obtained from the University of Venda. She is currently a Sociology PhD candidate at the University of Venda under the Department of Arts and Social Sciences. She is a scholar with research interests in rural development, community engagement, Indigenous knowledge systems, food security, water governance, smallholder farming, climate change, and just energy transition. She has published several papers in Department of Higher Education and Training

(DHET) accredited journals. She has also presented papers at several research conferences, such as the annual Africa Young Graduates and Scholars (AYGS) hosted by the Human Sciences Research Council (HSRC) and the annual conferences hosted by the Oral History Association of South Africa (OHASA). She is currently a Programme Officer for Climate Change, Environment and Energy at the Academy of Science of South Africa (ASSAF).

Sandile Tyrant Nhuvunga is a Master of Development Studies candidate at the University of Mpumalanga and an emerging development practitioner committed to translating research into actionable economic solutions. His work focuses on bridging academic insight with practical implementation, particularly within higher education institutions, township economies, and rural development contexts. As an aspiring economic architect, Sandile is driven by the belief that sustainable development requires institutional integrity, locally grounded knowledge systems, and enterprise-led growth.

Lindelwa Veronicah Nkosi holds a Bachelor of indigenous knowledge systems degree obtained in 2018 and a Master of Arts in African studies degree obtained from the university of Venda in 2022. She started her career as a tutor for the department of African studies at the university of Venda. in 2021 she joined the Human Science Research Council (HSRC) as an intern and has worked for the Department of Culture, Sport and Recreation in Mpumalanga under the Archives and records management division. She is currently a lecturer in the university of Mpumalanga in Culture and Heritage Studies in the school of Social Sciences. She intends to pursue a PhD in indigenous knowledge systems Focusing on the promotion and preservation of Indigenous languages.

Ananya Pandey is a Research Scholar in School of Law at Christ University, Bangalore. She has her research interests in the field of Constitutional Law, Environment Law, Human Rights, Food Security, Intellectual Property Law and Jurisprudence.

Delecta Rajendren is an Assistant professor, EDP in Talent Management at XLRI Jamshedpur, an experienced researcher, and first-time initiator in her university to assess and map the Cross-cultural competence of the middle-level professionals, working in the footwear industry in India and China. Extensively traveled in China and examined the cross-cultural behavior, and working environment, co-established the presence, while presenting, and attended International conferences across the continents. A Sr. HR professional with over 9+years of working experience, with an array to manage the various roles, handling the HR specimen roles, to giving an exclusivity, as starting the 'Campus Placement' drive, during the academic head for the MBA department in the J.M.C.E institution. Extensively researched over the years in Cross-Cultural competence, published papers, presented at various International and National conferences, and co-authored the book on Managerial Behavior and Effectiveness. An experienced soft-skill trainer, established a reputation in conducting the mock-drills sessions, for the upcoming professionals; to handling the grievance addressal systems in the manufacturing unit

Rick Rantz is a scholar and educator whose work focuses on educational equity, culturally responsive curriculum, and the structural dynamics that shape student experiences in educational institutions. His scholarship examines the intersections of culture, identity, and educational systems, with particular attention to the experiences of minoritized students. Through his research and writing, he seeks to advance more inclusive and responsive educational practices. His education background includes a teaching English as second language certificate, a master's degree in linguistics and bilingual education, and doctoral studies in higher education at Baylor University and the University of Houston. As a former K-12 teacher, he worked in California, Ecuador, and Argentina. At the higher education level, he has held executive positions at Feather River College, Sierra College, and Allan Hancock College.

Mani Sachdev is an esteemed academician currently serving as Professor and Head of the Institute of Behavioural and Allied Sciences at Amity University Rajasthan, Jaipur. Her academic journey includes earning an M.Phil. in Philosophy and completing her Doctorate at the University of Rajasthan, specializing in Aesthetics, Metaphysics, and Epistemology. With an impressive tenure spanning 37 years, she has excelled as an Associate Professor of Philosophy. Dr. Sachdev's pioneering roles include founding the Department of Philosophy at Vedic Kanya P. G. College, Jaipur, and establishing the Department of Philosophy at B.N.D. Arts Government P.G. College Chimanpura, Jaipur. Her scholarly contributions and administrative leadership underscore her commitment to advancing philosophical discourse and academic excellence in higher education. Her professional portfolio encompasses the design and development of innovative curricula tailored for students, ensuring adherence to rigorous quality assurance standards within the University framework. Additionally, she contributes to the pedagogical landscape by teaching ethics to medical professionals at a distinguished medical university in Jaipur, where she serves as a distinguished member of the ethics committee. Her scholarly contributions extend to numerous international and national conferences, seminars, and scholarly journals, where she has presented research papers, authored book chapters, and contributed book reviews. Beyond academia, she has demonstrated versatility in media production, having produced, directed, and scripted corporate films for hospitals and organizations, as well as documentaries for national broadcast on Doordarshan. Furthermore, she has conducted in-depth interviews with over 50 prominent global figures, spanning governors, philosophers, academics, and renowned personalities from the film industry. Her cultural engagement includes creating, directing, writing, and narrating the acclaimed 'Light and Sound Cultural Show' celebrating Rajasthan's art, culture, and heritage, showcased at numerous tourism festivals and international conferences.

Aahana Saha is an Assistant Professor of Psychology with over nine years of teaching experience. She holds a Master's degree, is UGC-NET qualified, and completed a Ph.D. on female police personnel's job satisfaction and happiness. Her work includes developing and validating various psychological scales, publishing 20+ research papers, and securing two patents on machine-learning applications for cyber defense and forensic behaviour prediction. She has guided more than 30 postgraduate dissertations, coordinated academic programs, and launched student initiatives. Currently she is pursuing a PG certificate in forensic psychology and criminal profiling and a Master's in Data Science from Purdue University.

sha Sharma is an Assistant Professor with academic specialization in constitutional and corporate law. Her research focuses on the role of Artificial Intelligence in the criminal justice system, examining issues of accountability, transparency, and constitutional safeguards in technology-driven adjudication. Her broader scholarly interests include sustainability, Indian Knowledge Systems (IKS), blockchain in corporate governance, economic analysis of Alternative Dispute Resolution (ADR), and corporate surveillance within regulatory frameworks. Through her academic engagement, she aims to contribute to progressive legal scholarship at the intersection of technology, governance, and constitutional values.

Shri. Jogen Sharma is a Ph.D. Research Scholar in Library and Information Science at KIIT Deemed to be University, Bhubaneswar, Odisha, India and Assistant Librarian at the India International Centre (IIC), New Delhi, India. His academic and professional work centers on bibliometric research, digital transformation, and sustainable library practices. He has contributed widely through publications in reputed journals, edited volumes and conferences. His expertise includes implementing advanced technologies to modernize academic libraries and supporting institutional research needs. He is actively involved in professional associations and is committed to advancing innovation and sustainability in Library and Information Science.

Neharshi Srivastava is an accomplished Assistant Professor at Amity University Rajasthan with a distinguished academic and professional background in psychology. A Gold Medalist from Lucknow University, she holds a license as a Rehabilitation Psychologist from the Rehabilitation Council of India (RCI). Her career spans over a decade, covering counseling, research, and industry practices, with a specialization in Cognitive Behavioral Therapy (CBT) from London. Dr. Srivastava's expertise lies in mindfulness, positive psychology, and yogic interventions. She has published over 40 research papers in prestigious international journals, including Scopus, Web of Science, and UGC-recognized publications. She has also authored eight books and contributed to several book chapters. Her outstanding work has earned her awards like the "Best Young Faculty Award" and the "Dhruv Ratan Award." As a Core Committee Member of the Indian Academy of Health Psychology, she plays a pivotal role in advancing the field. Additionally, she has around 14 years of experience in electronic media and serves as a Radio Jockey for AIR FM Rainbow 100.7 MHz. All India Radio, Gyanvani and various community radio.

Tri Suci, S.K.M., M.K.M. is a lecturer and researcher in Public Health, with expertise in health promotion, health advocacy and policy, and community empowerment. She completed her Bachelor's and Master's degrees in Public Health and earned a doctoral degree in the same field. Dr. Tri Suci is actively engaged in teaching, research, and community service as part of the university's tri dharma mission.

Dayawanti Tarmali serves as an Assistant in the Post Graduate Diploma Department at Management Development Institute Murshidabad, where she is responsible for departmental coordination and academic administrative duties. She contributes actively to the effective functioning of postgraduate academic programs through her organisational and managerial expertise. She has authored several scholarly articles in Library and Information Science and Management, reflecting her interdisciplinary academic engagement and research orientation. Her work demonstrates an interest in knowledge organisation, academic administration, and management practices in educational institutions. She holds a Master of Commerce (M.Com) and a Master of Business Administration (MBA), which provide a strong foundation for her professional role and research pursuits.

Livhuwani Tshikukuvhe is a lecturer in the Department of Indigenous Knowledge Systems and Heritage Studies at the University of Venda. She holds a master's degree in African studies, a bachelor of Applied Anthropology, a Bachelor of Development Studies and a Diploma in Women Development and micro finance from the university of Venda. She is also a motivational speaker and an entrepreneur .

Felix Sandile Thwala is a researcher with academic experience in both South Africa and Germany, developed through his interdisciplinary scholarly journey. His work is strongly research-driven, focusing on empowering communities, socio-economic development, indigenous knowledge systems, data science, and sustainable innovation. He is particularly interested in how indigenous knowledge can be integrated with modern scientific approaches to address local and global development challenges. Felix has authored and contributed to a range of academic outputs, including municipal policy frameworks, a conference paper, a Sabinet-indexed publication, and book chapters. His research reflects a commitment to producing socially relevant and impactful knowledge that bridges academia, policy, and practice. Through his work, he continues to contribute to emerging conversations on inclusive innovation and the role of indigenous systems in sustainable development.

Nidhi Tyagi is a Professor and researcher at Graphic Era (Deemed to be University) in Dehradun, serving within the Department of Humanities & Social Sciences. She has nearly two decades of experience in academia, specializing in Sociology and its intersection with modern technology and law. Dr. Nidhi Tyagi is a prominent academician and researcher whose work primarily explores the sociological dimensions of modern technology, particularly Artificial Intelligence, and its impact on traditional legal and social structures. As a Professor her research portfolio reflects a unique blend of Social Anthropology and contemporary legal-tech issues.

Index

A

acknowledge 25, 28, 115, 119, 134, 184, 194, 274, 297
African Indigenous Knowledge 65, 89, 108, 110, 111, 178, 179, 180, 187, 189, 190, 191, 192, 194, 195, 196, 199, 200, 221, 265, 304
African Indigenous Knowledge Systems 89, 108, 111, 199, 200, 265, 304
Afrikology 287, 288, 289, 290, 305
Afrocentric Ethical Guidelines 177, 178, 179, 190, 198, 221
Afrocentricity 109, 179, 200, 203, 204, 205, 214, 216, 217, 219, 223
Ancient Indian Ecological Governance 383, 387, 398, 407
Artificial Intelligence 1, 2, 3, 5, 7, 13, 14, 15, 16, 17, 20, 35, 38, 42, 92, 108, 117, 120, 123, 144, 151, 152, 153, 163, 171, 186, 286, 312, 324, 357, 358, 363, 366, 372, 376, 379, 380, 381, 413, 415, 432
Augmented Reality 128, 145, 153, 414, 415, 423, 428, 432

B

BCPs 249, 250, 251, 252, 253, 254, 255, 257, 258, 259, 260, 261, 262
Biocultural Rights 263, 265, 266, 330, 342, 348, 349, 354, 355
Biodiversity 1, 2, 4, 16, 17, 19, 28, 37, 45, 46, 47, 48, 49, 50, 51, 52, 53, 54, 55, 62, 69, 70, 72, 74, 78, 80, 81, 87, 110, 116, 118, 125, 126, 143, 148, 151, 239, 240, 243, 254, 255, 258, 260, 263, 264, 266, 270, 272, 273, 277, 282, 285, 292, 294, 296, 297, 303, 327, 328, 329, 330, 331, 332, 333, 334, 335, 336, 337, 339, 340, 341, 343, 344, 345, 346, 347, 348, 349, 351, 352, 354, 355, 384, 388, 391, 396, 399, 405, 406, 442, 450

C

Climate 1, 2, 3, 4, 5, 16, 17, 20, 28, 37, 39, 46, 47, 48, 50, 51, 52, 53, 54, 65, 70, 71, 79, 81, 83, 84, 85, 86, 109, 110, 111, 116, 118, 125, 126, 143, 144, 151, 182, 197, 235, 237, 239, 247, 253, 254, 270, 272, 273, 276, 279, 280, 282, 283, 285, 301, 310, 322, 330, 331, 332, 336, 340, 343, 344, 345, 348, 349, 351, 352, 353, 383, 384, 385, 386, 388, 395, 396, 399, 401, 404, 405, 406, 408, 409, 410, 411
Climate Resilience 4, 16, 47, 50, 125, 144, 151, 247, 330, 332, 336, 340, 383, 385, 404, 405
Collective 1, 2, 3, 4, 5, 6, 8, 16, 17, 21, 23, 29, 35, 37, 42, 46, 47, 48, 53, 55, 61, 62, 63, 115, 116, 118, 120, 128, 130, 132, 134, 182, 184, 203, 204, 213, 216, 217, 226, 229, 251, 269, 270, 271, 272, 273, 274, 275, 276, 277, 278, 281, 282, 283, 285, 288, 293, 295, 296, 297, 299, 300, 309, 315, 330, 332, 334, 335, 337, 338, 339, 340, 341, 342, 344, 346, 348, 349, 353, 354, 355, 360, 361, 384, 385, 389, 390, 391, 393, 394, 395, 398, 400, 404, 405, 406, 410, 420, 428
colonialism 44, 46, 47, 67, 68, 113, 114, 117, 129, 130, 134, 135, 175, 178, 180, 182, 197, 202, 304
Commons Governance 273, 275, 330, 383, 389, 390, 393, 395, 398, 404
Communication 13, 40, 68, 73, 77, 80, 113, 114, 128, 163, 165, 175, 199, 201, 202, 203, 204, 205, 206, 207, 208, 209, 210, 213, 214, 215, 216, 217, 219, 220, 221, 232, 245, 303, 324, 377, 395, 410, 413, 415, 417, 423, 424, 426, 428, 449
Communities 2, 4, 5, 7, 13, 14, 16, 17, 18, 21, 25, 26, 27, 28, 29, 30, 31, 32, 33, 35, 36, 37, 38, 39, 42, 43, 44, 45, 46, 47, 48, 49, 50, 51, 52, 53, 54, 55, 56, 57, 58, 59, 60, 61, 62, 63, 64, 67, 68, 69, 70, 71, 72, 74, 77, 79, 80, 81, 82, 84, 87, 89, 90, 91, 92, 93, 94, 95, 96, 97, 98, 99, 100, 101, 102, 103, 104, 105, 106, 107, 108, 109, 110, 111, 112, 114, 116, 118, 119, 120, 130, 131, 132, 135, 136, 140, 141, 143, 147, 151, 153, 154, 157, 161, 163, 164, 165, 166, 167, 168, 170, 171, 172, 173, 175, 178, 179, 180, 181, 182, 184, 185, 189, 191, 192, 194, 195, 196, 198, 201, 202, 203, 204, 205, 206, 207, 210, 211, 212, 215, 217, 219, 226, 227, 228, 229, 235, 236, 239, 241, 244, 249, 250, 251, 253, 254, 255, 257, 258, 259, 260, 261, 262, 263, 264, 265, 266, 267, 269, 270, 271, 272, 274, 275, 276, 277, 278, 279, 280, 281, 282, 289, 293, 294, 299, 303, 307, 308, 309, 311, 312, 313, 314, 316, 317, 318, 320, 322, 324, 325, 327, 331, 334, 338, 340, 341, 342, 343, 377, 378, 384, 386, 390, 393, 399, 402, 404, 406, 408, 420, 423, 435, 436, 437, 438, 439, 440, 441, 442, 443, 444, 445, 446, 447, 448, 449, 450, 451
community-based governance 342, 396
Community-Centered Design 157, 164
Community Empowerment 142, 165, 170, 227, 247, 250, 256, 269, 270, 283
Community Engagement 30, 34, 44, 105, 108, 114, 165, 197

Community Stewardship 355, 385, 389, 396, 404, 408
Conservation 1, 2, 16, 17, 28, 46, 49, 50, 52, 53, 54, 63, 69, 71, 72, 78, 80, 81, 83, 87, 114, 116, 118, 125, 128, 143, 144, 225, 226, 229, 230, 232, 233, 239, 240, 247, 254, 255, 266, 270, 272, 273, 277, 282, 292, 293, 294, 297, 310, 329, 333, 342, 343, 344, 351, 352, 353, 354, 383, 384, 386, 388, 390, 393, 394, 396, 398, 399, 405, 406, 408, 409, 410, 413, 414, 415, 416, 417, 418, 421, 422, 423, 426, 429, 430, 432, 435, 436, 437, 438, 440, 441, 442, 444, 445, 446, 447, 448, 449, 450, 451
Contestations 89, 90, 93, 94, 95, 96, 103, 104, 105, 106, 111
Cooperative Governance 330
cultural commodification 277, 441, 442, 444, 451
Cultural Heritage 17, 54, 58, 60, 61, 62, 63, 64, 107, 115, 130, 143, 148, 153, 175, 176, 188, 199, 202, 204, 226, 227, 228, 229, 237, 238, 239, 240, 241, 244, 245, 246, 247, 258, 277, 283, 301, 310, 314, 413, 414, 415, 416, 417, 418, 422, 423, 426, 427, 428, 430, 431, 432, 435, 438, 442
Cultural Preservation 13, 47, 52, 54, 63, 189, 204, 291, 316, 319, 428, 435, 436, 439
Cultural Resilience 281, 297

D

Data Governance 9, 35, 37, 40, 42, 113, 115, 116, 117, 132, 135, 148, 151, 152, 153, 155, 164, 166, 170, 172, 174, 182, 184, 198
Data Sovereignty 2, 3, 4, 9, 13, 14, 16, 19, 23, 32, 35, 38, 39, 40, 42, 116, 130, 131, 132, 135, 136, 146, 147, 148, 149, 151, 152, 153, 154, 155, 157, 158, 163, 166, 167, 168, 171, 172, 174, 175, 176, 183, 184, 185, 195, 198, 280, 303, 359
Decolonial Digital Practices 130
Decoloniality 289, 301, 305
Decolonization 4, 24, 25, 34, 41, 42, 43, 87, 117, 142, 263, 302, 324
Dharma 233, 246, 357, 358, 359, 361, 362, 363, 364, 366, 367, 370, 372, 373, 374, 376, 378, 380, 382, 383, 384, 387, 394, 396, 398, 401, 404, 408, 412
Digital Content Creation 110, 177, 178, 179, 185, 186, 187, 190, 191, 192, 193, 194, 195, 196, 198, 200, 314
Digital Heritage 114, 140, 146, 147, 154, 155, 166, 169, 170, 174
Digitalisation 107, 110, 199, 201
Digital Preservation 71, 79, 83, 85, 92, 100, 105, 106, 107, 109, 110, 111, 121, 132, 134, 147, 152, 153, 154, 163, 165, 166, 167, 170, 171, 172, 316, 325
Digital Technology Platforms 178
Digitisation 89, 90, 92, 93, 94, 95, 96, 97, 98, 99, 100, 101, 102, 103, 104, 105, 106, 107, 108, 111, 178, 187, 188, 189, 190, 194, 195, 197, 198, 199, 200
disciplinary knowledge 67, 77, 81

E

Ecological 1, 2, 4, 5, 15, 17, 23, 26, 27, 28, 33, 34, 37, 43, 45, 46, 47, 48, 49, 50, 51, 52, 53, 58, 59, 60, 61, 62, 64, 65, 111, 114, 115, 116, 118, 120, 126, 128, 144, 147, 148, 151, 152, 163, 171, 172, 174, 225, 226, 229, 233, 235, 266, 270, 271, 272, 273, 275, 276, 277, 280, 282, 283, 284, 286, 288, 291, 292, 293, 294, 295, 296, 297, 299, 300, 305, 310, 312, 327, 328, 329, 330, 331, 332, 333, 334, 335, 336, 337, 338, 339, 340, 341, 342, 343, 344, 345, 346, 347, 348, 349, 350, 351, 352, 354, 355, 383, 384, 385, 386, 387, 388, 389, 390, 393, 394, 395, 396, 398, 399, 400, 401, 402, 404, 405, 406, 407, 408, 409, 410, 415, 436, 451
Economic Development 52, 72, 202, 206, 249, 250, 251, 252, 253, 254, 258, 259, 260, 261, 262, 264, 266, 269, 270, 272, 278, 292, 308, 311, 312, 314, 319, 320, 324, 328, 347, 359, 436, 450, 451
Economic Prosperity 269, 270, 283
Emerging Technologies 4, 6, 128, 151, 152, 153, 155, 159, 163, 166, 167, 170, 219
Employment 60, 195, 236, 250, 253, 254, 259, 260, 261, 262, 277, 291, 292, 307, 308, 310, 311, 312, 313, 314, 316, 318, 320, 323, 324, 325
Environment 41, 48, 49, 50, 53, 58, 59, 60, 65, 66, 68, 69, 75, 80, 107, 110, 114, 116, 118, 120, 134, 144, 179, 180, 181, 188, 200, 201, 226, 228, 229, 231, 235, 240, 249, 257, 259, 263, 271, 273, 279, 303, 305, 308, 309, 311, 325, 328, 330, 331, 332, 343, 351, 355, 359, 384, 386, 388, 394, 395, 398, 407, 409, 411, 418, 419, 421, 422, 423, 424, 425, 428, 431, 439, 450
Epistemic Justice 14, 24, 36, 38, 42, 43, 143, 147, 251, 253, 262, 263, 295, 297, 298, 300, 305, 347, 355
Epistemic Sovereignty 4, 261, 262, 298, 305
Ethics 2, 3, 4, 5, 6, 8, 9, 15, 17, 19, 20, 21, 23, 24, 25, 29, 30, 31, 32, 35, 36, 37, 38, 39, 40, 41, 42, 43, 52, 96, 115, 116, 117, 129, 132, 136, 141, 142, 145, 146, 147, 153, 154, 164, 168, 175, 182, 184, 185, 195, 197, 198, 199, 200, 204, 353, 355, 358, 359, 360, 361, 362, 364, 367, 374, 375, 376, 379, 380, 381, 382, 385, 387, 388, 393, 395, 396, 401, 404, 405, 407, 410

F

Financial Innovation 371, 372

G

Governance 1, 2, 4, 5, 9, 13, 14, 15, 16, 17, 21, 23, 26, 30, 31, 32, 35, 36, 37, 38, 39, 40, 42, 43, 45, 46, 47, 49, 50, 61, 62, 63, 64, 107, 113, 114, 115, 116, 117, 120, 124, 126, 129, 130, 131, 132, 134, 135, 136, 141, 142, 143, 144, 145, 146, 147, 148, 151, 152, 153, 154, 155, 157, 158, 159, 163, 164, 165, 166, 167, 168, 169, 170, 171, 172, 173, 174, 182, 184, 198, 200, 225, 226, 227, 228, 229, 230, 233, 234, 235, 237, 241, 244, 250, 258, 264, 269, 270, 271, 272, 273, 275, 276, 277, 278, 279, 280, 281, 282, 283, 285, 288, 290, 293, 295, 296, 298, 299, 327, 328, 329, 330, 331, 332, 333, 334, 335, 336, 337, 338, 339, 340, 341, 342, 343, 344, 345, 346, 347, 348, 349, 350, 353, 354, 355, 357, 358, 359, 360, 361, 362, 363, 364, 365, 366, 367, 368, 369, 370, 371, 372, 373, 374, 375, 376, 377, 382, 383, 384, 385, 386, 387, 388, 389, 390, 391, 393, 394, 395, 396, 398, 399, 400, 401, 402, 404, 405, 406, 407, 408, 411, 412, 435, 436, 437, 438, 439, 440, 441, 442, 443, 444, 445, 446, 447, 450, 451
Grassroots Innovation 291, 292

H

Heritage 10, 11, 17, 21, 47, 52, 54, 58, 59, 60, 61, 62, 63, 64, 71, 79, 83, 89, 107, 110, 114, 115, 128, 130, 140, 143, 145, 146, 147, 148, 151, 153, 154, 155, 166, 169, 170, 174, 175, 176, 178, 188, 190, 197, 199, 202, 204, 205, 208, 219, 225, 226, 227, 228, 229, 230, 231, 232, 233, 234, 236, 237, 238, 239, 240, 241, 242, 243, 244, 245, 246, 247, 248, 255, 258, 264, 277, 283, 301, 302, 307, 308, 310, 311, 314, 316, 319, 320, 321, 399, 405, 413, 414, 415, 416, 417, 418, 422, 423, 426, 427, 428, 429, 430, 431, 432, 433, 435, 437, 438, 439, 441, 442, 444, 446, 447
Higher Education 23, 24, 25, 32, 36, 37, 38, 39, 40, 41, 42, 43, 67, 68, 70, 72, 84, 85, 86, 87, 107, 108, 301, 302, 318, 320, 448
Holistic 2, 4, 7, 24, 37, 42, 45, 46, 47, 48, 51, 55, 56, 59, 64, 74, 81, 118, 155, 201, 229, 292, 304, 311, 357, 358, 359, 361, 369, 371, 373, 375, 388, 411
Honorifics 204, 206, 207, 209, 217

I

IKS 1, 2, 3, 4, 5, 6, 7, 8, 11, 12, 13, 14, 15, 16, 17, 45, 46, 47, 48, 49, 51, 52, 53, 54, 55, 56, 58, 59, 60, 61, 62, 63, 64, 65, 67, 68, 69, 71, 72, 74, 75, 76, 77, 78, 79, 80, 81, 82, 86, 90, 91, 92, 93, 94, 96, 97, 98, 99, 100, 101, 102, 103, 104, 105, 106, 107, 109, 110, 113, 114, 115, 118, 119, 120, 124, 129, 130, 132, 134, 135, 136, 140, 142, 143, 144, 147, 188, 201, 202, 203, 204, 205, 206, 207, 208, 218, 226, 245, 254, 260, 262, 266, 287, 289, 290, 291, 292, 293, 294, 295, 296, 297, 298, 299, 308, 311, 312, 313, 314, 315, 316, 317, 320, 330, 331, 354, 357, 358, 359, 360, 362, 363, 364, 366, 367, 369, 370, 372, 373, 374, 375, 376, 377, 378, 435, 436, 439, 440, 443, 445, 446, 447, 451
Inclusive Development 16, 17, 45, 442
Indigenous Data Sovereignty 2, 3, 4, 9, 14, 16, 19, 32, 35, 38, 40, 116, 130, 131, 132, 136, 147, 148, 149, 153, 155, 157, 163, 166, 168, 174, 175, 176, 183, 184, 185, 195, 198, 280
Indigenous Ethics 29
Indigenous Knowledge 1, 2, 3, 4, 5, 6, 7, 11, 14, 15, 16, 17, 23, 24, 25, 26, 27, 28, 29, 30, 31, 32, 34, 35, 36, 37, 38, 39, 40, 41, 42, 43, 44, 45, 46, 47, 48, 49, 51, 52, 53, 54, 55, 56, 58, 59, 60, 61, 62, 63, 64, 65, 66, 67, 68, 70, 71, 72, 73, 74, 77, 78, 79, 80, 81, 83, 84, 85, 86, 87, 89, 90, 91, 92, 93, 95, 97, 98, 99, 100, 101, 102, 103, 104, 107, 108, 109, 110, 111, 112, 113, 114, 115, 116, 117, 118, 119, 120, 122, 129, 133, 134, 135, 136, 138, 143, 144, 146, 147, 148, 149, 151, 152, 153, 154, 155, 157, 159, 160, 163, 164, 166, 167, 169, 170, 171, 172, 174, 175, 177, 178, 179, 180, 181, 182, 184, 185, 187, 188, 189, 190, 191, 192, 193, 194, 195, 196, 197, 198, 199, 200, 201, 205, 206, 208, 209, 219, 220, 221, 222, 225, 226, 227, 235, 239, 241, 242, 243, 245, 248, 249, 250, 251, 253, 254, 255, 258, 259, 260, 262, 263, 264, 265, 266, 267, 269, 270, 271, 272, 273, 274, 275, 276, 277, 278, 279, 280, 281, 282, 283, 286, 287, 288, 289, 290, 291, 292, 293, 294, 295, 296, 297, 298, 299, 301, 302, 303, 304, 305, 307, 308, 309, 310, 311, 312, 313, 314, 315, 316, 317, 318, 319, 320, 321, 322, 323, 324, 325, 327, 329, 330, 333, 336, 341, 342, 347, 348, 349, 353, 354, 355, 381, 382, 384, 387, 404, 406, 408, 409, 411, 412, 435, 437, 438, 439, 440, 441, 443, 444, 445, 447, 448, 449, 451
Indigenous Knowledge Holders 30, 72, 122, 177, 178, 182, 185, 189, 192, 193, 194, 195, 196, 260, 280, 297, 298, 303, 313, 320, 445, 447

Indigenous Knowledge Systems 1, 2, 3, 4, 5, 7, 14, 15, 16, 17, 24, 25, 26, 37, 38, 40, 45, 46, 47, 48, 49, 51, 52, 53, 54, 55, 56, 58, 59, 60, 61, 62, 63, 64, 65, 66, 68, 83, 86, 89, 90, 91, 92, 93, 95, 103, 107, 108, 109, 110, 111, 112, 113, 114, 115, 117, 118, 120, 129, 133, 135, 143, 144, 146, 147, 149, 151, 152, 170, 181, 182, 188, 189, 190, 195, 197, 198, 199, 200, 201, 208, 209, 220, 221, 222, 245, 250, 259, 262, 264, 265, 266, 269, 270, 272, 274, 282, 283, 287, 288, 289, 291, 292, 293, 294, 295, 296, 297, 298, 299, 301, 302, 303, 304, 305, 307, 308, 309, 310, 312, 313, 317, 320, 322, 323, 324, 325, 327, 329, 330, 333, 336, 341, 342, 347, 348, 349, 354, 355, 384, 387, 411, 435, 437, 438, 439, 443, 444, 451
indigenous people 18, 31, 67, 68, 69, 70, 71, 72, 73, 74, 76, 77, 78, 79, 80, 81, 83, 91, 93, 95, 96, 97, 98, 99, 100, 101, 102, 105, 106, 112, 118, 119, 124, 125, 128, 134, 135, 143, 179, 220, 227, 237, 238, 239, 240, 241, 245, 254, 263, 308, 406
Indigenous Research Methodologies 23, 24, 25, 26, 32, 36, 37, 38, 39, 40, 41, 42, 44, 185, 198
Innovation 2, 3, 5, 14, 15, 16, 17, 38, 41, 71, 72, 74, 79, 83, 91, 92, 110, 117, 129, 144, 146, 147, 152, 166, 170, 174, 183, 184, 188, 221, 245, 250, 254, 263, 270, 274, 275, 277, 280, 281, 283, 287, 288, 289, 290, 291, 292, 293, 294, 295, 296, 297, 298, 299, 300, 301, 302, 303, 305, 307, 309, 310, 312, 313, 316, 318, 319, 321, 322, 328, 332, 346, 347, 349, 359, 370, 371, 372, 377, 382, 408, 418, 421, 423, 424, 430, 431
Internet of Things 413, 414, 415, 419, 422, 423, 424, 427, 428, 430, 431

J

Job opportunities 325

K

Karma 357, 358, 359, 360, 361, 362, 363, 364, 366, 367, 368, 369, 370, 372, 373, 374, 375, 376, 378, 380, 381, 382
Knowledge Preservation 3, 4, 7, 21, 38, 46, 47, 58, 62, 90, 91, 92, 103, 112, 116, 117, 152, 154, 155, 157, 159, 160, 163, 166, 167, 172, 307, 309, 310, 315, 316, 318, 325
Knowledge Sovereignty 270, 274, 278, 280, 283, 288

L

Land 1, 4, 6, 23, 24, 26, 27, 28, 29, 32, 33, 34, 36, 37, 38, 40, 41, 42, 43, 47, 49, 50, 51, 52, 54, 61, 62, 63, 64, 65, 66, 83, 115, 116, 118, 125, 126, 143, 172, 181, 199, 236, 239, 247, 253, 255, 259, 266, 273, 275, 277, 278, 279, 280, 282, 286, 327, 328, 329, 331, 332, 333, 334, 335, 336, 338, 341, 342, 343, 345, 346, 349, 381, 393, 394, 400, 405, 407, 435, 436, 437, 438, 439, 440, 441, 442, 443, 444, 445, 446, 447, 450, 451
land tenure 278, 329, 333, 342, 349, 394
Local Governance 136, 272, 299, 337, 343, 345, 385, 396, 406

M

Mixed-Methods 1, 3, 6, 7, 9, 15, 16

N

Nature 17, 19, 20, 45, 46, 47, 48, 49, 52, 55, 56, 58, 59, 60, 61, 64, 65, 67, 84, 90, 101, 104, 151, 152, 179, 180, 182, 187, 195, 213, 226, 228, 229, 233, 234, 239, 253, 254, 255, 261, 277, 279, 286, 292, 293, 295, 296, 299, 301, 302, 303, 304, 305, 319, 320, 330, 351, 352, 355, 358, 360, 384, 387, 388, 393, 394, 395, 398, 406, 407, 408, 410, 411, 418, 443, 449, 450, 451

P

Participatory Action Research 89, 90, 94, 110, 298
partnerships 61, 64, 67, 68, 141, 144, 197, 230, 258, 279, 281, 283, 291, 312, 317, 338, 339, 342, 343, 402, 436, 439, 441, 442, 446, 448
PLS-SEM 1, 3, 6, 8, 9, 15
Policy 2, 3, 15, 17, 19, 20, 23, 28, 38, 41, 45, 46, 61, 65, 67, 69, 71, 72, 76, 77, 78, 79, 81, 82, 83, 84, 86, 87, 110, 114, 115, 116, 117, 118, 119, 126, 129, 142, 143, 144, 146, 148, 154, 166, 175, 176, 178, 184, 188, 189, 190, 194, 197, 198, 199, 218, 220, 221, 225, 227, 228, 229, 231, 233, 237, 239, 241, 243, 245, 246, 249, 253, 254, 262, 263, 265, 266, 269, 270, 271, 278, 279, 281, 283, 285, 286, 287, 288, 289, 290, 293, 294, 295, 296, 297, 298, 299, 302, 304, 312, 313, 314, 315, 318, 320, 321, 327, 328, 329, 330, 331, 341, 346, 347, 358, 360, 364, 366, 368, 369, 372, 375, 376, 381, 383, 384, 385, 387, 395, 398, 399, 401, 402, 405, 408, 409, 414, 424, 431, 435, 436, 437, 438, 443, 446, 447,

449, 451
Post-Extractive Development 348, 355
Practices 2, 4, 6, 16, 21, 26, 27, 29, 30, 34, 36, 39, 40, 41, 44, 46, 47, 48, 49, 50, 51, 52, 53, 54, 55, 56, 57, 58, 59, 60, 61, 62, 63, 65, 66, 67, 68, 69, 70, 71, 72, 73, 74, 75, 77, 78, 79, 80, 81, 83, 85, 86, 87, 90, 92, 107, 108, 112, 113, 114, 116, 117, 130, 132, 134, 145, 147, 151, 152, 153, 163, 172, 175, 178, 179, 180, 182, 186, 187, 191, 194, 195, 199, 200, 201, 202, 203, 204, 205, 206, 209, 210, 211, 215, 223, 225, 226, 229, 231, 235, 236, 241, 249, 250, 251, 253, 254, 255, 256, 258, 259, 269, 270, 271, 272, 273, 274, 275, 276, 277, 279, 280, 281, 282, 283, 289, 290, 291, 292, 293, 294, 296, 298, 302, 303, 304, 305, 308, 309, 310, 311, 312, 314, 315, 316, 317, 318, 319, 320, 324, 325, 327, 328, 329, 330, 331, 332, 333, 335, 336, 339, 342, 344, 345, 351, 355, 358, 360, 361, 362, 367, 368, 372, 373, 377, 383, 384, 385, 386, 388, 389, 390, 391, 396, 398, 402, 404, 405, 407, 408, 411, 412, 424, 425, 426, 428, 437, 438, 439, 441, 442, 444, 445, 446, 451
Public Finance 357, 358, 359, 360, 361, 362, 363, 364, 366, 367, 368, 369, 370, 371, 372, 373, 374, 375, 376, 377

R

Relational Accountability 2, 4, 5, 9, 21, 23, 24, 26, 27, 33, 34, 38, 42, 43, 44, 117, 332
Resilience 4, 16, 17, 21, 25, 38, 45, 46, 47, 49, 50, 51, 52, 53, 54, 62, 84, 114, 118, 125, 144, 151, 155, 226, 233, 234, 235, 236, 241, 243, 244, 245, 247, 248, 269, 270, 272, 273, 274, 275, 276, 277, 279, 281, 283, 285, 286, 292, 296, 297, 304, 308, 310, 321, 327, 330, 331, 332, 333, 335, 336, 337, 340, 341, 343, 344, 345, 346, 347, 348, 349, 351, 354, 355, 358, 362, 383, 385, 387, 389, 394, 395, 396, 398, 402, 404, 405, 407, 408, 409, 418, 420
Resource 20, 28, 31, 32, 33, 37, 38, 43, 45, 46, 47, 48, 49, 50, 51, 52, 53, 54, 55, 62, 84, 124, 158, 161, 163, 168, 169, 170, 171, 172, 175, 187, 201, 226, 237, 239, 249, 250, 254, 258, 260, 264, 269, 270, 271, 272, 273, 274, 277, 278, 280, 282, 283, 285, 291, 292, 295, 304, 307, 308, 309, 310, 311, 313, 318, 320, 321, 327, 332, 333, 336, 337, 342, 347, 348, 360, 369, 370, 371, 372, 382, 383, 384, 385, 386, 389, 398, 400, 401, 408, 409, 410, 416, 421, 423, 447
Resource Management 28, 48, 50, 52, 54, 55, 187, 201, 226, 237, 250, 254, 264, 269, 270, 272, 273, 274, 292, 360, 383, 385, 398, 410
revive 71, 90, 91, 120, 135, 435, 438

S

Smart Cities 413, 415, 418, 419, 430
Smart Tourism 413, 414, 415, 416, 417, 422, 423, 426, 428, 430, 431, 433
Social media 35, 128, 186, 191, 200, 201, 202, 203, 204, 205, 206, 207, 208, 209, 210, 211, 212, 213, 214, 215, 216, 217, 218, 219, 220, 221, 222, 223, 265, 307, 312, 319, 324, 427, 430
Social Media Platforms 186, 191, 200, 202, 203, 204, 212, 217, 319
Sovereignty 2, 3, 4, 9, 13, 14, 16, 19, 23, 24, 30, 31, 32, 35, 37, 38, 39, 40, 41, 42, 43, 53, 113, 114, 116, 117, 121, 129, 130, 131, 132, 135, 136, 138, 143, 146, 147, 148, 149, 151, 152, 153, 154, 155, 157, 158, 163, 166, 167, 168, 170, 171, 172, 174, 175, 176, 183, 184, 185, 195, 198, 225, 226, 227, 228, 229, 241, 245, 261, 262, 270, 274, 276, 278, 280, 283, 288, 294, 298, 300, 303, 305, 337, 346, 359, 405, 406, 451
Spiritual 2, 4, 24, 26, 27, 28, 29, 37, 45, 46, 47, 48, 49, 50, 55, 56, 57, 58, 61, 62, 98, 108, 114, 124, 132, 172, 216, 228, 229, 234, 235, 335, 336, 360, 390, 405, 407
Sustainability 2, 3, 16, 20, 26, 27, 34, 37, 45, 46, 47, 48, 49, 50, 52, 54, 58, 61, 62, 64, 65, 68, 69, 72, 78, 79, 81, 84, 85, 109, 113, 114, 115, 116, 117, 118, 120, 122, 132, 136, 138, 144, 147, 153, 155, 157, 166, 168, 170, 171, 175, 226, 231, 235, 239, 241, 247, 248, 253, 254, 260, 262, 265, 266, 270, 271, 272, 273, 274, 275, 276, 279, 280, 283, 285, 286, 288, 290, 291, 292, 294, 295, 297, 299, 303, 305, 307, 308, 310, 311, 312, 320, 323, 327, 328, 331, 334, 337, 345, 347, 348, 349, 351, 355, 358, 359, 360, 367, 368, 369, 370, 374, 375, 376, 377, 383, 384, 385, 386, 387, 388, 389, 393, 395, 396, 398, 399, 400, 401, 404, 406, 407, 408, 409, 410, 412, 415, 416, 418, 423, 431, 432, 433, 438, 439, 442, 443, 444, 445, 447, 448, 449
Sustainable Development 45, 46, 47, 55, 58, 61, 62, 63, 65, 66, 67, 69, 70, 78, 81, 84, 107, 109, 110, 113, 114, 118, 135, 144, 146, 147, 180, 187, 197, 221, 239, 245, 246, 249, 250, 251, 253, 254, 259, 261, 263, 264, 265, 266, 272, 279, 283, 286, 290, 294, 297, 308, 311, 322, 323, 324, 325, 351, 359, 367, 380, 395, 409, 411, 448, 449, 450
Sustainable Livelihoods 45, 52, 54, 65, 276, 309, 311, 313, 314, 316, 447

T

Traditional 5, 11, 17, 21, 25, 27, 28, 32, 33, 34, 36, 45, 46, 47, 49, 52, 53, 54, 55, 56, 57, 58, 59, 60, 61, 62, 63, 65, 68, 71, 73, 78, 79, 91, 95, 96, 108, 109, 111, 114, 115, 118, 119, 120, 126, 129, 130, 133, 134, 136, 144, 149, 151, 158, 166, 167, 172, 175, 179, 180, 188, 189, 192, 194, 199, 201, 202, 203, 204, 206, 209, 210, 211, 212, 214, 216, 221, 226, 227, 229, 235, 236, 237, 238, 240, 243, 246, 249, 250, 253, 254, 258, 259, 261, 263, 264, 265, 266, 269, 270, 271, 272, 273, 275, 276, 277, 278, 280, 282, 285, 291, 293, 301, 304, 305, 307, 308, 309, 310, 312, 313, 315, 316, 317, 322, 341, 347, 351, 354, 358, 359, 376, 383, 384, 385, 388, 389, 390, 391, 396, 398, 399, 404, 405, 406, 407, 408, 409, 410, 411, 412, 420, 425, 426, 427, 428, 440, 441, 445, 447, 448

Traditional Ecological Knowledge 27, 28, 46, 49, 59, 65, 118, 272, 277, 384, 385, 399, 408, 409

W

Western countries 67, 74

Worldview 45, 46, 48, 195, 375, 384, 387, 388

Y

Youth Engagement 143, 157, 277, 309, 310, 311, 314, 315, 317, 320, 321, 322, 325, 446